T0358292

Fundamentals of General Equilibrium Analysis

Series on Mathematical Economics and Game Theory

ISSN: 2010-1953

Series Editor: Ezra Einy *(Ben Gurion University)*

Editorial Advisory Board

Tatsuro Ichiishi
Hitotsubashi University

James S. Jordan
The Penn State University

Ehud Kalai
Northwestern University

Semih Koray
Bilkent University

John O. Ledyard
California Institute of Technology

Richard P. McLean
Rutgers University

Dov Monderer
The Technion

Bezalel Peleg
The Hebrew University of Jerusalem

Stanley Reiter
Northwestern University

Dov E. Samet
Tel Aviv University

Timothy Van Zandt
INSEAD

Eyal Winter
The Hebrew University of Jerusalem

Itzhak Zilcha
Tel Aviv University

Published

Vol. 6: *Fundamentals of General Equilibrium Analysis*
by Takashi Suzuki

Vol. 5: *Fixed Points and Economic Equilibria*
by Ken Urai

Vol. 4: *General Equilibrium Analysis of Production and Increasing Returns*
by Takashi Suzuki

Vol. 3: *Cooperative Extensions of the Bayesian Game*
by Tatsuro Ichiishi & Akira Yamazaki

Vol. 2: *Theory of Conjectural Variations*
by C. Figuières, A. Jean-Marie, N. Quérou & M. Tidball

Vol. 1: *Theory of Regular Economies*
by Ryo Nagata

More information on this series can also be found at https://www.worldscientific.com/series/smegt

Series on Mathematical Economics and Game Theory

Vol. 6

Fundamentals of General Equilibrium Analysis

Takashi Suzuki

Meiji Gakuin University, Japan

World Scientific

NEW JERSEY • LONDON • SINGAPORE • BEIJING • SHANGHAI • HONG KONG • TAIPEI • CHENNAI • TOKYO

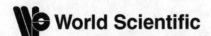

Published by

World Scientific Publishing Co. Pte. Ltd.
5 Toh Tuck Link, Singapore 596224
USA office: 27 Warren Street, Suite 401-402, Hackensack, NJ 07601
UK office: 57 Shelton Street, Covent Garden, London WC2H 9HE

Library of Congress Cataloging-in-Publication Data
Names: Suzuki, Takashi, 1959– author.
Title: Fundamentals of general equilibrium analysis / Takashi Suzuki (Meiji Gakuin University, Japan).
Description: USA : World Scientific, 2020. | Series: Series on mathematical economics and game theory, 2010-1953 ; vol 6 | Includes bibliographical references and index.
Identifiers: LCCN 2020022816 | ISBN 9789811219610 (hardcover) | ISBN 9789811219627 (ebook) | ISBN 9789811219634 (ebook other)
Subjects: LCSH: Equilibrium (Economics)--Mathematical models.
Classification: LCC HB145 .S889 2020 | DDC 339.501/5118--dc23
LC record available at https://lccn.loc.gov/2020022816

British Library Cataloguing-in-Publication Data
A catalogue record for this book is available from the British Library.

Copyright © 2021 by World Scientific Publishing Co. Pte. Ltd.

All rights reserved. This book, or parts thereof, may not be reproduced in any form or by any means, electronic or mechanical, including photocopying, recording or any information storage and retrieval system now known or to be invented, without written permission from the Publisher.

For photocopying of material in this volume, please pay a copying fee through the Copyright Clearance Center, Inc., 222 Rosewood Drive, Danvers, MA 01923, USA. In this case permission to photocopy is not required from the publisher.

For any available supplementary material, please visit
https://www.worldscientific.com/worldscibooks/10.1142/11808#t=suppl

Desk Editors: Balasubramanian Shanmugam/Shreya Gopi/Yulin Jiang

Typeset by Stallion Press
Email: enquiries@stallionpress.com

Printed in Singapore

To Tatsuro Ichiishi

Preface to the First Edition

The economies of modern advanced countries are characterized by the enormous extension of markets, the rapid development of technologies, and the ongoing monopolization of the markets by small numbers of ever larger companies. These market characteristics are mutually related, and at least, the last two belong to the same Marshallian tradition of economic thought. Unfortunately, modern general equilibrium theory, which has essentially grown out of the Walrasian tradition, is currently insufficient for considering the problems arising from markets with these characteristics. To tackle these problems successfully, I believe that one has to go back to Marshall and construct a theory that unifies the ideas of these two great masters of equilibrium analysis. The chapters in this book are a first step toward such a theory. I want to show in this book that the main body of Marshallian concepts, including increasing returns and monopolistic competition, indeed fit into the scheme of modern general equilibrium theory. At the same time, I trust that this book will be a useful introductory text on general equilibrium theory for graduate students. It will achieve this purpose if it becomes a good companion to those students who are ambitious to contribute to further development of the theory.

During my own time as a student and in writing this book, I have become indebted to many people for their assistance. I thank in particular Professors Kenjiro Ara, Marcus Berliant, Pierre Dehez,

Jacques Dreze, Jean-Michel Grandmont, Kazuya Kamiya, Toru Maruyama, Lionel McKenzie, Takashi Negishi, Kazuo Nishimura, Mitsunori Noguchi, Sin-ichi Takekuma, Akira Yamazaki, and Makoto Yano.

I am most grateful to Professor Tatsuro Ichiishi for inviting me to write this book. There is no doubt that it would not have been written if not for his kind invitation and generous encouragement. I dedicate this entire volume to him.

I also express my gratitude to my wife Nobuko as it is extremely doubtful whether the book would have been completed without her help and support. I thank my friend Mr. Susumu Sakuma for cheering me up with his music. Finally, my gratitude goes to our cat Kijiko (1991–2008) and our dog Tetsu (1997–2013) for granting my wife and me the joy of life.

Takashi Suzuki
Tokyo

Preface to the Second Edition

This is a revised and expanded edition of *General Equilibrium Analysis of Production and Increasing Returns* (World Scientific, 2009). The fundamental premise of the book has not changed, as it continues to seek to incorporate Marshallian ideas such as external increasing returns and monopolistic competition into a general equilibrium framework within the Walrasian tradition. However, this edition has included some new sections and two new chapters, of which one is of technical and of a Walrasian nature and the other of a political/philosophical nature that draws on the external increasing returns (i.e., the division of labor) of Adam Smith and Alfred Marshall.

More specifically, in Chapter 5, I present equilibrium existence and core equivalence theorems for an infinite-horizon economy with a measure space of consumers. These results are currently the focus of extensive studies by mathematical theorists concerning the application of an advanced mathematical concept called the *saturated* (*super-atomless*) *measure space*. I trust that this new chapter will be helpful for graduate students and researchers interested in this topic. The results in Chapter 5 have been obtained under research collaboration with Professors M. Ali Khan and Nobusumi Sagara. I would like to express my profound appreciation to them. Of course, I am solely responsible for any remaining errors and defects.

The second major change in this new edition is in Section 7.5 in which I present a simple toy model of a liberal society that implements the *difference principle* proposed by John Rawls as a principle of distributive justice. I trust that this new section opens up the possibility of connecting theoretical economics and political philosophy and that it will then be of interest to students and researchers in both areas. It is often alleged in an ideological context that neoclassical economics (general equilibrium theory) is an economic theory that extremely emphasizes the "efficiency of free markets" achieved by the "invisible hand" of the price mechanism, and thus, it only serves to justify neoliberalism or the libertarian standpoint of political philosophy. I would like to insist that the Smith/Marshall tradition of economic thought has conveyed not only the idea of a price mechanism working in closed markets but also the idea of *reciprocity*, which is incidentally a crucial concept in Rawls's philosophy of working outside markets through the background institutions of liberal societies. In my mind, the notion of external increasing returns manifestly exemplifies this.

My research partner Professor Khan strongly argued that the new edition must also include a section concerning the marginal cost pricing equilibrium; hence, I have included this in Section 7.1. Section 7.4, which discusses the welfare properties of external increasing returns, is also completely revised. I note that these also belong to the Marshall/Pigou tradition and the Cambridge School.

I have also included a mathematical appendix (Appendix C) that considers the basics of *singular homology theory*. The purpose of this appendix is to provide readers with an idea of a proof of Brower's fixed point theorem from the perspective of algebraic topology. Although the fixed point theorem is originally a theorem of algebraic topology, most economic students of our generation have learned it via the differential topological proof of Milnor or the proof based on a technical lemma due to Sperner. The advantage of these proofs is that they are short and require relatively small prerequisites. However, they are inclined to obscure the true background of the fixed point theorem. Considering the significance of the fixed point theorem and its key role in general equilibrium theory, I wanted to consider this from the "right place." However, no concepts and techniques such as homology groups or exact sequences in Appendix C will be not used in the text. Appendix E also includes Milnor's neat proof applying Sard's theorem and the implicit function theorem.

Typographical and technical errors in the previous edition have also been corrected, and I hope that any flaws in the previous edition are remedied by these revisions and corrections.

With this opportunity, I would like to acknowledge the many people to whom I am indebted for this work. In addition to those mentioned in the preface to the first edition, I thank Professors Chiaki Hara, Atsushi Kajii, M. Ali Khan, Jean F. Mertens, Konrad Podczeck, Heraklis M. Polemarchakis, Nobusumi Sagara, Yeneng Sun, and Nicholas Yannelis. The entire volume remains dedicated to Professor Tatsuro Ichiishi with gratitude and acknowledgment for his constant encouragement. I would also like to thank Ms. Shreya Gopi and Mr. Balasubramanian Shanmugam of World Scientific for their excellent works and my wife Nobuko for her help and support. The financial support of a Grant-in-Aid for Scientific Research (No. 15K03362) from the Ministry of Education, Culture, Sports, Science and Technology, Japan is also gratefully acknowledged. Finally, I express my gratitude to our cats and dog Umeko (2007–2017), Sakura-ko (2017–2020), and Kotetsu for giving both of us the joy of life.

Takashi Suzuki
Tokyo

About the Author

Takashi Suzuki obtained his B.A. with economics major from the Hitotsubashi University, Tokyo and Ph.D. from the University of Rochester, New York. After teaching economics at Yokohama National University for one year, he served as an assistant professor at Meiji Gakuin University in 1993, a visiting fellow of CORE (Center for Operation Research and Econometrics) between 1998 and 1999, and he has been a full professor from 2006 at the same university. Professor Suzuki has been sustaining his research interest for general equilibrium theory from his student years at Hitotsubashi, but from 2010 or so, he has been also interested in Political Philosophy, mainly the theory of "Justice as Fairness" by J. Rawls.

Contents

Chapter 1

A Brief History of Equilibrium Analysis

1.1 From the 19th Century to the 1940s

The first mathematical analysis of market equilibria was undertaken by Cournot [47], who constructed a mathematical model of an economic market with a downward sloping total demand curve and a finite number of firms whose technologies were given by cost functions. In such an economy, he established the concepts of monopolistic equilibrium and oligopolistic equilibrium, which are named after him.

Although limited within partial equilibrium analysis, these contributions are truly original, and the clarity and modernity of his analysis are impressive. Indeed, his profound insight is equal to that of modern theorists. He characterized the perfectly competitive equilibrium as the limit of a sequence of imperfectly competitive equilibria with an increasing number of producers. In other words, he realized the effect of a large number of economic agents in the market. This style of analysis was followed by Edgeworth [66] and Debreu and Scarf [55] in the study of the core of an exchange economy, which will be discussed in Section 2.6.[1]

[1]For modern discussions of the "limit theorem" of Cournot's imperfectly competitive equilibria, see Roberts [207], Novshek [181], Novshek and Sonnenschein [182], and Mas-Colell [147]. We are indebted for the descriptions of the work of Walras and Edgeworth to Arrow and Hahn [11, Chapter 1].

Furthermore, in the Cournot model, the total demand function $F(p)$ (of the price p) was not put into the model simply "by hands," but rather, he provided an intuitive argument that justifies the continuity of the function $F(p)$. He wrote

> We will assume that the function $F(p)$, which expresses the law of demand or of the market, is a continuous function, i.e., a function which does not pass suddenly from one value to another, but which takes in passing all intermediate values. It might be otherwise if the number of consumers were limited....
>
> ... But the wider the market extends, and the more the combinations of needs, of fortunes, or even caprices, and varied among consumers, the closer the function $F(p)$ will come to varying with price in a continuous manner [47, English translated version 1927, Section 22, pp. 49–50].

What Cournot was explaining above is the "smoothing effect of aggregation," which we will discuss in Section 3.5.

As is well known, the market demand function was derived from the individual demand functions via utility maximization behaviors by Menger [162], Jevons [101], Walras [264], and Marshall [141]. Among these "marginal revolutionists," credit for establishing the general equilibrium model is given to Walras. He defined the general competitive equilibrium as a solution of a system of equations; the system consists of the first-order conditions of the utility maximization problem of consumers, the first-order conditions of the profit maximization problem of firms, the budget constraints of consumers, and the market conditions of the supply and demand of each commodity.[2]

In Walras' system of equations, only the $\ell - 1$ relative prices of the ℓ commodities are determined, along with the budget constraints summed over all households, made up of an identity equation which we now call after his name. Consequently, the equations of the system are not independent, and the number of equations equals the number of unknowns in the model. Walras appeared satisfied with this feature of the model and believed it demonstrated the consistency of his

[2] We do not consider the markets of the intermediate (capital) commodities in his model.

model. As we will see in Section 1.2, the question of the existence of economically meaningful solutions of the Walras system was left as a legacy to mathematical theorists of the following century. Instead, Walras discussed the stability of his equilibrium. His stability theory, known as *tâtonnements*, is described next.

The price adjustment process occurs sequentially across markets. The first market equilibrium is achieved when the price adjusts to any change in demand and/or supply. Next, the second market adjusts in the same way. However, because the price in the second market affects the first market, the equilibrium in the first market is disturbed. Therefore, when price adjustment in the final market is completed, only this market is in equilibrium. Walras assumed that the impact of price changes is largest for the own market and believed that the price in each market will be closer to the equilibrium price than prior to the adjustment process, and if the price adjustment process is repeated, then the market prices will converge to the equilibrium prices.

The welfare economics of general equilibrium markets were studied by Edgeworth [66] and Pareto [186]. Pareto formulated maximum social welfare as the solution of the maximization problem of the social welfare function and proposed the criterion of the social optimum that is named after him.

Edgeworth considered a market model with two commodities and two traders (consumers), each of which has his/her own utility function and initial endowment. He illustrated the model using the Edgeworth Box diagram, which is used as a basic tool by modern theorists. He assumed that (a) the traders would not make a trade if another would be more beneficial for one or both of them and that (b) neither would make a trade that would make them worse off. He called the set of allocations satisfying conditions (a) and (b) the contract curve and observed that the competitive equilibrium allocation was contained in this set. He assumed that markets contained two types of traders rather than two individuals, and there are equal numbers of each type of trader, all with the same utility function and the same initial endowment. Edgeworth generalized the contract curve to this situation in which he assumed that no trade among any traders of any type would be completed if there existed a group of traders who could make a different trade among themselves, using only their own initial endowments in such a way that nobody in the group is worse off and at least one member of the group is better off than the initially proposed trade. It is clear that the set of allocations satisfying this condition is the core of the market game in modern terminology. Using the Edgeworth Box diagram, Edgeworth concluded that as the number of individuals of each type became large, the core shrank to the competitive equilibrium allocation. In the terminology of modern game theory, Cournot considered the competitive equilibrium to be the limit of the non-cooperative solutions of market games with an increasing number of producers, and Edgeworth considered it to be the limit of the cooperative solutions of market games with an increasing number of consumers.

Among the many contributions of Marshall [141],[3] we are especially interested in his idea of external economies and increasing returns. Marshall divided increasing returns into two types based on their sources and provided the following definition:

> We may divide the economies arising from an increase in the scale of production of any kind of goods, into two classes — firstly, those dependent on the general development of the industry; and secondly, those dependent on the resources of the individual houses of business engaged in it, on their organization and the efficiency of their management. We may call the former *external economies*, and the latter *internal economies* [141, Chapter IX, p. 221].

Of course, as a theorist in the nineteenth century, he believed that the limitations of production factors such as land naturally led to decreasing returns. For both external and internal economies, he assumed that increasing returns were related to the efficiency of human skill and technology.

---

[3] A.C. Pigou said, "It's all in Marshall."

> We say broadly that while the part which nature plays in production shows a tendency to diminishing return, the part which man plays shows a tendency to increasing return. The *law of increasing return* may be worded thus: — an increase of labour and capital leads generally to improved organization, which increases the efficiency of the work of labour and capital [141, Chapter XIII, p. 265].

Therefore, it seems fair to assume that (at least) most elements of "internal economies" in Marshall's sense correspond to non-convex production sets in modern terminology, and "external economies" to (positive) external effects between individual firms when one tries to express his concept of economies of scale within the context of modern equilibrium theory.

In any case, Marshall proposed these "economies of scale" as theoretical concepts, not simply as observations of reality. This means that he had assumed that increasing returns, both internal or external, are consistent with his system as a whole, or more specifically, he assumed that the increasing returns are compatible with a perfectly competitive equilibrium.

Marshall seemed to believe the consistency between the external economies and the competitive (or price-taking) behavior of individual firms. One reason for this is that his equilibrium concept of perfect competition is defined over a long period of time, with free entry and exit, and is therefore different from a temporary equilibrium. Invoking the famous metaphor of trees in the forest, he defined his equilibrium concept where each firm lives its own lifetime in the market.

> But here we may read a lesson from the young trees of the forest as they struggle upwards through the benumbing shade of their older rivals. Many succumb on the way, and a few only survive; those few become stronger with every year, they get a larger share of light and air with every increase of their height, and at last in their turn they tower above their neighbours, and seem as though they would grow on forever, and for ever become stronger as they grow. But they do not. One tree will last longer in full vigour and attain a greater size than another; but sooner or later age tells on them all. Though the taller ones have a better access to light and air than their

> rivals, they gradually lose vitality; and one after another they give place to others, which, though of less material strength, have on their side the vigour of youth[4] [141, Chapter XIII, p. 263].

Unfortunately, by the intuitive and verbatim nature of his exposition, the "Marshallian increasing returns (or equivalently decreasing marginal costs)" invited furious debate among theorists of younger generations. Indeed, one of the most critical opponents, Knight, wrote in his reply [125] to Graham [79], who supported external increasing returns, that they were "empty economic boxes." Regarding controversies on increasing returns, Chipman [42] reported the following:

> It is widely agreed that Marshall's theoretical treatment of external economies [141, Book IV, Ch. XIII; Book V, Ch. XII; and Appendix H]) was far from satisfactory; indeed it seems certain that Marshall was not satisfied with it himself. ... Marshall's only definition consists in the statement ([141, p. 266 (p. 221 in the new edition)]) that external economies are those economies of scale which are "dependent on the general development of the industry." The absence of any more elaborate formal definition in Marshall's writing is so conspicuous that it must be interpreted as deliberate; Robertson used the term "evasive" (cf. Newman [171, p. 601]). In an earlier skeptical paper, Robertson ([208, p. 26]), after enumerating the usual examples (including the inevitable trade journal), sighed: "we have all at some time tried to memorize and to reproduce the formidable list." In the same year, Knight ([124, p. 597]) set forth his famous objection to the concept of external economies in the words: "external economies in one business unit are internal economies in some other, within the industry." ... Knight's paper was largely a criticism of the concept of external economies — as was Robertson's 1924 paper — as used by both Pigou [189] and Graham [78]. Graham based his argument for protection on an analysis which took the compatibility of perfect competition and increasing returns for granted; this very assumption is what was challenged by Knight, and as long as Knight's objection stood, Graham's entire

[4]In many places of *The Principles*, Marshall emphasized the analogy between economics and biology. His attitude toward biology is sometimes compared with that of Walras toward mechanics.

> argument — whatever other defects it had, and there were several — was vitiated by having this as its premise. In his reply to Knight, Graham [79] failed to come to grips with the main issue; and Knight [125], in his rejoinder, fairly placed the burden of proof on those who believed that competitive conditions could be reconciled with increasing returns. In saying with respect to external economies that "I have never succeeded in picturing them in my mind," Knight [125, p. 323] was undoubtedly expressing a feeling that was widespread but suppressed, owing to the authority of Marshall and Pigou. ... [42, pp. 740–741].

In his paper of external increasing returns, Chipman summarized and concluded those discussions as follows:

> [The compatibility of increasing returns with perfectly competitive equilibrium] was once a lively subject of debate. The debate appears to have petered out in the 1930s, with nobody the apparent winner. That this was the outcome seems evident from later writings of some of the participants. Thus, Sir Dennis Robertson presented in 1957 an account which was substantially unaltered from his contribution to the 1930 Symposium on Increasing Returns, supporting the compatibility of increasing returns with perfect competition (Robertson [210, Vol. I, Ch. IX, pp. 114–123]). On the other hand, Sir Roy Harrod, in 1967, was able to state flatly, without any qualification, as to whether economies were internal or external, that: "Increasing returns can, of course only occur if competition is less than perfect (Harrod [89, p. 74])." In the contemporary international trade literature, some authors maintain that perfect competition can prevail under conditions of increasing returns, provided the economies of scale are external to individual firms (e.g., James Edward Mead [161, p. 33]); whereas others deny the compatibility of economies of scale with perfect competition under any circumstances, and with equal confidence (Cf. R.G. Lipsey [131, pp. 496–513]. He states on pp. 511–512 [p. 277 of the reprinted version]: "It is, of course, well known that exhausted economies of scale are incompatible with the existence of perfect competition, but it is equally well known that unexhausted economies of scale are compatible with the existence of imperfect competition as long as long-run marginal cost is declining faster [sic] than the marginal revenue.") [43, pp. 347–349].

Although they did not reach any definitive conclusion as to whether increasing returns were compatible with perfect competition, the controversies, from our point of view, led to two important consequences as by-products.

The first of these is the idea of Young [276] that (external) increasing returns are the force that drive economies to grow. This idea dates back to at least Adam Smith [233] and the division of labor in a pin factory. At first, Young [275] was skeptical about increasing returns. But in his 1928 paper, he observed the division of labor resulting in increasing returns is limited by the extent of the market, and conversely, that the extent of the market is in turn enlarged by the division of labor. As a consequence, he wrote "the division of labour depends in large part upon the division of labour," and added that "this is more than mere tautology." It is apparent from our point of view that his observations on increasing returns are the source of modern endogenous growth theory.

The second of these consequences is the theory of imperfect competition, which, using Newman's [171] words, "rose from the ashes." In one of his most influential papers, in attacking the idea of external increasing returns, Sraffa [239] pointed out that the very concept of an industrial supply function was illegitimate because it did not consider the interdependence of industries. He rejected the concept of competitive equilibrium and proposed an alternative equilibrium concept, namely the *monopolistically competitive equilibrium.*

Two influential monographs by Robinson [211] and Chamberlin [40] appeared in this line of research. An essential property of their theories was that firms in the market do not take prices as given when they make their production decisions, which means that, in the context of partial equilibrium analysis, each firm faces a downward sloping subjective demand curve. Robinson [211] pursued the study of the welfare properties of imperfectly competitive equilibrium using the concept of consumer surplus. Chamberlin [40] introduced commodity differentiation and free entry of firms into the model and clarified the concept of monopolistic competition.

The purpose of this book is to examine the Marshallian tradition of economic thought, increasing returns, and monopolistic competition from the perspective that has grown out of the Walrasian tradition of general equilibrium theory (see the following sections). The main goal of Chapter 7 in this book is to demonstrate the existence of a competitive equilibrium with external increasing returns in a dynamic infinite time horizon economy whose equilibrium consumption and production paths grow without bound. This model is considered to be a realization of Young's view of increasing returns.

In Chapter 8, we present a monopolistically competitive equilibrium of a market in which each firm perceives a (subjective) downward sloping demand curve and its technology exhibits a kind of (internal) increasing returns from large setup costs.

1.2 After the 1950s: Existence and Limit Theorems

The mathematical foundations of modern general equilibrium theory were established by Arrow and Debreu [10], McKenzie [154], and Nikaido [173]; these proved the existence of the competitive equilibrium in the Walrasian market with a finite number of commodities and economic agents.[5] It should be emphasized that the proof of the existence of the Nash equilibrium [166] had a close theoretical connection with Arrow and Debreu [10]. In fact, their method of proof applied a version of Nash's theorem (Debreu [49]) to a given market model constructed from an abstract game theoretic model.

In these papers, all basic concepts, such as the commodity space, prices, agents' characteristics, and so on, are represented rigorously by pure mathematical concepts, and since then, the general equilibrium theory has been a "geometry of agents and commodities." Indeed, theorems on competitive equilibrium and its welfare properties are stated and proven in terms of topology and convex analysis, and the key results used were fixed point and separation hyperplane theorems. Using these mathematical techniques, the question of the

[5]In a series of papers [260, 261], Wald studied the existence of competitive equilibria in alternative market models, including a linear technology model and a pure exchange model.

existence of equilibrium posed by Walras has been answered affirmatively and definitively in a more general way than Walras himself expected.

Edgeworth's conjecture on the core limit theorem was solved by Debreu and Scarf [55]. They generalized Edgeworth's result for two commodities and two consumers to the case of an arbitrarily finite number of commodities and types of consumers with strictly concave utility functions.[6] More specifically, they considered a sequence of economies, as did Edgeworth, over which the number of consumers of each type becomes large, and proved that the core allocations shrink to the competitive equilibrium in an appropriate sense. In a crucial step of their proof, the separation hyperplane theorem was used (Section 2.6).

In view of these results, the power of mathematical methods used in economic analysis is evident; the economic concepts and propositions have been clarified, and the open questions and conjectures held by theorists in the nineteenth century have been solved definitively. In later chapters of this book, we present several results obtained using modern mathematical techniques and pursue this line of research further.

1.3 Local Uniqueness and Stability of Equilibria

Once the existence of equilibrium has been established, we are interested in its stability and uniqueness. Walras' *tâtonnements*[7] was formulated by Samuelson [220] as a system of differential equations. This work was followed by Arrow and Hurwicz [13], Arrow *et al.* [12], and McKenzie [158] etc.

The uniqueness of equilibrium was first studied by Wald [262], who developed alternative sufficient conditions for the uniqueness; the weak axiom of revealed preferences holds for the market excess

[6] Nishino [176] is also an important contribution to the core limit theorem.

[7] For a textbook-level exposition of Walras' *tâtonnements*, see Arrow and Hahn [11, Chapters 11 and 12], McKenzie [160, Chapter 2].

demand functions, or all commodities are gross substitutes. Both these conditions are quite strong, but Wald's results stimulated almost all subsequent studies of this issue. Samuelson [221, 222] discussed the problem in the context of factor price equalization in world trade. Gale and Nikaido [74] pointed out an error in Samuelson's argument and generalized the condition of gross substitutes of Wald.[8] A general condition was finally obtained by Dierker [60] using differential topology.

The purpose of the aforementioned research was to develop sufficient conditions for the (global) uniqueness of equilibrium. It is clear from these studies that the globally unique equilibrium is very special and appears under very strong conditions.

Debreu [53] discussed the problem from a drastically different perspective. He studied the local uniqueness of equilibria, namely those that are discrete and locally stable, which means that they change continuously when the agents' characteristics are perturbed continuously. An economy in which the competitive equilibria are discrete and locally stable is now called a *regular economy*, or a *singular economy* otherwise. *Regularity* is an important property in economic analysis because most applications of theoretical models are performed using comparative statics, which perturbs the economic parameters slightly and examines the associated changes in the equilibria. Hence, these models will lose any predictive power for singular economies. The regularity of equilibria is a theoretical basis of comparative statics.

Debreu did not identify any sufficient conditions under which economies become regular. Rather, he asked *how generally* economies are regular, and his answer was that *almost all* economies are regular. We explain this point more precisely because this result has led to drastic changes, both methodologically and philosophically, in economic analysis. Recall that in their research on the core limit theorem, Edgeworth, Debreu, and Scarf already considered a *sequence* of economies. However, Debreu considered a *topological space* of

[8] For the comparative statics, see McKenzie [160, Chapter 4].

economies. He showed that under suitable differentiability assumptions, the set of singular economies is a closed subset with Lebesgue measure equal to zero of the whole space of economies. Before this paper appeared, economists had been interested in an *individual economy* and examined its economic properties. Debreu opened up another possibility to be explored, namely the mathematical structures of the *space of economies*. Methodologically, his paper introduced differential topology into economic analysis. Sard's theorem has been added to theorists' toolboxes (Appendix E). We review this theory in Section 2.8.

1.4 Markets with a Continuum of Traders

In a competitive equilibrium, market prices are given to economic agents by definition. This price-taking behavior is justified by a very large number of traders in the market so that the influence of each individual on prices is negligible. However, this is inconsistent with traditional economic models with a finite number of traders that have been discussed so far.

Aumann [14] introduced a model of an exchange economy with a continuum of consumers to present rigorously from a mathematical perspective the idea of perfect competition and showed that in such an economy, the set of core allocations coincides with the set of competitive allocations. Edgeworth's conjecture was realized as the core equivalence theorem (Section 3.4). The actual markets, however, only contain a finite number of traders. What is the economic meaning of a market with a continuum of traders? Aumann explained it using an analogy from physics. He wrote

> Actually, it is no stranger than a continuum of prices or of strategies or a continuum of "particles" in fluid mechanics. In all these cases, the continuum can be considered an approximation to the "true" situation in which there is a large but finite number of particles (or traders or possible prices)... There is perhaps a certain psychological difference between a fluid with a continuum of particles and a market with a continuum of traders. Though we are intellectually convinced that a fluid

> contains only finitely many particles, to the naked eyes it still looks quite continuous. The economic structure of a shopping center, on the other hand, does not look like continuous at all. But, for the economic policy maker in Washington, or for any professional macro economist, there is no such difference. He works with figures that are summarized for geographic regions, different industries, and so on; the individual consumer (or merchant) is as anonymous to him as the individual molecule is to the physicists [15, p. 41].

Hildenbrand [94] provided a more statistical explanation. As will be shown in Chapter 3, an economic model with a continuum of traders induces an atomless or continuous distribution over the set of agent's characteristics, such as the income distribution. Subsequently, he wrote

> To view the distribution of agent's characteristics of a *finite* set A of agents as atomless distribution means, strictly speaking, that the "actual" distribution is considered as a distribution of a *sample* of size $\sharp A$ drawn from a "hypothetical" population.[9] This statistical point of view is based on the well-known fact that the sample distributions converge with increasing sample size to the hypothetical distribution [94, p. 110].

Studies of general equilibrium models with a finite number of traders have relied heavily on convexity assumptions regarding the agents' characteristics. For instance, the existence of the competitive equilibria is ascertained under the assumption that every consumer has a convex preference defined on a convex consumption set. Aumann [17] showed that in an exchange economy with a continuum of consumers, the existence of equilibrium could be obtained without the convexity assumption on preferences. Mathematically speaking, this is a consequence of Lyapunov's theorem (Theorem F.1), which states that the integral of a measurable correspondence over an atomless measure space is convex-valued. Therefore, even if individual demand correspondences are not convex-valued, the market demand correspondence, which is defined as the integral of the individual demand correspondence over the atomless measure space

---

[9] $\sharp A$ means the number of elements of the set A.

of consumers, is convex-valued, and we can apply Kakutani's theorem.

In subsequent studies, Mas-Colell [146] proved the existence of equilibria without the convexity assumption on the consumption sets in an exchange economy with indivisible commodities, or the commodities that can be consumed in integer units. Hence, the consumption set in this case is $\mathbb{R}_+ \times \mathbb{N}^{\ell-1}$, where $\mathbb{R}_+$ is the coordinate of a divisible commodity and the other $\ell - 1$ commodities are indivisible, which is obviously not a convex set. Yamazaki [265] generalized this result to the consumption sets that are closed and bounded from below.

When one does not assume that the consumption sets are convex, individual demand will generally exhibit discontinuous behaviors for continuous changes in prices. The key observation of Mas-Colell and Yamazaki is that the discontinuous behaviors of the individual traders will be smoothed out when the distribution of agents' characteristics, such as the income distribution, is sufficiently dispersed. These phenomena are exactly what Cournot had already expected over 150 years ago. We explain the "smoothing effect of the aggregation" in Section 3.5 and discuss the problems that arise when one considers production in Section 6.4.

1.5 Markets with Infinitely Many Commodities

An economic model with an infinite time horizon first appeared in the context of optimal growth theory by Ramsey [198]. His problem was to find the level of savings that maximizes total future utility for a population. Von Neumann [169, 170] presented a general equilibrium model of growth in which there were no demand functions, only productions of linear activities. The economy is in equilibrium at each time period, and the equilibrium configurations are identical from period to period (stationary equilibrium). To prove the existence of equilibrium, he generalized a saddle point theorem of bilinear form that was used to show the existence of equilibrium in two-person zero-sum games. He deduced the saddle point theorem from a fixed point theorem, and it was the first time that the fixed point theorem appeared in economic analysis.

The need for infinite-dimensional commodity spaces in general equilibrium analysis was identified clearly by Debreu [51]. In Note 2 of Chapter 2, he wrote

> The assumption of a finite number of dates has the great mathematical convenience of enabling one to stay within a finite-dimensional commodity spaces. There are, however, conceptual difficulties in postulating a predetermined instant beyond which all economic activity either ceases or is outside the scope of the analysis. It is therefore worth noticing that many results of the following chapters can be extended to infinite dimensional commodity spaces. In general, the *commodity space* would be assumed to be a linear space L over the reals and instead of a price vector p, one would consider a linear form v on L defining for every action $a \in L$ its *value* $v(a)$. In this framework could also be studied cases where the date, the location, the quality of commodities are treated as continuous variables [51, pp. 35–36].

This analysis originated with Peleg and Yaari [188] and Bewley [27], both of whom proved the existence of equilibrium for infinite time-horizon economies. The commodity space of Peleg and Yaari is the space of all sequences endowed with the product topology, or the space $\mathbb{R}^\infty$. To prove the existence, they applied a Debreu–Scarf-type core limit theorem. The commodity space of Bewley's paper is the space of all bounded sequences (for the space of all essentially bounded measurable functions, see Appendix G), which is denoted by ℓ^∞ (or L^∞ for the function space). He considered a sequence of equilibria of a finite number, say ℓ, of dimensional subeconomies and showed that the limit of the finite-dimensional equilibria as $\ell \to \infty$ is indeed an equilibrium of the original infinite-dimensional economy.

Mas-Colell [145] and Jones [103] worked with the space of Borel measures on a compact set K, denoted by $ca(K, \mathcal{B}(K))$, where the set K is the set of commodity characteristics. The commodity vector $\boldsymbol{x}$ is postulated as a measure on K and the value (distribution) for a Borel set $B \subset K$, $\boldsymbol{x}(B)$ represents the quantity of characteristics contained in the set B. This commodity space represents commodity differentiation in the most general form. Mas-Colell [150] proved the existence of competitive equilibria for a general topological vector space in an exchange economy, and Zame [277] generalized his result to a production economy. These accomplishments of the past will be presented in Chapter 4.

The next natural step is to integrate the results obtained in Chapters 3 and 4. Indeed, current theoretical interest focuses on studies of market models with an infinite number of commodities and a continuum of traders. These models, however, are a technically formidable task and the research remains incomplete. In Chapter 5, we present an infinite horizon exchange economy with a measure space of consumers and establish the existence and core equivalence of the competitive equilibria.

1.6 Modern Developments of External Increasing Returns

The concept of external increasing returns of Marshall was clarified by Chipman [43]. He called it "parametric economies of scale," because in his formulation, each firm takes the scale parameter in its production function as given and believes that it operates under constant returns to scale, but actually, the parameter is affected by the total industry inputs, and consequently, the production functions of all firms exhibit increasing returns to scale. The following example illustrates this idea.

Suppose that there exist ν identical firms in an industry, each of which has the same production function $y = \varsigma z$, where y is the output and z is the input and the coefficient ς is a parameter. The firms take ς as given when they make their production decisions, so that their subjective production function displays constant returns to scale, but actually, ς is assumed to be determined endogenously based on total industry inputs, $\sum_{j=1}^{\nu} z_j$, where z_j is the input level of firm j. That is, assuming that all firms use the same amount of inputs z, $\varsigma = \nu z$ holds in equilibrium. Then, the firms' objective production function is $y = \nu z^2$, which exhibits increasing returns, and the resulting equilibrium concept is a competitive equilibrium with production externalities.

The idea of parametric economies of scale originated with Edgeworth. In the midst of the desperately confusing debates on the compatibility between competitive equilibria and increasing returns (decreasing costs), which we discussed in Section 1.1, he was looking at the truth. We quote Chipman:

> The essential idea put forward by Edgeworth [67, pp. 66–68] and [68, Papers III, pp. 140–141] was that marginal cost was a function of a particular firm's output, and also of aggregate industrial output; and that it might be rising with respect to the former and falling with respect to the latter. According to this conception, rising marginal cost curves for the individual firms would shift downwards with a rise in industrial output, leading to a falling supply curve for the industry [42, p. 739].

In a subsequent paragraph, Chipman also wrote:

> To illustrate the case, an expansion in a certain industry may make possible a further division of labor, and give rise to new categories of technicians. The contribution of each individual firm to this process may be so negligible that no single entrepreneur will take into account the effect of his own scale of operations on the development of new specialized skills. This element of cost therefore plays the same role as do market prices. It is curious indeed that Edgeworth, of all people, did not notice the analogy between this concept of external economies and his own limit theorem justifying the competitive price mechanism (cf. Edgeworth [66, pp. 240–243]). All we need to assume is that a firm's size has a small effect (negligible from its point of view) on the organization of the industry (especially the labor market), and that the firm consciously adjusts its organization to the changed condition of the industry [*ibid.*, p. 740].

For the first time, Chipman [43] showed that the external increasing returns were indeed compatible with a competitive equilibrium and even Pareto optimality in one consumer economy. In Section 7.4, we generalize his result to the case of several consumers.

As stated in Section 1.1, the idea of external increasing returns dates back to Adam Smith's idea of division of labor in the pin factory. The division of labor increases the productivity of the firm (factory) by dividing tasks and specializing work. However, each separate task makes no sense in and of itself. Although each part of the process is conducted by a single worker (or group of workers), all other parts are interdependent. The division of labor then presupposes cooperative relationships between the workers in the factory. Furthermore, these cooperative actions are generally not intentionally altruistic or benevolent; rather, they generally result from each person's rational behavior. That is to say, there is a form of mutual advantage or reciprocity within the factory. External increasing returns are then simply the generalization of the reciprocity within the factory to the entire industry and/or the whole economy. We must bear in mind that the problem is not that of the "efficiency of markets," but of the reciprocity incorporated into liberal societies. Put differently, it is not a problem within markets, but rather, outside markets, namely, the externalities. The crucial aspect is the basic structure of the liberal societies in which the *reciprocity* is incorporated. This is exactly what Rawls [202, 203] emphasized as the fundamental basis of social justice.

One generally recognizes the cooperative relationship in the division of labor in retrospect. For instance, there are generally no "central planners" in any factory to form such relationships between workers. The division of labor is self-organizing, and therefore, so are external increasing returns. In addition, they do not come from any particular human mind. In other words, the reciprocity embodied in external increasing returns is the work of the institutions in a liberal society that develop naturally to increase the welfare of the members of the society.[10] In this sense, external increasing returns exist everywhere in liberal society. The tendency to form such an institution comes from the fact that any single person in any society cannot do

[10]Of course, this is also the case for the formation of markets.

everything he/she wants or in fact needs. People just do what they can do and rely on others for the rest, that is, they have natural complementarities to one another. This is exactly what Rawls called *social union*, an idea that later deeply impressed Kenneth Arrow, who likewise pointed out its similarity with Adam Smith's pin factory

> Indeed, one of the most brilliant passages in Rawls' book is on what he calls "social union" (Rawls [202, pp. 520–530]). He argues that no human life can encounter more than a small fraction of the experiences needed for completeness, so that individuals have a natural complementarity with each other (a more mundane version of this idea is Adam Smith's emphasis of the importance of the division of labor) [9, p. 262].

In Section 7.5, we present a socioeconomic model in which the external increasing returns implement the *difference principle*, which was proposed by Rawls [202, 203] as the principle of distributive justice in liberal societies.

1.7 Modern Developments of Monopolistic Competition

Considerable effort has been devoted to generalizing the classical general equilibrium models of production economies to models, including (internal) increasing returns to scale technologies. These increasing returns, or non-convex production sets, are of course incompatible with the competitive behavior of firms. Instead, the firms are assumed to operate under some pricing rule, for instance, that which satisfies the first-order condition of profit maximization. Mathematically speaking, the pricing rule in this case is what assigns to each firm a normal vector of efficient production plans. This pricing rule is called the marginal cost pricing (MCP) rule. Under the MCP rule, a firm with increasing returns to scale technology possibly earns a negative profit in the equilibrium. Such losses are assumed to be covered by lump-sum taxes collected from households. Therefore, the firms with increasing returns technologies in this theory can be thought of as privately owned public utilities, which are regulated. Mantel [140] and Beato [23] proved the existence of the MCP equilibrium. We will discuss MCP equilibria in Section 7.1.

The study of the general equilibrium model of imperfect competition, however, has been relatively limited because of two fundamental difficulties, one conceptual and the other technical. As we have seen, most of the general equilibrium theory results have relied on convexity assumptions. Therefore, the theory only relates to firms with constant or decreasing returns to scale technologies, with a few exceptions, such as the MCP theory explained earlier. However, monopolistically competitive firms with constant or decreasing returns to scale technologies are rare, and the economic meaning of models with such firms seems to be restricted. Moreover, from a technical perspective, even if one assumes that the monopolistically competitive firms have convex technologies, the conditions on the fixed point map to which Kakutani's theorem can be effectively applied are not generally guaranteed, as Roberts and Sonnenschein [206] showed.

Nevertheless, we have two remarkable achievements by Negishi [168] and Gabszewicz and Vial [77]. Negishi introduced the downward sloping subjective demand function for each firm that passes through the equilibrium point, meaning that the firms observe equilibrium prices, and Gabszewicz and Vial [77] constructed a model in which the firms' behaviors are monopolistically competitive in the sense of Cournot, and they make their production decisions using the true market demand function. Both these papers proved the existence of equilibria under the assumption that both the production and profit functions are concave, which is necessary for the reason explained earlier.

In Chapter 8, we generalize the Negishi-type model of monopolistic competition to the case in which the firms are allowed to have non-convex production sets because of large setup costs. We prove the existence of equilibrium using techniques that have been developed by Dehez and Dreze [56] in a study of equilibrium with increasing returns and a pricing rule.

1.8 Organization of the Book

The organization of the book is outlined in the table of contents. Part I discusses the mathematical structure of general equilibrium models of exchange economies. We shall prove various theoretical results ranging from elementary proofs to those at the forefront of research. Therefore, Part I is (almost) purely mathematical and technical in nature.

Chapter 2 presents the classical results on exchange economies. This chapter, in which we introduce the notations and basic terminologies, serves as an introduction to the basics of general equilibrium theory for beginning graduate students. The concepts of individual and market demand are established, and the existence and the local uniqueness of competitive equilibria are discussed. The limit theorem of Debreu and Scarf, which is developed into the core equivalence theorem of Aumann in Chapter 3, is presented. The abstract game form of the market model that is applied to production economies with external increasing returns in Chapter 7 is also introduced.

Chapter 3 presents the theory of the large economy developed by Aumann, Hildenbrand, and others. We also discuss the smoothing effects on aggregated demand, which are particularly relevant for the model, with many consumers with non-convex consumption sets, which makes it possible to incorporate indivisible commodities.

Chapter 4 is devoted to the study of equilibrium theory with infinite-dimensional commodity spaces, which was developed by Bewley [27], Mas-Colell [145, 150], Jones [103], Zame [277], and many others. We focus on the existence of competitive equilibria for such economies and provide methods and techniques utilized in later chapters. In fact, the results of Chapters 3 and 4 are combined in Chapter 5 to prove the existence and core equivalence of infinite horizon exchange economies with a continuum of consumers. We introduce two distinctive types of formulations, those of individualized economies and those of distributionalized economies, and see that they play the distinctive roles from economic perspectives, but are equivalent mathematically. To overcome some mathematical problems, we introduce a stronger concept of standard (Lebesgue) measure spaces, called *saturated* or *super atomless* measure spaces (see Appendix F).

Part II examines the economic and philosophical aspects of the general equilibrium theory. In Chapter 6, we study production economies and present two formulations: private ownership economies and coalition production economies. The concept of classical competitive equilibria with decreasing or constant returns to scale is generalized to the concept of social equilibria proposed by Scarf [224] and to the case of (internal) increasing returns to scale in Section 7.2, and we prove the existence of the social equilibrium via the core existence theorem.

The main theme of Chapter 7, however, is the concept of external increasing returns. We will prove the existence of competitive equilibrium for an infinite horizon economy in the presence of external increasing returns in the most general form. This result is considered to be a final answer for the question discussed in the 1920–1930s. As stated in Section 1.6, we also present a model of liberal societies and show how external increasing returns operate as the exact mechanism for implementing the difference principle, which could not be explained in Rawls [202, 203]. Chapter 8 concludes by proving the existence of a Negishi-type monopolistic competitive equilibrium with a large setup cost.

As the title indicates, our theme in this book is restricted to the most basic topics of general equilibrium theory. There are evidently many other topics that are not covered by this book such as theories of optimal growth, overlapping generations economies, financial or incomplete markets with money, uncertainty, asymmetric information, and so on. The readers will be able to consult appropriate monographs or survey articles on these topics.

The author has attempted to consider the balance between mathematical generality and economic intuition carefully. Therefore, the results in the text are not necessarily presented in the most general forms from a mathematical perspective. Rather, the author provides illustrative proofs that provide readers with the essential mathematical foundations behind those theorems. This is achieved most often by the original proofs. For mathematically more general results in the literature, the readers can refer the notes of each chapter.

All mathematical techniques used in the text are outlined in Appendices A–G. Appendices A–D provide the mathematical foundations of the entire book, and Appendix E is relevant to the discussion of the local uniqueness in Section 2.8 and the differentiability of demand functions in Section 4.5, Appendices F and G to Chapters 3–5, respectively. References for the readers who are interested in greater detail, including the proofs, can be found in Appendix H.

Part I

Theory of Exchange

Chapter 2

Classical Exchange Economies

2.1 Commodities and Markets

In every economic phenomenon, we examine two categories of "actors," namely *economic agents* and *commodities*. The "stage" on which they play their roles is called a *market*. The market is an abstract "field" in which the economic agents trade the commodities with each other. However, if we continue this description, it will then become circulatory and not make any sense. In any case, the agents' exchange or trade commodities in the market and the market is something that consists of the agents and commodities. This suggests that "the agents," "the commodities," and "the market" are the most fundamental concepts, which are, to some extent, *given* to our science. Any completely formal, independent, and self-contained definitions of these concepts are probably impossible and perhaps not useful. Rather, our understanding of these fundamental concepts is essentially based on our experience in our actual everyday economic lives. Moreover, even if we succeeded in formalizing an abstract economic theory that is suitably constructed from a mathematical point of view, the theory will implicitly make presumptions regarding a variety of social institutions that are outside the economic area: for instance, laws, politics, history, and so on, which cannot be completely specified and are usually not even mentioned at all. We hope that the readers bear in mind these remarks when they read the following explanations of basic concepts.

We start with commodities. It is assumed that there exists an open market for each commodity. As is well known (see, e.g., Debreu [51]), commodities are distinguished from each other by their physical character and by the location and time at which they are available to traders. We can arrange the commodities in a suitable order in the form of finite series or vectors, such as

$$(x^1, x^2 \dots x^\ell),$$

when we assume that there are ℓ distinguishable commodities. In this case, the *commodity vectors*, which are typically denoted by $\boldsymbol{x} = (x^1 \dots x^\ell)$, are considered to be elements of ℓ-dimensional Euclidean space $\mathbb{R}^\ell$. Each coordinate x^t is sometimes called a *good*. A good that is made available *to* an economic agent is called an *input* for that agent. A good that is made available *by* an economic agent is called an *output* for that agent. For the economic agents called the *consumers*, which will be explained in Section 2.2, the input goods are called *consumption goods* and are represented by positive quantities of the corresponding coordinates of the commodity space. For the economic agents, labor is an example of an output good and is represented by a negative quantity. In Chapter 6, we introduce the economic agents called *firms*. The output goods of the firms are represented by positive quantities and the inputs by negative quantities. The convenience of this notational convention will become clear in subsequent chapters. In other chapters, particularly Chapter 4, we will assume that there are infinitely many goods. Hence, in this case the commodity vectors are infinite sequences such as

$$(x^0, x^1 \dots x^t \dots).$$

NB: Throughout the book, we define the set of natural numbers $\mathbb{N}$ as the set of non-negative integers, or $\mathbb{N} = \{0, 1, 2, \dots\}$. Then, it is notationally convenient to start the commodity index with 0 when there are infinitely many commodities in the market.

There is a need for such infinite vectors because goods should sometimes be indexed by an infinite number of indices. As a typical case, a commodity index indicates the time period over which a good is delivered to consumers. In such a theory, it seems natural not to specify the terminal period of trades (the infinite horizon model). We will generally use the index t for the coordinate of the commodity vectors even if the number of goods is finite.

We use L to denote the set of all commodity vectors (or *commodity bundles*) and refer to it as the *commodity space*. In this space L, the sum $\boldsymbol{x}+\boldsymbol{y}$ and the scalar multiple $a\boldsymbol{x}$, $a \in \mathbb{R}$ are defined in an obvious way. Then, the space L is a finite- or an infinite-dimensional vector space (Appendices A and E). For each commodity t, we can observe a market price p^t at which the commodity t is traded among the economic agents (traders). The prices will consist of a vector

$$(p^1, p^2 \dots p^\ell)$$

when ℓ commodities exist, and

$$(p^0, p^1 \dots p^t \dots)$$

when there are infinitely many goods. The (*market*) *value* of a commodity vector $(x^1 \dots x^\ell)$ evaluated at a price vector $(p^1 \dots p^\ell)$ is defined by $\sum_{t=1}^{\ell} p^t x^t$. When the number of goods is infinite, we can define the market value similarly as $\sum_{t=0}^{\infty} p^t x^t$ when this value is well defined.

Mathematically speaking, it is natural to define the price vector as a *linear functional* or a *linear form* (linear and real-valued function) on the vector space L of the commodity vectors. Let, $\boldsymbol{p}$ be a *price functional*. The value of $\boldsymbol{p}$ at $\boldsymbol{x} \in L$ is denoted by $\boldsymbol{p}(\boldsymbol{x})$ which is the *market value* of the commodity bundle $\boldsymbol{x}$. Then, for every $\boldsymbol{x}, \boldsymbol{y} \in L$ and every real number $a, b \in \mathbb{R}$, it follows by definition of linearity that

$$\boldsymbol{p}(a\boldsymbol{x} + b\boldsymbol{y}) = a\boldsymbol{p}(\boldsymbol{x}) + b\boldsymbol{p}(\boldsymbol{y}).$$

In mathematical terminology, the set of all linear and continuous functionals on L is called the *dual space* of L and denoted by L^*. For the topologies and the theory of linear functionals on the vector spaces, we refer readers who are not familiar with these mathematical concepts to Appendices E and G.

Let $\boldsymbol{e}_t = (0 \ldots 0, 1, 0 \ldots) \in L$ be the commodity vector that contains 1 unit of the good t and 0 unit of the good for all $s \neq t$. Let $\boldsymbol{p} \in L^*$ be a price functional and set $p^t = \boldsymbol{p}(\boldsymbol{e}_t)$. Then, the price functional $\boldsymbol{p}$ has a vector representation that is denoted by $\boldsymbol{p} = (p^0, p^1 \ldots p^t \ldots)$. In this case, we call $\boldsymbol{p} = (p^t)$ the *price vector.* Let $\boldsymbol{x} = (x^0, x^1 \ldots x^t \ldots) = \sum_{t=0}^{\infty} x^t \boldsymbol{e}_t$ be a commodity vector. For each T, we have

$$\boldsymbol{p}\left(\sum_{t=0}^{T} x^t \boldsymbol{e}_t\right) = \sum_{t=0}^{T} p^t x^t$$

from the linearity of $\boldsymbol{p}$. Hence,

$$\boldsymbol{p}(\boldsymbol{x}) = \boldsymbol{p}\left(\lim_{T\to\infty} \sum_{t=0}^{T} x^t \boldsymbol{e}_t\right) = \lim_{T\to\infty} \sum_{t=0}^{T} p^t x^t = \sum_{t=0}^{\infty} p^t x^t,$$

as it should be under the suitable continuity assumption regarding $\boldsymbol{p}$, which is not specified here.

To summarize: *The set of all commodity bundles form a finite- or an infinite-dimensional vector space L. We call L the commodity space. A linear functional on L is called a price functional. The set of all price functionals is what mathematicians call the dual space of L and is denoted by L^*.*

2.2 Preference of a Consumer

In the economy, consumers are economic agents who purchase consumption goods and supply production factors such as labor. Let A be the set of all consumers in the economy. The set A may be a finite or an infinite set. We usually denote an individual member of the set A by a.

A consumer $a \in A$ must determine his/her consumption plan vectors or, put more simply, consumption vectors $\boldsymbol{x}_a = (x_a^t) \in L$. As stated in Section 2.1, it is commonly assumed that if $x_a^t > 0$, then the commodity t is *demanded* by consumer a, whereas if $x_a^t < 0$, then the commodity t is *supplied* by the consumer a (for example, the commodity t could be labor, measured by hours). The set of all consumption vectors possible for consumer $a \in A$ is called the *consumption set* of a and is denoted by X_a. It is a subset of the commodity space L. It seems that there are few general conditions that we can impose on the consumption set from an economic point of view. We will return to this point in Chapter 3, but it seems reasonable that the consumption set is bounded from below, as the ability of consumers to supply labor is normally limited in time. That is to say, there exists a fixed vector (possibly depending on a) $\boldsymbol{b}_a \in L$ such that for all $\boldsymbol{x}_a \in X_a$, it follows that $\boldsymbol{b}_a \leq \boldsymbol{x}_a$. Furthermore, for a technical reason, we will assume that the consumption set is a closed subset of the commodity space L with respect to an appropriate topology that will be specified in each case. The specification of the topologies is important when the commodity spaces are infinite dimensional, as in Chapters 4 and 5. When the commodity space is finite dimensional and $L = \mathbb{R}^\ell$, then we will use the usual topology on $\mathbb{R}^\ell$.

To sum up; *for every consumer $a \in A$, his/her consumption set is a closed subset X_a of the commodity space L which is bounded below.*

Now, we introduce preferences of consumers. For a consumer $a \in A$, we assume that there exists a binary relation on X_a, denoted by $\prec_a$. Let $\boldsymbol{x}$, $\boldsymbol{y} \in X_a$ be consumption vectors. Then, $(\boldsymbol{x}, \boldsymbol{y}) \in \prec_a$ is read as "$\boldsymbol{y}$ is strictly preferred to $\boldsymbol{x}$ by the consumer a." $(\boldsymbol{x}, \boldsymbol{y}) \in \prec_a$ is usually written as $\boldsymbol{x} \prec_a \boldsymbol{y}$. We call $\prec_a$ the *preference relation* of the consumer a.

NB: Throughout this book, we postulate that the preference relations always satisfy the following three conditions unless otherwise stated:

(IR) (Irreflexivity): for all $\boldsymbol{x} \in X_a$, $(\boldsymbol{x}, \boldsymbol{x}) \notin \prec_a$,
(TR) (Transitivity): $\boldsymbol{x} \prec_a \boldsymbol{y}$ and $\boldsymbol{y} \prec_a \boldsymbol{z}$ imply $\boldsymbol{x} \prec_a \boldsymbol{z}$ for all $\boldsymbol{x}, \boldsymbol{y}$ and $\boldsymbol{z} \in X_a$,
(CT) (Continuity): The set $\{(\boldsymbol{x}, \boldsymbol{y}) \in X_a \times X_a | \, \boldsymbol{x} \prec_a \boldsymbol{y}\}$ is open in $X_a \times X_a$.

As already mentioned, the topology is specified when the space L is infinite dimensional, otherwise we mean the usual topology. We refer to the conditions (IR), (TR), and (CT) as the *basic postulates* for preferences. In this chapter, we will often assume that $\prec_a$ satisfies the following *negative transitivity* condition:

(NTR) (Negative transitivity): $(\boldsymbol{x}, \boldsymbol{y}) \notin \prec_a$ and $(\boldsymbol{y}, \boldsymbol{z}) \notin \prec_a$ imply that $(\boldsymbol{x}, \boldsymbol{z}) \notin \prec_a$ for all $\boldsymbol{x}, \boldsymbol{y}$ and $\boldsymbol{z} \in X_a$.

$(\boldsymbol{x}, \boldsymbol{y}) \notin \prec_a$ is written as $(\boldsymbol{x}, \boldsymbol{y}) \in \succsim_a$ or $\boldsymbol{x} \succsim_a \boldsymbol{y}$ and is read as "$\boldsymbol{x}$ is at least as desired as $\boldsymbol{y}$ by the consumer a." We define $\boldsymbol{x} \sim_a \boldsymbol{y}$ by $\boldsymbol{x} \succsim_a \boldsymbol{y}$ and $\boldsymbol{y} \succsim_a \boldsymbol{x}$ and read it as "$\boldsymbol{x}$ is *indifferent* with $\boldsymbol{y}$." The relations $\succsim_a$ and $\sim_a$ are referred to as the *weak preference relation* and the *indifference relation* of the consumer a, respectively. The irreflexivity is obviously equivalent with the *reflexivity* for $\succsim_a$:

(RF)(Reflexivity): For all $\boldsymbol{x} \in X_a$, $\boldsymbol{x} \succsim_a \boldsymbol{x}$,

and the irreflexivity and the transitivity conditions imply the completeness of $\succsim_a$:

(CP) (Completeness): For every $\boldsymbol{x}, \boldsymbol{y} \in X_a$, $\boldsymbol{x} \succsim_a \boldsymbol{y}$ or $\boldsymbol{y} \succsim_a \boldsymbol{x}$.

It is also evident that the negative transitivity is equivalent to the transitivity of $\succsim_a$:

(NTR) (Negative transitivity): $\boldsymbol{x} \succsim_a \boldsymbol{y}$ and $\boldsymbol{y} \succsim_a \boldsymbol{z}$ imply $\boldsymbol{x} \succsim_a \boldsymbol{z}$ for all $\boldsymbol{x}, \boldsymbol{y}$ and $\boldsymbol{z} \in X_a$.

A topological condition that is sometimes required is

(LNS) (Local non-satiation): for every $\boldsymbol{x} \in X_a$ and every neighborhood U of $\boldsymbol{x}$, there exists a $\boldsymbol{y} \in U \cap X_a$ such that $\boldsymbol{x} \prec_a \boldsymbol{y}$,

with the usual remark on the topology when the space is infinite dimensional. Occasionally, we will assume an even stronger condition, such as

(MT) (Monotonicity): $X_a = L_+$ and if $\boldsymbol{x}, \boldsymbol{y} \in X_a$ and $\boldsymbol{x} < \boldsymbol{y}$, then $\boldsymbol{x} \prec_a \boldsymbol{y}$.

Finally, we introduce the convexity assumption, which states that the consumers prefer mixed consumption vectors rather than each of the individual consumption bundles or, in other words, they prefer mixed rather than extreme commodities.

(CV) (Convexity): X_a is a convex subset of L and for every $\boldsymbol{x} \in X_a$, the set $\{\boldsymbol{y} \in X_a | \ \boldsymbol{y} \succsim_a \boldsymbol{x}\}$ is convex.

It is well known that before consumer preferences were represented by binary relations, they were given by real-valued functions on consumption sets called *utility function*,

$$u_a : X_a \to \mathbb{R}, \ a \in A,$$

which is related to the preference relation by $u_a(\boldsymbol{x}) < u_a(\boldsymbol{y})$ if and only if $\boldsymbol{x} \prec_a \boldsymbol{y}$. When this is the case, we say that the utility function u_a *represents* the preference relation $\prec_a$ (up to monotone transformations). Although some philosophical confusion existed with respect to the "meaning" of the values of the utility functions, we are not concerned with these controversies. We consider that the utility function remains a useful tool of economic analysis, and indeed, we will use it throughout this book. Note that when a preference relation is represented by a utility function, the irreflexivity, the transitivity, and the negative transitivity conditions are automatically satisfied and the continuity (CT) condition should be stated as

(UCT) (Continuity): The utility function $u_a : X_a \to \mathbb{R}$ is continuous,

which will be always assumed without explicitly stating it. The other assumptions in terms of the utility functions are

(LNS) (Local non-satiation): For every $\boldsymbol{x} \in X_a$ and every neighborhood U of $\boldsymbol{x}$, there exists a $\boldsymbol{y} \in U \cap X_a$ such that $u_a(\boldsymbol{x}) < u_a(\boldsymbol{y})$,

(UMT) (Monotonicity): $X_a = L_+$ and if $\boldsymbol{x}, \boldsymbol{y} \in X_a$ and $\boldsymbol{x} < \boldsymbol{y}$, then $u_a(\boldsymbol{x}) < u_a(\boldsymbol{y})$,

(QCV) (Quasi-concavity): X_a is a convex subset of L and for every $\boldsymbol{x} \in X_a$, the set $\{\boldsymbol{y} \in X_a | u_a(\boldsymbol{y}) \geq u_a(\boldsymbol{x})\}$ is convex.

Of course, regarding assumptions (UCT) and (LNS), the usual caveat on the topology is understood to apply.

To summarize: *The preference relation $\prec_a$ of the consumer a is a irreflexive, transitive, and continuous binary relation on the consumption set X_a. The weak preference relation $\succsim_a$ is defined by $\succsim_a = X_a \setminus \prec_a$. When a real-valued function on X_a, $u_a : X_a \to \mathbb{R}$ satisfies $u_a(\boldsymbol{x}) < u_a(\boldsymbol{y})$ if and only if $\boldsymbol{x} \prec_a \boldsymbol{y}$, then we refer to u_a as a utility function representing $\prec_a$.*

2.3 Demand of a Consumer

Until the end of this chapter, we assume that there exist m consumers in the economy, so that $A = \{1 \ldots m\}$. In addition, we assume that the commodity space L is the ℓ-dimensional Euclidean space $\mathbb{R}^\ell$ and the consumption set X_a of each consumer $a \in A$ is a closed and convex subset of $\mathbb{R}^\ell$, which is bounded from below. Each consumer a is assumed to hold initially a vector ω_a called an *initial endowment* vector. For $\boldsymbol{p} \in \mathbb{R}^\ell_+$ with $\boldsymbol{p} \neq \{\boldsymbol{0}\}$ given, the value $w = \boldsymbol{p}\omega_a$ of the endowment vector at $\boldsymbol{p}$ is the wealth of the consumer a paying his/her purchases.

Definition 2.1. Given $\boldsymbol{p} \in \mathbb{R}^\ell_+$, the *budget set* of the consumer a is defined by $\beta(a, \boldsymbol{p}) = \{\boldsymbol{x} \in X_a | \ \boldsymbol{p}\boldsymbol{x} \leq \boldsymbol{p}\omega_a\}$, and the *demand set* of a is

$$\phi(a, \boldsymbol{p}) = \{\boldsymbol{x} \in X_a | \ \boldsymbol{x} \in \beta(a, \boldsymbol{p}), \ \boldsymbol{x} \succsim_a \boldsymbol{z} \text{ for every } \boldsymbol{z} \in \beta(a, \boldsymbol{p})\}.$$

In the following, $\beta(a, \boldsymbol{p})$ and $\phi(a, \boldsymbol{p})$ will be sometimes written as $\beta_a(\boldsymbol{p})$ and $\phi_a(\boldsymbol{p})$ respectively, when there are finitely many consumers. The demand set describes consumer a's behavior, namely that he/she maximizes his/her utility subject to the budget constraint. From the definition of the set $\beta(a, \boldsymbol{p})$, it is clear that $\beta(\tau\boldsymbol{p}) = \beta(a, \boldsymbol{p})$ for all $\tau > 0$, or that $\beta(a, \boldsymbol{p})$ is *positively homogeneous of degree 0*. Hence, so is the set $\phi(a, \boldsymbol{p})$. Therefore, without loss of generality, we can restrict the price vectors $\boldsymbol{p}$ on the simplex (cf. Appendix A),

$$\Delta = \left\{ \boldsymbol{p} = (p^t) \in \mathbb{R}^\ell \,\middle|\, \sum_{t=1}^{\ell} p^t = 1,\ p^t \geq 0,\ t = 1 \ldots \ell \right\},$$

and we naturally consider that β and ϕ are correspondences from Δ to X_a,

$$\beta(a, \cdot) : \Delta \to X_a,\ \boldsymbol{p} \mapsto \beta(a, \boldsymbol{p}),$$

and

$$\phi(a, \cdot) : \Delta \to X_a,\ \boldsymbol{p} \mapsto \phi(a, \boldsymbol{p}).$$

Then, our first task is to check that $\beta(a, \boldsymbol{p})$ and $\phi(a, \boldsymbol{p})$ are non-empty, if we insist that they are correspondences, and we are also interested in their continuity properties. Unfortunately, it is difficult to ensure the non-emptiness and the continuity of these correspondences unless we restrict their domains and/or ranges. We chose to restrict the ranges. For a sufficiently large $k > 0$, we can obtain $\omega_a \ll k\mathbf{1}$, where $\mathbf{1} = (1 \ldots 1)$. We define that $\hat{K} = \{\boldsymbol{x} \in \mathbb{R}^\ell |\ \boldsymbol{x} \leq k\mathbf{1} = (k \ldots k)\}$ and set

$$\hat{X}_a = X_a \cap \hat{K} = \{\boldsymbol{x} \in X_a |\ \boldsymbol{x} \leq k\mathbf{1}\}$$

(see Figure 2.1). It is obvious that the set $\hat{X}_a$ is convex. As $\hat{X}_a \subset B(0, M)$, where $M > max\{\|\boldsymbol{b}_a\|, \sqrt{\ell}k\}$ and $\boldsymbol{b}_a$ is the lower bound of X_a, the set $\hat{X}_a$ is bounded. Let $\{\boldsymbol{x}_n\} \subset \hat{X}_a$ be a converging sequence in $\hat{X}_a$ with $\boldsymbol{x}_n \to \boldsymbol{x}$. For each n, $\boldsymbol{x}_n \leq k\mathbf{1}$. Passing to the limit, we have $\boldsymbol{x} \leq k\mathbf{1}$. Hence, K is closed, and so is $\hat{X}_a$ as the intersection of the closed sets. From Proposition B.13, $\hat{X}_a$ is compact.

We define the *restricted budget correspondence* $\hat{\beta}(a, \cdot) : \Delta \to \hat{X}_a$ and the *restricted demand correspondence* $\hat{\phi}(a, \cdot) : \Delta \to \hat{X}_a$ as follows:

$$\hat{\beta}(a, \boldsymbol{p}) = \{\boldsymbol{x} \in \hat{X}_a |\ \boldsymbol{p}\boldsymbol{x} \leq \boldsymbol{p}\omega_a\},$$

and

$$\hat{\phi}(a, \boldsymbol{p}) = \{\boldsymbol{x} \in \hat{X}_a |\ \boldsymbol{x} \in \hat{\beta}(a, \boldsymbol{p}), \boldsymbol{x} \succsim_a \boldsymbol{z} \text{ for every } \boldsymbol{z} \in \hat{\beta}(a, \boldsymbol{p})\},$$

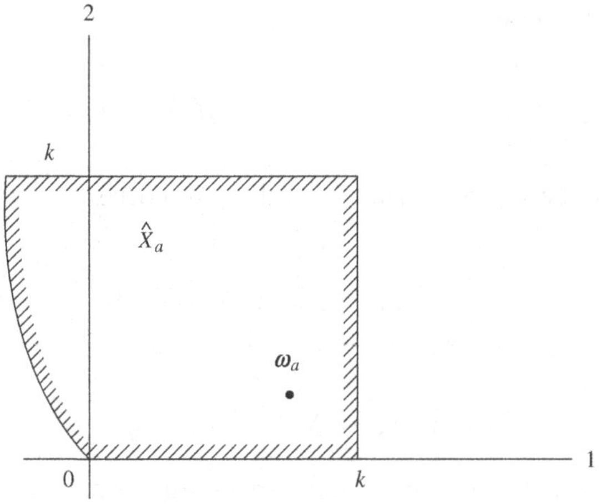

Figure 2.1. Truncated consumption set.

respectively. The set $\hat{\beta}(a, \boldsymbol{p})$ is closed. We fix a vector $\boldsymbol{p} \in \Delta$ and take a converging sequence $\{\boldsymbol{x}_n\} \subset \hat{\beta}(a, \boldsymbol{p})$ with $\boldsymbol{x}_n \to \boldsymbol{x}$. As $\hat{X}_a$ is closed, $\boldsymbol{x} \in \hat{X}_a$. For every n, we have $\boldsymbol{p}\boldsymbol{x}_n \leq \boldsymbol{p}\omega_a$. Passing to the limit, one obtains $\boldsymbol{p}\boldsymbol{x} \leq \boldsymbol{p}\omega_a$, namely that $\boldsymbol{x} \in \hat{\beta}(a, \boldsymbol{p})$. Hence, the set $\hat{\beta}(a, \boldsymbol{p})$ is a closed subset of $\hat{X}_a$ that is compact. Being a bounded and closed subset of $\mathbb{R}^\ell$, the set $\hat{\beta}(a, \boldsymbol{p})$ is compact for every $\boldsymbol{p} \in \Delta$. We now prove that the demand relation is non-empty valued. For later use, we prove the proposition in a general form, omitting the consumer index a.

Proposition 2.1. *The set $\hat{\phi} = \{\boldsymbol{x} \in B|\ \boldsymbol{x} \in B, \boldsymbol{x} \succsim \boldsymbol{z}$ for every $\boldsymbol{z} \in B\}$ is non-empty for a nonempty compact set $B \subset X$. When B is convex and $\succsim$ satisfies (NTR) and (CV), $\hat{\phi}$ is a convex set.*

Proof. Suppose that $\hat{\phi} = \emptyset$. For each $\boldsymbol{x} \in B$, the set $L(\boldsymbol{x}) = \{\boldsymbol{z} \in B|\boldsymbol{z} \prec \boldsymbol{x}\}$ is obviously open in B and we claim $B \subset \cup_{\boldsymbol{x} \in B} L(\boldsymbol{x})$. If not, then there exists $\boldsymbol{y} \in B$ such that $\boldsymbol{y} \notin L(\boldsymbol{x})$ for all $\boldsymbol{x}$. This means that $\boldsymbol{y} \succsim \boldsymbol{x}$ for all $\boldsymbol{x} \in B$, namely that $\boldsymbol{y} \in \hat{\phi}$, which is a contradiction. As B is compact, there exist a finite number of vectors $\{\boldsymbol{x}_1 \ldots \boldsymbol{x}_n\} \subset B$ such that $B \subset \cup_{i=1}^n L(\boldsymbol{x}_i)$. By the transitivity of $\prec$, we have a $\boldsymbol{x}* \in \{\boldsymbol{x}_1, \ldots, \boldsymbol{x}_n\}$ such that $\boldsymbol{x}* \succsim \boldsymbol{x}_i$, $i = 1, \ldots, n$. Hence,

$\boldsymbol{x}* \notin L(\boldsymbol{x}_i)$, $i = 1, \dots, n$, contradicting $B \subset \cup_{i=1}^{n} L(\boldsymbol{x}_i)$. Suppose that B is convex and $\succsim$ satisfies (NTR) and (CV) and take $\boldsymbol{x}_1, \boldsymbol{x}_2 \in \hat{\phi}$. As $\succsim$ is complete, we can assume without loss of generality $\boldsymbol{x}_1 \succsim \boldsymbol{x}_2$. Define $\boldsymbol{x}_\tau = \tau \boldsymbol{x}_1 + (1-\tau)\boldsymbol{x}_2$. Then, $\boldsymbol{x}_\tau \in B$ and $\boldsymbol{x}_\tau \succsim \boldsymbol{x}_2$ by the condition (CV). Hence, $\boldsymbol{x}_\tau \succsim \boldsymbol{z}$ whenever $\boldsymbol{z} \in B$ by the negative transitivity (NTR), or $\boldsymbol{x}_\tau \in \hat{\phi}$. □

We say that the consumer a satisfies the *minimum income condition* at $\boldsymbol{p}$ if the following condition is satisfied:

(MI): $\boldsymbol{p}\omega_a > \inf\{\boldsymbol{p}\boldsymbol{x} | \ \boldsymbol{x} \in X_a\}$.

If (MI) holds for every $\boldsymbol{p} \in \mathbb{R}_+^\ell \backslash \{\boldsymbol{0}\}$, we simply say that the condition (MI) holds.

Now, we show that the budget correspondence $\hat{\beta}(a, \cdot) : \Delta \to \hat{X}_a$ is continuous or that it is upper hemicontinuous and lower hemicontinuous. As $\hat{X}_a$ is compact, $\hat{\beta}(a, \cdot)$ is upper hemicontinuous if its graph

$$\text{Graph}\, \hat{\beta}(a, \cdot) = \{(\boldsymbol{p}, \boldsymbol{x}) \in \hat{X}_a | \ \boldsymbol{x} \in \hat{\beta}(a, \boldsymbol{p})\}$$

is closed, from Proposition D.2. Let $(\boldsymbol{p}_n, \boldsymbol{x}_n)_{n \in \mathbb{N}}$ be a sequence in $\Delta \times \hat{X}_a$ such that $\boldsymbol{p}_n \boldsymbol{x}_n \le \boldsymbol{p}_n \omega_a$ for all n and $(\boldsymbol{p}_n, \boldsymbol{x}_n) \to (\boldsymbol{p}, \boldsymbol{x})$. Passing to the limit, we have $\boldsymbol{p}\boldsymbol{x} \le \boldsymbol{p}\omega_a$. As $\Delta \times \hat{X}_a$ is closed, we have $(\boldsymbol{p}, \boldsymbol{x}) \in \Delta \times \hat{X}_a$. This proves the upper hemicontinuity of $\hat{\beta}(a, \cdot)$.

Next, we show that $\hat{\beta}(a, \cdot)$ is lower hemicontinuous *under the condition (MI)*. Let $(\boldsymbol{p}_n)_{n \in \mathbb{N}}$ be a sequence in Δ converging to $\boldsymbol{p}$, and let $\boldsymbol{x} \in \hat{\beta}(a, \cdot)$. From Propositions D.7 and D.8, it suffices to construct a sequence $(\boldsymbol{x}_n)_{n \in \mathbb{N}}$ converging to $\boldsymbol{x}$, and $\boldsymbol{x}_n \in \hat{\beta}(a, \boldsymbol{p}_n)$ for all n. We consider two cases, as follows. (Case 1): $\boldsymbol{p}\boldsymbol{x} < \boldsymbol{p}\omega_a$. Then, $\boldsymbol{p}_n \boldsymbol{x}_n < \boldsymbol{p}_n \omega_a$ for all sufficiently large n, say for all $n \ge n_0$. We pick an arbitrary $\boldsymbol{x}_n$ from $\hat{\beta}(a, \boldsymbol{p}_n)$ for $n = 1 \dots n_0$ and set $\boldsymbol{x}_n = \boldsymbol{x}$ for $n \ge n_0$. Then, we obtain a desired sequence. (Case 2): $\boldsymbol{p}\boldsymbol{x} = \boldsymbol{p}\omega_a$. From the assumption (MI) or $\boldsymbol{p}\omega_a > \inf \boldsymbol{p}X_a$, there exists a vector $\boldsymbol{z} \in X_a$ such that $\boldsymbol{p}\boldsymbol{z} < \boldsymbol{p}\omega_a$. Hence, $\boldsymbol{p}_n(\boldsymbol{x} - \boldsymbol{z}) > 0$ and $\boldsymbol{p}_n(\omega_a - \boldsymbol{z}) > 0$ for all sufficiently large n. We set $\tau_n = \boldsymbol{p}_n(\omega_a - \boldsymbol{z})/\boldsymbol{p}_n(\boldsymbol{x} - \boldsymbol{z})$. Then, $\tau_n > 0$ for all sufficiently large n and $\tau_n \to 1$. Let $\boldsymbol{y}_n = \tau_n \boldsymbol{x} + (1 - \tau_n)\boldsymbol{z}$, $n = 1 \dots$. We define that $\boldsymbol{x}_n = \boldsymbol{y}_n$ if $\boldsymbol{y}_n \in [\boldsymbol{z}, \boldsymbol{x}]$ and $\boldsymbol{x}_n = \boldsymbol{x}$ if $\boldsymbol{y}_n \notin [\boldsymbol{z}, \boldsymbol{x}]$.

(Note that it could occur that $\boldsymbol{y}_n \notin \hat{X}_a$.) As $\boldsymbol{p}_n\boldsymbol{x}_n \leq \boldsymbol{p}_n\omega_a$ and $\boldsymbol{x}_n \to \boldsymbol{x}$, we obtain a desired sequence. Therefore, the proof of lower hemicontinuity is established.

Finally, we note that for every $\boldsymbol{p}$, the set $\hat{\beta}(a, \boldsymbol{p})$ is convex. Indeed, denoting the *half space* determined by the vector $\boldsymbol{p} \in \mathbb{R}^\ell$ and the scalar $w \in \mathbb{R}$ by

$$H(\boldsymbol{p}, w) = \{\boldsymbol{x} \in \mathbb{R}^\ell | \, \boldsymbol{p}\boldsymbol{x} \leq w\}.$$

Clearly, the set $H(\boldsymbol{p}, w)$ is convex, and so is the set $\hat{\beta}(a, \boldsymbol{p}) = H(\boldsymbol{p}, \boldsymbol{p}\omega_a) \cap \hat{X}_a$, as the intersection of convex sets. Therefore, we have established the next proposition.

Proposition 2.2. *The restricted budget correspondence*

$$\hat{\beta}(a, \cdot) : \Delta \to \hat{X}_a, \hat{\beta}(a, \boldsymbol{p}) = \{\boldsymbol{x} \in \hat{X}_a | \, \boldsymbol{p}\boldsymbol{x} \leq \boldsymbol{p}\omega_a\}$$

is a (possibly empty) compact, convex valued. Moreover, if the consumer a satisfies (MI) at $\mathbf{p} \in \mathbb{R}^\ell \backslash \{\mathbf{0}\}$, *then it is continuous at* $\mathbf{p}$.

When the condition (MI) is fulfilled at some price $\boldsymbol{p} \in \Delta$, then the set $\hat{\beta}(a, \boldsymbol{p})$ is non-empty at $\boldsymbol{p}$. Otherwise, we have to check $\hat{\beta}(a, \boldsymbol{p}) \neq \emptyset$. In Proposition 2.2, a remark on the role of the minimum income condition (MI) is required. In Figure 2.2, $X_a = \mathbb{R}^2_+$, the endowment vector is $\omega = (1, 0)$, and the condition (MI) is violated at $\boldsymbol{p} = (0, 1)$. Consider the strictly positive price vectors $\boldsymbol{p}_n$ converging to $\boldsymbol{p} = (0, 1)$. As long as $\boldsymbol{p}_n \neq \boldsymbol{p}$, the budget set $\hat{\beta}(a, \boldsymbol{p})$ is the shaded triangle area with a limit of segment $[0, 1]$. However, actually in the limit, $\hat{\beta}(a, \boldsymbol{p}) = \mathbb{R}_+$. The segment $[1, +\infty]$ suddenly appears in the limit, and we cannot construct any sequence $(\boldsymbol{x}_n)_{n \in \mathrm{N}}$ in $\hat{\beta}(a, \boldsymbol{p}_n)$ converging to $\boldsymbol{x} = (x, 0)$ with $x > 1$. Hence, the lower hemicontinuity of $\hat{\beta}(a, \cdot)$ breaks down at $\boldsymbol{p} = (0, 1)$.

The upper hemicontinuity of the restricted demand correspondence is now immediate.

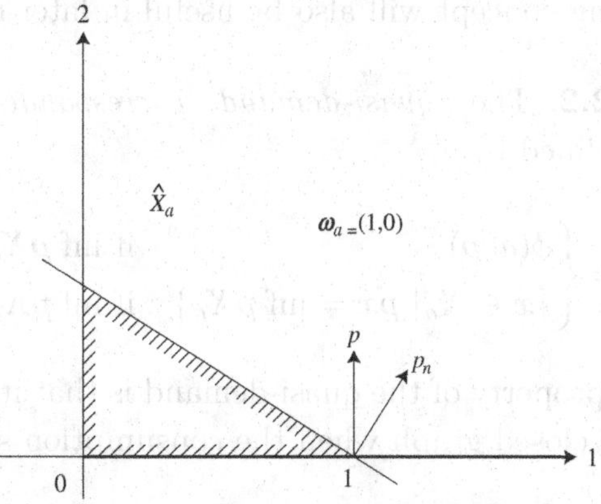

Figure 2.2. Violation of lower hemicontinuity.

Proposition 2.3. *Suppose that the consumer a satisfies (MI) at* $\boldsymbol{p} \in \Delta$. *Then, the restricted demand correspondence*

$$\hat{\phi}(a,\cdot) : \Delta \to \hat{X}_a,\ \boldsymbol{p} \mapsto \hat{\phi}(a,\boldsymbol{p})$$
$$= \{\boldsymbol{x} \in \hat{\beta}(a,\boldsymbol{p}) |\ \boldsymbol{x} \succsim_a \boldsymbol{y}\ \textit{whenever}\ \boldsymbol{y} \in \hat{\beta}(a,\boldsymbol{p})\}$$

is upper hemicontinuous at $\boldsymbol{p}$.

Proof. Let $\boldsymbol{x}_n \in \hat{\phi}(a,\boldsymbol{p}_n)$, $n \in \mathbb{N}$, $\boldsymbol{p}_n \to \boldsymbol{p}$, and $\boldsymbol{x}_n \to \boldsymbol{x}$. As the range of the correspondence $\hat{\phi}$ is compact, it is sufficient to show that $\boldsymbol{x} \in \hat{\phi}(a,\boldsymbol{p})$. From Proposition 2.2, we have $\boldsymbol{x} \in \hat{\beta}(a,\boldsymbol{p})$. Let $\boldsymbol{z} \in \hat{\beta}(a,\boldsymbol{p})$. Again, from Proposition 2.2, we have a sequence $\boldsymbol{z}_n \in \hat{\beta}(a,\boldsymbol{p}_n)$ for all n and $\boldsymbol{z}_n \to \boldsymbol{z}$ as $n \to \infty$. Hence, $\boldsymbol{x}_n \succsim_a \boldsymbol{z}_n$ for all n and, therefore from (CT), $\boldsymbol{x} \succsim_a \boldsymbol{z}$ in the limit. This proves $\boldsymbol{x} \in \hat{\phi}(a,\boldsymbol{p})$. □

We notice that Proposition 2.3 is proved without assuming (NTR) or (CV). Readers may not be satisfied with Propositions 2.1 and 2.3 because they claim the non-emptiness and the (upper hemi-) continuity only for restricted (truncated) correspondence $\hat{\beta}(a,\cdot)$ and $\hat{\phi}(a,\cdot)$, rather than for $\beta(a,\cdot)$ and $\phi(a,\cdot)$. However, the purpose of these results is to apply them to prove the existence of the competitive equilibrium demonstrated in Section 2.4, and for that purpose, our results for the truncated correspondences are adequate.

The following concept will also be useful in later discussions.

Definition 2.2. The *quasi-demand correspondence* $\tilde{\phi}(a, \cdot) : \mathbb{R}^\ell_+ \to X_a$ is defined by

$$\tilde{\phi}(a, \boldsymbol{p}) = \begin{cases} \phi(a, \boldsymbol{p}) & \text{if } \inf \boldsymbol{p}X_a < \boldsymbol{p}\omega_a, \\ \{\boldsymbol{x} \in X_a | \, \boldsymbol{px} = \inf \boldsymbol{p}X_a\} & \text{if } \inf \boldsymbol{p}X_a \geq \boldsymbol{p}\omega_a. \end{cases}$$

A pleasant property of the quasi-demand is that it is a correspondence with the closed graph when the consumption set is convex.

Proposition 2.4. *Suppose that X_a is a convex subset of $\mathbb{R}^\ell$. Then, the graph $\{(\boldsymbol{p}, \boldsymbol{x}) \in \mathbb{R}^\ell_+ \times X_a | \, \boldsymbol{x} \in \tilde{\phi}(a, \boldsymbol{p})\}$ is closed in $\mathbb{R}^\ell_+ \times X_a$. When the preference $\prec_a$ satisfies (NTR) and (CV), then $\tilde{\phi}(a, \cdot)$ is convex valued.*

Proof. By definition, it is clear that $\tilde{\phi}(a, \boldsymbol{p})$ is convex for each $\boldsymbol{p} \in \mathbb{R}^\ell_+$. Let $(\boldsymbol{p}_n, \boldsymbol{x}_n)$ be a sequence converging to $(\boldsymbol{p}, \boldsymbol{x})$ and $\boldsymbol{x}_n \in \tilde{\phi}(a, \boldsymbol{p}_n)$ for all n. If $\inf \boldsymbol{p}X_a < \boldsymbol{p}\omega_a$, then $\inf \boldsymbol{p}_n X_a < \boldsymbol{p}_n \omega_a$ for all sufficiently large n. Hence, by definition, $\tilde{\phi}(a, \boldsymbol{p}_n) = \phi(a, \boldsymbol{p}_n)$ and we can show that $\boldsymbol{x} \in \tilde{\phi}(a, \boldsymbol{p})$ in the same way as the proof of Proposition 2.3. If $\inf \boldsymbol{p}X_a \geq \boldsymbol{p}\omega_a$, then we have only to show that $\boldsymbol{px} = \inf \boldsymbol{p}X_a$. As $\boldsymbol{x} \in X_a$, it is obvious that $\boldsymbol{px} \geq \inf \boldsymbol{p}X_a$. However, it is impossible that $\boldsymbol{px} > \inf \boldsymbol{p}X_a$ because, if this was the case, then $\boldsymbol{p}_n \boldsymbol{x}_n > \inf \boldsymbol{p}_n X_a$ for all sufficiently large n. Hence, $\boldsymbol{x}_n \in \phi(a, \boldsymbol{p}_n) \subset \beta(a, \boldsymbol{p}_n)$ and $\boldsymbol{p}_n \omega_a \geq \boldsymbol{p}_n \boldsymbol{x}_n > \inf \boldsymbol{p}_n X_n = \boldsymbol{p}_n \omega_a$ for all sufficiently large n. In the limit, we have $\boldsymbol{p}\omega_a \geq \boldsymbol{px} \geq \inf \boldsymbol{p}X_a \geq \boldsymbol{p}\omega_a$. The convex valuedness under (NTR) and (CV) is clear from Proposition 2.1. □

We conclude this section by presenting the somewhat surprising result obtained by Sonnenschein [235]. We have assumed that the preference relations are transitive, which is considered to represent that consumers are rational enough to make their consumption decisions possible. However, Sonnenschein proved that transitivity is not necessary for the demand correspondence to be non-empty when the preference is irreflexive and convex. As the convexity of preferences seems to represent the consumers' "taste" rather than their "rationality," namely it represents that they prefer "mixed" commodities to "extreme" commodities, his result is not intuitively obvious at all.

Let $\prec$ be a binary relation on a consumption set $X \subset \mathbb{R}^\ell$ that is irreflexive. We do not assume it to be transitive or continuous. The continuity condition will resurrect in a suitable form (see (CT_{NT}) below). We define the (strict) *preference correspondence* $P : X \to X$, $\boldsymbol{x} \mapsto P(\boldsymbol{x}) = \{\boldsymbol{y} \in X |\ \boldsymbol{x} \prec \boldsymbol{y}\}$. The value $P(\boldsymbol{x})$ is called the *preferred set* of $\boldsymbol{x}$. We also define a correspondence $L : X \to X$, $\boldsymbol{x} \mapsto L(\boldsymbol{x}) = \{\boldsymbol{y} \in X |\ \boldsymbol{x} \in P(\boldsymbol{y})\}$. The value $L(\boldsymbol{x})$ is called the *lower section* of P at $\boldsymbol{x}$. In Figure 2.3(a), the situation where $\boldsymbol{y} \in L(\boldsymbol{x})$ is illustrated and Figure 2.3(b) illustrates the situation $\boldsymbol{y} \notin L(\boldsymbol{x})$. Finally, we define the correspondence

$$Q : X \to X,\ \boldsymbol{x} \mapsto Q(\boldsymbol{x}) = \{\boldsymbol{y} \in X |\ \boldsymbol{x} \notin P(\boldsymbol{y})\}.$$

Note that the value $Q(\boldsymbol{x})$ is the complement of the lower section $L(\boldsymbol{x})$ in X. We have now succeeded in discarding the transitivity of consumer preferences, and the preference relation of consumers is represented by the (strict) preference correspondence $P(\cdot)$ until the end of this section. We postulate the following assumptions on P:

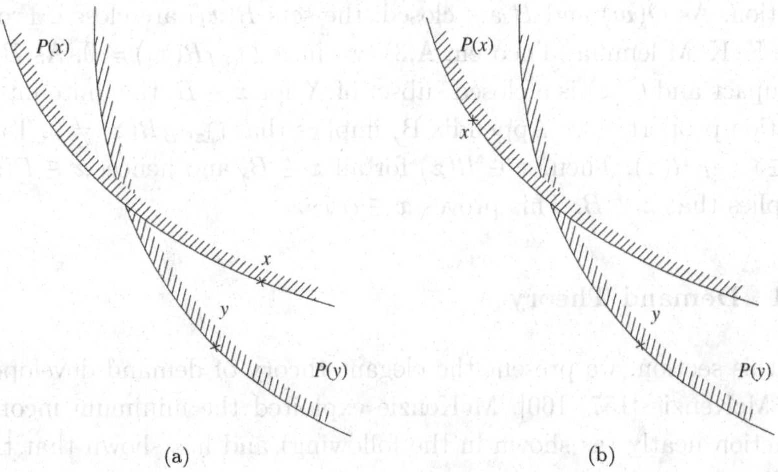

Figure 2.3. (a) $\boldsymbol{y} \in L(\boldsymbol{x})$. (b) $\boldsymbol{y} \notin L(\boldsymbol{x})$.

(IRCV$_{NT}$) (Irreflexivity and convexity): X is a convex subset of $\mathbb{R}^\ell$ and for all $\boldsymbol{x} \in X$, $\boldsymbol{x} \notin coP(\boldsymbol{x})$, where coZ denotes the convex hull of a set Z, and

(CT$_{NT}$) (Continuity): $L(\boldsymbol{x})$ is open relative to X for every $\boldsymbol{x} \in X$.

As the set $Q(\boldsymbol{x})$ is the complement of the set $L(\boldsymbol{x})$, we have from (CT_{NT}) that $Q(\boldsymbol{x})$ is closed for every $\boldsymbol{x}$. Let B be a nonempty, compact and convex subset of $\mathbb{R}^\ell$. Then, the demand set is defined by

$$\phi_{NT} = \{\boldsymbol{x} \in B | \; \boldsymbol{z} \in P(\boldsymbol{x}) \text{ implies } \boldsymbol{z} \notin B\}.$$

Sonnenschein's theorem now reads as follows:

Theorem 2.1. *Under the assumptions* (IRCV$_{NT}$) *and* (CT$_{NT}$), *the demand set is non-empty,* $\phi_{NT} \neq \emptyset$.

Proof. Take $\{\boldsymbol{x}_0 \ldots \boldsymbol{x}_r\} \subset B$ and $J \subset I = \{0, 1, \ldots, r\}$. We define for $\boldsymbol{w} \in B$ the set $R(\boldsymbol{w}) = Q(\boldsymbol{w}) \cap B$. $R(\boldsymbol{w})$ is the set of commodities in B that are as good as $\boldsymbol{w}$. Note that $\boldsymbol{z} \notin R(\boldsymbol{w})$ implies $\boldsymbol{w} \in P(\boldsymbol{z})$. We claim that $co\{\boldsymbol{x}_i\}_{i \in J} \subset \cup_{i \in J} R(\boldsymbol{x}_i)$. Suppose that this is not the case. Then, there exists a bundle $\boldsymbol{y} \in co\{\boldsymbol{x}_i\}_{i \in J}$ such that $\boldsymbol{y} \notin \cup_{i \in J} R(\boldsymbol{x}_i)$. This implies that $\boldsymbol{x}_i \in P(\boldsymbol{y})$ for all $i \in J$. Hence, $co\{\boldsymbol{x}_i\}_{i \in J} \subset coP(\boldsymbol{y})$, and $\boldsymbol{y} \in coP(\boldsymbol{y})$, which represents a contradiction. As $Q(\boldsymbol{w})$ and B are closed, the sets $R(\boldsymbol{x}_i)$ are closed. From the K–K–M lemma (Theorem A.3), we have $\cap_{i \in I} R(\boldsymbol{x}_i) \neq \emptyset$. As B is compact and $Q(\boldsymbol{z})$ is a closed subset of X for $\boldsymbol{z} \in B$, the finite intersection property (see Appendix B) implies that $\cap_{\boldsymbol{z} \in B} R(\boldsymbol{z}) \neq \emptyset$. Take $\boldsymbol{x} \in \cap_{\boldsymbol{z} \in B} R(\boldsymbol{z})$. Then, $\boldsymbol{x} \in R(\boldsymbol{z})$ for all $\boldsymbol{z} \in B$, and hence, $\boldsymbol{z} \in P(\boldsymbol{x})$ implies that $\boldsymbol{z} \notin B$. This proves $\boldsymbol{x} \in \phi_{NT}$. □

2.4 Demand Theory

In this section, we present the elegant theory of demand developed by McKenzie [157, 160]. McKenzie exploited the minimum income function neatly (as shown in the following) and has shown that the utility functions are not necessary to derive the main result (the Slutsky equation, Theorem 2.3) of the demand theory. He has also discarded the convexity assumption on the consumption set from the classical demand theory of Hicks [93] and Samuelson [220].

In this section, we omit the consumer's index a. The consumption set X is assumed to be a closed subset of $\mathbb{R}^\ell$ that is bounded from below. We do *not* assume it to be convex. We also assume that the preference $\prec$ satisfies (NTR) and (LNS). Of course, the assumptions (IR), (TR), and (CT) are maintained.

For every $\boldsymbol{p} \in \mathbb{R}^\ell_+$ and $w \in \mathbb{R}$, recall that the budget set and the demand set are defined by

$$\beta(\boldsymbol{p}, w) = \{\boldsymbol{x} \in X |\, \boldsymbol{p}\boldsymbol{x} \leq w\},$$

and

$$\phi(\boldsymbol{p}, w) = \{\boldsymbol{x} \in X |\, \boldsymbol{x} \in \beta(\boldsymbol{p}, w) \text{ and } \boldsymbol{z} \in \beta(\boldsymbol{p}, w) \text{ implies } \boldsymbol{x} \succsim \boldsymbol{z}\},$$

respectively. Then, for $\boldsymbol{x} \in X$ and $\boldsymbol{p} \in \mathbb{R}^\ell_+$, we define the *minimum income function* $\mu_{\boldsymbol{x}}$ by

$$\mu_{\boldsymbol{x}} : \mathbb{R}^\ell_+ \to \mathbb{R},\ \boldsymbol{p} \mapsto \mu_{\boldsymbol{x}}(\boldsymbol{p}) = \inf\{\boldsymbol{p}\boldsymbol{z} |\, \boldsymbol{z} \succsim \boldsymbol{x}\}.$$

The basic properties of the minimum income function are

Proposition 2.5. *The function $\mu_{\boldsymbol{x}}(\cdot)$ is positive homogeneous of degree 1 and concave.*

Proof. It is clear from the definition that $\mu_{\boldsymbol{x}}(\tau\boldsymbol{p}) = \tau\mu_{\boldsymbol{p}}(\boldsymbol{p})$ for all $\tau \geq 0$. To prove the concavity, let $\boldsymbol{p} = \tau\boldsymbol{p}_0 + (1-\tau)\boldsymbol{p}_1$, $0 \leq \tau \leq 1$. Given $\epsilon > 0$, we obtain $\boldsymbol{z} \in X$ with $\boldsymbol{z} \succsim \boldsymbol{x}$ and $\mu_{\boldsymbol{x}}(\boldsymbol{p}) > \boldsymbol{p}\boldsymbol{z} - \epsilon$, namely that

$$\mu_{\boldsymbol{x}}(\boldsymbol{p}) > \tau\boldsymbol{p}_0\boldsymbol{z} + (1-\tau)\boldsymbol{p}_1\boldsymbol{z} - \epsilon.$$

Hence, from the definition of $\mu_{\boldsymbol{x}}$, we obtain

$$\mu_{\boldsymbol{x}}(\boldsymbol{p}) > \tau\mu_{\boldsymbol{x}}(\boldsymbol{p}_0) + (1-\tau)\mu_{\boldsymbol{x}}(\boldsymbol{p}_1) - \epsilon.$$

As this holds for any $\epsilon > 0$, it follows that

$$\mu_{\boldsymbol{x}}(\boldsymbol{p}) \geq \tau\mu_{\boldsymbol{x}}(\boldsymbol{p}_0) + (1-\tau)\mu_{\boldsymbol{x}}(\boldsymbol{p}_1).$$

This proves the concavity of $\mu_{\boldsymbol{x}}(\cdot)$. □

From Theorem A.2, the concave function is continuous on the interior of its domain. Therefore, the function $\mu_{\boldsymbol{x}}(\cdot)$ is continuous for a strictly positive price vector $\boldsymbol{p} \gg \mathbf{0}$.

Proposition 2.6. *For any $\boldsymbol{p} \gg \mathbf{0}$ and any $\boldsymbol{x} \in X$, $\mu_{\boldsymbol{x}}(\boldsymbol{p}) = \boldsymbol{pz}$ for some $\boldsymbol{z} \in X$ with $\boldsymbol{z} \succsim \boldsymbol{x}$.*

Proof. First, we show that for $\boldsymbol{p} \gg \mathbf{0}$, the budget set $\beta(\boldsymbol{p}, w) = \{\boldsymbol{x} \in X | \ \boldsymbol{px} \leq w\}$ is compact. It is closed at the intersection of closed sets X and the half-space determined by $\boldsymbol{p}$ and w, $HS(\boldsymbol{p}, w) \equiv \{\boldsymbol{x} \in \mathbb{R}^{\ell} | \ \boldsymbol{px} \leq w\}$. As X is bounded from below, so is $\beta(\boldsymbol{p}, w)$. Suppose that $\beta(\boldsymbol{p}, w)$ is not bounded above, and let $\{\boldsymbol{x}_n\}$ be a sequence in $\beta(\boldsymbol{p}, w)$ with $\|\boldsymbol{x}_n\| \to \infty$. Then, as $\boldsymbol{px}_n \leq w$ for all n, one obtains $\boldsymbol{px}_n/\|\boldsymbol{x}_n\| \leq w/\|\boldsymbol{x}_n\|$ for all n. As $\{\boldsymbol{x}_n/\|\boldsymbol{x}_n\|\}$ is a sequence in the unit sphere that is compact, we can assume that $\boldsymbol{x}_n/\|\boldsymbol{x}_n\| \to \boldsymbol{y}$ and $\|\boldsymbol{y}\| = 1$. As the sequence $\{\boldsymbol{x}_n\}$ is bounded from below, $\boldsymbol{y} \geq \mathbf{0}$. As $\boldsymbol{p} \gg \mathbf{0}$, we have $\boldsymbol{px}_n/\|\boldsymbol{x}_n\| \to \boldsymbol{py} > 0$. However, $w/\|\boldsymbol{x}_n\| \to 0$, which is a contradiction.

Let $R(\boldsymbol{x}) = \{\boldsymbol{z} \in X | \ \boldsymbol{z} \succsim \boldsymbol{x}\} \cap \beta(\boldsymbol{p}, \boldsymbol{px})$. Then, $R(\boldsymbol{x}) \neq \emptyset$ because $\boldsymbol{x} \in R(\boldsymbol{x})$. Therefore, the function $\boldsymbol{pz}$ achieves its minimum value at some point in $R(\boldsymbol{x})$. □

A counterexample of Proposition 2.6 for the case lacking the condition $\boldsymbol{p} \gg \mathbf{0}$ is given by the utility function $u(x^1, x^2) = x^1 x^2$ defined on $X = \mathbb{R}^2_+$ for $\boldsymbol{x} = (1, 1)$ and $\boldsymbol{p} = (0, 1)$. Consider the sequence $\boldsymbol{x}_n = (n, 1/n)$. It follows from $u(\boldsymbol{x}_n) = u(\boldsymbol{x}) = 1$ for all n and $\boldsymbol{px}_n = 1/n \to 0$ that $\mu_{\boldsymbol{x}}(\boldsymbol{p}) = 0$. However, we have no $\boldsymbol{z} \in X$ such that $u(\boldsymbol{z}) \geq 1$ and $\boldsymbol{pz} = 0$.

Proposition 2.7. *If $\boldsymbol{x} \in \phi(\boldsymbol{p}, w)$, then $w = \boldsymbol{px} = \mu_{\boldsymbol{x}}(\boldsymbol{p})$.*

Proof. From the definition of $\phi(\boldsymbol{p}, w)$, we have $\boldsymbol{px} \leq w$. Suppose that $\boldsymbol{px} < w$. Then, from (LNS), there exists a $\boldsymbol{z} \in X$ that is close enough to $\boldsymbol{x}$ so that $\boldsymbol{pz} < w$ and $\boldsymbol{x} \prec \boldsymbol{z}$, which contradicts $\boldsymbol{x} \in \phi(\boldsymbol{p}, w)$. Hence, $\boldsymbol{px} = w$. If there was a $\boldsymbol{z} \in X$ such that $\boldsymbol{z} \succsim \boldsymbol{x}$ and $\boldsymbol{pz} < w$, then, in the same way as above, we would obtain a $\boldsymbol{w} \in X$ such that $\boldsymbol{z} \prec \boldsymbol{w}$ and $\boldsymbol{pw} < w$. Suppose that $\boldsymbol{x} \succsim \boldsymbol{w}$. Then, together with $\boldsymbol{z} \succsim \boldsymbol{x}$, it follows that $\boldsymbol{z} \succsim \boldsymbol{w}$ from the negative transitivity, which contradicts $\boldsymbol{z} \prec \boldsymbol{w}$. Thus, we obtain $\boldsymbol{x} \prec \boldsymbol{w}$. Together with $\boldsymbol{p}\boldsymbol{w} < w$, we have a contradiction to $\boldsymbol{x} \in \phi(\boldsymbol{p}, w)$. Therefore, $w = \mu_{\boldsymbol{x}}(\boldsymbol{p})$. □

Let the consumption bundle $\boldsymbol{x} \in X$ and the price vector $\boldsymbol{p} \gg \boldsymbol{0}$ be given. We define the *compensated demand set* $\phi_{\boldsymbol{x}}(\boldsymbol{p})$ by

$$\phi_{\boldsymbol{x}}(\boldsymbol{p}) = \phi(\boldsymbol{p}, \mu_{\boldsymbol{x}}(\boldsymbol{p})).$$

In the following, we assume that $\phi(\boldsymbol{p}, w)$ and $\phi_{\boldsymbol{x}}(\boldsymbol{p})$ are functions over $\boldsymbol{p} \gg \boldsymbol{0}$ and $w > 0$.

Proposition 2.8. *Suppose that $\phi_{\boldsymbol{x}}(\boldsymbol{p})$ is differentiable at $\boldsymbol{p}(\gg \boldsymbol{0})$. Then, we have*

$$\boldsymbol{p}\partial\phi_{\boldsymbol{x}}(\boldsymbol{p}) = \boldsymbol{0},$$

where $\partial\phi_{\boldsymbol{x}}(\boldsymbol{p}) = (\partial_s\phi^t_{\boldsymbol{x}}(\boldsymbol{p}))^{\ell}_{s,t=1}$ is called the substitution matrix (see below).

Proof. Let $\boldsymbol{y} = \phi_{\boldsymbol{x}}(\boldsymbol{p})$ and take a $\boldsymbol{q}$ close enough to $\boldsymbol{p} \gg \boldsymbol{0}$ so that $\boldsymbol{q} \gg \boldsymbol{0}$ and let $\boldsymbol{z} = \phi_{\boldsymbol{x}}(\boldsymbol{q})$. From Proposition 2.6, there exists a $\boldsymbol{w}$ such that $\mu_{\boldsymbol{x}}(\boldsymbol{q}) = \boldsymbol{q}\boldsymbol{w}$ and $\boldsymbol{w} \succsim \boldsymbol{x}$. We claim that $\boldsymbol{z} \succsim \boldsymbol{x}$ because, if it is not the case, then $\boldsymbol{z} \prec \boldsymbol{x}$. As $\boldsymbol{w} \succsim \boldsymbol{x}$, it follows from (NTR) that $\boldsymbol{z} \prec \boldsymbol{w}$ because otherwise we have $\boldsymbol{z} \succsim \boldsymbol{x}$. However, as $\boldsymbol{q}\boldsymbol{w} = \mu_{\boldsymbol{x}}(\boldsymbol{q})$ and $\boldsymbol{z} \prec \boldsymbol{w}$, one has $\boldsymbol{z} \neq \phi_{\boldsymbol{x}}(\boldsymbol{q})$, which is a contradiction. Thus, $\boldsymbol{z} \succsim \boldsymbol{x}$.

$\boldsymbol{y} = \phi_{\boldsymbol{x}}(\boldsymbol{p}) = \phi(\boldsymbol{p}, \mu_{\boldsymbol{x}}(\boldsymbol{p}))$ implies that $\boldsymbol{p}\boldsymbol{y} \leq \mu_{\boldsymbol{x}}(\boldsymbol{p})$. On the other hand, $\boldsymbol{z} \succsim \boldsymbol{x}$ implies that $\mu_{\boldsymbol{x}}(\boldsymbol{p}) \leq \boldsymbol{p}\boldsymbol{z} = \boldsymbol{p}\phi_{\boldsymbol{x}}(\boldsymbol{q})$. From Proposition 2.7, we have $\boldsymbol{p}\boldsymbol{y} = \mu_{\boldsymbol{x}}(\boldsymbol{p})$. Hence, $\boldsymbol{p}\boldsymbol{y} = \boldsymbol{p}\phi_{\boldsymbol{x}}(\boldsymbol{p}) \leq \boldsymbol{p}\phi_{\boldsymbol{x}}(\boldsymbol{q})$ for all $\boldsymbol{q} \gg \boldsymbol{0}$ near $\boldsymbol{p}$. Differentiating $\boldsymbol{p}\phi_{\boldsymbol{x}}(\boldsymbol{q})$ with respect to $\boldsymbol{q}$ and evaluating at $\boldsymbol{p}$, we have the necessary condition for the minimum,

$$\boldsymbol{p}\partial\phi_{\boldsymbol{x}}(\boldsymbol{p}) = \boldsymbol{0}.$$

This proves the proposition. □

The following proposition is the key to McKenzie's demand theory.

Proposition 2.9.

$$\frac{\partial}{\partial p^t}\mu_{\boldsymbol{x}}(\mathbf{p}) = \phi^t_{\boldsymbol{x}}(\mathbf{p}), \quad t = 1 \ldots \ell,$$

where we write $\phi_{\boldsymbol{x}}(\mathbf{p}) = (\phi^1_{\boldsymbol{x}}(\mathbf{p}) \ldots \phi^\ell_{\boldsymbol{x}}(\mathbf{p}))$.

Proof. Differentiating $\mu_{\boldsymbol{x}}$, we obtain

$$\frac{\partial}{\partial p^t}\mu_{\boldsymbol{x}}(\mathbf{p}) = \frac{\partial}{\partial p^t}(\mathbf{p}\phi_{\boldsymbol{x}}(\mathbf{p})) = \phi^t_{\boldsymbol{x}}(\mathbf{p}) + \mathbf{p}\partial_t\phi_{\boldsymbol{x}}(\mathbf{p}) = \phi^t_{\boldsymbol{x}}(\mathbf{p})$$

by Proposition 2.8. □

As a corollary, we have under (LNS),

$$\frac{\partial^2}{\partial p^t \partial p^s}\mu_{\boldsymbol{x}}(\mathbf{p}) = \frac{\partial}{\partial p^s}\phi^t_{\boldsymbol{x}}(\mathbf{p}), \quad t, s = 1 \ldots \ell,$$

if the derivative exists. The main result of the demand theory is now immediate. We define the substitution matrix $S_{\boldsymbol{x}}(\mathbf{p})$ by

$$S_{\boldsymbol{x}}(\mathbf{p}) = \partial\phi_{\boldsymbol{x}}(\mathbf{p}) = \left(\frac{\partial}{\partial p^s}\phi^t_{\boldsymbol{x}}(\mathbf{p})\right)^{\ell}_{s,t=1} = \left(\frac{\partial^2}{\partial p^t \partial p^s}\mu_{\boldsymbol{x}}(\mathbf{p})\right)^{\ell}_{s,t=1}.$$

Theorem 2.2. *For each* $\boldsymbol{x} \in X$, *a* $S_{\boldsymbol{x}}(\mathbf{p})$ *exists for almost all* $\mathbf{p} \gg \mathbf{0}$, *and it satisfies the following*:

(a) $S_{\boldsymbol{x}}(\mathbf{p})$ *is symmetric, and*
(b) $S_{\boldsymbol{x}}(\mathbf{p})$ *is negative semi-definite.*

Proof. From Proposition 2.5, the function $\mu_{\boldsymbol{x}}(\mathbf{p})$ is concave. Therefore,

$$S_{\boldsymbol{x}}(\mathbf{p}) = \left(\frac{\partial^2}{\partial p^t \partial p^s}\mu_{\boldsymbol{x}}(\mathbf{p})\right)$$

is symmetric and negative semi-definite. □

Theorem 2.3. (*Slutsky equation*) *If the derivative exists for* $\mathbf{p} \gg \mathbf{0}$,

$$\frac{\partial}{\partial p^s}\phi^t_{\boldsymbol{x}}(\mathbf{p}) = \frac{\partial}{\partial p^s}\phi^t(\mathbf{p}, w) + \phi^s(\mathbf{p}, w)\frac{\partial}{\partial w}\phi^t(\mathbf{p}, w), \ s, t = 1 \ldots \ell.$$

Proof. From Proposition 2.7, $\boldsymbol{p}\boldsymbol{x} = w = \mu_{\boldsymbol{x}}(\boldsymbol{p})$. Differentiating the compensated demand function $\phi^t_{\boldsymbol{x}}(\boldsymbol{p}) = \phi^t(\boldsymbol{p}, \mu_{\boldsymbol{x}}(\boldsymbol{p}))$, $t = 1 \ldots \ell$, we obtain

$$\begin{aligned}\frac{\partial}{\partial p^s}\phi^t_{\boldsymbol{x}}(\boldsymbol{p}) &= \frac{\partial}{\partial p^s}\phi^t(\boldsymbol{p}, \mu_{\boldsymbol{x}}(\boldsymbol{p})) = \frac{\partial}{\partial p^s}\phi^t(\boldsymbol{p}, w) + \frac{\partial}{\partial w}\phi^t(\boldsymbol{p}, w)\frac{\partial \mu_{\boldsymbol{x}}(\boldsymbol{p})}{\partial p^s} \\ &= \frac{\partial}{\partial p^s}\phi^t(\boldsymbol{p}, w) + \phi^s(\boldsymbol{p}, w)\frac{\partial}{\partial w}\phi^t(\boldsymbol{p}, w), \quad s, t = 1 \ldots \ell.\end{aligned}$$

□

2.5 Competitive Equilibria

In this section, we maintain the assumptions that the commodity space L is an ℓ-dimensional Euclidean space $\mathbb{R}^\ell$ and that the set of consumers A is finite, $A = \{1 \ldots m\}$. Following McKenzie [159, 160], we call such an economy *classical.* Moreover, we restrict ourselves to an exchange economy in which there are no producers, so that all trades of commodities occur among the consumers. This means that every consumer a owns his/her initial endowment vector $\omega_a \in \mathbb{R}^\ell$, and the trades undertaken by the consumers are limited by the initial endowment. Hence, the total demand (and supply) is bounded by the total resource vector $\sum_{a=1}^m \omega_a$. Recall that the consumption set X_a is a closed subset of $\mathbb{R}^\ell$ with a lower bound $\boldsymbol{b}_a$. In this section, we also assume that X_a is convex. Let $\prec_a$ be a preference relation of consumer a. An *exchange economy* is a $2m$-tuple of the preference relations and the initial endowments $(\prec_a, \omega_a)_{a=1}^m$ and is denoted by $\mathcal{E}$.

An m-tuple of the consumption vectors $(\boldsymbol{x}_1 \ldots \boldsymbol{x}_m) \in \prod_{a=1}^m X_a$ is called an *allocation*, which is said to be *feasible* if

$$\sum_{a=1}^m \boldsymbol{x}_a \leq \sum_{a=1}^m \omega_a.$$

A feasible allocation is called *exactly feasible* if $\sum_{a=1}^m \boldsymbol{x}_a = \sum_{a=1}^m \omega_a$. A feasible allocation $(\boldsymbol{x}_1 \ldots \boldsymbol{x}_m) \in \prod_{a=1}^m X_a$ is called *Pareto-optimal* if and only if there exists no other feasible allocation $(\boldsymbol{y}_1 \ldots \boldsymbol{y}_m) \in \prod_{a=1}^m X_a$, such that $\boldsymbol{y}_a \succsim_a \boldsymbol{x}_a$ for all $a \in A$ and $\boldsymbol{x}_a \prec \boldsymbol{y}_a$ holds for at least one $a \in A$. Now, we state the definition of the fundamental equilibrium concept.

Definition 2.3. A pair $(\boldsymbol{p}, \boldsymbol{x}_a)$ of a non-zero price vector $\boldsymbol{p} \in L^*$ and an allocation $(\boldsymbol{x}_a) \in \prod_{a=1}^m X_a$ is said to consist of a *competitive equilibrium* or a *Walrasian equilibrium* if and only if

(E-1) $\boldsymbol{p}\boldsymbol{x}_a \leq \boldsymbol{p}\omega_a$ and $\boldsymbol{x}_a \succsim_a \boldsymbol{y}$ whenever $\boldsymbol{y} \in X_a$ and $\boldsymbol{p}\boldsymbol{y} \leq \boldsymbol{p}\omega_a$, $a = 1 \dots m$, and

(E-2) $\sum_{a=1}^m \boldsymbol{x}_a \leq \sum_{a=1}^m \omega_a$.

Recall that $\Delta = \{\boldsymbol{p} = (p^t) \in \mathbb{R}_+^\ell |\ \sum_{t=1}^\ell p^t = 1\}$ is the standard ℓ-simplex. The following easy proposition is known as the *first fundamental theorem of welfare economics*.

Proposition 2.10. *Let* $(\boldsymbol{p}, \boldsymbol{x}_1 \dots \boldsymbol{x}_m) \in \Delta \times \prod_{a=1}^m X_a$ *be a competitive equilibrium of an economy* $(\prec_a, \omega_a)_{a=1}^m$ *that satisfies (LNS) for all* $a \in A$. *Then, the allocation* $(\boldsymbol{x}_1 \dots \boldsymbol{x}_m)$ *is Pareto-optimal.*

Proof. Suppose that $(\boldsymbol{x}_1 \dots \boldsymbol{x}_m)$ was not Pareto-optimal. Then, there exists a feasible allocation $(\boldsymbol{y}_1 \cdots \boldsymbol{y}_m) \in \prod_{a=1}^m X_a$ such that $\boldsymbol{y}_a \succsim_a \boldsymbol{x}_a$ for all $a \in A$ and $\boldsymbol{x}_a \prec \boldsymbol{y}_a$ holds for at least one $a \in A$. Then, it follows that $\boldsymbol{p}\boldsymbol{y}_a \geq \boldsymbol{p}\omega_a$ for all $a \in A$. For a such that $\boldsymbol{p}\omega_a \leq \inf \boldsymbol{p}X_a$, the inequality follows trivially. Suppose that $\boldsymbol{p}\omega_a > \inf \boldsymbol{p}X_a$ and $\boldsymbol{p}\boldsymbol{y}_a < \boldsymbol{p}\omega_a$ for some a. From condition (E-1) of Definition 2.3, we have $\boldsymbol{x}_a \succsim_a \boldsymbol{y}_a$, hence $\boldsymbol{x}_a \sim_a \boldsymbol{y}_a$. Then, from (LNS), there exists a bundle $\boldsymbol{z} \in X_a$ that is close enough to $\boldsymbol{y}_a$ such that $\boldsymbol{y}_a \prec_a \boldsymbol{z}$ and $\boldsymbol{p}\boldsymbol{z} < \boldsymbol{p}\omega_a$. Therefore, $\boldsymbol{x}_a \prec_a \boldsymbol{z}$ and $\boldsymbol{p}\boldsymbol{z} < \boldsymbol{p}\omega_a$, contradicting the condition (E-1). Furthermore, for a such that $\boldsymbol{x}_a \prec_a \boldsymbol{y}_a$, we have $\boldsymbol{p}\boldsymbol{y}_a > \boldsymbol{p}\omega_a$. Summing these inequalities over a, one obtains $\boldsymbol{p}\sum_{a=1}^m \boldsymbol{y}_a > \boldsymbol{p}\sum_{a=1}^m \omega_a$. On the other hand, as the allocation $(\boldsymbol{y}_1 \dots \boldsymbol{y}_m)$ is feasible, we have $\sum_{a=1}^m \boldsymbol{y}_a \leq \sum_{a=1}^m \omega_a$. As $\boldsymbol{p} \geq \boldsymbol{0}$, we have $\boldsymbol{p}\sum_{a=1}^m \boldsymbol{y}_a \leq \boldsymbol{p}\sum_{a=1}^m \omega_a$, which is a contradiction. □

Note that in the above proof, we did not use the basic postulates but (LNS). Now, we state our first result regarding the existence of the equilibrium. It is the most fundamental theorem of this book.

Indeed, most parts of the book are, in a sense, variations on the theme of the following theorem.

Theorem 2.4. *Suppose that an economy* $\mathcal{E} = (\prec_a, \omega_a)_{a=1}^m$ *satisfies (NTR), (CV) and (MI) for every* $a \in A$*. Then, there exists a competitive equilibrium* $(\boldsymbol{p}^*, \boldsymbol{x}_1^* \dots \boldsymbol{x}_m^*) \in \Delta \times \prod_{a=1}^m X_a$ *for* $\mathcal{E}$.

Proof. Take a positive number k such that $\| \sum_{a=1}^m \omega_a \| < k$, and set $\hat{K} = \{\boldsymbol{x} \in \mathbb{R}^\ell | \, \boldsymbol{x} \leq k\mathbf{1}\}$, where $\mathbf{1} = (1 \dots 1)$. As before, we define the restricted consumption sets,

$$\hat{X}_a(k) = X_a \cap \hat{K}, \quad a = 1 \dots m,$$

and we consider the (restricted) budget correspondence $\hat{\beta}(a, \cdot) : \Delta \to \hat{X}_a(k)$,

$$\hat{\beta}(a, \boldsymbol{p}) = \{\boldsymbol{x} \in \hat{X}_a(k) | \, \boldsymbol{p}\boldsymbol{x} \leq \boldsymbol{p}\omega_a\}, \quad a = 1 \dots m.$$

The (restricted) demand correspondence $\hat{\phi}(a, \cdot) : \Delta \to \hat{X}_a(k)$ is given by

$$\hat{\phi}(a, \boldsymbol{p}) = \{\boldsymbol{x} \in \hat{\beta}(a, \boldsymbol{p}) | \, \boldsymbol{x} \succsim_a \boldsymbol{y} \text{ for all } \boldsymbol{y} \in \hat{\beta}(a, \boldsymbol{p})\}, \quad a = 1 \dots m.$$

Then, we define the *aggregate excess demand correspondence* $\hat{\zeta} : \Delta \to \mathbb{R}^\ell$ by

$$\hat{\zeta}(\boldsymbol{p}) = \sum_{a=1}^m \hat{\phi}(a, \boldsymbol{p}) - \sum_{a=1}^m \omega_a.$$

Set $\hat{X}(k) = \sum_{a=1}^m \hat{X}_a(k) - \sum_{a=1}^m \omega_a$. As $\hat{X}_a(k) - \{\omega_a\}$ is a compact and convex subset of $\mathbb{R}^\ell$ for each a, the set $\hat{X}(k)$ is also compact and convex, and $\hat{\zeta}(\boldsymbol{p}) \subset \hat{X}(k)$ for every $\boldsymbol{p} \in \Delta$. From Proposition 2.3 and the assumptions (NTR) and (CV), the correspondence $\hat{\zeta}$ is upper hemicontinuous and convex-valued and satisfies $\boldsymbol{p}\hat{\zeta}(\boldsymbol{p}) \leq 0$ by the budget condition of each a. Applying the Gale–Nikaido–Debreu lemma (Theorem D.3), we obtain $\boldsymbol{p}(k) \in \Delta$ and

$\boldsymbol{x}_a(k) \in \hat{\phi}(a, \boldsymbol{p}(k)),\ a = 1 \ldots m$, such that

$$\sum_{a=1}^{m} \boldsymbol{x}_a(k) - \sum_{a=1}^{m} \omega_a \leq \boldsymbol{0}.$$

As $\boldsymbol{p}(k) \in \Delta$ and $\boldsymbol{b}_a \leq \boldsymbol{x}_a(k) \leq \sum_a \omega_a - \sum_{c \neq a} \boldsymbol{b}_c$ for all k, we can assume that $\boldsymbol{p}(k) \to \boldsymbol{p}^* \in \Delta$ and $\boldsymbol{x}_a(k) \to \boldsymbol{x}_a^*$. We claim that $(\boldsymbol{p}^*, \boldsymbol{x}_1^* \ldots \boldsymbol{x}_m^*)$ is a desired competitive equilibrium. Passing to the limit in the above market condition for $(\boldsymbol{x}_a(k))$, we obtain $\sum_{a=1}^m \boldsymbol{x}_a^* \leq \sum_{a=1}^m \omega_a$. Suppose there exists $\boldsymbol{y} \in X_a$ such that $\boldsymbol{p}^*\boldsymbol{y} \leq \boldsymbol{p}^*\omega_a$ and $\boldsymbol{x}_a^* \prec_a \boldsymbol{y}$. For a sufficiently large k, we have $\boldsymbol{y} \in \hat{X}_a(k)$. From (CT) and (MI), we can take $\boldsymbol{z} \in \hat{X}_a(k)$, which is close enough to $\boldsymbol{y}$ such that $\boldsymbol{x}_a^* \prec_a \boldsymbol{z}$ and $\boldsymbol{p}^*\boldsymbol{z} < \boldsymbol{p}^*\omega_a$. Hence, for a large enough k, we have $\boldsymbol{x}_a(k) \prec_a \boldsymbol{z}$ and $\boldsymbol{p}(k)\boldsymbol{z} < \boldsymbol{p}(k)\omega_a$, which contradicts $\boldsymbol{x}_a(k) \in \hat{\phi}(a, \boldsymbol{p}(k))$. Therefore, the condition (E-1) is met. □

When we assume the monotonicity of preferences, the minimum income condition for each individual is replaced by the following *survival condition*[1]:

(SV): $\omega_a \in X_a$ for each a,

and the *positive total endowment*:

(PTE): $\sum_{a=1}^m \omega_a \gg \boldsymbol{0}$.

Theorem 2.5. *Suppose that an economy $\mathcal{E} = (\prec_a, \omega_a)_{a=1}^m$ satisfies (NTR), (CV), (MT) and (SV) for every $a \in A$. Suppose further that $\mathcal{E}$ satisfies (PTE). Then, there exists a competitive equilibrium $(\boldsymbol{p}^*, \boldsymbol{x}_1^* \ldots \boldsymbol{x}_m^*)$ for $\mathcal{E}$.*

Proof. Let $\hat{X}_a(k)$ and $\hat{\beta}(a, \boldsymbol{p})$ be defined as in Theorem 2.4. The aggregate excess demand correspondence $\tilde{\zeta} : \Delta \to \mathbb{R}^\ell$ is now

[1]When the conditions (SV) and (MT) are imposed simultaneously, the former simply means $\omega_a \geq \boldsymbol{0}$.

defined by

$$\tilde{\zeta}(\boldsymbol{p}) = \sum_{a=1}^{m} \tilde{\phi}(a, \boldsymbol{p}) - \sum_{a=1}^{m} \omega_a,$$

where $\tilde{\phi}(a, \cdot) : \Delta \to \hat{X}_a$ is the quasi-demand correspondence restricted to $\hat{X}_a$. $\tilde{\phi}(a, \boldsymbol{p}) \neq \emptyset$ because $\hat{X}_a(k)$ is compact.

Applying the Gale–Nikaido–Debreu lemma to $\tilde{\zeta}(\boldsymbol{p})$, we obtain $\boldsymbol{p}(k) \in \Delta$ and $\boldsymbol{x}_a(k) \in \tilde{\phi}(a, \boldsymbol{p}(k))$, $a = 1 \dots m$. As in Theorem 2.4, we obtain the limit $\boldsymbol{p}(k) \to \boldsymbol{p}^* \in \Delta$ and $\boldsymbol{x}_a(k) \to \boldsymbol{x}_a^*$ such that $\sum_{a=1}^{m} \boldsymbol{x}_a^* \leq \sum_{a=1}^{m} \omega_a$. It follows from $\sum_{a=1}^{m} \omega_a \gg \boldsymbol{0}$ that $\boldsymbol{p}^* \omega_a > 0$ for some a. For such a, $\boldsymbol{x}^* \in \phi(a, \boldsymbol{p}^*)$ by definition, and hence, (E-1) holds for a such that $\boldsymbol{p}^* \omega_a > 0$. From the (MT) condition, we have $\boldsymbol{p}^* \gg \boldsymbol{0}$. For if $p^{*t} = 0$ for some t, then $\boldsymbol{p}^*(\boldsymbol{x}_a^* + \epsilon \boldsymbol{e}_t) = \boldsymbol{p}^* \boldsymbol{x}_a^*$ and $\boldsymbol{x}_a^* \prec_a \boldsymbol{x}_a^* + \epsilon \boldsymbol{e}_t$ for $\epsilon > 0$, where $\boldsymbol{e}_t = (0 \dots 0, 1, 0 \dots 0)$ (1 at the tth position), which is a contradiction. Then, if $\boldsymbol{p}^* \omega_a = 0$, the budget set $\{\boldsymbol{x} \in \mathbb{R}_+^\ell | \, \boldsymbol{p}^* \boldsymbol{x} \leq \boldsymbol{p}^* \omega_a = 0\}$ is a singleton $\{\boldsymbol{0}\}$, and hence, the condition (E-1) holds trivially. We conclude that $(\boldsymbol{p}^*, \boldsymbol{x}_a^*)$ is a competitive equilibrium. □

Remark 2.1. In Theorem 2.5, the monotonicity implies the strict positivity of the equilibrium price vector; $\boldsymbol{p}^* \gg \boldsymbol{0}$, or otherwise, one can obtain a strictly preferred consumption vector within the budget set. Hence, the resource constraint holds with the equality, or the equilibrium allocation is exactly feasible. We won't repeat the same remark for theorems in which (MT) is assumed in subsequent sections or chapters.

Suppose that the economy satisfies (LNS) and the *strong convexity* conditions in addition to all of the assumptions of Theorem 2.4.

(SCV): Let X_a be a convex subset of L and $\boldsymbol{x}, \boldsymbol{y} \in X_a$ commodity bundles, such that $\boldsymbol{x} \neq \boldsymbol{y}$ and $\boldsymbol{x} \succsim_a \boldsymbol{y}$. Then, $\boldsymbol{y} \prec_a \tau \boldsymbol{x} + (1 - \tau)\boldsymbol{y}$ for every $0 < \tau < 1$.

Then, the (restricted) demand $\hat{\phi}(a, \boldsymbol{p})$ would be a continuous function and satisfy $\boldsymbol{p}\hat{\phi}(a, \boldsymbol{p}) = \boldsymbol{p}\omega_a$. This implies that the feasibility condition $\sum_{a=1}^{m} \hat{\phi}^t(a, \boldsymbol{p}) \leq \sum_{a=1}^{m} \omega_a^t$, $t = 1 \dots \ell$, holds with equality unless $p^t = 0$. Theorem 2.4 in this case is reduced to the following lemma.

Lemma 2.1. *If a continuous function* $\zeta(\boldsymbol{p}) = (\zeta^t(\boldsymbol{p}))$ *from* Δ *to* $\mathbb{R}^\ell$ *satisfies* $\boldsymbol{p}\zeta(\boldsymbol{p}) = 0$, *then there exists a vector* $\hat{\boldsymbol{p}}$ *such that* $\zeta^t(\hat{\boldsymbol{p}}) \leq 0$ *with equality unless* $\hat{p}^t = 0$.

Lemma 2.1 is proved via Brower's fixed point theorem (Theorem C.5) in exactly the same way as the proof of Gale–Nikaido lemma (Theorem D.3) is shown using Kakutani's theorem. Uzawa [257] observed that this lemma implies Brower's fixed point theorem, and hence, they are equivalent propositions.

Proposition 2.11. *Suppose that Lemma* 2.1 *holds and let* $f : \Delta \to \Delta$ *be a continuous function. Then, there exists a vector* $\hat{\boldsymbol{p}} \in \Delta$ *with* $\hat{\boldsymbol{p}} = f(\hat{\boldsymbol{p}})$.

Proof. Let $f(\boldsymbol{p}) = (f^t(\boldsymbol{p}))$ be a continuous function from Δ to itself. We define an "excess demand" function ζ from Δ to $\mathbb{R}^\ell$ by

$$\zeta(\boldsymbol{p}) = f(\boldsymbol{p}) - \left(\frac{\boldsymbol{p}f(\boldsymbol{p})}{\boldsymbol{p}\boldsymbol{p}}\right)\boldsymbol{p}.$$

It can be easily verified that ζ satisfies the assumptions of Lemma 2.1, and hence, there exists a vector $\hat{\boldsymbol{p}}$ with

$$f^t(\hat{\boldsymbol{p}}) - \left(\frac{\hat{\boldsymbol{p}}f(\hat{\boldsymbol{p}})}{\hat{\boldsymbol{p}}\hat{\boldsymbol{p}}}\right)\hat{p}^t \leq 0, \quad t = 1\ldots\ell$$

with equality unless $\hat{p}^t = 0$. Summing over t and noticing that $\hat{\boldsymbol{p}}$, $f(\hat{\boldsymbol{p}}) \in \Delta$, we obtain $\hat{\boldsymbol{p}}f(\hat{\boldsymbol{p}}) = \hat{\boldsymbol{p}}\hat{\boldsymbol{p}}$ or $f^t(\hat{\boldsymbol{p}}) \leq \hat{p}^t$ with equality unless $\hat{p}^t = 0$. As $\hat{\boldsymbol{p}}, f(\hat{\boldsymbol{p}}) \in \Delta$, we conclude $\hat{\boldsymbol{p}} = f(\hat{\boldsymbol{p}})$. □

Now, we investigate the welfare properties of the competitive equilibria. In Proposition 2.10, we saw that the competitive equilibrium is Pareto-optimal. A partial converse of this result, which is known as the second fundamental theorem of welfare economics, has been obtained. To present this result, we state a new equilibrium concept.

Definition 2.4. Suppose that a price vector $\boldsymbol{p} \in \mathbb{R}^{\ell}_{+}\backslash\{\mathbf{0}\}$ is given. An allocation $(\boldsymbol{x}_a)_{a=1}^{m}$ is said to be an *equilibrium relative to* $\boldsymbol{p}$ if and only if

(R-1) $\boldsymbol{p}\boldsymbol{y} \leq \boldsymbol{p}\boldsymbol{x}_a$ implies that $\boldsymbol{x}_a \succsim_a \boldsymbol{y}$ for all $a \in A$, and
(E-2) $\sum_{a=1}^{m} \boldsymbol{x}_a \leq \sum_{a=1}^{m} \omega_a$.

It is obvious from the definitions that if the $m+1$-tuple $(\boldsymbol{p}^*, \boldsymbol{x}_a^*)$ is a competitive equilibrium of the economy $\mathcal{E} = (\prec_a, \omega_a)_{a=1}^{m}$, then the allocation $(\boldsymbol{x}_a^*)$ is the equilibrium relative to $\boldsymbol{p}^*$ of the economy $\mathcal{E}$. On the other hand, let $(\boldsymbol{x}_a^*)$ be an equilibrium relative to the price vector $\boldsymbol{p}^*$ in the economy $\mathcal{E}$. Then, $(\boldsymbol{p}^*, \boldsymbol{x}_a^*)$ is a competitive equilibrium of the economy $\mathcal{E}' = (\prec_a, \omega_a')$, where $\omega_a' = \boldsymbol{x}_a^*$, $a = 1 \ldots m$. For this reason, some authors call the equilibrium relative to $\boldsymbol{p}$ is the *competitive equilibrium with redistribution* in the economy $\mathcal{E}$. A somewhat weaker equilibrium concept is the following.

Definition 2.5. Suppose that a price vector $\boldsymbol{p} \in \mathbb{R}^{\ell}_{+}\backslash\{\mathbf{0}\}$ is given. An allocation $(\boldsymbol{x}_a)_{a=1}^{m}$ is said to be a *quasi-equilibrium relative to the price vector* $\boldsymbol{p}$ if and only if:

(Q-1) $\boldsymbol{x}_a \prec_a \boldsymbol{y}$ implies that $\boldsymbol{p}\boldsymbol{x}_a \leq \boldsymbol{p}\boldsymbol{y}$ for all $a \in A$, and
(E-2) $\sum_{a=1}^{m} \boldsymbol{x}_a \leq \sum_{a=1}^{m} \omega_a$.

Clearly, if an allocation $(\boldsymbol{x}_a)$ is an equilibrium relative to $\boldsymbol{p}$, then it is the quasi-equilibrium. However, the converse does not hold as the following counter example from Arrow shows. In Figure 2.4, the consumption bundles $\boldsymbol{x}$ and $\boldsymbol{z}$ have the same value when evaluated by the price vector $\boldsymbol{p}$, or $\boldsymbol{p}\boldsymbol{x} = \boldsymbol{p}\boldsymbol{z}$. However, $\boldsymbol{z}$ is strictly preferred to $\boldsymbol{x}$, namely $\boldsymbol{x} \prec \boldsymbol{z}$. Therefore, the condition (Q-1) holds but not the condition (R-1). This kind of phenomenon appears when the allocation bundle $\boldsymbol{x}$ satisfies $\boldsymbol{p}\boldsymbol{x} = \inf\{\boldsymbol{p}\boldsymbol{z} | \, \boldsymbol{z} \in X\}$.

The *second fundamental theorem of welfare economics* reads as follows:

Theorem 2.6. *Suppose that for every consumer $a \in A$, the preference satisfies (NTR), (CV) and (MT). Then, every Pareto-optimal allocation is a quasi-equilibrium relative to some price vector* $\boldsymbol{p} \in \mathbb{R}^{\ell}_{+}\backslash\{\mathbf{0}\}$.

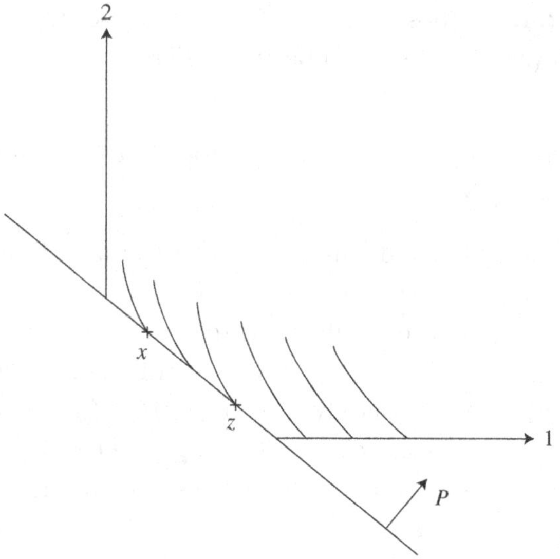

Figure 2.4. Arrow's example.

Proof. Let $(\boldsymbol{x}_a)_{a=1}^m$ be a Pareto-optimal allocation. Then, because it is a feasible allocation, the condition (E-2) is met. For each consumer a, define the strictly preferred set $P_a(\boldsymbol{x})$ by

$$P_a(\boldsymbol{x}) = \{\boldsymbol{z} \in X_a |\ \boldsymbol{x} \prec_a \boldsymbol{z}\}.$$

The set $P_a(\boldsymbol{x})$ is non-empty for all $\boldsymbol{x} \in X_a$ from the assumption (MT). We claim that the set $P_a(\boldsymbol{x})$ is convex for all $\boldsymbol{x} \in X_a$. Let $\boldsymbol{z}_1, \boldsymbol{z}_2 \in P_a(\boldsymbol{x})$ and take $0 \leq \tau \leq 1$. Without loss of generality, we may assume that $\boldsymbol{z}_1 \succsim_a \boldsymbol{z}_2$ from the completeness (CP). From the assumption (CV), the set $\{\boldsymbol{z} \in X_a |\ \boldsymbol{z} \succsim_a \boldsymbol{x}\}$ is convex for every $\boldsymbol{x} \in X_a$. Hence, $\tau\boldsymbol{z}_1 + (1-\tau)\boldsymbol{z}_2 \succsim_a \boldsymbol{z}_2$ and $\boldsymbol{x} \prec_a \boldsymbol{z}_2$, and therefore, $\boldsymbol{x} \prec_a \tau\boldsymbol{z}_1 + (1-\tau)\boldsymbol{z}_2$. If this is not the case, we have $\boldsymbol{x} \succsim_a \tau\boldsymbol{z}_1 + (1-\tau)\boldsymbol{z}_2$. Then, it follows that $\boldsymbol{x} \succsim_a \boldsymbol{z}_2$ from (NTR). This contradicts $\boldsymbol{z}_2 \in P_a(\boldsymbol{x})$. Therefore, $\tau\boldsymbol{z}_1 + (1-\tau)\boldsymbol{z}_2 \in P_a(\boldsymbol{x})$ and $P_a(\boldsymbol{x})$ is convex.

Let $Q = \sum_{a=1}^m P_a(\boldsymbol{x}_a)$. Then, Q is convex, as it is the sum of the convex sets. As the allocation $(\boldsymbol{x}_a)_{a=1}^m$ is Pareto-optimal, $\sum_{a=1}^m \omega_a \notin Q$. Applying the separation theorem (Theorem A.1), we obtain a vector $\boldsymbol{p} \in \mathbb{R}^\ell$ with $\boldsymbol{p} \neq \mathbf{0}$ and

$$\boldsymbol{p}\sum_{a=1}^{m} \omega_a \leq \boldsymbol{p}\sum_{a=1}^{m} \boldsymbol{z}_a$$

for every $\boldsymbol{z}_a \in P_a(\boldsymbol{x}_a)$, $a = 1 \ldots m$. By the (MT) condition, it follows that $\boldsymbol{p} = (p^t) \geq \mathbf{0}$. If $p^t < 0$ for some t, $\boldsymbol{z}_a + (0 \ldots 0, \tau, 0 \ldots 0) \in P_a(\boldsymbol{x})$ (τ is at the tth position) for all $\tau > 0$ and $\boldsymbol{p}(\boldsymbol{z}_a + (0 \ldots 0, \tau, 0 \ldots 0)) \to -\infty$ ($\tau \to +\infty$), which is a contradiction.

Now, fix an a. For $b \neq a$, we can take a $\boldsymbol{z}_b$ arbitrarily close to $\boldsymbol{x}_b$. Hence,

$$\boldsymbol{p}\sum_{a=1}^{m} \omega_a \leq \boldsymbol{p}\sum_{b\neq a} \boldsymbol{x}_b + \boldsymbol{p}\boldsymbol{z}_a \leq \boldsymbol{p}\sum_{a=1}^{m} \omega_a - \boldsymbol{p}\boldsymbol{x}_a + \boldsymbol{p}\boldsymbol{z}_a,$$

and from this, we obtain

$$\boldsymbol{p}\boldsymbol{x}_a \leq \boldsymbol{p}\boldsymbol{z}_a \text{ for every } \boldsymbol{z}_a \in P_a(\boldsymbol{x}_a), \quad a = 1 \ldots m.$$

Therefore, the condition (Q-1) is met, and the theorem is proved. □

If the situation that $\boldsymbol{p}\boldsymbol{x}_a = \inf \boldsymbol{p}X_a$ is excluded, then we have an equilibrium rather than a quasi-equilibrium.

Corollary 2.1. *Under the assumptions of Theorem 2.6, if the Pareto-optimal allocation* $(\boldsymbol{x}_a)_{a=1}^m$ *satisfies* $\boldsymbol{p}\boldsymbol{x}_a > \inf \boldsymbol{p}X_a$, *then it is an equilibrium relative to* $\boldsymbol{p}$.

Proof. Suppose the above is not the case. Then, there exists a consumption bundle $\boldsymbol{z}$ such that $\boldsymbol{p}\boldsymbol{z} \leq \boldsymbol{p}\boldsymbol{x}_a$ and $\boldsymbol{x}_a \prec_a \boldsymbol{z}$. We can take $\boldsymbol{y} \in X_a$ such that $\boldsymbol{x}_a \prec_a \boldsymbol{y}$ and $\boldsymbol{p}\boldsymbol{y} < \boldsymbol{p}\boldsymbol{x}_a$. As the set $Q_a(\boldsymbol{x}_a) = \{\boldsymbol{z} \in X_a | \ \boldsymbol{x}_a \prec_a \boldsymbol{z}\}$ is convex, it follows that

$$\boldsymbol{x}_a \prec_a \tau\boldsymbol{z} + (1-\tau)\boldsymbol{y} \quad \text{for every } 0 \leq \tau \leq 1.$$

However, $\boldsymbol{p}(\tau\boldsymbol{z} + (1-\tau)\boldsymbol{y}) < \boldsymbol{p}\boldsymbol{x}_a$, which contradicts that $(\boldsymbol{x}_a)$ is a quasi-equilibrium. □

2.6 A Limit Theorem of the Core

In this section, we assume that the consumer $a(= 1 \dots m)$ has the consumption set $X_a = \mathbb{R}^\ell_+$ and the non-negative orthant of $\mathbb{R}^\ell$ and that the preferences of the consumer are represented by a continuous utility function,

$$u_a : X_a \to \mathbb{R}, \ \boldsymbol{x} \mapsto u_a(\boldsymbol{x}).$$

Let $A = \{1 \dots m\}$ be the set of consumers. A non-empty subset C of A is called a *coalition*. Let $C \subset A$ be a coalition. A feasible allocation $(\boldsymbol{x}_a) \in \prod_{a=1}^m X_a$ is said to be *blocked* or *improved upon* by the coalition C if and only if there exists an allocation $(\boldsymbol{y}_a) \in \prod_{a=1}^m X_a$ such that

$$u_a(\boldsymbol{y}_a) \geq u_a(\boldsymbol{x}_a) \quad \text{for all } a \in C, > \text{for some } a \in C,$$

and

$$\sum_{a \in C} \boldsymbol{y}_a \leq \sum_{a \in C} \omega_a.$$

Definition 2.6. A feasible allocation is called a *core allocation* if no coalition can block it. The set of all core allocations is called the *core*.

By definition, it is clear that the core allocations are Pareto-optimal because they cannot be blocked by the grand coalition $C = A$. Indeed, the first theorem of welfare economics Proposition 2.10 is extended to the following proposition.

Proposition 2.12. *Under (LNS), every competitive equilibrium allocation is in the core.*

Proof. Let $(\boldsymbol{x}_1 \dots \boldsymbol{x}_m) \in \prod_{a=1}^m X_a$ be a competitive equilibrium allocation, such that there exists a price vector $\boldsymbol{p} \in \Delta$ such that $m+1$-tuple $(\boldsymbol{p}, \boldsymbol{x}_a) \in \Delta \times \prod_{a=1}^m X_a$ is a competitive equilibrium. Hence, $u_a(\boldsymbol{y}) > u_a(\boldsymbol{x}_a)$ implies that $\boldsymbol{py} > \boldsymbol{p}\omega_a$. We claim that $u_a(\boldsymbol{y}) \geq u_a(\boldsymbol{x}_a)$ implies that $\boldsymbol{py} \geq \boldsymbol{p}\omega_a$. Otherwise, $\boldsymbol{py} < \boldsymbol{p}\omega_a$ and, from (LNS), we can find a $\boldsymbol{z} \in X_a$ close enough to $\boldsymbol{y}$ so that $\boldsymbol{pz} < \boldsymbol{p}\omega_a$ and $u_a(\boldsymbol{z}) > u_a(\boldsymbol{y}) \geq u_a(\boldsymbol{x}_a)$, which would contradict condition (E-1) of Definition 2.3 for consumer a. Suppose that $C \subset A$ was a blocking coalition, so that one has an allocation $(\boldsymbol{y}_a)$ such that $\sum_{a \in C} \boldsymbol{y}_a \leq \sum_{a \in C} \omega_a$, and $u_a(\boldsymbol{y}_a) \geq u_a(\boldsymbol{x}_a)$ for all $a \in C$ and $u_a(\boldsymbol{y}_a) > u_a(\boldsymbol{x}_a)$ for at least one $a \in C$. Therefore, we have $\boldsymbol{p}\boldsymbol{y}_a \geq \boldsymbol{p}\omega_a$ for all $a \in C$ and $\boldsymbol{p}\boldsymbol{y}_a > \boldsymbol{p}\omega_a$ for at least one $a \in C$. Hence, $\sum_{a \in C} \boldsymbol{p}(\boldsymbol{y}_a - \omega_a) > 0$, contradicting $\sum_{a \in C} \boldsymbol{p}(\boldsymbol{y}_a - \omega_a) \leq 0$. □

Note that only the (LNS) condition was used in the proof of Proposition 2.12, as in the proof of Proposition 2.10. The intuitive relationship between Pareto-optimal allocations, core allocations, and competitive equilibrium allocations can be depicted in Figure 2.5.

In the rest of this section, we will study the theoretical relationship between the competitive equilibria and the core. As we saw in the competitive equilibrium, consumers watch the market price signal only and do not pay attention to other consumers' consumption decisions. That is to say, they act independent of other consumers'

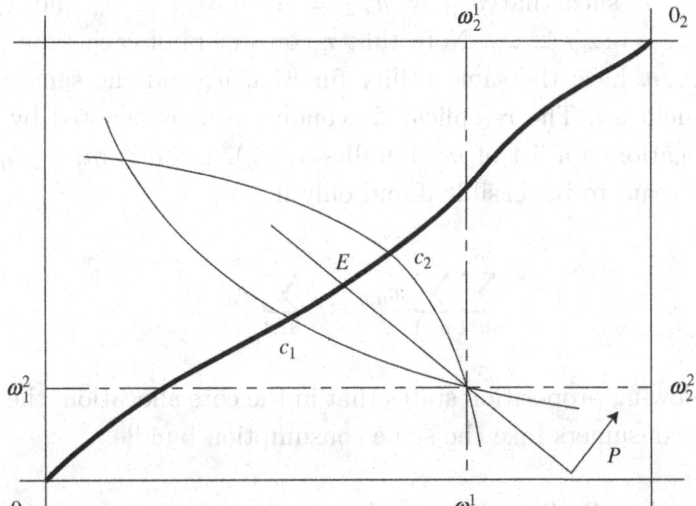

Figure 2.5. Edgeworth diagram. E: Equilibrium, c_1–c_2: core, O_1–O_2: optimal allocations.

buying and selling actions or, in game theoretic terms, they behave non-cooperatively with each other. On the other hand, in the concept of the core (Definition 2.6), we consider bargaining among individual consumers. The bargaining process is not finalized until there is no coalition that prefers a new bargain to their existing bargains. In game theoretic terms, the concept of the core is a cooperative one. Edgeworth [66] showed that in the case of two commodities and two types of consumers, the core allocation will approach the competitive equilibrium allocation as the number of consumers becomes larger and larger, in such a way that identical consumers of each type increase proportionally, or each type of consumer is "replicated." Debreu and Scarf [55] then generalized this result to the case of arbitrary (finite) numbers of commodities and consumers. We present their result in Theorem 2.7.

We consider m types of consumers, indexed by $a = 1 \dots m$, with r consumers of each type indexed by $q = 1 \dots r$. Therefore, each individual consumer is indexed by a pair of integers (a, q), and there exist mr consumers in the economy. The r-replicated economy of the economy $\mathcal{E} = (u_a, \omega_a)_{a=1}^{m}$ is an mr-tuple $(u_{a,q}, \omega_{a,q})$, $1 \leq a \leq m$, $1 \leq q \leq r$, such that $u_{a,1} = u_{a,2} = \dots = u_{a,r} \equiv u_a$ and $\omega_{a,1} = \omega_{a,2} = \dots = \omega_{a,r} \equiv \omega_a$. Note that r consumers of each type (a, q), $q = 1 \dots r$, have the same utility function u_a and the same initial endowment ω_a. The r-replicated economy of $\mathcal{E}$ is denoted by $\mathcal{E}(r)$. An allocation is a list of mr bundles, $(\boldsymbol{x}_{a,q})$, $1 \leq a \leq m$, $1 \leq q \leq r$, and it is said to be feasible if and only if

$$\sum_{a=1}^{m} \sum_{q=1}^{r} \boldsymbol{x}_{a,q} \leq r \sum_{a=1}^{m} \omega_a.$$

The following proposition states that in the core allocation, the same type of consumers take the same consumption bundle.

Proposition 2.13. *Let* $(\boldsymbol{x}_{a,q})$, $1 \leq a \leq m$, $1 \leq q \leq r$, *be a core allocation of the economy* $\mathcal{E}(r)$. *Then, under (SCV),* $\boldsymbol{x}_{a,1} = \dots \boldsymbol{x}_{a,r}$ *for each* $a = 1 \dots m$.

Proof. For any type of fixed a, let the q_ath individual take the least desired bundle among r consumers, namely

$$u_a(\boldsymbol{x}_{a,q_a}) \le u_a(\boldsymbol{x}_{a,q}) \quad \text{for all } q = 1 \dots r.$$

and suppose that for some b, $\boldsymbol{x}_{b,s} \neq \boldsymbol{x}_{b,t}$, $1 \le s,\ t \le r$. We denote $\boldsymbol{x}_{a,q_a}$ by $\boldsymbol{x}_a$. Then, from the assumption (SCV),

$$u_a\left(\frac{1}{r}\sum_{q=1}^{r} \boldsymbol{x}_{a,q}\right) \ge u_a(\boldsymbol{x}_a) \quad \text{for all } a$$

and

$$u_b\left(\frac{1}{r}\sum_{q=1}^{r} \boldsymbol{x}_{b,q}\right) > u_b(\boldsymbol{x}_b).$$

However, we have

$$\sum_{a=1}^{m}\left(\frac{1}{r}\sum_{q=1}^{r} \boldsymbol{x}_{a,q} - \omega_a\right) = \mathbf{0}$$

and hence the coalition, which consists of one consumer of each type a who receives $\boldsymbol{x}_a = \boldsymbol{x}_{a,q_a}$, would be a blocking coalition. □

From Proposition 2.13, we can describe the core allocations of the economy $\mathcal{E}(r)$, $r \ge 1$, by m-tuples of consumption bundles $(\boldsymbol{x}_1 \dots \boldsymbol{x}_m)$ such that $\sum_{a=1}^{m} \boldsymbol{x}_a \le \sum_{a=1}^{m} \omega_a$. Also from Proposition 2.12, every competitive equilibrium allocation of the economy $\mathcal{E}(r)$ is a core allocation of $\mathcal{E}(r)$. Therefore, competitive equilibrium allocations are also described by the m-tuples of feasible consumption bundles.

Let $\mathscr{W}(r)$ and $\mathscr{C}(r)$ be the set of equilibrium allocations and the core allocations of the economy $\mathcal{E}(r)$, respectively. We have shown that $\mathscr{W}(r) \subset \mathscr{C}(r)$ for each r. Furthermore, we have

Proposition 2.14. *$\mathscr{C}(r+1) \subset \mathscr{C}(r)$ for all $r \ge 1$.*

Proof. Let $A(r) = \{(1,1) \dots (1,r), (2,1) \dots (m,r)\}$ be the set of consumers in the economy $\mathcal{E}(r)$. Obviously, we have

$$A = A(1) \subset A(2) \subset \dots \subset A(r) \subset A(r+1) \subset \cdots.$$

If an allocation of the economy $\mathcal{E}(r)$ is blocked by a coalition $C \subset A(r)$, then C will block that allocation in the economy $\mathcal{E}(r+1)$. Therefore, if an allocation does not belong to $\mathscr{C}(r)$, it does not belong to $\mathscr{C}(r+1)$. Hence, the proposition follows. □

If $(\boldsymbol{p}, \boldsymbol{x}_a)_{a=1}^m$ is a competitive equilibrium of the economy $\mathcal{E} = (u_a, \omega_a)_{a=1}^m$, then it is also a competitive equilibrium for all $\mathcal{E}(r)$, $r \geq 1$. Under the equilibrium price $\boldsymbol{p}$ in $\mathcal{E}$, consumer (a, s) and consumer (a, t), $1 \leq s,\ t \leq r$, have the same utility function u_a, which is strictly convex, and the same initial endowment ω_a. Thus, they demand the same consumption bundle $\boldsymbol{x}_a (= \boldsymbol{x}_{a,s} = \boldsymbol{x}_{a,t})$. In other words, we have $\mathscr{W} = \mathscr{W}(1) = \cdots \mathscr{W}(r) = \cdots$. On account of Proposition 2.14, it follows that $\mathscr{W} \subset \cdots \mathscr{C}(r+1) \subset \mathscr{C}(r) \subset \cdots \subset \mathscr{C}$, or $\mathscr{W} \subset \cap_{r=1}^{\infty} \mathscr{C}(r)$, where $\mathscr{W}$ and $\mathscr{C}$ are the set of competitive equilibrium allocations and the core allocations of the economy $\mathcal{E}$, respectively. The limit theorem of Debreu and Scarf asserts the converse.

Theorem 2.7. *Let $\mathscr{W}$ be the set of competitive equilibrium allocations of an economy $\mathcal{E} = (u_a, \omega_a)_{a=1}^m$, and let $\mathscr{C}(r)$ be the set of core allocations of the r-replicated economy $\mathcal{E}(r)$. Then, under the assumptions (SCV), (MT), and (PTE),*

$$\mathscr{W} = \cap_{r=1}^{\infty} \mathscr{C}(r).$$

Proof. $\mathscr{W} \subset \cap_{r=1}^{\infty} \mathscr{C}(r)$ has already been shown. We prove the converse. Let $(\boldsymbol{x}_1 \dots \boldsymbol{x}_m) \in \cap_{r=1}^{\infty} \mathscr{C}(r)$, and define the set $\mathscr{Z}_a$ by

$$\mathscr{Z}_a = \{\boldsymbol{z} \in \mathbb{R}^\ell \mid u_a(\boldsymbol{z} + \omega_a) > u_a(\boldsymbol{x}_a)\}, \quad a = 1 \dots m,$$

and the set $\mathscr{Z}$ by

$$\mathscr{Z} = co\,(\cup_{a=1}^m \mathscr{Z}_a),$$

where coA denotes the convex hull of a set A. Note that the set $\mathscr{Z}_a$ is convex by the convexity assumption on the utility function u_a, so that $\mathscr{Z}$ can be written as

$$\mathscr{Z} = \left\{ \sum_{a=1}^m \alpha_a \boldsymbol{z}_a \,\middle|\, \alpha_a \geq 0, \sum_{a=1}^m \alpha_a = 1, u_a(\boldsymbol{z}_a + \omega_a) > u_a(\boldsymbol{x}_a) \right\}.$$

We claim that $\mathbf{0} \notin \mathscr{Z}$. Suppose that this is not the case. Then, we have bundles $\boldsymbol{z}_1, \ldots, \boldsymbol{z}_m$ such that $\sum_{a=1}^{m} \alpha_a \boldsymbol{z}_a = \mathbf{0}$ with $\alpha_a \geq 0$, $\sum_{a=1}^{m} \alpha_a = 1$, $u_a(\boldsymbol{z}_a + \omega_a) > u_a(\boldsymbol{x}_a)$. Let $k \in \mathbb{N}$ be an integer and let α_a^k be the smallest integer such that $\alpha_a^k \geq k\alpha_a$. We denote $I = \{a |\ \alpha_a > 0\}$. For each $a \in I$, let $\boldsymbol{z}_a^k = (k\alpha_a/\alpha_a^k)\boldsymbol{z}_a$. Then, it follows that $\omega_a \leq \boldsymbol{z}_a^k + \omega_a \leq \boldsymbol{z}_a + \omega_a$, and $\boldsymbol{z}_a^k \to \boldsymbol{z}_a$ as $k \to \infty$. By the continuity of u_a, $u_a(\boldsymbol{z}_a^k + \omega_a) > u_a(\boldsymbol{x}_a)$ for k is sufficiently large, and

$$\sum_{a \in I} \alpha_a^k \boldsymbol{z}_a^k = k \sum_{a \in I} \alpha_a \boldsymbol{z}_a = \mathbf{0}.$$

Consider the coalition that consists of α_a^k members of the type $a \in I$, each one of whom receives $\omega_a + \boldsymbol{z}_a^k$. This coalition would block the core allocation $(\boldsymbol{x}_1 \ldots \boldsymbol{x}_m)$ of the economy $\mathcal{E}(\max\{\alpha_a^k |\ a \in I\})$, which is a contradiction. Hence, $\mathbf{0} \notin \mathscr{Z}$. Then, by the separation hyperplane Theorem A.1, one obtains the non-zero vector $\boldsymbol{p}$ such that $\boldsymbol{p}\boldsymbol{z} \geq 0$ for all $\boldsymbol{z} \in \mathscr{Z}$. As in the proof of Theorem 2.6, we can show that $\boldsymbol{p} \geq \mathbf{0}$.

If $u_a(\boldsymbol{y}) > u_a(\boldsymbol{x}_a)$, then $\boldsymbol{y} - \omega_a \in \mathscr{Z}_a \subset \mathscr{Z}$, so that $\boldsymbol{p}\boldsymbol{y} \geq \boldsymbol{p}\omega_a$, $a = 1 \ldots m$. From (MT), we have $\boldsymbol{y}$, which is strictly preferred to $\boldsymbol{x}_a$ and arbitrarily close to $\boldsymbol{x}_a$. Thus, it follows that $\boldsymbol{p}\boldsymbol{x}_a \geq \boldsymbol{p}\omega_a$, $a = 1 \ldots m$. However,

$$\sum_{a=1}^{m} (\boldsymbol{x}_a - \omega_a) \leq \mathbf{0},$$

and $\boldsymbol{p} \geq \mathbf{0}$, so that $\boldsymbol{p}\boldsymbol{x}_a = \boldsymbol{p}\omega_a$, $a = 1 \ldots m$. Therefore, we have demonstrated that the vector $\boldsymbol{p}$ and the allocation $(\boldsymbol{x}_a)$ consist of the quasi-equilibrium, namely the following conditions:

(Q-1) $u_a(\boldsymbol{x}_a) < u_a(\boldsymbol{y})$ implies that $\boldsymbol{p}\boldsymbol{x}_a \leq \boldsymbol{p}\boldsymbol{y}$ for all $a = 1 \ldots m$, and

(E-2) $\sum_{a=1}^{m} \boldsymbol{x}_a \leq \sum_{a=1}^{m} \omega_a$.

As $\sum_{a=1}^{m} \omega_a \gg \mathbf{0}$, we have $\boldsymbol{p} \sum_{a=1}^{m} \omega_a > 0$, and hence, $\boldsymbol{p}\omega_a > 0$ for some a. Now, we show that for $a \in A$ with positive income $\boldsymbol{p}\omega_a > 0$, $\boldsymbol{p}\boldsymbol{y} \leq \boldsymbol{p}\omega_a$ implies $u_a(\boldsymbol{x}_a) \geq u_a(\boldsymbol{y})$. For such a consumer, we can take a sequence $\{\boldsymbol{y}_n\}$ with $\boldsymbol{y}_n \to \boldsymbol{y}$ and $\boldsymbol{p}\boldsymbol{y}_n < \boldsymbol{p}\omega_a$ for all n. It follows from the condition (Q-1) that $u_a(\boldsymbol{x}_a) \geq u_a(\boldsymbol{y}_n)$ for all n.

From (CT), we obtain $u_a(\boldsymbol{x}_a) \geq u_a(\boldsymbol{y})$ in the limit. From (MT), we have $\boldsymbol{p} \gg \boldsymbol{0}$. Otherwise, if $p^t = 0$ for some t, then $\boldsymbol{p}(\boldsymbol{x}_a + \epsilon \boldsymbol{e}_t) = \boldsymbol{p}\boldsymbol{x}_a$ and $u_a(\boldsymbol{x}_a) < u_a(\boldsymbol{x}_a + \epsilon \boldsymbol{e}_t)$ for $\epsilon > 0$, where $\boldsymbol{e}_t = (0 \dots 0, 1, 0 \dots 0)$ (1 at the tth position), which is a contradiction. Then, if $\boldsymbol{p}\omega_a = 0$, the budget set $\{\boldsymbol{x} \in \mathbb{R}^\ell_+ | \ \boldsymbol{p}\boldsymbol{x} \leq \boldsymbol{p}\omega_a = 0\}$ is a singleton $\{\boldsymbol{0}\}$, and hence, the condition (E-1) holds trivially. We conclude that $(\boldsymbol{p}, \boldsymbol{x}_a)$ is a competitive equilibrium, or $(\boldsymbol{x}_a) \in \mathscr{W}$. □

2.7 Nash Equilibria and the Core of Games

In this section, we will explain some applications of fixed point theory to game theory. A (strategic form of a) non-cooperative N person game $\mathcal{G}$ is formally described as follows. Let $A = \{1 \dots N\}$ be the set of *players*. The player $a \in A$ has a *choice set* X_a that is a non-empty subset of $\mathbb{R}^\ell$, and we denote $X = \prod_{a=1}^N X_a$. The player $a \in A$ also has a *constraint correspondence* $\mathcal{C}_a : X \to X_a$ and a *preference correspondence* $P_a : X \to X_a$. The $3N$-tuple $\mathcal{G} = (X_a, \mathcal{C}_a, P_a)_{a=1}^N$ is called an *N person (non-cooperative) abstract game*. An interpretation of the correspondence $P_a(\boldsymbol{x})$ is the set of *actions* that is preferred by a to the action $\boldsymbol{x}_a$ when the profile of actions $\boldsymbol{x} = (\boldsymbol{x}_1 \dots \boldsymbol{x}_a \dots \boldsymbol{x}_N)$ is given.

Definition 2.7. An N-tuple of strategies $\boldsymbol{x}^* = (\boldsymbol{x}_1^* \dots \boldsymbol{x}_N^*) \in X$ is called a *Nash equilibrium* of the abstract game $\mathcal{G} = (X_a, \mathcal{C}_a, P_a)_{a=1}^N$ if and only if for each $a = 1 \dots N$,

$$\boldsymbol{x}_a^* \in \mathcal{C}_a(\boldsymbol{x}^*) \text{ and } P_a(\boldsymbol{x}^*) \cap \mathcal{C}_a(\boldsymbol{x}^*) = \emptyset.$$

Suppose that an exchange economy $\mathcal{E} = (X_a, u_a, \omega_a)_{a=1}^{N-1}$ with $N-1$ consumers is given. We can interpret $\mathcal{E}$ using a fictitious "market player," which is explained in what follows as an N person game $\mathcal{G}$, and the Nash equilibrium of $\mathcal{G}$ is reduced to the competitive equilibrium of $\mathcal{E}$ as follows.

The first $N-1$ players have the choice sets X_a, which are the consumption sets of the consumer $a = 1 \dots N-1$, and the last player N, called the market player, has the choice set $X_N = \Delta$, the unit simplex. The player $a(= 1 \dots N - 1)$ has the constraint correspondence

$$\mathcal{C}_a : X \to X_a,\ \mathcal{C}_a((\boldsymbol{x}_b), \boldsymbol{p}) = \{\boldsymbol{x}' \in X_a |\ \boldsymbol{p}\boldsymbol{x}' \leq \boldsymbol{p}\omega_a\},$$

which is the budget correspondence of consumer a, and the market player's constraint correspondence is given by $\mathcal{C}_N((\boldsymbol{x}_b), \boldsymbol{p}) = \Delta$.

The first $N - 1$ players' preference is given by the preference correspondence (see Section 2.3),

$$P_a : X \to X_a,\ P_a((\boldsymbol{x}_b), \boldsymbol{p}) = \{\boldsymbol{x}' \in X_a |\ u_a(\boldsymbol{x}') > u_a(\boldsymbol{x}_a)\},$$

and the market player's preference correspondence $P_N : X \to \Delta$ is given by

$$P_N((\boldsymbol{x}_b), \boldsymbol{p}) = \left\{ \boldsymbol{q} \in \Delta \,\middle|\, \boldsymbol{q} \sum_{a=1}^{N-1} (\boldsymbol{x}_a - \omega_a) > \boldsymbol{p} \sum_{a=1}^{N-1} (\boldsymbol{x}_a - \omega_a) \right\}.$$

It is easy to check that a Nash equilibrium of the game $\mathcal{G}$ is a competitive equilibrium of the economy $\mathcal{E}$. Shafer and Sonnenschein [228] proved the existence of the Nash equilibrium based on the work of Nash [166] and Debreu [49].

Theorem 2.8. *Let $\mathcal{G} = (X_a, \mathcal{C}_a, P_a)_{a=1}^N$ be an N person game satisfying the following for each $a = 1 \dots N$:*

(a) *X_a is a non-empty compact and convex subset of $\mathbb{R}^\ell$,*
(b) *$\mathcal{C}_a : X \to X_a$ is a continuous correspondence,*
(c) *for each $\boldsymbol{x} = (\boldsymbol{x}_b) \in X$, the set $\mathcal{C}_a(\boldsymbol{x})$ is non-empty and convex,*
(d) *$Graph(P_a) \equiv \{(\boldsymbol{x}, \boldsymbol{y}) \in X \times X_a |\ \boldsymbol{x} \in X,\ \boldsymbol{y} \in P_a(\boldsymbol{x})\}$ is open in $X \times X_a$,*
(e) *for each $\boldsymbol{x} = (\boldsymbol{x}_a) \in X$, $\boldsymbol{x}_a \notin coP_a(\boldsymbol{x})$.*

Then, $\mathcal{G}$ has a Nash equilibrium.

Proof. Define a real valued function u_a on $X \times X_a$ by

$$u_a(\boldsymbol{y}, \boldsymbol{x}_a) = \inf\{d((\boldsymbol{y}, \boldsymbol{x}_a), \boldsymbol{z}) |\ \boldsymbol{z} \notin \mathrm{Graph}(P_a)\},$$

where $d(\boldsymbol{z}_1, \boldsymbol{z}_2) = \|\boldsymbol{z}_1 - \boldsymbol{z}_2\|$ for $\boldsymbol{z}_1, \boldsymbol{z}_2 \in X \times X_a$. Then, u_a is continuous and $u_a(\boldsymbol{y}, \boldsymbol{x}_a) > 0$ if and only if $\boldsymbol{x}_a \in P_a(\boldsymbol{y})$. Indeed, take a converging sequence $(\boldsymbol{y}(n), \boldsymbol{x}_a(n)) \to (\boldsymbol{y}, \boldsymbol{x}_a)$. As $X \times X_a$ is compact from assumption (a) and the graph of P_a is open from the assumption (d), the infimum of the distance between $(\boldsymbol{y}(n), \boldsymbol{x}_a(n))$ and a compact set $X \times X_a \backslash Graph P_a$ is achieved by $\boldsymbol{z}(n)$. We can assume $\boldsymbol{z}(n) \to \boldsymbol{z}$. It is easy to verify that $u_a(\boldsymbol{y}(n), \boldsymbol{x}_a(n)) = d((\boldsymbol{y}(n), \boldsymbol{x}_a(n)), \boldsymbol{z}(n)) \to d((\boldsymbol{y}, \boldsymbol{x}_a), \boldsymbol{z}) = u_a(\boldsymbol{y}, \boldsymbol{x}_a)$. This proves the continuity of u_a. The second claim that $u_a(\boldsymbol{y}, \boldsymbol{x}_a) > 0$ if and only if $\boldsymbol{x}_a \in P_a(\boldsymbol{y})$ follows from the definition of u_a.

For each a, we define the correspondence $F_a : X \to X_a$ by

$$F_a(\boldsymbol{y}) = \{\boldsymbol{x}_a \in \mathcal{C}_a(\boldsymbol{y}) |\ u_a(\boldsymbol{y}, \boldsymbol{x}_a) \geq u_a(\boldsymbol{y}, \boldsymbol{z}_a) \text{ for every } \boldsymbol{z}_a \in \mathcal{C}_a(\boldsymbol{y})\}.$$

As u_a is continuous and $\mathcal{C}_a$ is a continuous and non-empty compact valued correspondence, it follows from Theorem D.2 that F_a is upper hemicontinuous and a compact-valued correspondence. We define the correspondence $G : X \to X$, $\boldsymbol{y} = (\boldsymbol{y}_a) \mapsto G(\boldsymbol{y})$ by

$$G(\boldsymbol{y}) = \prod_{a=1}^{N} coF_a(\boldsymbol{y}_a).$$

Then, from Propositions D.5 and D.6, G is a upper hemicontinuous, compact- and convex-valued correspondence. From the Kakutani fixed point theorem (Theorem D.1), there exists a fixed point $\boldsymbol{x}^* \in G(\boldsymbol{x}^*)$ or $\boldsymbol{x}_a^* \in coF_a(\boldsymbol{x}_a^*)$ for every a. We claim that $\boldsymbol{x}^*$ is a Nash equilibrium of $\mathcal{G}$.

As $F_a(\boldsymbol{x}^*) \subset \mathcal{C}_a(\boldsymbol{x}^*)$ and $\mathcal{C}_a(\boldsymbol{x}^*)$ is convex from assumption (b),

$$coF_a(\boldsymbol{x}^*) \subset \mathcal{C}_a(\boldsymbol{x}^*),$$

hence $\boldsymbol{x}_a^* \in \mathcal{C}_a(\boldsymbol{x}^*)$. If there was a point $\boldsymbol{z}_a \in P_a(\boldsymbol{x}^*) \cap \mathcal{C}_a(\boldsymbol{x}^*)$, then $u_a(\boldsymbol{x}^*, \boldsymbol{z}_a) > 0$. Thus, $u_a(\boldsymbol{x}^*, \boldsymbol{y}_a) > 0$ for all $\boldsymbol{y}_a \in F_a(\boldsymbol{x}^*)$. Therefore, $\boldsymbol{z}_a \in P_a(\boldsymbol{x}^*) \cap \mathcal{C}_a(\boldsymbol{x}^*)$ implies that $F_a(\boldsymbol{x}^*) \subset P_a(\boldsymbol{x}^*)$, which yields that $\boldsymbol{x}_a^* \in coF_a(\boldsymbol{x}_a^*) \subset coP_a(\boldsymbol{x}^*)$, a contradiction. □

It should be noted that the non-empty values of the correspondence P_a or, $P_a(\boldsymbol{x}) \neq \emptyset$ for all $\boldsymbol{x} \in X$, is not required in Theorem 2.8. Two more remarkable points of the theorem are that (1) the transitivity of preferences is not assumed, and (2) the correspondences $\mathcal{C}_a$

and P_a admit externalities, as they include other players' actions in their variables. We will apply the Shafer–Sonnenschein theorem in Section 6.2.

Next, we will discuss the core of a game and its relations with that of an exchange economy. In Section 2.6, we saw that the competitive equilibria are in the core and, as we proved the existence of the equilibrium in Section 2.5, we have also obtained the non-emptiness of the core of the exchange economies. However, the competitive equilibrium and the core are mutually independent concepts, so that the proof of the existence of the core is important and interesting in its own right. We will present the classical existence result from Scarf [223]. First, we will provide definitions of a (nonside-payment) game and its core and explain how the exchange economy can be interpreted as the game. We will see that the core of the economy is reduced to that of the game. The proof of the existence of the core will follow.

Let $A = \{1 \dots N\}$ be the set of players and $\mathcal{A} = \{C \subset A |\, C \neq \emptyset\}$ be the family of non-empty subsets of A, which we call coalitions. For each coalition $C \in \mathcal{A}$, we define the subspace $\mathbb{R}^C$ of $\mathbb{R}^N$ by

$$\mathbb{R}^C = \{\boldsymbol{x} = (x^a) \in \mathbb{R}^N \,|\, x^a = 0 \text{ for } a \notin C\}.$$

A *nonside-payment game* in *characteristic function form* is a correspondence $V : \mathcal{A} \to \mathbb{R}^N$ such that $V(C) \subset \mathbb{R}^C$ for every $C \in \mathcal{A}$. The intended interpretation is that each element $(v^a) \in V(C)$ represents the *utility allocation* of the members of C, which results from their cooperation. In the following, however, we embed $V(C)$ into $\mathbb{R}^N$ as a cylinder to simplify our analysis. Thus, for each $C \in \mathcal{A}$, we define the subset $\mathcal{V}(C)$ of $\mathbb{R}^N$ by

$$\mathcal{V}(C) = \left\{ \boldsymbol{v} = (v^a) \in \mathbb{R}^N \middle| \, proj_C(\boldsymbol{v}) \in V(C) \right\},$$

where $proj_C(\boldsymbol{v}) = (w^a)$ is the *projection* of the vector $\boldsymbol{v}$ to $\mathbb{R}^C$, that is, $w^a = v^a$ for $a \in C$ and $w^a = 0$ otherwise. From now on, we will work with $\mathcal{V}$ rather than V and call it a *game*.

A utility allocation $(v^a)_{a=1}^N \in \mathbb{R}^N$ is called feasible if $(v^a) \in \mathcal{V}(A)$. A coalition $C \in \mathcal{A}$ is said to *block* or *improve upon* a utility allocation

(v^a) if there exists a utility allocation $(w^a) \in \mathcal{V}(C)$ such that $v^a < w^a$ for all $a \in C$. The definition of the core of the game $\mathcal{V}$, denoted by $\mathscr{C}(\mathcal{V})$, is now immediate.

Definition 2.8. The *core* of a game $\mathcal{V}$, $\mathscr{C}(\mathcal{V})$, is the set of feasible utility allocations that are not improved upon by any coalition.

Now, we explain the relationship between the game $\mathcal{V}$ and the exchange economy $\mathcal{E} = (u^a, \omega_a)_{a=1}^m$ with $m = N$. For each coalition $C \in \mathcal{A}$, we define the following:

$$\mathcal{V}(C) = \left\{ \boldsymbol{v} = (v^a) \in \mathbb{R}^N \,\middle|\, \text{There exists an allocation } (\boldsymbol{y}_a) \in \prod_{a=1}^N X_a \right.$$
$$\left. \text{such that } v^a \leq u_a(\boldsymbol{y}_a) \text{ for all } a \in C \text{ and } \sum_{a \in C} \boldsymbol{y}_a = \sum_{a \in C} \omega_a \right\}.$$

The economic meaning of the set $\mathcal{V}(C)$ is now clear; it is the set of utility allocations that can be obtained by redistribution of the endowments among the members of C, and it is obvious that under the assumptions of (CT) and (MT), the core of the exchange economy $\mathcal{E}$ coincides with that of the reduced game $\mathcal{V}$.

A family $\mathcal{B}$ of subsets of $\mathcal{A}$ is *balanced* if there exist non-negative "balanced weights" w_C for $C \in \mathcal{B}$ such that $\sum_{C \in \mathcal{B}_a} w_C = 1$ for every $a \in A$, where $\mathcal{B}_a = \{C \in \mathcal{B} | \ a \in C\}$ is the family of subsets of A containing a. A (nonside-payment) game $\mathcal{V}$ is called *balanced* if, for every balanced subfamily $\mathcal{B}$ of $\mathcal{A}$, it follows that $\cap_{C \in \mathcal{B}} \mathcal{V}(C) \subset \mathcal{V}(A)$. We will show that essentially, the balancedness condition is sufficient for non-emptiness of the core of the game $\mathcal{V}$. Before we state the main theorem of the existence of the core, we show that the game derived from the exchange economy is balanced.

Proposition 2.15. *Let $\mathcal{E} = (u_a, \omega_a)$ be an exchange economy and suppose that for every $a \in A$, the consumption set X_a is a convex subset of $\mathbb{R}^\ell$, $\omega_a \in X_a$ and u_a is quasi-concave, or satisfies the condition (QCV). Then, the associated game $\mathcal{V}$ is balanced.*

Proof. Let $\mathcal{B} \subset \mathcal{A}$ be a balanced subfamily of $\mathcal{A}$ with the weights $\{w_C |\ C \in \mathcal{B}\}$ and take any $\boldsymbol{v} = (v^a) \in \cap_{C\in\mathcal{B}}\mathcal{V}(C)$. Then, for each $C \in \mathcal{B}$, there exists an allocation $(\boldsymbol{y}_a) \in \prod_{a=1}^m X_a$ such that $\sum_{a\in C} \boldsymbol{y}_a = \sum_{a\in C} \omega_a$ and $v^a \leq u_a(\boldsymbol{y}_a)$ for all $a \in C$. Define $\boldsymbol{x}_a = \sum_{C\in\mathcal{B}_a} w_C \boldsymbol{y}_a$ for each $a \in C$. As X_a is convex and $u_a(\cdot)$ is quasi-concave, we have $\boldsymbol{x}_a \in X_a$ and $v^a \leq u_a(\boldsymbol{x}_a)$. Moreover, it follows that

$$\sum_{a=1}^m \boldsymbol{y}_a = \sum_{a=1}^m \sum_{C\in\mathcal{B}_a} w_C \boldsymbol{y}_a = \sum_{C\in\mathcal{B}_a} \sum_{a=1}^m w_C \boldsymbol{y}_a$$

$$= \sum_{C\in\mathcal{B}_a} \sum_{a=1}^m w_C \omega_a = \sum_{a=1}^m \sum_{C\in\mathcal{B}_a} w_C \omega_a = \sum_{a=1}^m \omega_a.$$

This proves that $\boldsymbol{v} = (v^a) \in \mathcal{V}(A)$. □

We define the real number $b_a = \sup\{v^a \in \mathbb{R}|\ (v^a) \in \mathcal{V}(\{a\})\}$ for each $a \in A$. Scarf's core existence theorem now reads as follows:

Theorem 2.9. *Let $\mathcal{V} : \mathcal{A} \to \mathbb{R}^N$ be a (nonside-payment) game. The core of $\mathcal{V}$ is non-empty if*

(a) *$\mathcal{V}(C) - \mathbb{R}^N_+ = \mathcal{V}(C)$ for every $C \in \mathcal{A}$,*
(b) *there exists an $M \in \mathbb{R}$ such that for each $C \in \mathcal{A}$, if $\boldsymbol{u} = (u^a) \in \mathcal{V}(C) \cap \{(b^a) + \mathbb{R}^N_+\}$, then $u^a < M$ for all $a \in C$,*
(c) *$\mathcal{V}(C)$ is a closed subset of $\mathbb{R}^N$ for every $C \in \mathcal{A}$, and*
(d) *the game $\mathcal{V}$ is balanced.*

Proof. Without loss of generality, we may assume that $b^a = 0$ for all $a \in A$. Given the positive number M in assumption (b), we define

$$\Delta^C = co\{-MN\boldsymbol{e}_a |\ a \in C\},$$

for every $C \in \mathcal{A}$, where $\boldsymbol{e}_a = (0\ldots0,1,0\ldots0) \in \mathbb{R}^N$. We also define the function $\tau : \Delta^A \to \mathbb{R}$ by $\tau(\boldsymbol{u}) = \sup\{\tau \in \mathbb{R}|\ \boldsymbol{u} + \tau\mathbf{1} \in \cup_{C\in\mathcal{A}}\mathcal{V}(C)\}$, where $\mathbf{1} = (1\ldots1) \in \mathbb{R}^N$. From Theorem D.2 (a), the function τ is continuous. Hence, the function $f : \Delta^A \to \mathbb{R}^N$ defined by $f(\boldsymbol{u}) = \boldsymbol{u} + \tau(\boldsymbol{u})\mathbf{1}$ is also continuous. Note that $f(\boldsymbol{u}) \geq 0$. Let $G_C = \{\boldsymbol{u} \in \Delta^A |\ f(\boldsymbol{u}) \in \mathcal{V}(C)\}$. As the set G_C is the inverse image under f of the closed set $\mathcal{V}(C)$, it is closed.

We claim that if $\Delta^C \cap G_B \neq \emptyset$ for $B, C \in \mathcal{A}$, then $B \subset C$. This claim trivially holds when $C = A$. Thus, assume that $\sharp C < N$, and take any $\boldsymbol{u} \in \Delta^C \cap G_B$. As $\sum_{a \in C} u^a = -MN$, there exists an $a \in C$ such that $u^a \leq -MN/\sharp C < -M$. It follows from $f(\boldsymbol{u}) = \boldsymbol{u} + \tau(\boldsymbol{u})\mathbf{1} \geq 0$ that $u^a + \tau(\boldsymbol{u}) \geq 0$. Hence, $\tau(\boldsymbol{u}) > M$. On the other hand, $f(\boldsymbol{u}) \in \mathcal{V}(B)$, so that for every $a \in B$, $u^a + \tau(\boldsymbol{u}) < M$. Thus, $u^a < 0$ for every $a \in B$. As $\boldsymbol{u} = (u^a) \in \Delta^C$ implies that $u^a = 0$ for every $a \notin C$, one obtains that $B \subset C$, and the claim is established (Figure 2.6).

Then, the conditions of the K–K–M–S lemma (Theorem A.4) are satisfied. Hence, there exists a balanced family $\mathcal{B}$ and a point $\boldsymbol{u}^* \in \Delta^A$ such that $\boldsymbol{u}^* \in \cap_{C \in \mathcal{B}} G_C$. Thus, $f(\boldsymbol{u}^*) \in \cap_{C \in \mathcal{B}} \mathcal{V}(C) \subset \mathcal{V}(A)$. By the definition of τ, $f(\boldsymbol{u}^*)$ is on the boundary of $\cap_{C \in \mathcal{A}} \mathcal{V}(C)$ and, according to (a), it cannot be improved upon by any coalition. Therefore, $f(\boldsymbol{u}^*) \in \mathscr{C}(\mathcal{V})$. □

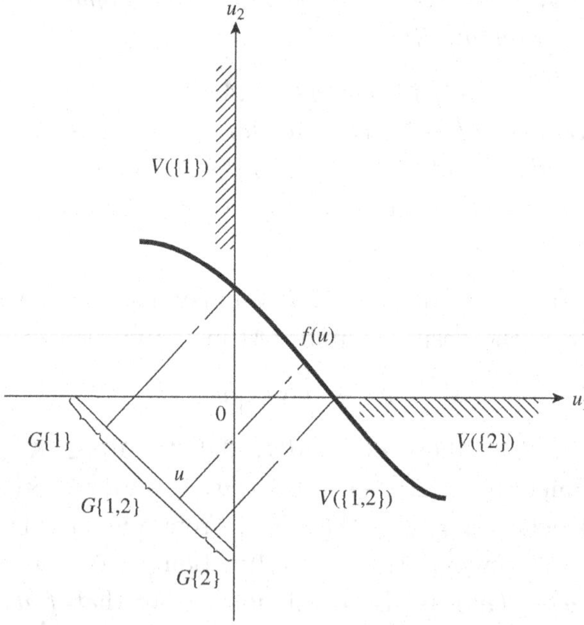

Figure 2.6. The set $\mathcal{V}(C)$ and G_C.

From Scarf's theorem, the non-emptiness of the core of the exchange economy $\mathcal{E} = (u_a, \omega_a)_{a=1}^m$ is obtained immediately.

Theorem 2.10. *Let* $\mathcal{E} = (u_a, \omega_a)_{a=1}^m$ *be a pure exchange economy such that for every* $a \in A$, *the consumption set* X_a *is a closed and convex subset of* $\mathbb{R}^\ell$, *which is bounded from below,* $\omega_a \in X_a$, *and the utility function* u_a *is quasi-concave. Then, the core of the economy* $\mathcal{E}$, $\mathscr{C}(\mathcal{E})$ *is non-empty.*

Proof. We will show that conditions (a)–(d) of Theorem 2.9 are satisfied. Condition (a) follows from the definition of $\mathcal{V}(C)$. To show conditions (b) and (c), we define the set $\mathscr{A}(C)$ of attainable allocations for the coalition C by

$$\mathscr{A}(C) = \left\{ (\boldsymbol{x}_a)_{a\in C} \in \prod_{a\in C} X_a \middle| \sum_{a\in C} \boldsymbol{x}_a = \sum_{a\in C} \omega_a \right\}.$$

As X_a is bounded from below and closed, $\mathscr{A}(C)$ is bounded and closed and, hence, it is compact. Then,

$$\mathcal{V}(C) = \{\boldsymbol{v} = (v^a) \in \boldsymbol{R}^N \mid v^a \le u_a(\boldsymbol{x}_a) \text{ for some } (\boldsymbol{x}_a)_{a\in C} \in \mathscr{A}(C)\}.$$

Conditions (b) and (c) follow from the compactness of $\mathscr{A}(C)$. Finally, condition (d) follows from Proposition 2.15. □

2.8 The Local Uniqueness of Equilibria

Consider an exchange economy with m consumers. The commodity space of our economy is an ℓ-dimensional Euclidean space $\mathbb{R}^\ell$. Consumer a's consumption set X_a is defined by

$$X_a = \{\boldsymbol{x} = (x^t) \in \mathbb{R}^\ell \mid x^t > 0,\ t = 1 \dots \ell\} \equiv \mathbb{R}^\ell_{++}, \quad a = 1 \dots m,$$

and consumer a's preference is represented by a utility function

$$u_a : X_a \to \mathbb{R},\ \boldsymbol{x} \mapsto u_a(\boldsymbol{x}), \quad a = 1 \dots m.$$

NB: In this section, we deal with a consumption set that is not closed, because we will work with smooth utility functions on it, which make the open domain suitable to work with. Furthermore, we do not distinguish the row vectors and the column vectors.

As usual, the consumer a's initial endowment vector is denoted by $\omega_a \in \mathbb{R}^\ell$ and the survival condition (SV) is assumed, $\omega \in X_a$. An economy is given by the list $(u_a, \omega_a)_{a=1}^m$ and denoted by $\mathcal{E}$. From now on, we fix the utility functions and identify an economy $\mathcal{E}$ with its initial endowment profile $\boldsymbol{\omega} \equiv (\omega_1 \dots \omega_m) \in \prod_{a=1}^m X_a$. Hence, the set $\Omega \equiv \prod_a X_a$ is the *space of all economies.* In the following, we will often refer to an endowment profile $\boldsymbol{\omega} = (\omega_a)$ simply as an economy.

Let $\mathscr{W}(\boldsymbol{\omega})$ be the set of equilibrium of the economy $\boldsymbol{\omega}$. In Section 2.5, we gave a sufficient condition for $\mathscr{W}(\boldsymbol{\omega})$ to be non-empty. The next property that we expect for $\mathscr{W}(\boldsymbol{\omega})$ is that it is a discrete set, that is, the equilibrium of $\boldsymbol{\omega}$ is locally unique. Furthermore, we hope that the set $\mathscr{W}(\boldsymbol{\omega})$ changes continuously when the economy $\boldsymbol{\omega}$ changes continuously. In other words, the equilibrium of $\boldsymbol{\omega}$ is locally stable. An economy in which all equilibria are locally unique and locally stable is called a *regular economy* (Definition 2.11). The main result of this section is that "almost all" economies are regular (Theorem 2.13). We assume that the utility function satisfies the following conditions.

(U-1): u_a is twice continuously differentiable, namely of class C^2 on X_a.

Let $Du_a(\boldsymbol{x}) = (\partial_1 u_a(\boldsymbol{x}) \dots \partial_\ell u_a(\boldsymbol{x}))$ be the derivative (tangent map) of u_a at $\boldsymbol{x} \in X_a$, where

$$\partial_t u_a(\boldsymbol{x}) = \lim_{h \to 0} \frac{u_a(x^1 \dots x^t + h \dots x^\ell) - u_a(x^1 \dots x^t \dots x^\ell)}{h},$$

and let

$$D^2 u_a(\boldsymbol{x}) = \begin{pmatrix} \partial_1 \partial_1 u_a(\boldsymbol{x}) & \dots & \partial_1 \partial_\ell u_a(\boldsymbol{x}) \\ \dots & \dots & \dots \\ \partial_\ell \partial_1 u_a(\boldsymbol{x}) & \dots & \partial_\ell \partial_\ell u_a(\boldsymbol{x}) \end{pmatrix}$$

be the second derivative. For every $\boldsymbol{x} \in X_a$, the Hessian $D^2u_a(\boldsymbol{x})$ is considered to be a linear map from $\mathbb{R}^\ell$ to itself. We assume for every a that

(U-2): u_a is *strictly differentiably monotone*, i.e., $Du_a(\boldsymbol{x}) \gg 0$ for every $\boldsymbol{x} \in X_a$,

(U-3): u_a is *strictly differentiably concave.*

The condition (U-3) means that $D^2u_a(\boldsymbol{x})$ is a non-degenerate, negative definite bilinear form on $\mathbb{R}^\ell$, or

$$\boldsymbol{y}D^2u_a(\boldsymbol{x})\boldsymbol{y} = \sum_{s,t=1}^{\ell} \partial_s\partial_t u_a(\boldsymbol{x})y^s y^t \le 0$$

for every $\boldsymbol{y} = (y^t) \in \mathbb{R}^\ell$ and the equality holds only when $\boldsymbol{y} = \boldsymbol{0}$. This implies that the linear map $D^2u_a(\boldsymbol{x})$ is a linear isomorphism of $\mathbb{R}^\ell$ to itself. Finally, we assume that

(U-4): For every sequence $\boldsymbol{x}_n = (x_n^t) \in X_a$ such that $x_n^t \to 0$ for some t, it follows that $u_a(\boldsymbol{x}_n) \to -\infty$.

The assumption (U-4) is sometimes called the *Inada condition.*

A price vector in the economies is an ℓ-vector $\boldsymbol{p} = (p^t) \gg 0$. In this section, the feasibility of an allocation $(\boldsymbol{x}_a) \in \prod_{a=1}^m X_a$ is meant to be in the exact sense, or $\sum_{a=1}^m \boldsymbol{x}_a = \sum_{a=1}^m \omega_a$. The concept of the Pareto optimality can be reformulated as follows.

Definition 2.9. An allocation $(\boldsymbol{x}_a)$ is said to be Pareto-optimal if it is a solution of the following *social welfare maximization problem:*

Given $\boldsymbol{\lambda} = (\lambda_2, \ldots, \lambda_m) \in \mathbb{R}_{++}^{m-1}$,

$$\text{maximize } u_1(\boldsymbol{x}_1) + \sum_{a=2}^{m} \lambda_a u_a(\boldsymbol{x}_a) \text{ subject to } \sum_a \boldsymbol{x}_a = \sum_a \omega_a.$$

It is easy to verify that the solution of this maximization problem exists. As will be shown in the proof of Proposition 8.1, the solution $\boldsymbol{x}_a$ associated with $\boldsymbol{\lambda}$ is a smooth function of $\boldsymbol{\lambda} = (\lambda_2 \ldots \lambda_m)$. Therefore, we will denote it by $\boldsymbol{x}_a(\boldsymbol{\lambda})$. On account of the second fundamental theorem of welfare economics (Theorem 2.6 and its corollary), a Pareto-optimal allocation that satisfies the budget constraints of all consumers is a competitive equilibrium. The following definition from Balasko [18] expresses this property of the competitive equilibrium.

Definition 2.10. A pair $(\boldsymbol{\lambda}, \boldsymbol{\omega}) \in \mathbb{R}^{m-1}_{++} \times \Omega$ is a $\boldsymbol{\lambda}$-*equilibrium* if it is a solution of the *Negishi equations*

$$\nu_a(\boldsymbol{\lambda}, \boldsymbol{\omega}) \equiv Du_a(\boldsymbol{x}_a(\boldsymbol{\lambda}))(\boldsymbol{x}_a(\boldsymbol{\lambda}) - \omega_a) = 0, \quad a = 2 \dots m.$$

Note that the first consumer's budget equation follows from the feasibility condition $\sum_{a=1}^{m}(\boldsymbol{x}_a - \omega_a) = 0$ and the first-order conditions of the maximization problem, namely that $Du_1(\boldsymbol{x}_1) = \lambda_a Du_a(\boldsymbol{x}_a)$, $a = 2, \dots, m$; see the proof of Proposition 2.16.

Example 2.1. There exist two consumers. The utility functions are of Cobb–Douglas or log-linear form,

$$u_a(\boldsymbol{x}_a) = \sum_{t=1}^{\ell} \rho_a^t \log x_a^t, \quad \rho_a^t > 0 \;\; a = 1, 2.$$

Then, we have

$$Du_a(\boldsymbol{x}_a) = \left(\frac{\rho_a^1}{x_a^1} \cdots \frac{\rho_a^\ell}{x_a^\ell}\right), \quad a = 1, 2,$$

and

$$D^2 u_a(\boldsymbol{x}_a) = -\begin{pmatrix} \frac{\rho_a^1}{(x_a^1)^2} & 0 & 0 & \dots & 0 \\ 0 & \frac{\rho_a^2}{(x_a^2)^2} & 0 & \dots & 0 \\ \dots & \dots & \dots & \dots & \\ 0 & 0 & 0 & \dots & \frac{\rho_a^\ell}{(x_a^\ell)^2} \end{pmatrix}, \quad a = 1, 2.$$

Let $\mathscr{E} \subset \mathbb{R}^{m-1}_{++} \times \Omega$ denote the set of $\boldsymbol{\lambda}$-equilibria. The projection map $\pi : \mathscr{E} \to \Omega$ is the restriction to $\mathscr{E}$ of the natural projection $(\boldsymbol{\lambda}, \boldsymbol{\omega}) \to \boldsymbol{\omega}$.

Proposition 2.16. *The Negishi map defined by*

$$\boldsymbol{\nu} : \mathbb{R}^{m-1}_{++} \times \Omega \to \mathbb{R}^{m-1}, \; (\boldsymbol{\lambda}, \boldsymbol{\omega}) \mapsto \boldsymbol{\nu}(\boldsymbol{\lambda}, \boldsymbol{\omega}) = (\nu_2(\boldsymbol{\lambda}, \boldsymbol{\omega}) \dots \nu_m(\boldsymbol{\lambda}, \boldsymbol{\omega}))$$

is of class C^1.

Proof. First, we will show that the map $\boldsymbol{\lambda} \to \boldsymbol{x}_a(\boldsymbol{\lambda})$, $a = 2 \dots m$, is smooth. Note that the function is determined implicitly by the first-order condition of the social welfare maximization problem with a resource constraint

$$\sum_{a=1}^{m} x_a - \sum_{a=1}^{m} \omega_a = 0,$$

$$Du_1(\boldsymbol{x}_1) - \lambda_b Du_b(\boldsymbol{x}_b) = 0, \quad b = 2 \dots m.$$

It is sufficient to check that the Jacobian matrix of the above equation system is invertible (Theorem E.5)

$$\begin{pmatrix} I & I & I & \dots & I \\ -D^2u_1 & \lambda_2 D^2 u_2 & 0 & \dots & 0 \\ -D^2u_1 & 0 & \lambda_3 D^2 u_3 & \dots & 0 \\ \dots & \dots & \dots & \dots & \dots \\ \dots & \dots & \dots & \dots & \dots \\ -D^2u_1 & 0 & 0 & \dots & \lambda_m D^2 u_m \end{pmatrix}.$$

This matrix is equivalent to the following matrix:

$$\begin{pmatrix} I & 0 & 0 & \dots & 0 \\ -D^2u_1 & D^2u_1 + \lambda_2 D^2 u_2 & D^2u_1 & \dots & D^2u_1 \\ -D^2u_1 & D^2u_1 & D^2u_1 + \lambda_3 D^2 u_3 & \dots & D^2u_1 \\ \dots & \dots & \dots & \dots & \dots \\ \dots & \dots & \dots & \dots & \dots \\ -D^2u_1 & D^2u_1 & D^2u_1 & \dots & D^2u_1 + \lambda_m D^2 u_m \end{pmatrix}.$$

Therefore, it suffices to show that the following matrix is invertible:

$$M = \begin{pmatrix} D^2u_1 + \lambda_2 D^2 u_2 & D^2u_1 & \dots & D^2u_1 \\ D^2u_1 & D^2u_1 + \lambda_3 D^2 u_3 & \dots & D^2u_1 \\ \dots & \dots & \dots & \dots \\ \dots & \dots & \dots & \dots \\ D^2u_1 & D^2u_1 & \dots & D^2u_1 + \lambda_m D^2 u_m \end{pmatrix}.$$

Let $\boldsymbol{v} = (\boldsymbol{v}_2 \dots \boldsymbol{v}_m) \in (\mathbb{R}^\ell)^{m-1}$ and consider the equation $M\boldsymbol{v} = 0$. Then we have

$$(D^2u_1 + \lambda_b D^2 u_b)\boldsymbol{v}_b + D^2u_1 \left(\sum_{c\neq b} \boldsymbol{v}_c\right) = 0, \quad b = 2 \dots m.$$

Then, making the inner product with the vector $\boldsymbol{v}_b$ yields

$$\boldsymbol{v}_b \left(D^2u_1 + \lambda_b D^2 u_b\right) \boldsymbol{v}_b + \boldsymbol{v}_b D^2u_1 \left(\sum_{c\neq b} \boldsymbol{v}_c\right) = 0, \quad b = 2 \dots m,$$

which simplifies into

$$\lambda_b \boldsymbol{v}_b D^2 u_b \boldsymbol{v}_b + \boldsymbol{v}_b D^2 u_1 \left(\sum_{c=2}^{m} \boldsymbol{v}_c\right) = 0, \quad b = 2 \dots m.$$

We add up these equations for b from 2 to m,

$$\sum_{b=2}^{m} \lambda_b \boldsymbol{v}_b D^2 u_b \boldsymbol{v}_b + \left(\sum_{b=2}^{m} \boldsymbol{v}_b\right) D^2 u_1 \left(\sum_{b=2}^{m} \boldsymbol{v}_b\right) = 0.$$

As D^2u_b is negative definite, each term in the sum of the left-hand side of the above equation is ≤ 0, and hence, each term must be 0

$$\left(\sum_{b=2}^{m} \boldsymbol{v}_b\right) D^2 u_1 \left(\sum_{b=2}^{m} \boldsymbol{v}_b\right) = 0, \ \lambda_b \boldsymbol{v}_b D^2 u_b \boldsymbol{v}_b = 0, \ b = 2 \dots m.$$

Therefore, we have $\boldsymbol{v}_2 = \cdots = \boldsymbol{v}_m = 0$ as desired. By construction, the Negishi map

$$\nu_b(\boldsymbol{\lambda}, \boldsymbol{\omega}) = Du_b(\boldsymbol{x}_b(\boldsymbol{\lambda}))(\boldsymbol{x}_b(\boldsymbol{\lambda}) - \omega_b), \quad b = 2 \dots m,$$

is of class C^1 with respect to $(\boldsymbol{\lambda}, \boldsymbol{\omega}) = ((\lambda_b), (\omega_a))$, $a = 1 \dots m$, $b = 2 \dots m$. □

Example 2.2. The utility functions for the consumers $a = 1, 2$ are given as in Example 2.1. Then, the first-order conditions for the social optimization are

$$x_1^t + x_2^t = \omega_1^t + \omega_2^t \equiv \omega^t,$$
$$\frac{\rho_1^t}{x_1^t} - \lambda\frac{\rho_2^t}{x_2^t} = 0, \quad t = 1 \ldots \ell.$$

From this, we obtain

$$x_1^t = \frac{\rho_1^t \omega^t}{\rho_1^t + \lambda\rho_2^t} \text{ and } x_2^t = \frac{\lambda\rho_2^t \omega^t}{\rho_1^t + \lambda\rho_2^t}.$$

Certainly, $\boldsymbol{x}_1 = (x_1^t)$ and $\boldsymbol{x}_2 = (x_2^t)$ are smooth functions of λ.

Proposition 2.17. *The set of equilibria $\mathscr{E}$ is a smooth submanifold of $\mathbb{R}^{m-1}\times\Omega$ of codimension $m-1$.*

Proof. We will prove this proposition by applying the regular value theorem (Theorem E.6), and it is sufficient to prove that $\mathbf{0} \in \mathbb{R}^{m-1}$ is a regular value of the map $\boldsymbol{\nu}$ that is smooth from Proposition 2.16.

Let $D\boldsymbol{\nu}(\boldsymbol{\lambda}, \boldsymbol{\omega}) : \mathbb{R}^{m-1}\times T_{\boldsymbol{\omega}}\Omega \to \mathbb{R}^{m-1}$ denote the tangent map (derivative) of map $\boldsymbol{\nu}$ at $(\boldsymbol{\lambda}, \boldsymbol{\omega}) \in \mathscr{E}$, where $T_{\boldsymbol{\omega}}\Omega$ is the tangent space of Ω at $\boldsymbol{\omega}$. We denote the tangent map $D\boldsymbol{\nu}(\boldsymbol{\lambda}, \boldsymbol{\omega})$ as $(D\nu_b(\boldsymbol{\lambda}, \boldsymbol{\omega}))_{b=2}^m$, which is defined by the $m-1$ coordinate mappings of the derivatives of ν_b with respect to $(\lambda_2 \ldots \lambda_m, \omega_1 \ldots \omega_m)$. Note that the partial derivative of ν_b with respect to ω_c $(c \neq b)$ is 0. Then, the derivative of ν_b is written as

$$D\nu_b(\boldsymbol{\lambda}, \boldsymbol{\omega})(\dot{\boldsymbol{\lambda}}, (\dot{\omega}_a)) = \sum_{c=2}^{m} \frac{\partial \nu_b(\boldsymbol{\lambda}, \boldsymbol{\omega})}{\partial \lambda_c}\dot{\lambda}_c + \sum_{t=1}^{\ell} \frac{\partial \nu_b(\boldsymbol{\lambda}, \boldsymbol{\omega})}{\partial \omega_b^t}\dot{\omega}_b^t,$$

where $\dot{\boldsymbol{\lambda}} = (\dot{\boldsymbol{\lambda}}_c) \in \mathbb{R}^{m-1}$ and $\dot{\omega}_b = (\dot{\omega}_b^t) \in \mathbb{R}^\ell$ are the tangent vectors. In order to show that the $D\boldsymbol{\nu}(\boldsymbol{\lambda}, \boldsymbol{\omega})$ is onto, it suffices to prove that a restriction of this map to some subspace of $\mathbb{R}^{m-1}\times T_{\boldsymbol{\omega}}\Omega$ is onto. Let us pick an arbitrary period t. The Jacobian matrix of $D\boldsymbol{\nu}$ with respect to $(\omega_2^t \dots \omega_m^t)$ is

$$\begin{pmatrix} \partial_t u_2(\boldsymbol{x}_2(\boldsymbol{\lambda})) & 0 & \dots & 0 \\ 0 & \partial_t u_3(\boldsymbol{x}_3(\boldsymbol{\lambda})) & \dots & 0 \\ \dots & \dots & \dots & \\ \dots & \dots & \dots & \dots \\ 0 & 0 & \dots & \partial_t u_m(\boldsymbol{x}_m(\boldsymbol{\lambda})) \end{pmatrix},$$

which is obviously of rank $m-1$ from assumption (U-2). □

From Proposition 2.17, it follows that the projection map $\pi : \mathscr{E} \to \Omega, (\boldsymbol{\lambda}, \boldsymbol{\omega}) \mapsto \boldsymbol{\omega}$ is smooth, as it is a composition of smooth maps, the *canonical embedding* $\mathscr{E} \to \mathbb{R}^{m-1}_{++} \times \Omega$ and the canonical projection map $\mathbb{R}^{m-1}_{++} \times \Omega \to \Omega$. We call $\mathscr{E}$ the *equilibrium manifold.*

Now, we determine the definition of our main subject in this section.

Definition 2.11. A $\boldsymbol{\lambda}$-equilibrium $(\boldsymbol{\lambda}, \boldsymbol{\omega}) \in \mathscr{E}$ is called the *regular* if it is a regular point of the projection map $\pi : \mathscr{E} \to \Omega$. The regular value $\boldsymbol{\omega} \in \Omega$ is called a *regular economy.* An economy that is not regular is called *critical.*

An important property of the map π is the following:

Proposition 2.18. *The projection map $\pi : \mathscr{E} \to \Omega$ is proper, that is, its inverse $\pi^{-1}(\mathcal{K})$ of every compact set $\mathcal{K} \subset \Omega$ is compact.*

Proof. Let $\mathcal{K}$ be a compact subset of Ω and let $\mathcal{K}_a$ be the image of $\mathcal{K}$ by the restriction of the natural projection map $(\boldsymbol{x}_a) \to \boldsymbol{x}_a$. Therefore, the set $\mathcal{K}_a$ is a compact subset of $X_a = \mathbb{R}^{\ell}_{++}$. The utility function $u_a : X_a \to \mathbb{R}$ is smooth and hence continuous, so that the image $u_a(\mathcal{K}_a)$ is compact and hence bounded by some $b_a \in \mathbb{R}$, $a = 1 \dots m$. We define the set $\mathcal{F} \subset \mathbb{R}^{\ell m}_{++}$ of the feasible allocations by

$$\mathcal{F} = \left\{ (\boldsymbol{x}_a) \in \mathbb{R}^{\ell m}_{++} \;\middle|\; \sum_{a=1}^{m} \boldsymbol{x}_a = \sum_{a=1}^{m} \omega_a,\ u_a(\boldsymbol{x}_a) \geq b_a \right\}.$$

We claim that the set $\mathcal{F}$ is compact. First, for each $a = 1 \ldots m$, we define the set $\mathcal{F}_a \subset \mathbb{R}^\ell_{++}$ by $\mathcal{F}_a = \{\boldsymbol{x}_a \in \mathbb{R}^\ell_{++} | \boldsymbol{x}_a \leq \sum_a \omega_a, u_a(\boldsymbol{x}_a) \geq b_a\}$. The set $\mathcal{F}_a$ is bounded as $\mathbf{0}$ is a lower bound of $\mathcal{F}_a$ and $\sum_a \omega_a$ is an upper bound. Next, the set $\mathcal{F}_a$ is closed. Let $(\boldsymbol{x}_n)$ $n = 1, 2 \ldots$ be a sequence in $\mathcal{F}_a$ with $\boldsymbol{x}_n \to \boldsymbol{x}_*$. Clearly, $\boldsymbol{x}_* \leq \sum_{a=1}^m \omega_a$ and, from the continuity of u_a, we have $u_a(\boldsymbol{x}_*) \geq b_a$. It remains to show that $\boldsymbol{x}_* \in \mathbb{R}^\ell_{++}$. Obviously, $\boldsymbol{x}_* \geq 0$. As $u_a(\boldsymbol{x}_*) \geq b_a$, it is impossible that $x^t_* = 0$ for any t by assumption (U-2). Therefore, $\mathcal{F}_a$ is compact, and hence, the product $\prod_{a=1}^m \mathcal{F}_a$ is also compact by the Tychonoff theorem (Theorem B.1). As a closed subset of $\prod_{a=1}^m \mathcal{F}_a$, the set $\mathcal{F}$ is compact.

The individual rationality means that if $\boldsymbol{x} \in \Omega$ is an equilibrium allocation associated with $\boldsymbol{\omega} \in \Omega$, then the inequality $u_a(\boldsymbol{x}_a) \geq u_a(\omega_a)$ is satisfied for all $a = 1 \ldots m$. This implies that for all equilibrium allocations $(\boldsymbol{x}_a)$ associated with endowments $\boldsymbol{\omega} \in \mathcal{K}$, the inequality $u_a(\boldsymbol{x}_a) \geq u_a(\omega_a) \geq b_a$ is satisfied for $a = 1 \ldots m$. In other words, if $(\omega_a) \in \mathcal{K}$, then $(\boldsymbol{x}_a) \in \mathcal{F}$. We now claim that the set of Pareto-optimal allocations associated with endowments $(\omega_a) \in \mathcal{K}$ is a closed subset of a compact set $\mathcal{F}$ and, hence, it is compact.

Let $(\boldsymbol{x}^n_a)$, $n = 1, 2, \ldots$, be a sequence of solutions of the social welfare maximization problem associated with the endowments (ω^n_a) in $\mathcal{K}$, such that $\boldsymbol{x}^n_a \to \boldsymbol{x}^*_a$. Then, for each n, there exists a social welfare weight $(\lambda^n_a)^m_{a=2}$, which satisfies the first-order conditions

$$\sum_{a=1}^m \boldsymbol{x}^n_a - \sum_{a=1}^m \omega^n_a = 0,$$

$$Du_1(\boldsymbol{x}^n_1) - \lambda^n_a Du_a(\boldsymbol{x}^n_a) = 0, \quad a = 2 \ldots m, \quad n = 1, 2, \ldots.$$

As $\mathcal{K}$ is compact, we may assume that $\omega^n_a \to \omega^*_a$, $a = 1 \ldots m$. For each n, we have $\lambda^n_a = \partial_1 u_1(\boldsymbol{x}^n_1)/\partial_1 u_a(\boldsymbol{x}^n_a)$. As the partial derivatives are continuous and $\boldsymbol{x}^n_a \to \boldsymbol{x}^*_a$, it follows that $\lambda^n_a \to \lambda^*_a = \partial_1 u_1(\boldsymbol{x}^*_1)/\partial_1 u_a(\boldsymbol{x}^*_a)$ from assumption (U-2). Clearly, we have

$$\sum_{a=1}^m \boldsymbol{x}^*_a - \sum_{a=1}^m \omega^*_a = 0,$$

$$Du_1(\boldsymbol{x}^*_1) - \lambda^*_a Du_a(\boldsymbol{x}^*_a) = 0, \quad a = 2 \ldots m.$$

This shows that the set of Pareto-optimal allocations associated with endowments in $\mathcal{K}$ is a closed subset of $\mathcal{F}$, as desired. The set of associated welfare weights is then necessarily a compact subset H of $\mathbb{R}^{m-1}_{++}$. As $\pi^{-1}(\mathcal{K}) = (H \times \mathcal{K}) \cap \mathscr{E}$ and $\mathscr{E}$ is closed in $\mathbb{R}^{m-1} \times \Omega$, it follows that $\pi^{-1}(\mathcal{K})$ is compact. □

The following proposition, which characterizes the regular and/or singular equilibria, will be useful for the subsequent analysis.

Proposition 2.19. *The* $\boldsymbol{\lambda}$*-equilibrium* $(\boldsymbol{\lambda}, \boldsymbol{\omega}) \in \mathscr{E}$ *is critical if and only if*

$$\det \frac{D\nu(\boldsymbol{\lambda}, \boldsymbol{\omega})}{D\boldsymbol{\lambda}} = 0,$$

where $\det A$ *means the determinant of a matrix* A.

Proof. The tangent space of $\mathscr{E}$ at $(\boldsymbol{\lambda}, \boldsymbol{\omega})$ and the derivative map of the projection π are written as

$$T_{(\boldsymbol{\lambda},\boldsymbol{\omega})}\mathscr{E} = \left\{ (\dot{\boldsymbol{\lambda}}, \dot{\boldsymbol{\omega}}) \in \mathbb{R}^{m-1} \times \mathbb{R}^{\ell m} \,\middle|\, \frac{D\nu(\boldsymbol{\lambda}, \boldsymbol{\omega})}{D\boldsymbol{\lambda}} \dot{\boldsymbol{\lambda}} + \frac{D\nu(\boldsymbol{\lambda}, \boldsymbol{\omega})}{D\boldsymbol{\omega}} \dot{\boldsymbol{\omega}} = 0 \right\},$$

and

$$D\pi : T_{(\boldsymbol{\lambda},\boldsymbol{\omega})}\mathscr{E} \to \mathbb{R}^{\ell m}, (\dot{\boldsymbol{\lambda}}, \dot{\boldsymbol{\omega}}) \mapsto \dot{\boldsymbol{\omega}},$$

where $\dot{\boldsymbol{\lambda}} \in \mathbb{R}^{m-1}$ and $\dot{\boldsymbol{\omega}} \in \mathbb{R}^{\ell m}$ are the tangent vectors.

As *dimension* $T_{(\boldsymbol{\lambda},\boldsymbol{\omega})}\mathscr{E} = \ell m$ from Theorem E.6, the map $D\pi$ fails to be onto if and only if the kernel $D\pi^{-1}(\mathbf{0}) = \{(\dot{\boldsymbol{\lambda}}, \dot{\boldsymbol{\omega}}) \in T_{(\boldsymbol{\lambda},\boldsymbol{\omega})}\mathscr{E} |\ \dot{\boldsymbol{\omega}} = \mathbf{0}\}$ contains linear space other than $\{0\}$, as $T_{(\boldsymbol{\lambda},\boldsymbol{\omega})}\mathscr{E}/D\pi^{-1}(\mathbf{0}) \approx D\pi(T_{(\boldsymbol{\lambda},\boldsymbol{\omega})}\mathscr{E})$. The necessary and sufficient condition for this is that the linear equation $\frac{D\nu(\boldsymbol{\lambda},\boldsymbol{\omega})}{D\boldsymbol{\lambda}} \dot{\boldsymbol{\lambda}} = 0$ has a non-zero solution $\dot{\boldsymbol{\lambda}} \neq 0$, namely that $det \frac{D\nu(\boldsymbol{\lambda},\boldsymbol{\omega})}{D\boldsymbol{\lambda}} = 0$. □

The fundamental property of regular economies is that they have finitely many (regular) equilibria, which are locally stable and unique.

Theorem 2.11. *If* $\boldsymbol{\omega}$ *is a regular economy, then the set of* $\boldsymbol{\lambda}$*-equilibria associated with* $\boldsymbol{\omega}$ *is finite.*

Proof. By Proposition 2.19, a $\boldsymbol{\lambda}$-equilibrium $(\boldsymbol{\lambda}_0, \boldsymbol{\omega}_0) \in \mathscr{E}$ is regular if and only if $det \frac{D\boldsymbol{\nu}(\boldsymbol{\lambda},\boldsymbol{\omega})}{D\boldsymbol{\lambda}} \neq 0$. Then, from the implicit function theorem (Theorem E.5), we can take a neighborhood U_0 of $\boldsymbol{\omega}_0$ such that $\boldsymbol{\lambda}$ is a C^1 function of $\boldsymbol{\omega}$ and $\boldsymbol{\nu}(\boldsymbol{\lambda}(\boldsymbol{\omega}), \boldsymbol{\omega}) = 0$ for all $\boldsymbol{\omega} \in U_0$. Let $\mathscr{E}_0 = \{(\boldsymbol{\lambda}, \boldsymbol{\omega}) \in \mathscr{E} |\ \boldsymbol{\omega} \in U_0\}$ and define the map $\rho : U_0 \to \mathscr{E}_0$ by $\rho(\boldsymbol{\omega}) = (\boldsymbol{\lambda}(\boldsymbol{\omega}), \boldsymbol{\omega})$. Then, we have a $\pi \circ \rho =$ identity on U_0, which means that π restricted to $\mathscr{E}_0$ is a diffeomorphism between U_0 and $\mathscr{E}_0$.

Let $\boldsymbol{\omega}$ be a regular value of π. As π is a proper map from Proposition 2.18, the set $\pi^{-1}(\boldsymbol{\omega})$ is compact. To show that the set $\pi^{-1}(\boldsymbol{\omega})$ is finite, it suffices to show that $\pi^{-1}(\boldsymbol{\omega})$ is discrete. Take open sets U_0 and $\mathscr{E}_0$ from the preceding discussion. We claim that $\pi^{-1}(\boldsymbol{\omega}) \cap \mathscr{E}_0$ is one-point set $\{(\boldsymbol{\lambda}(\boldsymbol{\omega}), \boldsymbol{\omega})\}$. If not, $\{(\boldsymbol{\lambda}(\boldsymbol{\omega}), \boldsymbol{\omega})\}$ contains at least two distinct points both of which are mapped to $\boldsymbol{\omega}$ by the projection π. This contradicts the fact that π restricted to $\mathscr{E}_0$ is bijective. □

Let $\mathscr{R}$ be the set of regular economies. We will show in Theorem 2.13 that $\mathscr{R}$ is an open and dense subset of $\Omega = \mathbb{R}^{\ell m}_{++}$. The following theorem shows that the regular equilibrium is locally unique and moves continuously when the economy changes continuously (locally stable).

Theorem 2.12. *For every $\boldsymbol{\omega} \in \mathscr{R}$, there exists an open neighborhood V of $\boldsymbol{\omega}$ in $\mathscr{R}$ such that the preimage $\pi^{-1}(V)$ is the disjoint union of a family of a finite number of open subsets U_i of $\pi^{-1}(\mathscr{R})$, $i = 1 \ldots k$ and the restriction $\pi_i : U_i \to V$ of π to each U_i is a homeomorphism.*

Proof. From Theorem 2.11, $\pi^{-1}(\boldsymbol{\omega})$ is a finite set $\{(\lambda_1, \boldsymbol{\omega}) \ldots (\lambda_k, \boldsymbol{\omega})\}$. As in the proof of Theorem 2.11, we can take open disjoint neighborhoods $U'_1 \ldots U'_k$ of regular equilibria $(\lambda_1, \boldsymbol{\omega}) \ldots (\lambda_k, \boldsymbol{\omega})$ such that the restriction of π to U'_i is a diffeomorphism with $V_i = \pi(U'_i)$, $i = 1 \ldots k$. We claim that the image of a closed set by the proper map π is closed. Let C be a closed set and take a converging sequence $\{\boldsymbol{y}_n\}$ in $\pi(C)$, $\boldsymbol{y}_n \to \boldsymbol{y}_*$. The set $Y = \{\boldsymbol{y}_n\} \cup \{\boldsymbol{y}_*\}$ being compact, $\pi^{-1}(Y)$ is a compact subset of C by the properness of π. Take $\boldsymbol{x}_n \in \pi^{-1}(\boldsymbol{y}_n)$ for each n. Then, we can assume that

$\boldsymbol{x}_n \to \boldsymbol{x}_* \in \pi^{-1}(Y) \subset C$. As the projection is continuous, we have $\boldsymbol{y}_* = \pi(\boldsymbol{x}_*) \in \pi(C)$, as desired.

As the set $\mathscr{E}\backslash(U_1' \cup \ldots \cup U_k')$ is closed in $\mathscr{E}$, its image by the map π is closed. We define the set V as follows:

$$V = (V_1 \cap \ldots \cap V_k)\backslash\pi(\mathscr{E}\backslash(U_1' \cup \ldots \cup U_k')).$$

Clearly, the set V is open in Ω. We show that $\boldsymbol{\omega} \in V$. As $\boldsymbol{\omega} \in \bigcap_{i=1}^k V_i$, it suffices to show that $\boldsymbol{\omega} \notin \pi(\mathscr{E}\backslash(U_1' \cup \ldots \cup U_k'))$. This follows from the fact that $\pi^{-1}(\boldsymbol{\omega}) \subset U_1' \cup \ldots \cup U_k'$.

We define that $U_i = U_i' \cap \pi^{-1}(V)$. Then, the restriction $\pi_i = \pi|_{U_i}$ is a homeomorphism between U_i and $\pi(U_i) = V$. It remains to be proven that $\pi^{-1}(V) = \bigcup_{i=1}^k U_i$. That $\bigcup_{i=1}^k U_i \subset \pi^{-1}(V)$ is clear. Suppose that $\pi^{-1}(V) \subset \bigcup_{i=1}^k U_i$ does not hold. Then, there exists a $\boldsymbol{x}' \in \pi^{-1}(V)$ such that $\boldsymbol{x}' \notin U_i$ for all i. Then, $\boldsymbol{x}'$ must belong to $\mathscr{E}\backslash(U_1' \cup \ldots \cup U_k')$, which implies that $\boldsymbol{\omega}' = \pi(x') \in \pi(\mathscr{E}\backslash(U_1' \cup \ldots \cup U_k'))$. Therefore, $\boldsymbol{\omega}' \notin V$, contradicting the choice of $\boldsymbol{x}' \in \pi^{-1}(V)$. □

Finally, we prove that the size of the regular economies is "big" in the space of all economies or, in other words, that the size of singular economies is "small" in the set Ω.

Theorem 2.13. *The set of singular economies is a closed subset of Ω that has measure zero.*

Proof. It follows from Sard's theorem (Theorem E.8) that the set of critical economies, as the set of critical values of a smooth map $\pi : \mathscr{E} \to \Omega$, is of measure zero, as *dimension* $\mathscr{E}$ = *dimension* Ω. From Proposition 2.19, the set of critical equilibria is a closed subset of $\mathscr{E}$, as the function $\det \frac{D\boldsymbol{\nu}(\boldsymbol{\lambda},\boldsymbol{\omega})}{D\boldsymbol{\lambda}}$ is continuous. Therefore, the set of critical economies is closed, as it is the image of a proper map π. □

2.9 Notes

Classical monographs of the general equilibrium theory are discussed in the references [11, 51, 160, 174]. These monographs are written by the founders of the theory and any serious graduate student should have a copy of them. Among the more recent textbooks, we refer to those in references [31, 71, 151, 240]. We now provide some remarks regarding each of the sections.

Section 2.1: The basic reference is Debreu [51, Chapter 2]. Chapter 7 of his book began by describing a market structure with uncertainty, which is outside the scope of this book. In this book, we first introduce the commodity space, including the infinite-dimensional cases, and we do not restrict the set of economic agents to be finite, in contrast to the classical monographs discussed earlier. We hope that the readers do not take this as pedantic.

Section 2.2: We refer to Debreu [51, Chapter 4] and Hildenbrand [94, Sections 1.1 and 1.2] as the basic references relating to this section. In particular, following Hildenbrand, we postulate the strict preference relations $\prec$ for which the negative transitivity is not assumed, rather than using (weak) preference relations $\succsim$ as the basic concept. This is because the space of strict preference relations has a simpler topological structure, which will be clear in Section 3.2. For various kinds of convexity of preferences and their relationships, see [51, Chapter 4].

Section 2.3: The basic references for this section are Debreu [51, Chapter 4], Arrow and Hahn [11, Chapter 4], Hildenbrand [94, Section 1.2], and McKenzie [160, Chapter 1]. For the idea of the proof of Theorem 3.1, in which the K–K–M lemma was applied, we followed [160, Section 1.2].

A surprising result, which was originally due to Sonnenschein [236, 237] and subsequently developed by Mantel [139] and Debreu [54], should be mentioned. In Section 2.3, we saw that an individual excess demand correspondence is upper hemicontinuous and hence continuous if it is a function (Proposition 2.3) and that it satisfies the budget equation (under local non-satiation); $\boldsymbol{p}(\phi(a,\boldsymbol{p})-\omega_a)=0$. Therefore, the aggregate excess demand $\sum_a(\phi(a,\boldsymbol{p})-\omega_a)$ is also continuous and fulfills Walras' law; $\boldsymbol{p}\sum_a(\phi(a,\boldsymbol{p})-\omega_a)=0$. Sonnenschein showed that these are the only properties that all aggregate excess demand functions should necessarily have, namely that for each continuous function (market excess demand function) which satisfies Walras' law, one can construct a finite exchange economy in which the sum of the individual excess demand functions coincides on the unit sphere (up to $\epsilon > 0$, see below) with the given excess demand function. Debreu [54] provided an elegant geometric proof for this result.

Theorem. *Let* $S_\epsilon = \{\boldsymbol{p} = (p^t) \in \mathbb{R}^\ell \mid \|\boldsymbol{p}\| = 1, p^t \geq \epsilon \text{ for all } t\}$ *and* $\Phi : S_\epsilon \to \mathbb{R}^\ell$ *be a continuous function satisfying* $\boldsymbol{p}\Phi(\boldsymbol{p}) = 0$. *Then, for every* $\epsilon > 0$, *there exist* ℓ *consumers* $(\precsim_a, \omega_a)_{a=1}^\ell$ *such that the sum of their individual excess demand functions is equal to* $\Phi(\boldsymbol{p})$ *on* S_ϵ, $\sum_{a=1}^\ell (\phi(a, \boldsymbol{p}) - \omega_a) = \Phi(\boldsymbol{p})$ *for* $\boldsymbol{p} \in S_\epsilon$.

An educational and readable exposition of this result can be found in Mas-Colell, Whinston and Green [151].

Section 2.4: For all of the expositions, we follow McKenzie [157, 160].

Section 2.5: A basic reference is Debreu [51, Chapter 5]. See also Arrow and Hahn [11, Chapter 5], McKenzie [160, Chapter 6], and Nikaido [174, Chapter 5]. A general procedure to prove the existence of equilibria is to construct some devices (a mapping or a game) to which Kakutani's fixed point theorem will be applied. At least three alternative constructions are known which are as follows:

(a) Construct mapping from the market excess demand correspondences, to which the Kakutani fixed point theorem is applied (*excess demand approach*): This was initiated by McKenzie [154] and Nikaido [173]. In this book, we have applied this method for Theorems 2.4, 2.5, and Theorems 3.3, 3.8, 6.2, 6.4, 6.6 will also be proved by it.

(b) Construct an abstract game from a market model and invoke an existence theorem of the Nash equilibrium (*game theoretic approach*): This method was initiated by Arrow and Debreu [10], who used a result of Debreu [49] that generalized the original result of Nash [166], and we will follow this procedure in Theorem 6.3 by using Theorem 2.8.

(c) Construct fixed point mapping that exploits the utility possibility frontier and the welfare theorems: This strategy of proof was initiated by Negishi [167] (*Negishi method*). Theorems 4.6, 7.1, and 8.1 will be obtained through this method.

Mas-Colell [144] observed that the demand correspondences are not generally convex-valued if the preferences are not complete or transitive; nevertheless, the competitive equilibrium exists without completeness or transitivity of the preferences. His proof is very sophisticated. Gale and Mas-Colell [75] (corrections [76]) presented a more readable proof. The monotonicity assumption (MT) implies that the commodities under consideration are goods. For an equilibrium existence theorem with bads, see Hara [88].

The concept of the (quasi) equilibrium relative to the price $\boldsymbol{p}$ is in Debreu [51, Chapter 6]. McKenzie [160] called it the *equilibrium with redistribution.* For the proof of the second welfare theorem (Theorem 2.6), we followed [51, Chapter 6]. See also Arrow and Hahn [11, Chapter 4].

Section 2.6: The proof of Theorem 2.7 followed the original paper of Debreu and Scarf [55]. See also Arrow and Hahn [11, Chapter 8] and McKenzie [160, Section 5.2].

Section 2.7: For the proof of Theorem 2.8, we followed Shafer and Sonnenschein [228]. We followed Ichiishi [99] for the presentation of the model and the proof of Theorem 2.9.

Section 2.8: A clear and readable textbook for this topic is Dierker [61]. Advanced readers will be interested in [21, 149]. Kehoe [106–108] examined extensively regular production economies. The exposition of the present section follows Balasko [18, 19]. The advantage of Balasko's approach, which parameterizes an economy by social welfare weights (Negishi method), will be apparent in the infinite-dimensional cases; see Balasko [20]. The most far-reaching result on this subject to date is that of Shannon and Zame [230] which discussed the problems of regularity on a topological vector lattice (see Appendix G); see also [41, 109, 110, 197].

Chapter 3

Economies with a Continuum of Traders

3.1 Markets with a Measure Space of Consumers

In this chapter, we introduce equilibrium models on the commodity space $\mathbb{R}^\ell$ with a measure space of economic agents. In Chapter 2, we discussed the competitive equilibria of exchange economies in which each consumer behaves as a price taker, namely that consumers have no power to affect market prices, and they take market prices as given when they make decisions about their consumption plans. From a logical point of view, however, this hypothesis of price-taking behavior by each consumer is not consistent with the assumption that there exist finitely many consumers in the market. Even if the number of consumers is very large, each of them has some size or weight in the market, and changes in their behavior will have some effect on equilibrium market prices as long as the population is finite.

Perfectly competitive equilibria can be achieved in an equilibrium model with a continuum of traders. In such a model, each individual would have a negligible effect, with a 0 weight compared with the whole market. To give a precise meaning to "weight" or "mass" of traders, we assume that the set of traders is a complete and atomless measure space $(A, \mathcal{A}, \lambda)$ (see Appendix F for the definition of the measure space). Then integrating with respect to the measure λ of individual demand $\phi(a, \boldsymbol{p})$, which is obtained from the price-taking and utility-maximizing behavior of the consumer $a \in A$, we obtain the total or aggregated demand $\Phi(\boldsymbol{p}) = \int_A \phi(a, \boldsymbol{p}) d\lambda$, and the competitive equilibrium is obtained by postulating that the total demand is equal to the total supply $\int_A \omega(a) d\lambda$, where $\omega(a)$ is the initial endowment of the consumer a. Sometimes we call the economy with the measure space of consumers a *large economy*. In the large economy, we can observe remarkable effects that come from the aggregation of the individual behaviors. One of these effects is the convexing effect of total demand correspondence that will be discussed in Section 3.3. This says that in markets with an atomless measure space of consumers (the atom is a subset of consumers that is indivisible and has a measure greater than 0; see Appendix F), the total demand correspondence is convex-valued even if the individual demand correspondences are not. Mathematically, this is a straightforward consequence of Lyapunov's theorem (Theorem F.1). Because of this result, we can prove the existence of equilibrium of exchange economies with an atomless measure space of consumers without the convexity of preferences (Theorem 3.3).

In Chapter 2, we discussed the limit theorem of the core, which says that the core of the sequence of the replica economies converges to the competitive equilibrium of the original economy as long as the number of consumers approaches infinity. In the economy with a continuum of consumers, an equivalence theorem of the core and the competitive equilibria holds, namely that in the large economy, the sets of the core allocations and the competitive equilibrium allocations are equal. This theorem will be discussed in Section 3.4.

There is one more effect of the aggregation that is relevant to the existence of equilibria. As we will see in Section 3.5, when the consumption set of a consumer is not convex, the demand correspondence of the consumer exhibits discontinuous behaviors with respect to the continuous change of the price system. If the set of the consumers with these discontinuous behaviors has a positive measure, the total demand correspondence will also behave discontinuously. However, we will observe that if the distribution of the consumers' characteristics (for example, the distribution of the initial endowment vectors) is *dispersed* in some appropriate sense, the total demand correspondence will be upper hemicontinuous even if the individual demand correspondences are not.

Finally, in Section 3.6 we will discuss the effect of aggregation on the law of demand, namely that the demand curve will be downward sloping under a suitable condition on income distribution, even if the individual demand curves are not.

Summing up; *In this chapter, the commodity space is* $\mathbb{R}^\ell$ *and the set of economic agents is a complete atomless measure space* $(A, \mathcal{A}, \lambda)$. *For each measurable subset* $C \in \mathcal{A}$, $\lambda(C)$ *is a mass or weight of the consumers belonging to the set* C.

3.2 Spaces of Preferences

Let $\mathcal{X}$ be the set of all *admissible consumption sets*, which means that there exist, a vector $\boldsymbol{b} \in \mathbb{R}^\ell$ and a compact set $H \subset \mathbb{R}^\ell$ such that for all $X \in \mathcal{X}$,

$$\boldsymbol{b} \leq X \text{ and } X \cap H \neq \emptyset.$$

Let $\prec$ be the strict preference relation on $X \in \mathcal{X}$ that satisfies the basic postulates (see Section 2.2); irreflexivity (IR), transitivity (TR), and continuity (CT). A pair $(X, \prec)$ consisting of the consumption set and the preference relation is called the *preference–consumption set pair*. We sometimes abuse the term and notation, in such a way that we call the pair $(X, \prec)$ simply the (strict) preference (on X) and denote it by $\prec$ when the consumption set is understood.

Definition 3.1. $\mathcal{P}$ is the set of all preference–consumption set pairs $(X, \prec)$ such that $\prec \subset \mathbb{R}^\ell \times \mathbb{R}^\ell$ is an irreflexive (IR) and transitive (TR) binary relation on $X \in \mathcal{X}$ and $\prec$ is open relative to $X \times X$, or satisfies the condition (CT).

The first task of this section is to endow some nice topology with $\mathcal{P}$. Recall that to every preference relation $(X, \prec) \in \mathcal{P}$, we associate the relation $(X, \precsim)$ or simply $\precsim$ defined by $(X, \precsim) = \{(\boldsymbol{x}, \boldsymbol{y}) \in X \times X |\ (\boldsymbol{x}, \boldsymbol{y}) \notin \prec\} = X \times X \setminus \prec$. Because the relations $\prec$ and $\precsim$ are complements, they are in one-to-one correspondences and $\precsim$ is a closed subset of $\mathbb{R}^{2\ell}$ if X is closed in $\mathbb{R}^{\ell}$ and $\prec$ is open in $X \times X$. Consequently, the set $\mathcal{P}$ of preference relations can be considered a subset of $\mathcal{F}(\mathbb{R}^{2\ell})$, the set of all closed subsets of $\mathbb{R}^{2\ell}$ endowed with the topology of closed convergence τ_c (see Appendix D). We then have the following theorem:

Theorem 3.1. *The set $\mathcal{P}$ is compact and metrizable, and a sequence $(X_n, \prec_n)$ converges to $(X, \prec)$ in the topology τ_c if and only if $L_i(\precsim_n) = \precsim = L_s(\precsim_n)$, where $\precsim_n$ and $\precsim$ are the associated relations to $(X_n, \prec_n)$ and $(X, \prec)$, respectively. Moreover, the topology τ_c is the weakest Hausdorff topology on $\mathcal{P}$ such that the set*

$$\{(X, \prec, \boldsymbol{x}, \boldsymbol{y}) \in \mathcal{P} \times \mathbb{R}^{\ell} \times \mathbb{R}^{\ell} |\ \boldsymbol{x}, \boldsymbol{y} \in X \text{ and } \boldsymbol{x} \precsim \boldsymbol{y}\}$$

is closed.

Proof. By Theorem D.6, the set $\mathcal{F}(\mathbb{R}^{2\ell})$ endowed with the topology τ_c is compact and metrizable. To show that $(\mathcal{P}, \tau_c)$ is compact, it suffices from Proposition B.5 that $\mathcal{P}$ is a closed subset of $(\mathcal{F}(\mathbb{R}^{2\ell}), \tau_c)$. Let $\{(X_n, \prec_n)\}_{n \in \mathbb{N}}$ be a sequence in $\mathcal{P}$ and $\{\precsim_n\}_{n \in \mathbb{N}}$ be a sequence of the corresponding complement relations $\precsim_n = \{(\boldsymbol{x}, \boldsymbol{y}) \in X_n \times X_n |\ (\boldsymbol{x}, \boldsymbol{y}) \in X_n \times X_n \setminus \prec_n\}$ such that $\precsim_n \to \precsim$ in the topology τ_c, where $\precsim = \{(\boldsymbol{x}, \boldsymbol{y}) \in X \times X |\ (\boldsymbol{x}, \boldsymbol{y}) \in X \times X \setminus \prec\}$.

We have to show that $(X, \prec)$ belongs to $\mathcal{P}$. First, we show that $\prec$ is irreflexive. Let $\boldsymbol{x} \in X$. Because $\prec_n$ is irreflexive for each n, $(\boldsymbol{x}, \boldsymbol{x}) \in \precsim_n$ for all n. Hence, $(\boldsymbol{x}, \boldsymbol{x}) \in \precsim = L_i(\precsim_n)$, so that $(\boldsymbol{x}, \boldsymbol{x}) \notin \prec$. Therefore, $\prec$ is irreflexive. Next, we show that $\prec$ is transitive. Let $\boldsymbol{x} \prec \boldsymbol{y}$ and $\boldsymbol{y} \prec \boldsymbol{z}$. Suppose that $(\boldsymbol{x}, \boldsymbol{z}) \notin \prec$, or $(\boldsymbol{x}, \boldsymbol{z}) \in \precsim$. Because $L_i(\precsim_n) = \precsim$, there exists a sequence $(\boldsymbol{x}_n, \boldsymbol{z}_n) \in \precsim_n$ with $(\boldsymbol{x}_n, \boldsymbol{z}_n) \to (\boldsymbol{x}, \boldsymbol{z})$. Now for n large enough, we have $(\boldsymbol{x}_n, \boldsymbol{y}_n) \notin \precsim_n$ and $(\boldsymbol{y}_n, \boldsymbol{z}_n) \notin \precsim_n$, where $\{\boldsymbol{y}_n\}$ is a sequence in X converging to $\boldsymbol{y}$. For if not, it would follow that $(\boldsymbol{x}, \boldsymbol{y}) \in L_s(\precsim_n) = \precsim$ or $(\boldsymbol{y}, \boldsymbol{z}) \in \precsim$, contradicting that $\boldsymbol{x} \prec \boldsymbol{y}$ and $\boldsymbol{y} \prec \boldsymbol{z}$. Hence, by the transitivity of $\prec$, we obtain $(\boldsymbol{x}_n, \boldsymbol{z}_n) \notin \precsim_n$, a contradiction. By Theorem D.6, we have that $\prec_n \to \prec$ in the topology τ_c if and only if $L_i(\precsim_n) = \precsim = L_s(\precsim_n)$.

Finally, because $(\mathcal{P}, \tau_c)$ is a compact space, the identity map on $\mathcal{P}$, $\iota_{\mathcal{P}} : (\mathcal{P}, \tau_c) \to (\mathcal{P}, \sigma)$ is continuous if and only if the topology σ is weaker than τ_c. Hence, by Proposition B.7, every Hausdorff topology σ on $\mathcal{P}$ that is weaker than τ_c coincides with τ_c.

It remains to show that the set $\{(X, \prec, \boldsymbol{x}, \boldsymbol{y}) \in \mathcal{P}\times\mathbb{R}^\ell\times\mathbb{R}^\ell |\, \boldsymbol{x}, \boldsymbol{y} \in X \text{ and } \boldsymbol{x} \succsim \boldsymbol{y}\}$ is closed in $\mathcal{P}\times\mathbb{R}^\ell\times\mathbb{R}^\ell$. Let $(X_n, \prec_n, \boldsymbol{x}_n, \boldsymbol{y}_n) \to (X, \prec, \boldsymbol{x}, \boldsymbol{y})$, where $\boldsymbol{x}_n$, $\boldsymbol{y}_n \in X_n$ and $\boldsymbol{x}_n \succsim_n \boldsymbol{y}_n$. Hence, $(\boldsymbol{x}_n, \boldsymbol{y}_n) \in \succsim_n$, which implies that $(\boldsymbol{x}, \boldsymbol{y}) \in L_i(\succsim)$, or $\boldsymbol{x}, \boldsymbol{y} \in X$ and $\boldsymbol{x} \succsim \boldsymbol{y}$. □

We shall often denote $(X_n, \prec_n) \to (X, \prec)$ by $\prec_n \to \prec$ or even simply by $\succsim_n \to \succsim$ when they should not be misunderstood. Recall that for a topological space, the Borel σ field on X, denoted by $\mathcal{B}(X)$, is the σ-algebra generated by open sets in X; see Appendix F for details. A set $B \in \mathcal{B}(X)$ is called a *Borel set* or *Borelian.* Because the space $\mathcal{P}$ is compact (and metrizable), it is Borelian.

Let $\mathcal{P}*$ be the subset of $\mathcal{P}$ of all negative transitive (NTR) preference relations. The other conditions will occasionally be used in the subsequent analysis; for example, we denote by $\mathcal{P}_{lns} \subset \mathcal{P}$ the set of all locally non-satiated preferences. We also denote by $\mathcal{P}_{mo} \subset \mathcal{P}$ all monotonic preference relations, and finally, let $\mathcal{P}_{co} \subset \mathcal{P}$ be all convex preference relations.

We are interested in the topological and measurable properties of $\mathcal{P}*$, $\mathcal{P}_{lns}$, $\mathcal{P}_{mo}$, and $\mathcal{P}_{co}$. A set Q is said to be G_δ if it is an intersection of countably many open sets. It is called F_σ if it is a union of countably many closed sets. Obviously, the G_δ sets and the F_σ sets are Borel measurable sets.

Theorem 3.2. *The sets* $\mathcal{P}*, \mathcal{P}_{lns}, \mathcal{P}_{mo}, \mathcal{P}_{co}$ *are* G_δ *sets.*

Proof. First we show that the set $\mathcal{P}*$ is G_δ. For $S, T \subset X$, we denote by $S \prec T$ that for all $\boldsymbol{x} \in S$ and all $\boldsymbol{y} \in T$, $\boldsymbol{x} \prec \boldsymbol{y}$. For positive integers m and n, we define the set $\mathcal{P}*^{m,n} \subset \mathcal{P}$ by

$$\mathcal{P}*^{m,n} = \Big\{ (X, \prec) \in \mathcal{P} \Big|\ \text{there exist } \boldsymbol{x}, \boldsymbol{y}, \boldsymbol{z} \in X \text{ such that } \|\boldsymbol{x}\| \leq m, \\ \|\boldsymbol{y}\| \leq m,\ \|\boldsymbol{z}\| \leq m,\ \boldsymbol{z} \succsim \boldsymbol{y},\ \boldsymbol{y} \succsim \boldsymbol{x},\ B(\boldsymbol{z}, 1/n) \cap X \\ \prec B(\boldsymbol{x}, 1/n) \cap X \Big\}.$$

Then clearly we have $\mathcal{P}\backslash\mathcal{P}* = \cup_{m=1}^{\infty}\cup_{n=1}^{\infty}\mathcal{P}*^{m,n}$. Therefore, it suffices to show that $\mathcal{P}*^{m,n}$ is closed for all m,n. Let $\{(X_k,\prec_k)\}_{k\in\mathbb{N}}$ be a sequence in $\mathcal{P}*^{m,n}$ such that $(X_k,\prec_k)\to(X,\prec)$ in the topology τ_c. We want to show that $\prec\in\mathcal{P}*^{m,n}$. For each of the positive integers m,n, there exists a sequence $(\boldsymbol{x}_k,\boldsymbol{y}_k,\boldsymbol{z}_k)_{k\in\mathbb{N}}$ such that $\boldsymbol{x}_k,\boldsymbol{y}_k,\boldsymbol{z}_k\in X_k$ for all k, $\|\boldsymbol{x}_k\|\le m$, $\|\boldsymbol{y}_k\|\le m$, $\|\boldsymbol{z}_k\|\le m$, $\boldsymbol{z}_k\precsim_k\boldsymbol{y}_k$, $\boldsymbol{y}_k\precsim_k\boldsymbol{x}_k$, and $B(\boldsymbol{z}_k,1/n)\cap X_k\prec_k B(\boldsymbol{x}_k,1/n)\cap X_k$. Because the sequence $(\boldsymbol{x}_k,\boldsymbol{y}_k,\boldsymbol{z}_k)_{k\in\mathbb{N}}$ is bounded, we can assume that $\boldsymbol{x}_k\to\boldsymbol{x}\in X$, $\boldsymbol{y}_k\to\boldsymbol{y}\in X$ and $\boldsymbol{z}_k\to\boldsymbol{z}\in X$. Obviously, we have $\|\boldsymbol{x}\|\le m$, $\|\boldsymbol{y}\|\le m$, $\|\boldsymbol{z}\|\le m$, and $\boldsymbol{z}\precsim\boldsymbol{y}$, $\boldsymbol{y}\precsim\boldsymbol{x}$. Suppose that $B(\boldsymbol{z},1/n)\cap X\prec B(\boldsymbol{x},1/n)\cap X$ does not hold. Then there exist $\boldsymbol{x}'\in B(\boldsymbol{z},1/n)\cap X$ and $\boldsymbol{z}'\in B(\boldsymbol{x},1/n)\cap X$ such that $\boldsymbol{z}'\precsim\boldsymbol{x}'$. Because $L_i(\precsim_k)=\precsim$, there exists a sequence $(\boldsymbol{z}'_k,\boldsymbol{x}'_k)\in\precsim_k$ with $(\boldsymbol{z}'_k,\boldsymbol{x}'_k)\to(\boldsymbol{z}',\boldsymbol{x}')\in\precsim$. For k sufficiently large, $\|\boldsymbol{x}'_k-\boldsymbol{x}_k\|\le 1/n$, $\|\boldsymbol{z}'_k-\boldsymbol{z}_k\|\le 1/n$, and $\boldsymbol{z}'_k\precsim\boldsymbol{x}'_k$. This contradicts the fact that $B(\boldsymbol{z}_k,1/n)\cap X\prec B(\boldsymbol{x}_k,1/n)\cap X$. This proves that $\mathcal{P}*^{m,n}$ is closed.

Similarly, for each of the positive integers m,n, define $\mathcal{P}_{lns}^{m,n}\subset\mathcal{P}$ by

$$\mathcal{P}_{lns}^{m,n}=\left\{(X,\prec)\in\mathcal{P}\,\middle|\,\text{there exists } \boldsymbol{x}\in X \text{ such that } \|\boldsymbol{x}\|\le m \text{ and } \boldsymbol{x}\precsim\boldsymbol{y} \text{ whenever } \boldsymbol{y}\in B(\boldsymbol{x},1/n)\cap X\right\}.$$

Then it is obvious that $\mathcal{P}\backslash\mathcal{P}_{lns}=\cup_{m=1}^{\infty}\cup_{n=1}^{\infty}\mathcal{P}_{lns}^{m,n}$. To prove that $\mathcal{P}_{lns}$ is G_δ, it suffices to show that $\mathcal{P}_{lns}^{m,n}$ is closed for all m,n. Let $\{(X_k,\prec_k)\}_{k\in\mathbb{N}}$ be a sequence in $\mathcal{P}_{lns}^{m,n}$ such that $(X_k,\prec_k)\to(X,\prec)$ in the topology τ_c. Then for each k, there exists $\boldsymbol{x}_k\in X_k$ such that $\|\boldsymbol{x}_k\|\le m$ and if $\boldsymbol{y}\in B(\boldsymbol{x}_k,1/n)\cap X$, then $\boldsymbol{x}_k\precsim\boldsymbol{y}$. We can assume that $\boldsymbol{x}_k\to\boldsymbol{x}\in X$. Then $\|\boldsymbol{x}\|\le m$. Suppose that for some $\boldsymbol{y}\in B(\boldsymbol{x},1/n)\cap X$, $\boldsymbol{x}\prec\boldsymbol{y}$. Because $\boldsymbol{x}_k\to\boldsymbol{x}$, $\|\boldsymbol{x}_k-\boldsymbol{y}\|<1/n$ for k large enough. For such a k, we have $\boldsymbol{y}\in B(\boldsymbol{x}_k,1/n)\cap X$ and $\boldsymbol{x}_k\prec_k\boldsymbol{y}$, a contradiction. This proves that $\prec\in\mathcal{P}_{lns}^{m,n}$.

To prove that $\mathcal{P}_{mo}$ is G_δ, for positive integers m, n, we define

$$\mathcal{P}_{mo}^{m,n} = \left\{ (X, \prec) \in \mathcal{P} \,\middle|\, \text{there exists } \boldsymbol{x}, \boldsymbol{y} \in X \text{ such that } \|\boldsymbol{x}\| \le m, \|\boldsymbol{y}\| \le m, \boldsymbol{y} \succsim \boldsymbol{x}, \boldsymbol{x} \ge \boldsymbol{y} \text{ and } \|\boldsymbol{x} - \boldsymbol{y}\| \ge 1/n \right\}.$$

Because $\mathcal{P}\backslash\mathcal{P}_{mo} = \cup_{m=1}^{\infty} \cup_{n=1}^{\infty} \mathcal{P}_{mo}^{m,n}$, $\mathcal{P}_{mo}$ is G_δ if each $\mathcal{P}_{mo}^{m,n}$ is closed. Let $\{(X_k, \prec_k)\}_{k\in\mathbb{N}}$ be a sequence in $\mathcal{P}_{mo}^{m,n}$ such that $(X_k, \prec_k) \to (X, \prec)$ in the topology τ_c. Then for each k, there exist $\boldsymbol{x}_k, \boldsymbol{y}_k \in X_k$ such that $\|\boldsymbol{x}_k\| \le m, \|\boldsymbol{y}\| \le m$, $\boldsymbol{y}_k \succsim_k \boldsymbol{x}_k$, $\boldsymbol{x}_k \ge \boldsymbol{y}_k$ and $\|\boldsymbol{y}_k - \boldsymbol{x}_k\| \ge 1/n$. We can assume that $\boldsymbol{x}_k \to \boldsymbol{x} \in X$ and $\boldsymbol{y}_k \to \boldsymbol{y} \in X$. Then passing to the limit, we have $\|\boldsymbol{x}\|, \|\boldsymbol{y}\| \le m$, $\boldsymbol{y} \succsim \boldsymbol{x}$, $\boldsymbol{y} \ge \boldsymbol{x}$ and $\|\boldsymbol{x} - \boldsymbol{y}\| \ge 1/n$. This proves that $\prec \in \mathcal{P}_{mo}^{m,n}$; hence, $\mathcal{P}_{mo}^{m,n}$ is a closed set.

Finally, to show that $\mathcal{P}_{co}$ is a G_δ set, we define for each of the positive integers m, n, r, s, the set $\mathcal{P}_{co}^{m,n,r,s} \subset \mathcal{P}$ by

$$\mathcal{P}_{co}^{m,n,r,s} = \left\{ (X, \prec) \in \mathcal{P} \,\middle|\, \begin{array}{l} \text{there exists } \boldsymbol{x}, \boldsymbol{y}, \boldsymbol{z} \in X \text{ and } \tau \in \mathbb{R} \text{ such that} \\ \|\boldsymbol{x}\| \le m, \|\boldsymbol{y}\| \le m, \|\boldsymbol{z}\| \le m, \|\boldsymbol{x} - \boldsymbol{y}\|, \|\boldsymbol{x} - \boldsymbol{z}\| \ge 1/n, \\ \text{and } (1/r) \le \tau \le 1 - (1/r), \boldsymbol{x} \succsim \tau\boldsymbol{y} + (1-\tau)\boldsymbol{z}, \\ B(\boldsymbol{x}, 1/s) \cap X \prec B(\boldsymbol{y}, 1/s) \cap X \text{ and} \\ B(\boldsymbol{x}, 1/s) \cap X \prec B(\boldsymbol{z}, 1/s) \cap X \end{array} \right\}.$$

By definition, $\mathcal{P}\backslash\mathcal{P}_{co} = \cup_{m=1}^{\infty} \cup_{n=1}^{\infty} \cup_{r=1}^{\infty} \cup_{s=1}^{\infty} \mathcal{P}_{co}^{m,n,r,s}$. It remains to show that each $\mathcal{P}_{co}^{m,n,r,s}$ is a closed set. Let $(X_k, \prec_k) \in \mathcal{P}_{m,n,r,s}$ and $(X_k, \prec_k) \to (X, \prec)$. Then for each k, there exist $\boldsymbol{x}_k, \boldsymbol{y}_k, \boldsymbol{z}_k \in X_k$ and $\tau_k \in \mathbb{R}$ such that $\|\boldsymbol{x}_k\| \le m, \|\boldsymbol{y}_k\| \le m, \|\boldsymbol{z}_k\| \le m, \|\boldsymbol{x}_k - \boldsymbol{y}_k\|, \|\boldsymbol{x}_k - \boldsymbol{z}_k\| \ge 1/n$, and $(1/r) \le \tau_k \le 1 - (1/r)$, $\boldsymbol{x}_k \succsim_k \tau_k\boldsymbol{y}_k + (1 - \tau_k)\boldsymbol{z}_k$, $B(\boldsymbol{x}_k, 1/s) \cap X_k \prec B(\boldsymbol{y}_k, 1/s) \cap X_k$, and $B(\boldsymbol{x}_k, 1/s) \cap X_k \prec B(\boldsymbol{z}_k, 1/s) \cap X_k$.

We can assume that $(\boldsymbol{x}_k, \boldsymbol{y}_k, \boldsymbol{z}_k, \tau_k) \to (\boldsymbol{x}, \boldsymbol{y}, \boldsymbol{z}, \tau)$. Then we have $\boldsymbol{x}, \boldsymbol{y}, \boldsymbol{z} \in X$ and $\|\boldsymbol{x}\| \le m$, $\|\boldsymbol{y}\| \le m$, $\|\boldsymbol{z}\| \le m$, $\|\boldsymbol{x} - \boldsymbol{y}\|, \|\boldsymbol{x} - \boldsymbol{z}\| \ge 1/n$, and $(1/r) \leq \tau \leq 1-(1/r), \boldsymbol{x} \succsim \tau \boldsymbol{y}+(1-\tau)\boldsymbol{z}$. Suppose that $B(\boldsymbol{x}, 1/s) \cap X \prec B(\boldsymbol{y}, 1/s) \cap X$ does not hold. Then there exist $\boldsymbol{x}' \in B(\boldsymbol{x}, 1/s) \cap X$ and $\boldsymbol{y}' \in B(\boldsymbol{y}, 1/s) \cap X$ such that $\boldsymbol{x}' \succsim \boldsymbol{y}'$. Because $L_i(\succsim_k) = \succsim$ and $(\boldsymbol{x}', \boldsymbol{y}') \in \succsim$, we have a sequence $(\boldsymbol{x}'_k, \boldsymbol{y}'_k) \in \succsim_k$ with $(\boldsymbol{x}'_k, \boldsymbol{y}'_k) \to (\boldsymbol{x}', \boldsymbol{y}')$. Because $\|\boldsymbol{x}_k - \boldsymbol{x}'_k\| \leq \|\boldsymbol{x}_k - \boldsymbol{x}\| + \|\boldsymbol{x} - \boldsymbol{x}'_k\|$, one has $\|\boldsymbol{x}_k - \boldsymbol{x}'_k\| \leq 1/s$ for k large enough. Similarly, we have $\|\boldsymbol{y}_k - \boldsymbol{y}'_k\| \leq 1/s$ for k sufficiently large. This contradicts the fact that $B(\boldsymbol{x}_k, 1/s) \cap X_k \prec B(\boldsymbol{y}_k, 1/s) \cap X_k$. Using similar arguments, we can show that $B(\boldsymbol{x}, 1/s) \cap X \prec B(\boldsymbol{z}, 1/s) \cap X$. Hence, $\prec \in \mathcal{P}_{co}^{m,n,r,s}$ and the proof is complete. □

Summing up; *The sets $\mathcal{P}*$, $\mathcal{P}_{lns}$, $\mathcal{P}_{mo}$, $\mathcal{P}_{co}$ are Borel measurable sets.*

3.3 Existence of Competitive Equilibria

Let $(A, \mathcal{A}, \lambda)$ be a complete atomless measure space of consumers. We assume that the measure λ is a probability measure, or $\lambda(A) = 1$. The economic implication will be explained subsequently.

Definition 3.2. An *exchange economy* is a Borel measurable mapping

$$\mathcal{E} : (A, \mathcal{A}, \lambda) \to (\mathcal{P} \times \mathbb{R}^\ell, \mathcal{B}(\mathcal{P} \times \mathbb{R}^\ell)), \ a \mapsto ((X_a, \prec_a), \omega_a)$$

such that $\int_A \omega(a) d\lambda < +\infty$.

A measurable mapping f from A to $\mathbb{R}^\ell$ is called an *allocation* if $f(a) \in X_a$, *a.e.* The allocation $f : A \to \mathbb{R}^\ell$ is said to be *feasible* if and only if it is integrable and satisfies the following resource constraint condition:

$$\int_A f(a) d\lambda \leq \int_A \omega(a) d\lambda.$$

The allocation $f(a)$ is often denoted by $\boldsymbol{x}(a)$. The allocation is *exactly feasible* if the equality holds.

We now remark on the economic interpretation of the integration of an allocation over the set of consumers A. First, consider the case in which the set A is finite, or $\sharp A < +\infty$, where $\sharp A$ means the number of elements of the set A. Then we have a natural probability measure on A, namely the counting measure λ defined by

$$\lambda(B) = \sharp B / \sharp A \text{ for every } B \subset A.$$

Therefore, the integration of an allocation $\phi : A \to \mathbb{R}^\ell$ with respect to the counting measure λ is

$$\int_A f(a)d\lambda = \left(\frac{1}{\sharp A}\right) \sum_{a \in A} f(a),$$

which is the mean or average of the allocation f, in other words the per capita total allocation.

By analogy with the finite case, we interpret the integration of the allocation $f : A \to \mathbb{R}^\ell$ over the atomless probability measure space of the traders $(A, \mathcal{A}, \lambda)$,

$$\int_A f(a)d\lambda,$$

as the mean (per capita total) allocation. Hence, the value of the integration is represented by the per capita term, rather than the simple aggregated term. For the same reason, we call the integral $\int_A \omega(a)d\lambda \in \mathbb{R}^\ell$ the *mean endowment vector*, or simply the mean endowment.

A feasible allocation $f : A \to \mathbb{R}^\ell$ is said to be *improved upon* or *blocked* if there exists another feasible allocation $g : A \to \mathbb{R}^\ell$ such that

$$g(a) \succsim_a f(a) \ a.e,$$

and

$$\lambda(\{a \in A | \ f(a) \prec_a g(a)\}) > 0.$$

In other words, the allocation f is improved upon by a feasible allocation g that is at least as desired as f by almost all consumers, and it is strictly preferred to f by a group of consumers with positive measure. A feasible allocation f is said to be *Pareto-optimal* if it is not improved upon by any other feasible allocations. We now state the definition of the competitive equilibrium of an exchange economy with the measure space of consumers.

Definition 3.3. A pair $(\boldsymbol{p}, \boldsymbol{x}(a))$ consisting of a price vector and an allocation is called a *competitive equilibrium* of an (exchange) economy $\mathcal{E} : a \mapsto ((X_a, \prec_a), \omega_a)$ if the following conditions hold.

(LE-1) $\boldsymbol{p}\boldsymbol{x}(a) \leq \boldsymbol{p}\omega(a)$ and $\boldsymbol{x}(a) \succsim_a \boldsymbol{y}$ whenever $\boldsymbol{p}\boldsymbol{y} \leq \boldsymbol{p}\omega(a)$ a.e.,
(LE-2) $\int_A \boldsymbol{x}(a)d\lambda \leq \int_A \omega(a)d\lambda$.

The condition (LE-1) says that almost all consumers "maximize his/her utility" under the budget constraint, and the condition (LE-2) says that the resource condition is met. We have a straightforward generalization of the first fundamental theorem of welfare economics, Proposition 2.10.

Proposition 3.1. *Let $(\boldsymbol{p}, \boldsymbol{x}(\cdot))$ be a competitive equilibrium of an exchange economy $\mathcal{E} : (A, \mathcal{A}, \lambda) \to (\mathcal{P}\times\mathbb{R}^\ell, \mathcal{B}(\mathcal{P}\times\mathbb{R}^\ell))$ such that $\mathcal{E}(a) \subset \mathcal{P}_{lns}\times\mathbb{R}^\ell$ a.e. Then the allocation $\boldsymbol{x}(\cdot)$ is Pareto-optimal.*

Proof. Suppose on the contrary that $\boldsymbol{x}(\cdot)$ is not Pareto-optimal. Then there exists a feasible allocation $\boldsymbol{y}(\cdot)$, which is an integrable map of A to $\mathbb{R}^\ell$ such that $\boldsymbol{y}(a) \in X_a$ a.e., and $\int_A \boldsymbol{y}(a)d\lambda \leq \int_A \omega(a)d\lambda$, and satisfies $\boldsymbol{y}(a) \succsim_a \boldsymbol{x}(a)$ *a.e.*, and

$$\lambda(\{a \in A | \boldsymbol{x}(a) \prec_a \boldsymbol{y}(a)\}) > 0.$$

Then it follows that $\boldsymbol{p}\boldsymbol{y}(a) \geq \boldsymbol{p}\omega(a)$ a.e. Because the inequality follows trivially for $a \in A$ with $\boldsymbol{p}\omega_a \leq \inf \boldsymbol{p}X_a$, suppose that $\boldsymbol{p}\omega_a > \inf \boldsymbol{p}X_a$ and $\boldsymbol{p}\boldsymbol{y}(a) < \boldsymbol{p}\omega_a$ on a set B of positive measure. By (LNS), there exists a consumption vector $\boldsymbol{z}(a) \in X_a$ close enough to $\boldsymbol{y}(a)$ such that $\boldsymbol{y}(a) \prec_a \boldsymbol{z}(a)$, and $\boldsymbol{p}\boldsymbol{z}(a) < \boldsymbol{p}\omega(a)$ on B. This contradicts the condition (LE-1) of Definition 3.3. Furthermore, for $a \in A$ such that $\boldsymbol{x}(a) \prec_a \boldsymbol{y}(a)$, we have $\boldsymbol{p}\boldsymbol{y}(a) > \boldsymbol{p}\omega(a)$. Integrating these inequalities over A, we have $\boldsymbol{p}\int_A \boldsymbol{y}(a)d\lambda > \boldsymbol{p}\int_A \omega(a)d\lambda$. On the other hand, by the feasibility condition (LE-2) of Definition 3.3 and $\boldsymbol{p} \geq \boldsymbol{0}$, it follows that $\boldsymbol{p}\int_A \boldsymbol{y}(a)d\lambda \leq \boldsymbol{p}\int_A \omega(a)d\lambda$, a contradiction. □

Recall that $\mathcal{P}$ is the set of all irreflexive, transitive, and continuous preference-consumption set pairs $(X, \prec)$. Consider the exchange economy

$$\mathcal{E} : (A, \mathcal{A}, \lambda) \to (\mathcal{P}\times\mathbb{R}^\ell, \mathcal{B}(\mathcal{P}\times\mathbb{R}^\ell)),\ \mathcal{E}(a) = ((X_a, \prec_a), \omega_a).$$

Let X be a closed and convex subset of $\mathbb{R}^\ell$, which is bounded from below. In the following theorem, we assume that (almost) all consumers have the same consumption set X.

Theorem 3.3. *Let* $\mathcal{E} : (A, \mathcal{A}, \lambda) \to (\mathcal{P}\times\mathbb{R}^\ell, \mathcal{B}(\mathcal{P}\times\mathbb{R}^\ell))$ *be an exchange economy such that* $X_a = X$ *a.e.,* $\omega(a) \in \mathbb{R}^\ell_+$ *a.e., and* $\int_A \omega(a)d\lambda \gg \mathbf{0}$. *Moreover, we assume (MI) for almost all* $a \in A$. *Then there exists a competitive equilibrium for* $\mathcal{E}$.

Before proving this theorem, we note that the convexity or the negative transitivity (transitivity of $\precsim$) of preferences is not assumed, because the convex-valuedness of the *individual* demand correspondences will not be required in Theorem 3.3 (cf. Proposition 2.1). As we pointed out in Section 3.1, this is the first aggregation effect in large markets. We can illustrate this with the following simple example.

Example 3.1. Consider a sequence of finite exchange economies $\mathcal{E}(m)$ with two commodities and the m consumers, $A(m) = \{1 \dots m\}$, $m = 1 \dots$. For each m, all consumers $a \in A(m)$ are assumed to have the same preference relation represented by a convex (not concave!) utility function of the form

$$u(x, y) = x^2 + y^2 \quad \text{for } \boldsymbol{x} = (x, y) \in \mathbb{R}^2_+$$

defined on the consumption set $X = \mathbb{R}^2_+$. All consumers also have the same endowment vector $\omega = (1, 1)$ (see Figure 3.1). Because the indifference curves are circular or concave to the origin, for every non-negative price vector $\boldsymbol{p} = (p, q)$, the consumer maximizes his/her utility with vectors on the x-axis or the y-axis. More specifically, from

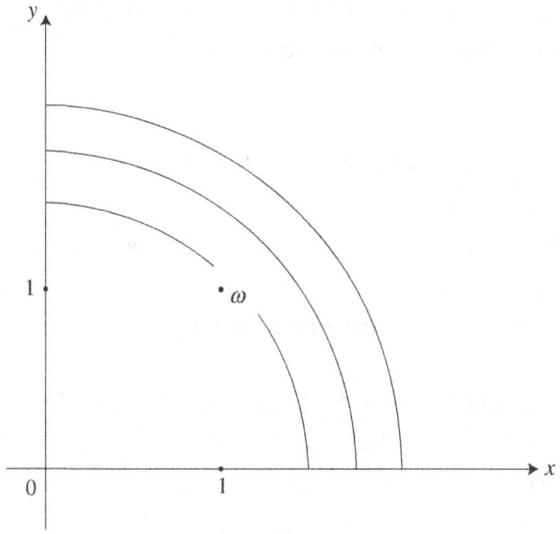

Figure 3.1. Example 3.1.

the budget equation

$$px + qy = p + q,$$

the consumer demands either $\boldsymbol{x} = \left(\frac{p+q}{p}, 0\right)$ or $\boldsymbol{x} = \left(0, \frac{p+q}{q}\right)$. Therefore, there does not exist any equilibrium for the one-consumer economy $\mathcal{E}(1)$ with $A(1) = \{1\}$.

However, there exists an equilibrium with the equilibrium price vector $\boldsymbol{p} = (1/2, 1/2) \in \Delta$ for the two-person economy $\mathcal{E}(2)$ with $A(2) = \{1, 2\}$. An equilibrium allocation is given by $\{\boldsymbol{x}_1, \boldsymbol{x}_2\} = \{(2,0),(0,2)\}$. The market condition $(1/2)(\boldsymbol{x}_1 + \boldsymbol{x}_2) = (1/2)((2,0)+(0,2)) = (1,1)$ is certainly met.

The reason for the existence of equilibrium in Example 3.1 is, of course, as the number of the consumers m increases, the number of the values in the mean demand correspondence $\Phi(\boldsymbol{p}) = (1/m)\sum_{a=1}^{m} \phi(a, \boldsymbol{p})$ increases (see Figures 3.2(a) and 3.2(b)). We expect that in the limit of $m \to +\infty$, the segment $co\{(2,0),(0,2)\}$ is full of the values of the mean demand, and indeed it is, thanks to the following theorem.

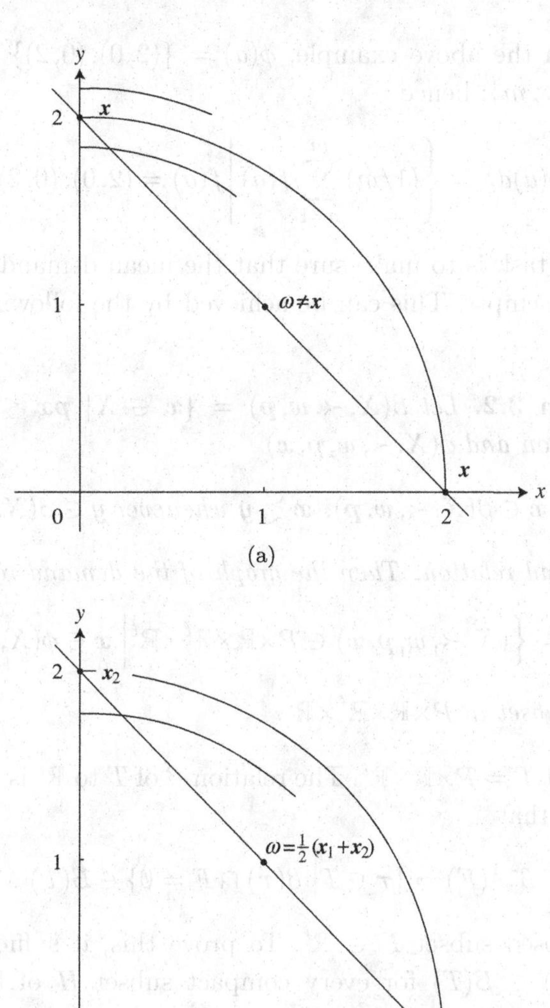

Figure 3.2. (a) One-consumer case. (b) Two-consumer case.

Theorem F.12. *Let ϕ be a correspondence of an atomless measure space $(A, \mathcal{A}, \lambda)$ to $\mathbb{R}^\ell$. Then the integral $\int_\Omega \phi(\omega)d\lambda$ is a convex subset of $\mathbb{R}^\ell$.*

Note that in the above example, $\phi(a) = \{(2,0),(0,2)\}$ for all $a \in A(m) = \{1 \dots m\}$; hence

$$\int_A \phi(a)d\lambda = \left\{(1/m)\sum_{a=1}^{m} f(a) \,\middle|\, f(a) = (2,0),(0,2)\right\}.$$

Our first task is to make sure that the mean demand correspondence is non-empty. This can be achieved by the following proposition:

Proposition 3.2. *Let* $\beta(X, \prec, w, \boldsymbol{p}) = \{\boldsymbol{x} \in X |\ \boldsymbol{px} \leq w\}$ *be the budget relation and* $\phi(X, \prec, w, \boldsymbol{p}, \boldsymbol{x})$

$$= \{\boldsymbol{x} \in X |\ \boldsymbol{x} \in \beta(X, \prec, w, \boldsymbol{p}),\ \boldsymbol{x} \succsim \boldsymbol{y} \textit{ whenever } \boldsymbol{y} \in \beta(X, \prec, w, \boldsymbol{p})\}$$

be the demand relation. Then the graph of the demand relation

$$Graph(\phi) = \left\{(X, \prec, w, \boldsymbol{p}, \boldsymbol{x}) \in \mathcal{P}\times\mathbb{R}\times\mathbb{R}^\ell\times\mathbb{R}^\ell \,\middle|\, \boldsymbol{x} \in \phi(X, \prec, w, \boldsymbol{p})\right\}$$

is a Borel subset in $\mathcal{P}\times\mathbb{R}\times\mathbb{R}^\ell\times\mathbb{R}^\ell$.

Proof. Let $T = \mathcal{P}\times\mathbb{R}\times\mathbb{R}^\ell$. The relation β of T to $\mathbb{R}^\ell$ is measurable in the sense that

$$\beta^{-1}(F) = \{\boldsymbol{\tau} \in T |\ \beta(\boldsymbol{\tau}) \cap F \neq \emptyset\} \in \mathcal{B}(T)$$

for every closed subset $F \subset \mathbb{R}^\ell$. To prove this, it suffices to show that $\beta^{-1}(H) \in \mathcal{B}(T)$ for every compact subset H of $\mathbb{R}^\ell$. Indeed, because the set F is separable, there exists a countable dense subset $\{x_0, x_1 \dots\}$ of F. Let $H_n = B(x_n, 1) \cap F$. Then $F = \cup_{n=0}^{\infty} H_n$ and $\beta^{-1}(F) = \cup_{n=0}^{\infty}\beta^{-1}(H_n)$. Let $\{\boldsymbol{\tau}_n\} = \{(X_n, \prec_n, w_n, \boldsymbol{p}_n)\}$ be a sequence in $\beta^{-1}(H)$ with $\boldsymbol{\tau}_n \to \boldsymbol{\tau} = (X, \prec, w, \boldsymbol{p})$. Then for each n, we have $\boldsymbol{x}_n \in \beta(\boldsymbol{\tau}_n) \cap H$. Because H is compact, we can assume that $\boldsymbol{x}_n \to \boldsymbol{x} \in H$. Because $\boldsymbol{x}_n \in \beta(\boldsymbol{\tau}_n) = \beta(X_n, \prec_n, w_n, \boldsymbol{p}_n)$, one has $\boldsymbol{x}_n \in X_n$ and $\boldsymbol{p}_n\boldsymbol{x}_n \leq w_n$ for all n. Passing to the limit, we have $\boldsymbol{x} \in X$ and $\boldsymbol{px} \leq w$. Hence, $\boldsymbol{x} \in \beta(\boldsymbol{\tau}) \cap H$ and $\beta^{-1}(H)$ is closed; hence, it belongs to $\mathcal{B}(T)$.

Let $\hat{T} = \{\boldsymbol{\tau} \in T |\ \beta(\boldsymbol{\tau}) \neq \emptyset\}$. Obviously, $\hat{T} \in \mathcal{B}(T)$.

By Proposition F.9, there exists a sequence $\{f_n\}_{n\in\mathbb{N}}$ of measurable functions of $\hat{T}$ to $\mathbb{R}^\ell$ such that

$$\beta(\boldsymbol{\tau}) = \overline{\{f_n(\boldsymbol{\tau})|\ n = 0, 1, \dots\}}$$

for every $\boldsymbol{\tau} \in \hat{T}$. Define

$$\phi_n(\boldsymbol{\tau}) = \{\boldsymbol{x} \in \beta(\boldsymbol{\tau})|\ \boldsymbol{x} \succsim f_n(\boldsymbol{\tau})\},\ n \in \mathbb{N}.$$

We will show that the graph G_n of the relation ϕ_n is a Borel subset of $T\times\mathbb{R}^\ell$. By Theorem 3.1, the set

$$\{(X, \prec, \boldsymbol{x}, \boldsymbol{y}) \in \mathcal{P}\times\mathbb{R}^\ell\times\mathbb{R}^\ell|\ \boldsymbol{x}, \boldsymbol{y} \in X \text{ and } \boldsymbol{y} \succsim \boldsymbol{x}\}$$

is closed. As for the argument showing the closedness of $\beta^{-1}(H)$, it is easy to show that the set

$$G = \{(\boldsymbol{\tau}, \boldsymbol{x}, \boldsymbol{y}) \in \hat{T}\times\mathbb{R}^\ell\times\mathbb{R}^\ell|\ \boldsymbol{x}, \boldsymbol{y} \in \beta(\boldsymbol{\tau}) \text{ and } \boldsymbol{y} \succsim_{\boldsymbol{\tau}} \boldsymbol{x}\}$$

is closed, where the preference $\succsim_{\boldsymbol{\tau}}$ corresponds to $\boldsymbol{\tau} = (X, \prec_{\boldsymbol{\tau}}, w, \boldsymbol{p}) \in \hat{T}$.

Let g_n be the mapping of $\hat{T}\times\mathbb{R}^\ell$ to $\hat{T}\times\mathbb{R}^\ell\times\mathbb{R}^\ell$ defined by

$$(\boldsymbol{\tau}, \boldsymbol{x}) \mapsto g_n(\boldsymbol{\tau}, \boldsymbol{x}) = (\boldsymbol{\tau}, f_n(\boldsymbol{\tau}), \boldsymbol{x}).$$

Then we have $G_n = g_n^{-1}(G)$. Because $f_n(\cdot)$ is a measurable function of $\hat{T}$ to $\mathbb{R}^\ell$, the mapping g_n is Borel measurable by Propositions F.3 and F.4; hence, $G_n \in \mathcal{B}(\hat{T}\times\mathbb{R}^\ell) \subset \mathcal{B}(T\times\mathbb{R}^\ell)$.

Finally, we show that $\phi(\boldsymbol{\tau}) = \cap_{n=0}^{\infty}\phi_n(\boldsymbol{\tau})$. Obviously, $\phi(\boldsymbol{\tau}) \subset \cap_{n=0}^{\infty}\phi_n(\boldsymbol{\tau})$. On the other hand, let $\boldsymbol{x} \in \cap_{n=0}^{\infty}\phi_n(\boldsymbol{\tau})$ but $\boldsymbol{x} \notin \phi(\boldsymbol{\tau})$. Then there exists a vector $\boldsymbol{y} \in \beta(\boldsymbol{\tau})$ with $\boldsymbol{x} \prec_{\boldsymbol{\tau}} \boldsymbol{y}$. Because the set $\{\boldsymbol{z} \in \beta(\boldsymbol{\tau})|\ \boldsymbol{x} \prec_{\boldsymbol{\tau}} \boldsymbol{z}\}$ is open relative to $\beta(\boldsymbol{\tau})$, there exists an integer N such that $\boldsymbol{x} \prec f_N(\boldsymbol{\tau})$, or $\boldsymbol{x} \notin \phi_N(\boldsymbol{\tau})$, a contradiction. □

We now prove Theorem 3.3. For the initial endowment vector $\omega(a) = (\omega^t(a))$, let M be a non-negative number defined by

$$M = \int_A \omega^1(a)d\lambda + \cdots \int_A \omega^\ell(a)d\lambda.$$

Because $\int_A \omega(a)d\lambda \gg \mathbf{0}$, $M > 0$. Define the set $A_k \subset A$ by

$$A_k = \{a \in A |\, \omega(a) \leq kM\mathbf{1}\},$$

where $\mathbf{1} = (1\ldots1)$. Then the set A_k is measurable by the measurability of the map $\omega : A \to \mathbb{R}^\ell$, and by the definition of M, we have $A_1 \neq \emptyset$. For if not, there exists $t \in \{1\ldots\ell\}$ such that $\int_A \omega^t(a)d\lambda > \int_A Md\lambda = M = \sum_{t=1}^{\ell} \int_A \omega^t(a)d\lambda \geq \int_A \omega^t(a)d\lambda$, a contradiction. Because $A_1 \subset A_2 \subset \cdots \subset A_k \subset \ldots$, we have $A_k \neq \emptyset$ for all $k \geq 1$. Define the sub σ-field $\mathcal{A}_k$ of $\mathcal{A}$ by

$$\mathcal{A}_k = \{C \cap A_k |\, C \in \mathcal{A}\},$$

and let λ_k be the restriction of λ to $(A_k, \mathcal{A}_k)$. Because the measure space $(A, \mathcal{A}, \lambda)$ is atomless, so are the measure spaces $(A_k, \mathcal{A}_k, \lambda_k)$, $k = 1, 2, \ldots$. Now for each positive integer k, let

$$X_k = \{\boldsymbol{x} \in X |\, \boldsymbol{x} \leq k(M\ldots M)\},$$

and for each $a \in A_k$, define the restricted budget sets and demand sets,

$$\beta_k(a, \boldsymbol{p}) = \{\boldsymbol{x} \in X_k |\, \boldsymbol{p}\boldsymbol{x} \leq \boldsymbol{p}\omega(a)\},$$

and

$$\phi_k(a, \boldsymbol{p}) = \{\boldsymbol{x} \in \beta_k(a, \boldsymbol{p}) |\, \boldsymbol{x} \succsim_a \boldsymbol{z} \text{ for every } \boldsymbol{z} \in \beta_k(a, \boldsymbol{p})\},$$

respectively. Finally, the restricted excess mean demand is defined by

$$\zeta_k(\boldsymbol{p}) = \int_{A_k} \phi_k(a, \boldsymbol{p})d\lambda_k - \int_{A_k} \omega(a)d\lambda_k.$$

We will show that the correspondence ζ_k of Δ to $Z_k \equiv \int_{A_k} X_k d\lambda_k - \{\int_{A_k} \omega(a)d\lambda_k\}$ satisfies the conditions that are needed to apply the Gale–Nikaido lemma (Theorem D.3),

(i) $\zeta_k : \Delta \to Z_k$ is an upper hemicontinuous correspondence from Δ to a compact and convex set Z_k,

(ii) for every $\boldsymbol{p} \in \Delta$, $\zeta_k(\boldsymbol{p})$ is a non-empty convex set,

(iii) for every $\boldsymbol{p} \in \Delta$, $\boldsymbol{p}\zeta_k(\boldsymbol{p}) \leq 0$.

First note that the set $\int_{A_k} X_k d\lambda_k$ is compact and convex by Theorem F.12 and Corollary F.1. Hence, so is the set Z_k. To show that $\zeta_k(\boldsymbol{p}) \neq \emptyset$, it is sufficient from Theorem F.11 (measurable selection theorem) to show that

$$\phi_k(a, \boldsymbol{p}) \neq \emptyset \ a.e \text{ for every } \boldsymbol{p} \in \Delta,$$

$$Graph(\phi_k) = \{(a, \boldsymbol{x}) \in A_k \times X_k | \ \boldsymbol{x} \in \phi_k(a, \boldsymbol{p})\} \in \mathcal{A}_k \times \mathcal{B}(X_k).$$

For almost all $a \in A_k$, it follows from (MI) that $\beta(a, \boldsymbol{p}) \neq \emptyset$. Because the set $\beta_k(a, \boldsymbol{p})$ is compact and the preference $\prec_a$ is continuous, we have $\phi_k(a, \boldsymbol{p}) \neq \emptyset$ by Proposition 2.1. By Proposition 3.2, the graph of the demand correspondence ϕ_k is measurable. This proves $\int_{A_k} \phi_k(a, \boldsymbol{p}) d\lambda_k \neq \emptyset$, and hence $\zeta_k(\boldsymbol{p}) \neq \emptyset$.

By Theorem F.12, we have $\int_{A_k} \phi_k(a, \boldsymbol{p}) d\lambda_k$ is a convex set, so is $\zeta_k(\boldsymbol{p})$. From the definition of the demand set, $\boldsymbol{p}\phi_k(a, \boldsymbol{p}) \leq \boldsymbol{p}\omega(a)$ a.e. in A_k, therefore integrating over A_k, one sees that

$$\boldsymbol{p}\zeta_k(\boldsymbol{p}) = \boldsymbol{p}\left(\int_{A_k} (\phi_k(a, \boldsymbol{p}) - \omega(a))\, d\lambda_k\right) \leq 0.$$

Finally by Proposition 2.3, the correspondence $\phi_k(a, \boldsymbol{p})$ is closed for almost all $a \in A_k$, hence so is $\int_{A_k} \phi(a, \boldsymbol{p}) d\lambda_k$. Therefore, the correspondence $\boldsymbol{p} \mapsto \zeta_k(\boldsymbol{p}) \subset Z_k$ is upper hemicontinuous at every $\boldsymbol{p} \in \Delta$ by Corollary F.2.

We now apply the Gale–Nikaido lemma to the correspondence $\zeta_k(\boldsymbol{p})$ and obtain the price vector $\boldsymbol{p}_k \in \Delta$ and an integrable function f_k of A_k to $\mathbb{R}^\ell$ such that

$$f_k(a) \in \phi(a, \boldsymbol{p}_k) \ a.e \text{ in } A_k, \text{ and } \int_{A_k} f_k(a) d\lambda_k \leq \int_{A_k} \omega(a) d\lambda_k.$$

Define a sequence of measurable functions $\{g_k(a)\}_{k \in \mathbb{N}}$ of A to $\mathbb{R}^\ell$ by

$$g_k(a) = \begin{cases} f_k(a) & \text{if } a \in A_k, \\ \omega(a) & \text{otherwise.} \end{cases}$$

Then by the construction, it follows that

$$g_k(a) \in \phi(a, \boldsymbol{p}_k) \text{ a.e. in } A_k, \text{ and } \int_A g_k(a) d\lambda \leq \int_A \omega(a) d\lambda.$$

Without loss of generality, we can assume that the sequence $\{\boldsymbol{p}_k\}$ is convergent, say, $\lim_{k\to\infty} \boldsymbol{p}_k = \boldsymbol{p} \in \Delta$. Let $\boldsymbol{b} = (b^t)$ be a lower bound of the consumption set X. Because the function $g_k(a)$ is bounded from below by $\min\{\boldsymbol{0}, \boldsymbol{b}\}$ a.e. in A for all k, the sequence $\{\int_A g_k(a)d\lambda\}$ is bounded from below. It is bounded from above by the vector $\int_A \omega(a)d\lambda$. Then it follows from Fatou's lemma in ℓ-dimensions (Theorem F.13) that there exists an integrable function f of A to $\mathbb{R}^\ell$ such that $f(a)$ is a cluster point of the sequence $\{g_k(a)\}$ and $\int_A f(a)d\lambda \leq \int_A \omega(a)d\lambda$. Let $k(a)$ be a positive vector with $k(a) \geq \|\omega(a)/M\|$. Then one sees that

$$0 \leq \omega^t(a) \leq \|\omega(a)\| \leq k(a)M, \quad t = 1\ldots\ell,$$

hence $a \in A_k$ for $k \geq k(a)$. Then it follows that

$$k > k(a) \text{ implies that } g_k(a) \in \phi_k(a, \boldsymbol{p}_k).$$

From this we can show that

$$f(a) \in \phi(a, \boldsymbol{p}) \text{ a.e. in } A.$$

Indeed, by the definition of $g_k(a)$, one sees that $f(a) \in \beta(a, \boldsymbol{p})$ a.e., because $f(a)$ is a cluster point of $\{g_k(a)\}$. If $\boldsymbol{x} \in X$ and $\boldsymbol{px} < \boldsymbol{p}\omega(a)$, then $\boldsymbol{p}_k\boldsymbol{x} < \boldsymbol{p}_k\omega(a)$ for all k sufficiently large and $g_k(a) \succsim_a \boldsymbol{x}$. Hence, by (CT), one has $f(a) \succsim_a \boldsymbol{x}$ a.e. in A. If $\boldsymbol{x} \in X$ and $\boldsymbol{px} = \boldsymbol{p}\omega(a)$, then by the minimum income condition (MI), we can take a sequence $\boldsymbol{x}_n$ converging to $\boldsymbol{x}$ and $\boldsymbol{px}_n < \boldsymbol{px}$, $n \in \mathbb{N}$. For each n, $g_k(a) \succsim_a \boldsymbol{x}_n$. Again by (CT), we have $f(a) \succsim_a \boldsymbol{x}$ a.e. in A. This proves condition (LE-1). Condition (LE-2) has already been proved. Therefore, $(\boldsymbol{p}, f(a))$ is a competitive equilibrium. □

From a technical point of view, in the above proof the role of Fatou's lemma in ℓ-dimensions is apparent. However, under the assumption of monotone preferences, we have a simpler proof of the theorem as in the finite-dimensional case applying the next concept.

Definition 3.4. The *quasi-demand relation* is defined by

$$\tilde{\phi}(X, \prec, w, \boldsymbol{p}) = \begin{cases} \phi(X, \prec, w, \boldsymbol{p}) & \text{if } \inf \boldsymbol{p}X < w, \\ \{\boldsymbol{x} \in X \mid \boldsymbol{px} = \inf \boldsymbol{p}X\} & \text{if } \inf \boldsymbol{p}X \geq w. \end{cases}$$

The following proposition is a generalization of Proposition 2.4.

Proposition 3.3. *Let* $T = \mathcal{P}\times\mathbb{R}\times\mathbb{R}^\ell$ *and*

$$T_c = \{\boldsymbol{\tau} = (X, \prec, w, \boldsymbol{p}) \in T \mid X \text{ is convex}\}.$$

Then graph of the quasi-demand relation

$$\{(X, \prec, w, \boldsymbol{p}, \boldsymbol{x}) \in T_c\times\mathbb{R}^\ell \mid \boldsymbol{x} \in \tilde{\phi}(x, \prec, w, \boldsymbol{p})\}$$

is closed.

Proof. Proposition 3.3 is proved in the same manner as Proposition 2.4; hence, we skip the proof. □

We can prove the existence of equilibrium without using Fatou's lemma apparently[1] when preferences are monotone. Note that the condition (MI) of Theorem 3.3 is replaced by the *survival condition* $\omega(a) \in X$ a.e. in Theorem 3.4.

Theorem 3.4. *Let* $\mathcal{E} :(A, \mathcal{A}, \lambda) \to (\mathcal{P}\times\mathbb{R}^\ell, \mathcal{B}(\mathcal{P}\times\mathbb{R}^\ell))$ *be an exchange economy such that* $\mathcal{E}(a) \subset \mathcal{P}_{mo} \times \mathbb{R}^\ell_+$ *a.e., and* $\int_A \omega(a)d\lambda \gg \mathbf{0}$. *Then there exists a competitive equilibrium for* $\mathcal{E}$.

Proof. Let $\tilde{\phi}(a, \boldsymbol{p})$ be the quasi-demand relation and we define the quasi-mean demand by

$$\tilde{\zeta}(\boldsymbol{p}) = \int_A \tilde{\phi}(a, \boldsymbol{p})d\lambda - \int_A \omega(a)d\lambda,$$

for every $\boldsymbol{p} \in \mathring{\Delta}$. We show that there exists a price vector $\boldsymbol{p}^* \in \mathring{\Delta}$ such that $\tilde{\zeta}(\boldsymbol{p}^*) = \mathbf{0}$. The existence of such a price vector is a consequence of Theorem D.4, and we want to verify that

(i) for every strictly positive vector $\boldsymbol{p} \gg \mathbf{0}$, it follows that $\boldsymbol{p}\tilde{\zeta}(\boldsymbol{p}) \leq \mathbf{0}$,
(ii) the correspondence $\tilde{\zeta} : \mathring{\Delta} \to \mathbb{R}^\ell$ is non-empty, compact, and convex-valued, bounded from below and upper hemicontinuous,
(iii) for every sequence $\{\boldsymbol{p}_n\} \in \mathring{\Delta}$ converging to $\boldsymbol{p} \in \partial\Delta$, it follows that $\inf\{\sum_{t=1}^{\ell} z^t \mid \boldsymbol{z} = (z^t) \in \tilde{\zeta}(\boldsymbol{p}_n)\} > 0$ for n large enough.

[1]Fatou's lemma is used in the proofs of Corollaries F.1 and F.2.

Because preferences are monotone, it is obvious that $\boldsymbol{px} = \boldsymbol{p}\omega(a)$ for every $\boldsymbol{x} \in \tilde{\phi}(a, \boldsymbol{p})$ when $\inf \boldsymbol{p}X < \boldsymbol{p}\omega(a)$. If $\inf \boldsymbol{p}X \geq \boldsymbol{p}\omega(a)$, then because $\inf \boldsymbol{p}X = \inf \boldsymbol{p}\mathbb{R}^\ell_+ = 0$, it follows that $0 \leq \boldsymbol{px} \leq \boldsymbol{p}\omega(a) = 0$. Hence, (i) is proved.

Let $\bar{\boldsymbol{p}} \gg \boldsymbol{0}$. Then there exists a compact neighborhood U of $\bar{\boldsymbol{p}}$ consisting of strictly positive vectors. For fixed $a \in A$, the correspondence $\tilde{\phi}(a, \cdot)$ is non-empty valued on U since the budget set is compact, and it is closed at $\bar{\boldsymbol{p}}$ by Proposition 3.3. Moreover, setting $h(a) = (1/\bar{\pi})(\sum_{t=1}^{\ell} \omega^t(a))$, where $\bar{\pi} = \min\{p^t | \boldsymbol{p} = (p^t) \in U\}$, we obtain an integrable function such that $\|\tilde{\phi}(a, \boldsymbol{p})\| \leq h(a)$ for every $\boldsymbol{p} \in U$ a.e. Then by Corollary F.2, the correspondence $\tilde{\Phi}(\boldsymbol{p}) = \int_A \tilde{\phi}(a, \boldsymbol{p}) d\lambda$ is closed at $\bar{\boldsymbol{p}}$. Because the correspondence $\tilde{\Phi}(\boldsymbol{p})$ is bounded on the neighborhood U of $\bar{\boldsymbol{p}}$, it follows from Corollary F.1 and Proposition D.2 that $\tilde{\Phi}(\boldsymbol{p})$ is compact-valued and upper hemicontinuous at $\bar{\boldsymbol{p}}$. By Theorem F.12, it is convex valued. Therefore, the property (ii) is verified. To prove (iii), we will show that

$$\inf \left\{ \|\boldsymbol{x}\| \,\middle|\, \boldsymbol{x} \in \tilde{\Phi}(\boldsymbol{p}_n) \right\} \to \infty \text{ as } n \to \infty$$

for every sequence $\{\boldsymbol{p}_n\}$ of strictly positive vectors converging to $\boldsymbol{p} \in \partial\Delta$. Suppose not. Then there exists a bounded set $B \subset \mathbb{R}^\ell$ such that $\tilde{\Phi}(\boldsymbol{p}_n) \cap B \neq \emptyset$ for infinitely many ns. Because $\int_A \omega(a) d\lambda \gg \boldsymbol{0}$, we have $\inf \boldsymbol{p}\mathbb{R}^\ell_+ = 0 < \boldsymbol{p}\omega(a)$ on a subset of A with positive measure. Hence, we can assume that $\phi(a, \boldsymbol{p}) \cap B \neq \emptyset$ for infinitely many ns on the non-null set. Let $\boldsymbol{x}_n(a) \in \phi(a, \boldsymbol{p}_n) \cap B$. There is a converging subsequence, say $\boldsymbol{x}_n(a) \to \boldsymbol{x}(a)$. Because $\inf \boldsymbol{p}\mathbb{R}^\ell_+ < \boldsymbol{p}\omega(a)$, it follows from Proposition 3.3 that $\boldsymbol{x} \in \phi(a, \boldsymbol{p})$, which is impossible. Indeed, this set is empty because $\prec_a \in \mathcal{P}_{mo}$ and the budget set $\beta(a, \boldsymbol{p})$ is unbounded for $\boldsymbol{p} \in \partial\Delta$.

By Theorem D.4, there exists a vector $\boldsymbol{p}^* \gg \boldsymbol{0}$ such that $\boldsymbol{0} \in \tilde{\zeta}(\boldsymbol{p}^*)$. Then there exists a measurable map $\boldsymbol{x}(a) \in \tilde{\phi}(a, \boldsymbol{p}^*)$ such that $\int_A \boldsymbol{x}(a) d\lambda = \int_A \omega(a) d\lambda$. Hence, condition (LE-2) is met. For $a \in A$ such that $\inf \boldsymbol{p}^*\mathbb{R}^\ell_+ < \boldsymbol{p}^*\omega(a)$, one has $\tilde{\phi}(a, \boldsymbol{p}^*) = \phi(a, \boldsymbol{p}^*)$; hence, condition (LE-1) holds for such an $a \in A$. For $a \in A$ such that $0 = \inf \boldsymbol{p}^*\mathbb{R}^\ell_+ \geq \boldsymbol{p}^*\omega(a)$, it follows from $\boldsymbol{p}^* \gg \boldsymbol{0}$ that the budget set is a singleton $\beta(a, \boldsymbol{p}^*) = \{\boldsymbol{0}\}$. Therefore, condition (LE-1) trivially holds for such an a. □

As in the classical finite (or simple) exchange economies in Chapter 2, we can prove the second fundamental theorem of welfare economics or the converse of Proposition 3.1.

Definition 3.5. An allocation f for the economy $\mathcal{E} : A \to \mathcal{P} \times \mathbb{R}^\ell$ is called an *equilibrium relative to a price vector* $\boldsymbol{p} > \boldsymbol{0}$ if

(R-1) $f(a) \succsim_a \boldsymbol{x}$ whenever $\boldsymbol{x} \in X_a$ and $\boldsymbol{px} \leq \boldsymbol{p}f(a)$ a.e. in A,
(LE-2) $\int_A f(a)d\lambda \leq \int_A \omega(a)d\lambda$.

The allocation f is called a *quasi-equilibrium relative to a price vector* $\boldsymbol{p} > \boldsymbol{0}$ if condition (R-1) is replaced by

(Q-1) $f(a) \succsim_a \boldsymbol{x}$ whenever $\boldsymbol{x} \in X_a$ and $\boldsymbol{px} < \boldsymbol{p}f(a)$ a.e. in A.

Recall that $\mathcal{P}_{mo}$ is the set of all monotone preference relations. The second fundamental theorem of welfare economics now reads as follows:

Theorem 3.5. *Let $\mathcal{E} : A \to \mathcal{P} \times \mathbb{R}^\ell$ be an economy which satisfies $\mathcal{E}(a) \subset \mathcal{P}_{mo}$. Then every Pareto-optimal allocation is a quasi-equilibrium relative to some price vector $\boldsymbol{p} > \boldsymbol{0}$.*

Proof. Let $f : A \to \mathbb{R}^\ell_+$ be a Pareto-optimal allocation and define $\psi : A \to X$ by $\psi(a) = \{\boldsymbol{x} \in X|\ f(a) \prec_a \boldsymbol{x}\}$. We now prove the following lemma.

Lemma 3.1. *Graph*$(\psi) \in \mathcal{B}(A \times \mathbb{R}^\ell)$.

Proof. By Theorem 3.1, Theorem 3.2 and Proposition F.1, the set

$$G = \{(X, \prec, \boldsymbol{x}, \boldsymbol{y}) \in \mathcal{P}_{mo} \times \mathbb{R}^\ell_+ \times \mathbb{R}^\ell_+|\ \boldsymbol{y} \prec \boldsymbol{x}\}$$

is a Borel set of $\mathcal{P}_{mo} \times \mathbb{R}^\ell_+ \times \mathbb{R}^\ell_+$. Defining a map $\Psi : A \times \mathbb{R}^\ell_+ \to \mathcal{P}_{mo} \times \mathbb{R}^\ell_+ \times \mathbb{R}^\ell_+$ by $\Psi(a, \boldsymbol{x}) = (\prec_a, \boldsymbol{x}, f(a))$, it follows that *Graph*$(\psi) = \Psi^{-1}(G)$. Because the map Ψ is measurable by Proposition F.6, the graph of ψ is measurable. □

Now we can define the integral $\int_A \psi(a)d\lambda$ and it is non-empty by the (MT) condition. By Theorem F.12, $\int_A \psi(a)d\lambda$ is a convex subset of $\mathbb{R}^\ell$. Because f is Pareto-optimal, $\int_A \omega(a)d\lambda \notin \int_A \psi(a)d\lambda$. Then we can apply the separation hyperplane theorem (Theorem A.1) and obtain a vector $\boldsymbol{p} \neq \boldsymbol{0}$ such that

$$\boldsymbol{p}\int_A \omega(a)d\lambda \leq \inf\left\{\boldsymbol{p}\boldsymbol{z} \in \mathbb{R} \,\middle|\, \boldsymbol{z} \in \int_A \psi(a)d\lambda\right\}.$$

We have $\boldsymbol{p} > \boldsymbol{0}$ by (MT), and thus, $\boldsymbol{p}\int_A f(a)d\lambda \leq \boldsymbol{p}\int_A \omega(a)$ by the feasibility of f. Therefore, it follows from Proposition F.11 that

$$\boldsymbol{p}\int_A f(a)d\lambda \leq \inf \boldsymbol{p}\int_A \psi(a)d\lambda = \int_A \inf \boldsymbol{p}\psi(a)d\lambda.$$

Because $\inf \boldsymbol{p}\psi(a) \leq \boldsymbol{p}f(a)$ a.e. in A, we obtain $\inf \boldsymbol{p}\psi(a) = \boldsymbol{p}f(a)$, a.e. in A. We have shown that

$$\boldsymbol{p}\boldsymbol{x} < \boldsymbol{p}f(a) \text{ implies } f(a) \succsim_a \boldsymbol{x} \text{ a.e. in } A.$$

This proves condition (Q-1). □

As in the case of classical economies, the quasi-equilibrium is reduced to the equilibrium relative to $\boldsymbol{p}$, if the situation $\boldsymbol{p}f(a) = \inf \boldsymbol{p}X_a = 0$ is excluded.

Corollary 3.1. *Under the assumptions of Theorem 3.5, if the Pareto-optimal allocation $f(a)$ satisfies $\boldsymbol{p}f(a) > 0$ a.e. in A, then it is an equilibrium relative to $\boldsymbol{p}$.*

Proof. Let $\boldsymbol{x} \in X_a$ be such that $\boldsymbol{p}\boldsymbol{x} = \boldsymbol{p}f(a)$ and $\boldsymbol{p}f(a) > 0$. Then we obtain a sequence $\{\boldsymbol{x}_n\}$ such that $\boldsymbol{p}\boldsymbol{x}_n < \boldsymbol{p}f(a)$ and $\boldsymbol{x}_n \to \boldsymbol{x}$. Because $f(a)$ is a quasi-equilibrium, $f(a) \succsim_a \boldsymbol{x}_n$ for all n. By (CT), we have $f(a) \succsim_a \boldsymbol{x}$ in the limit. □

3.4 Equivalence of Core and Equilibria

Let $(A, \mathcal{A}, \lambda)$ be a complete atomless measure space of consumers. Every measurable subset $C \in \mathcal{A}$ is called a *coalition*. When the set of consumers A is finite, every subset C of A is a coalition. If it is λ-measure zero, or $\lambda(C) = 0$, the coalition C is called *null*. The null coalitions are considered to have no influence on the market. The empty coalition $C = \emptyset$ is an example of the null coalition. The core of an economy $\mathcal{E} : A \to \mathcal{P} \times \mathbb{R}^\ell$ is now defined in a natural way.

Definition 3.6. An allocation $f : A \to \mathbb{R}^\ell$, $f(a) \in X_a$ a.e. in A, is *blocked* by a coalition $C \in \mathcal{A}$ if there exists an allocation $g : A \to \mathbb{R}^\ell$, $g(a) \in X_a$ a.e. in A such that

$$f(a) \prec_a g(a) \text{ for almost all } a \in C,$$

and

$$\int_C g(a) d\lambda \leq \int_C \omega(a) d\lambda.$$

The set of feasible allocations that are not blocked by any non-null coalition is called the *core* and denoted by $\mathscr{C}(\mathcal{E})$.

Let $\mathscr{W}(\mathcal{E})$ be the set of competitive allocations of the economy $\mathcal{E}$,

$$\mathscr{W}(\mathcal{E}) = \{f : A \to X_a |\ (\boldsymbol{p}, f) \text{ is a competitive equilibrium for some } \boldsymbol{p} \in \mathbb{R}^\ell_+\}.$$

The following proposition is a generalization of Proposition 2.12.

Proposition 3.4. *Let $\mathcal{E} : A \to \mathcal{P} \times \mathbb{R}^\ell$ be an atomless exchange economy such that $\mathcal{E}(a) \subset \mathcal{P}_{mo} \times \mathbb{R}^\ell_+$ and the mean endowment is strictly positive or $\int_A \omega(a) d\lambda \gg \mathbf{0}$. Then $\mathscr{W}(\mathcal{E}) \subset \mathscr{C}(\mathcal{E})$.*

Proof. Suppose $\boldsymbol{x} \in \mathscr{W}(\mathcal{E})$. Then there exists a price vector $\boldsymbol{p} \in \mathbb{R}^\ell_+$ such that $(\boldsymbol{p}, \boldsymbol{x})$ is a competitive equilibrium for $\mathcal{E}$. Suppose that $\boldsymbol{x} \notin \mathscr{C}(\mathcal{E})$, hence the allocation $\boldsymbol{x}$ is blocked by a non-null coalition C through an allocation $\boldsymbol{y}$ such that $\boldsymbol{x}(a) \prec_a \boldsymbol{y}(a)$ for almost all $a \in C$ and $\int_C \boldsymbol{y}(a) d\lambda \leq \int_C \omega(a) d\lambda$. Then by condition (LE-1) of Definition 3.3, we have $\boldsymbol{p}\boldsymbol{y}(a) > \boldsymbol{p}\omega(a)$ a.e. in C. Hence, $\boldsymbol{p} \int_C \boldsymbol{y}(a) d\lambda = \int_C \boldsymbol{p}\boldsymbol{y}(a) > \int_C \boldsymbol{p}\omega(a) d\lambda = \boldsymbol{p} \int_C \omega(a) d\lambda$. On the other hand, because $\boldsymbol{y}$ is feasible in C, we have $\boldsymbol{p} \int_C \boldsymbol{y}(a) d\lambda \leq$

$\boldsymbol{p}\int_C \omega(a)d\lambda$, a contradiction. Therefore, $\boldsymbol{x} \in \mathscr{C}(\mathcal{E})$ and this proves $\mathscr{W}(\mathcal{E}) \subset \mathscr{C}(\mathcal{E})$. □

In Section 2.6, we proved the core limit theorem (Theorem 2.7) which says that the core approaches the equilibrium when the population of the economy becomes large proportionally. In economies with an atomless measure space of consumers, we can prove the *core equivalence theorem*, which asserts the exact equality of core and equilibria.

Theorem 3.6. *Let $\mathcal{E} : A \to \mathcal{P}\times\mathbb{R}^\ell$ be an atomless exchange economy such that $\mathcal{E}(a) \subset \mathcal{P}_{mo} \times \mathbb{R}^\ell_+$ and the mean endowment is strictly positive or $\int_A \omega(a)d\lambda \gg \mathbf{0}$. Then $\mathscr{W}(\mathcal{E}) = \mathscr{C}(\mathcal{E})$.*

Proof. $\mathscr{W}(\mathcal{E}) \subset \mathscr{C}(\mathcal{E})$ has been proved in Proposition 3.4. It suffices to show $\mathscr{C}(\mathcal{E}) \subset \mathscr{W}(\mathcal{E})$. Let $f \in \mathscr{C}(\mathcal{E})$ and define $\psi : A \to \mathbb{R}^\ell$ by

$$\psi(a) = \{\boldsymbol{z} \in \mathbb{R}^\ell |\ f(a) \prec_a (\boldsymbol{z} + \omega(a))\} \cup \{\mathbf{0}\}.$$

In exactly the same way as Lemma 3.1, we can prove $Graph(\psi) \in \mathcal{B}(A \times \mathbb{R}^\ell)$. Then we can define the integral $\int_A \psi(a)d\lambda$ and it is nonempty, because $\mathbf{0} \in \int_A \psi(a)d\lambda$. We now show that $\int_A \psi(a)d\lambda \cap \mathbb{R}^\ell_- = \{\mathbf{0}\}$. Suppose not. Then there exists a function $h : A \to \mathbb{R}^\ell$ such that $h(a) \in \psi(a)$ a.e., and $\int_A h(a)d\lambda < \mathbf{0}$. Set $C = \{a \in A|\ h(a) \neq \mathbf{0}\}$. Obviously, $\lambda(C) > 0$. Define an integrable function $g : A \to \mathbb{R}^\ell$ by

$$g(a) = h(a) + \omega(a) - \frac{\int_A h(a)d\lambda}{\lambda(C)}.$$

It is easy to see that $f(a) \prec_a g(a)$ a.e. in C and $\int_C g(a)d\lambda = \int_C \omega(a)d\lambda$, contradicting $f \in \mathscr{C}(\mathcal{E})$. We can apply the separation hyperplane theorem (Theorem A.1) to the convex sets $\int_A \psi(a)d\lambda$ and $\mathbb{R}^\ell_-$ and obtain a vector $\boldsymbol{p} > \mathbf{0}$ such that

$$0 \leq \boldsymbol{p}\zeta \text{ whenever } \zeta \in \int_A \psi(a)d\lambda.$$

It follows from Proposition F.11 that

$$\inf\left\{\boldsymbol{p}\zeta \in \mathbb{R} \,\middle|\, \zeta \in \int_A \psi(a)d\lambda\right\} = \int_A \inf\{\boldsymbol{p}\boldsymbol{z} \in \mathbb{R}|\ \boldsymbol{z} \in \psi(a)\}d\lambda,$$

hence, $0 \le \int_A \inf\{\boldsymbol{pz} \in \mathbb{R} |\ \boldsymbol{z} \in \psi(a)\} d\lambda$. Because $\mathbf{0} \in \psi(a)$, we have $\inf \boldsymbol{p}\psi(a) \le 0$ a.e., hence $\inf \boldsymbol{p}\psi(a) = 0$ a.e. Thus, we have shown that

$$\boldsymbol{p}\omega(a) \le \boldsymbol{px} \text{ for every } \boldsymbol{x} \in X \text{ such that } f(a) \prec_a \boldsymbol{x}$$

for almost all $a \in A$. We now claim $\boldsymbol{p}\omega(a) = \boldsymbol{p}f(a)$ a.e. in A. By (CT), it follows immediately from the above that $\boldsymbol{p}\omega(a) \le \boldsymbol{p}f(a)$. If $\boldsymbol{p}\omega(a) < \boldsymbol{p}f(a)$ on a subset of A with a positive measure, then $\boldsymbol{p}\int_A \omega(a)d\lambda < \boldsymbol{p}\int_A f(a)d\lambda$, contradicting $\int_A \omega(a)d\lambda \ge \int_A f(a)d\lambda$.

We now show that for $a \in A$ with positive income $\boldsymbol{p}\omega(a) > 0$, $\boldsymbol{px} \le \boldsymbol{p}\omega(a)$ implies $f(a) \precsim_a \boldsymbol{x}$. For such a consumer, we can take a sequence $\{\boldsymbol{x}_n\}$ with $\boldsymbol{x}_n \to \boldsymbol{x}$ and $\boldsymbol{px}_n < \boldsymbol{p}\omega(a)$ for all n. It follows from the above condition that $f(a) \succsim_a \boldsymbol{x}_n$ for all n. By (CT), we obtain $f(a) \succsim_a \boldsymbol{x}$ in the limit. It follows from $\int_A \omega(a)d\lambda \gg \mathbf{0}$ that $\lambda(\{a \in A |\ \boldsymbol{p}\omega(a) > 0\}) > 0$.

By (MT), we have $\boldsymbol{p} \gg \mathbf{0}$. For if $p^t = 0$ for some t, then $\boldsymbol{p}(f(a) + \epsilon \boldsymbol{e}_t) = \boldsymbol{p}f(a)$ and $f(a) \prec_a f(a) + \epsilon \boldsymbol{e}_t$ for $\epsilon > 0$, where $\boldsymbol{e}_t = (0 \ldots 0, 1, 0 \ldots 0)$ (1 at the tth position), a contradiction. Then if $\boldsymbol{p}\omega(a) = 0$, the budget set $\{\boldsymbol{x} \in \mathbb{R}^\ell_+ |\ \boldsymbol{px} \le 0\}$ is a singleton $\{\mathbf{0}\}$; hence, condition (LE-1) holds trivially. We conclude that $(\boldsymbol{p}, f(a))$ is a competitive equilibrium or $f \in \mathscr{W}(E)$. □

3.5 Existence Theorem for Non-Convex Economies

We have assumed so far that the consumption sets of consumers are convex. In this section and Section 6.4, we will discuss economies with a non-convex consumption set. We will do so not simply as a mathematical generalization, but also with significant economic motivations. The first example of non-convex consumption sets is the case of *indivisible commodities*. In Figure 3.3, the commodity x is assumed to be perfectly divisible, hence the quantity consumed can be any real number value, as it has been so far. However, commodity y is an indivisible commodity, so that it is consumed only in integer units, and its coordinate axis is represented by $\mathbb{N}$, the set of natural numbers. Therefore, the consumption set X is given by

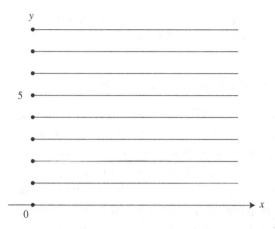

Figure 3.3. Indivisible goods.

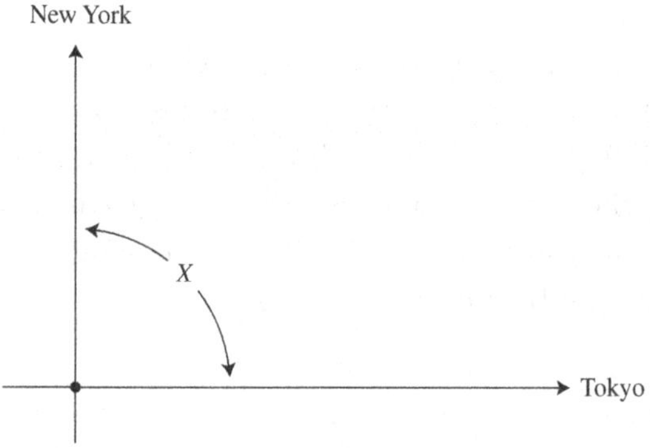

Figure 3.4. Mutually exclusive goods.

$X = \mathbb{R}_+ \times \mathbb{N}$. The second example is given by the case of mutually exclusive commodities.

In Figure 3.4, the commodity x is a good that is consumed in Tokyo and the commodity y is consumed in New York at the same time. Usually, any individual consumer can consume a commodity in either Tokyo or New York, but not both simultaneously, the consumption set X that represents this situation is something like

$$X = \{(x, 0) |\ x \in \mathbb{R}\} \cup \{(0, y) |\ y \in \mathbb{R}\},$$

which is obviously not convex. Recall that commodities are distinguished not only by their physical properties but also by their consumption characteristics. Some of the characteristics of commodities, say location or date, will induce this kind of situation; the consumption of a commodity by an individual consumer necessarily excludes the possibility of consumption by the same individual of another commodity.

These examples indicate that the economic models in which the consumption sets are assumed to be convex are not adequate for some applications. However, a technical problem arises from the models with non-convex consumption sets; the demand correspondences generally exhibit discontinuous behaviors as the following example shows.

Example 3.2. Consider a non-convex consumption set X that is defined by

$$X = \{(x, y) \in \mathbb{R}^2 | \, x \geq 1, \; y \geq 0\} \cup \{(x, y) \in \mathbb{R}^2 | \, x \geq 0, \; y \geq 3\}.$$

The set of consumers A is given by the unit interval $A = [0, 1]$, and all consumers $a \in [0, 1]$ have the same preferences, which are represented by the convex (not concave!) utility function

$$u(x, y) = x^2 + y^2, \quad (x, y) \in X.$$

Suppose that the endowment vector $\omega(a)$ is the same for every consumer $a \in [0, 1]$ and given by $\omega(a) = (1, 1)$ for $a \in [0, 1]$ (see Figure 3.5).

Because the indifference curves are circular, centered at the origin, for every price vector $\boldsymbol{p} = (p, q)$, the consumer maximizes his/her utility at the consumption bundle on the x-axis or the y-axis, or $\boldsymbol{x} = (x, 0)$ or $= (0, y)$. In the following, we normalize the price vector as $\boldsymbol{p} = (p, 1 - p)$, $0 \leq p \leq 1$. Hence, the budget equation of the consumer is given by

$$px + (1 - p)y = 1.$$

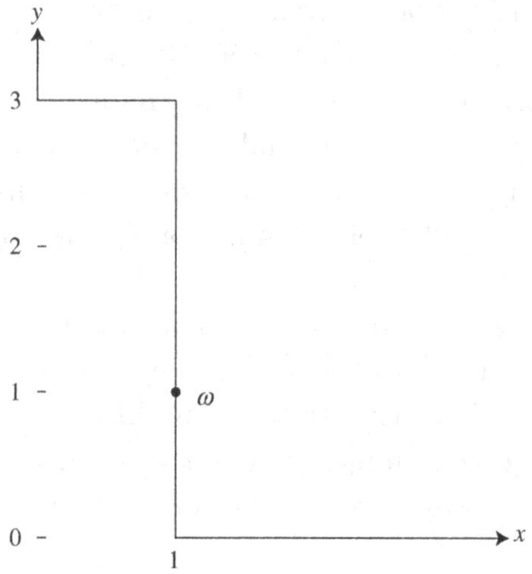

Figure 3.5. Example 3.2.

From the budget equation, we see that the consumer's demand $a \in [0, 1]$ is determined by

$$\phi(a, p) = \begin{cases} \left(\frac{1}{p}, 0\right) & \text{for } 0 \leq p < 2/3, \\ \left(0, \frac{1}{p-1}\right) & \text{for } 2/3 \leq p \leq 1, \end{cases}$$

which is discontinuous at $p = 2/3$ for every $a \in [0, 1]$. Because all consumers have the same demand function, the mean excess demand function can be simply calculated as

$$\begin{aligned} \zeta(p) &= \int_0^1 \phi(a, p) d\lambda - \int_0^1 \omega(a) d\lambda \\ &= \begin{cases} \left(\frac{1}{p} - 1, -1\right) & \text{for } 0 \leq p < 2/3, \\ \left(-1, \frac{1}{p-1} - 1\right) & \text{for } 2/3 \leq p \leq 1, \end{cases} \end{aligned}$$

and the mean demand function is also discontinuous at $p = 2/3$, and it stays away from 0. Therefore, the economy does not have an equilibrium. Figure 3.6 illustrates the graph of the x-coordinate of the mean demand function $\zeta(p)$.

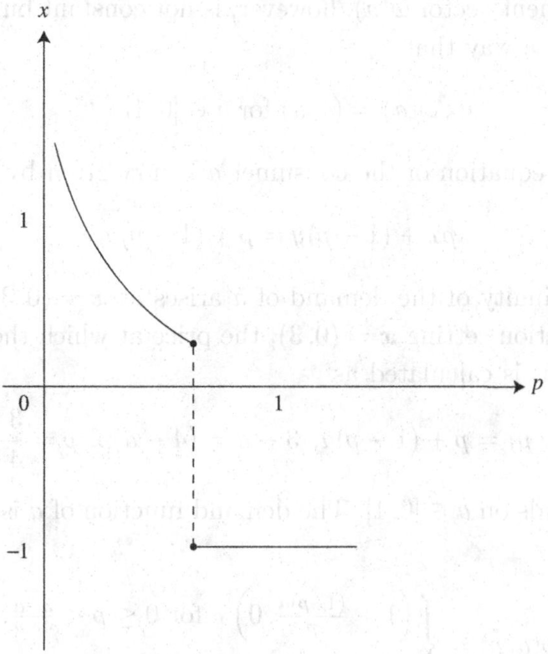

Figure 3.6. Graph of $\zeta(p)$.

The reason for the discontinuous behavior of $\zeta(p)$ is of course that the consumers on a set of positive measure (actually whole space $[0,1]$ in this example) behave discontinuously at a particular price. Therefore, we expect that, if for each price vector, the set of the discontinuous consumers is of measure 0, the mean demand $\Phi(\boldsymbol{p}) = \int_A \phi(a, \boldsymbol{p})d\lambda$ will be continuous, and so is the mean demand correspondence $\zeta(\boldsymbol{p}) = \Phi(\boldsymbol{p}) - \int_A \omega(a)d\lambda$. This is indeed the case, as the following example shows.

Example 3.3. For each consumer $a \in [0,1]$, the consumption set X and the utility function are the same as in Example 3.2,

$$X = \{(x,y) \in \mathbb{R}^2 |\ x \geq 1,\ y \geq 0\} \cup \{(x,y) \in \mathbb{R}^2 |\ x \geq 0,\ y \geq 3\},$$

$$u(x,y) = x^2 + y^2,\ (x,y) \in X.$$

The endowment vector $\omega(a)$, however, is not constant but it depends on a in such a way that

$$\omega(a) = (1, a) \text{ for } a \in [0, 1].$$

The budget equation of the consumer a is now given by

$$px + (1 - p)y = p + (1 - p)a.$$

The discontinuity of the demand of a arises at $\boldsymbol{x} = (0, 3)$. From the budget equation setting $\boldsymbol{x} = (0, 3)$, the price at which the demand is discontinuous is calculated as

$$3(1 - p) = p + (1 - p)a,\ 3 - a = (4 - a)p,\ p = \frac{3 - a}{4 - a},$$

which depends on $a \in [0, 1]$. The demand function of a is then determined by

$$\phi(a, p) = \begin{cases} \left(1 + \frac{(1-p)a}{p}, 0\right) & \text{for } 0 \leq p < \frac{3-a}{4-a}, \\ \left(0, \frac{p}{1-p} + a\right) & \text{for } \frac{3-a}{4-a} \leq p \leq 1. \end{cases}$$

Now we compute the mean excess demand function for commodity x, say. The consumer who "jumps" at the price p is given by

$$3(1 - p) = p + (1 - p)a,\ a = \frac{3 - 4p}{1 - p}.$$

From this, the mean excess demand function for x can be calculated as

$$\begin{aligned} \zeta(p) &= \int_0^{\frac{3-4p}{1-p}} \left(1 + \frac{(1 - p)a}{p}\right) d\lambda - \int_0^1 d\lambda = \left(a + \frac{1 - p}{2p} a^2\right)_0^{\frac{3-4p}{1-p}} - 1 \\ &= \frac{3 - 4p}{1 - p} + \frac{(3 - 4p)^2}{2p(1 - p)} - 1 \\ &= \frac{1}{2p(1 - p)} \{2p(3 - 4p) + (3 - 4p)^2 - 2p(1 - p)\} \\ &= \frac{1}{2p(1 - p)} (10p^2 - 20p + 9), \end{aligned}$$

which is continuous for $0 < p < 1$, and the equilibrium price for which $\zeta(p) = 0$ is given by

$$10p^2 - 20p + 9 = 0,\ p = \frac{10 - \sqrt{10}}{10}.$$

From this observation, we are naturally led to

Definition 3.7. The endowment distribution of the economy $\mathcal{E} : A \to \mathcal{P} \times \mathbb{R}^\ell$, $\mathcal{E}(a) = (\prec_a, \omega(a))$ is said to be *dispersed* at $\boldsymbol{p} \in \mathbb{R}^\ell_+$, $\boldsymbol{p} \neq \boldsymbol{0}$ if and only if the measure $\nu_{\boldsymbol{p}}$ on $\mathbb{R}$ defined by

$$\nu_{\boldsymbol{p}}(B) = \lambda(\{a \in A |\ \boldsymbol{p}\omega(a) \in B\})$$

for every Borel set $B \in \mathcal{B}(\mathbb{R})$ does not give positive value to any particular value $w \in \mathbb{R}$, or

$$\nu_{\boldsymbol{p}}(\{w\}) = 0$$

for every $w \in \mathbb{R}$. The endowment distribution is said to be dispersed if it is dispersed at every $\boldsymbol{p} \in \mathbb{R}^\ell_+$, $\boldsymbol{p} \neq \boldsymbol{0}$.

Note that in Example 3.3, the endowment distribution is dispersed at every $\boldsymbol{p} = (p, q) \in \mathbb{R}^2_+$ but $\boldsymbol{p} = (p, 0)$. When the consumption set is not convex, the monotonicity of preferences is too strong, because it implicitly assumes that $X_a + \mathbb{R}^\ell_+ \subset X_a$. Instead, we assume *weak desirability*.

(WD) (Weak desirability): for every $\boldsymbol{x} = (x^t) \in X_a$ and $t = 1 \ldots \ell$, there exists a vector $\boldsymbol{y} \in X_a$ such that $y^t > x^t$ and $y^s \leq x^s$ for $s \neq t$, and $\boldsymbol{x} \prec_a \boldsymbol{y}$.

Let $\mathcal{P}_{wd}$ denote the set of all preference relations that satisfy the weak desirability, and $\mathcal{P}_{wd,lns} = \mathcal{P}_{wd} \cap \mathcal{P}_{lns}$, where $\mathcal{P}_{lns}$ is the set of all preferences that satisfy (LNS). In the remainder of this section, we assume that all consumers in the economy have the same consumption set $X \subset \mathbb{R}^\ell$, which is closed and bounded from below. We do not assume X to be convex. The *regularizing effect* for the mean demand by a large number of consumers now reads as follows:

Theorem 3.7. *Let $\mathcal{E} : A \to \mathcal{P} \times \mathbb{R}^\ell$ be an economy such that $\mathcal{E}(a) \subset \mathcal{P}_{wd,lns} \times X$ a.e. Suppose that the endowment distribution is dispersed. Then we have*

(a) *for every $\boldsymbol{p} \in \mathring{\Delta}$, there exists a λ-null set $A_p \subset A$ such that for all $a \in A \backslash A_p$, the individual demand correspondence $\phi(a, \cdot)$ of $\mathring{\Delta}$ to $\mathbb{R}^\ell$ is upper hemicontinuous at $\boldsymbol{p}$,*

(b) *the mean demand correspondence $\Phi : \mathring{\Delta} \to \mathbb{R}^\ell$, $\Phi(\boldsymbol{p}) = \int_A \phi(a, \boldsymbol{p}) d\lambda$ is compact, convex-valued, and upper hemicontinuous.*

Proof. Let $\boldsymbol{p} \neq \boldsymbol{0}$ in $\mathbb{R}^\ell_+$ be given. The open half-ball centered at $\boldsymbol{x} \in \mathbb{R}^\ell$ with radius $\delta > 0$ is defined by

$$HB_{\boldsymbol{p}}(\boldsymbol{x}, \delta) = \{\boldsymbol{z} \in \mathbb{R}^\ell |\ \|\boldsymbol{x} - \boldsymbol{z}\| < \delta,\ \boldsymbol{pz} < \boldsymbol{px}\}.$$

Define the subset $NC_{\boldsymbol{p}}$ of X and $CW_{\boldsymbol{p}}$ of $\mathbb{R}$ by

$$NC_{\boldsymbol{p}} = \{\boldsymbol{x} \in X |\ HB_{\boldsymbol{p}}(\boldsymbol{x}, \delta) \cap X = \emptyset \text{ for some } \delta > 0\},$$

and

$$CW_{\boldsymbol{p}} = \{w \in \mathbb{R} |\ w = \boldsymbol{px} \text{ for some } \boldsymbol{x} \in NC_{\boldsymbol{p}}\},$$

respectively. We call a point $\boldsymbol{z}$ in $HB_{\boldsymbol{p}}(\boldsymbol{x}, \delta) \cap X$ the *local cheaper point* of $\boldsymbol{x}$. $NC_{\boldsymbol{p}}$ is the set of consumption vectors that have no local cheaper points. In Example 3.3, the vector $(0, 3)$ has no local cheaper points. Then the consumer a such that $\boldsymbol{p}\omega(a) \in CW_{\boldsymbol{p}}$ possibly behaves discontinuously. The w in $CW_{\boldsymbol{p}}$ is called the *critical wealth level.* The fundamental lemma is the following.

Lemma 3.2. *The set $CW_{\boldsymbol{p}}$ is a countable set for every $\boldsymbol{p} \in \mathbb{R}^\ell$ with $\boldsymbol{p} \neq \boldsymbol{0}$.*

Proof. Suppose that $CW_{\boldsymbol{p}}$ is an uncountable set $CW_{\boldsymbol{p}} = \{w_\alpha\}$. Then there are uncountably many points $\{\boldsymbol{x}_\alpha\}$ in $NC_{\boldsymbol{p}}$ such that $\boldsymbol{px}_\alpha = w_\alpha$ for each $w_\alpha \in CW_{\boldsymbol{p}}$. Denote $C = \{\boldsymbol{x}_\alpha\}$ and let $\epsilon > 0$ be a fixed positive number. Since the space $\mathbb{R}^\ell$ is second countable, it can be covered by countably many closed balls with radius ϵ. Because C

is an uncountable subset of $\mathbb{R}^\ell$, at least one of these balls, say B_0, must contain uncountably many points in C. Set $C_0 = C \cap B_0$.

We now claim that there exists a point $\boldsymbol{x}$ in C_0 and a sequence $\{\boldsymbol{x}_n\}$ converging to $\boldsymbol{x}$ such that $\boldsymbol{p}\boldsymbol{x}_n < \boldsymbol{p}\boldsymbol{x}$ for all $n \in \mathbb{N}$. Suppose not. Then for every $\boldsymbol{x} \in C_0 \subset NC_{\boldsymbol{p}}$, there exists a positive number $\delta_{\boldsymbol{x}}$ such that $HB_{\boldsymbol{p}}(\boldsymbol{x}, \delta_{\boldsymbol{x}}) \cap C_0 = \emptyset$. For each positive integer $n \geq 1$, define a subset C_n of C_0 by

$$C_n = \{\boldsymbol{x} \in C_0 |\ HB_{\boldsymbol{p}}(\boldsymbol{x}, 1/n) \cap C_0 = \emptyset\}.$$

Then $C_0 = \cup_{n=1}^{\infty} C_n$. Because C_0 is uncountable, at least one of the C_ns, say C_N contains uncountably many points of C_0. Therefore, one can take a sequence of district points $\{\boldsymbol{z}_n\}$ in C_N such that $\boldsymbol{p}\boldsymbol{z}_n \neq \boldsymbol{p}\boldsymbol{z}_m$ for $n \neq m$. Because the set $C_N \subset C_0$ is bounded, we can take a converging subsequence of $\{\boldsymbol{z}_n\}$, hence without loss of generality, we can assume that for sufficiently large j and k with $j \neq k$, one has $\|\boldsymbol{z}_j - \boldsymbol{z}_k\| < 1/2N$ and $\boldsymbol{p}\boldsymbol{z}_j \neq \boldsymbol{p}\boldsymbol{z}_k$, say $\boldsymbol{p}\boldsymbol{z}_j < \boldsymbol{p}\boldsymbol{z}_k$. Hence, one obtains $\boldsymbol{z}_j \in HB_{\boldsymbol{p}}(\boldsymbol{z}_k, 1/N)$. This contradicts the fact that both $\boldsymbol{z}_j$ and $\boldsymbol{z}_k$ are distinct elements of the set C_N. Therefore, our claim is verified.

Take a point $\boldsymbol{x} \in C_0$ and a sequence $\{\boldsymbol{x}_n\}$ in C_0 that converges to $\boldsymbol{x}$ and $\boldsymbol{p}\boldsymbol{x}_n < \boldsymbol{p}\boldsymbol{x}$ for each n. Because $\boldsymbol{x} \in NC_{\boldsymbol{p}}$, there is a positive number δ such that $HB_{\boldsymbol{p}}(\boldsymbol{x}, \delta) \cap X = \emptyset$. But for n sufficiently large, $\boldsymbol{x}_n \in HB_{\boldsymbol{p}}(\boldsymbol{x}, \delta)$, a contradiction (Figure 3.7). □

We now continue the proof of Theorem 3.7. For given $\boldsymbol{p} \neq \boldsymbol{0}$ in $\mathbb{R}^\ell_+$, define a set $A_{\boldsymbol{p}} \subset A$ by

$$A_{\boldsymbol{p}} = \{a \in A |\ \boldsymbol{p}\omega(a) \in CW_{\boldsymbol{p}}\}.$$

Then by Lemma 3.2, $CW_{\boldsymbol{p}}$ is a countable set $\{w_0, w_1 \dots\}$. Hence, $\lambda(A_{\boldsymbol{p}}) = \sum_{n=0}^{\infty} \lambda(\{a \in A |\ \boldsymbol{p}\omega(a) = w_n\}) = 0$, because the endowment distribution is dispersed. We will show that the relation $\{(\boldsymbol{q}, \boldsymbol{x}) \in \mathbb{R}^\ell \times \mathbb{R}^\ell |\ \boldsymbol{x} \in \phi(a, \boldsymbol{q})\}$ is closed at $\boldsymbol{p}$ for each $a \in A \backslash A_{\boldsymbol{p}}$. Let $(\boldsymbol{p}_n.\boldsymbol{x}_n)$ be a sequence such that $\boldsymbol{p}_n \to \boldsymbol{p}$, $\boldsymbol{x}_n \to \boldsymbol{x}$ and $\boldsymbol{x}_n \in \phi(a, \boldsymbol{p}_n)$ for all $n \in \mathbb{N}$. For each $n \in \mathbb{N}$, $\boldsymbol{p}_n\boldsymbol{x}_n \leq \boldsymbol{p}_n\omega(a)$. Passing to the limit, we have $\boldsymbol{p}\boldsymbol{x} \leq \boldsymbol{p}\omega(a)$. For $\boldsymbol{z} \in X$ such that $\boldsymbol{p}\boldsymbol{z} < \boldsymbol{p}\omega(a)$, one has $\boldsymbol{p}_n\boldsymbol{z} < \boldsymbol{p}_n\omega(a)$ for all n large enough. Thus, $\boldsymbol{x}_n \succsim_a \boldsymbol{z}$ for all n large

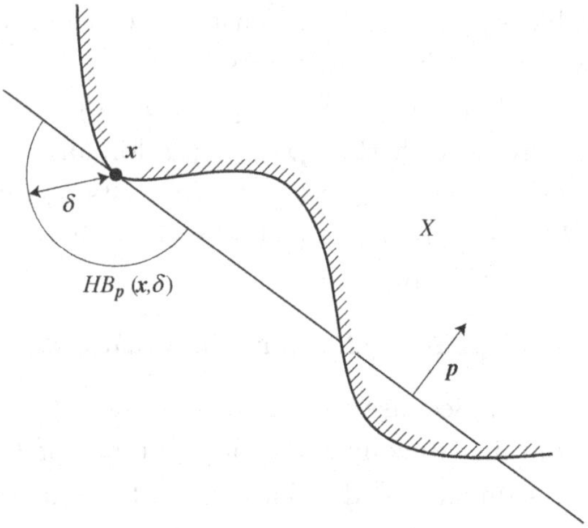

Figure 3.7. $\boldsymbol{x} \in NC_p$.

enough. By the continuity of $\succsim_a$, it follows that $\boldsymbol{x} \succsim_a \boldsymbol{z}$. For $\boldsymbol{z} \in X$ with $\boldsymbol{pz} = \boldsymbol{p}\omega(a)$, there exists a sequence $\{\boldsymbol{z}_n\}$ such that $\boldsymbol{z}_n \to \boldsymbol{z}$ and $\boldsymbol{pz}_n < \boldsymbol{p}\omega(a)$, because $a \notin A_{\boldsymbol{p}}$. For every $n \in \mathbb{N}$, we have $\boldsymbol{x} \succsim_a \boldsymbol{z}_n$. Hence, again by the continuity of $\succsim_a$, one sees that $\boldsymbol{x} \succsim_a \boldsymbol{z}$. Therefore, the relation $\phi(a, \cdot)$ is closed at $\boldsymbol{p}$.

Let $\boldsymbol{p} \in \mathring{\Delta}$ be given. Because $\omega(a) \in X$, the budget set $\beta(a, \boldsymbol{p}) = \{\boldsymbol{x} \in X|\ \boldsymbol{px} \leq \boldsymbol{p}\omega(a)\}$ is non-empty and compact. By Proposition 2.1, we have $\phi(a, \boldsymbol{p}) \neq \emptyset$. Let V be a compact neighborhood of $\boldsymbol{p} \gg \boldsymbol{0}$ with strictly positive vectors and define $\bar{\pi} > 0$ and $\underline{\pi} > 0$ by $\bar{\pi} = \max\{q^t|\ \boldsymbol{q} = (q^t) \in V,\ t = 1 \dots \ell\}$ and $\underline{\pi} = \min\{q^t|\ \boldsymbol{q} = (q^t) \in V,\ t = 1 \dots \ell\}$, respectively. Let $b < 0$ be a lower bound of X, or $(b \dots b) \leq \boldsymbol{x}$ for all $\boldsymbol{x} \in X$. Define a function $h : A \to \mathbb{R}$ by $h(a) = \underline{\pi}^{-1}\left(\sum_{t=1}^{\ell} \omega^t(a) - \ell\bar{\pi}b\right)$. Then $b \leq \phi^t(a, \boldsymbol{q}) \leq h(a)$ for all $\boldsymbol{q} \in V$, $t = 1 \dots \ell$. Because $\phi(a, \cdot)$ is closed at $\boldsymbol{p} \in V$, it is upper hemicontinuous. This proves (a).

By Proposition 3.2, the correspondence $\phi(a, \boldsymbol{p})$ has a measurable graph and $h(a)$ is clearly integrable. Therefore, by Theorems F.11 and F.12 and Corollaries F.1 and F.2, the mean demand correspondence $\int_A \phi(a, \boldsymbol{p}) d\lambda$ is non-empty, compact, and convex-valued and upper hemicontinuous. This proves (b). □

Now we can present the existence of equilibrium without the convexity assumptions.

Theorem 3.8. *Let $\mathcal{E} : A \to \mathcal{P} \times \mathbb{R}^\ell$ be an economy such that $\mathcal{E}(a) \subset \mathcal{P}_{wd,lns} \times X$ a.e. Suppose that the endowment distribution is dispersed. Then there exists a competitive equilibrium for $\mathcal{E}$.*

Proof. The proof goes exactly as that of Theorem 3.4. Consider the mean excess demand correspondence from $\mathring{\Delta}$ to $\mathbb{R}^\ell$,

$$\zeta(\boldsymbol{p}) = \int_A \phi(a, \boldsymbol{p}) d\lambda - \int_A \omega(a) d\lambda,$$

for every $\boldsymbol{p} \in \mathring{\Delta}$. We will show that the correspondence ζ satisfies conditions (i), (ii), and (iii), which are needed to apply Theorem D.4. Because preferences are locally non-satiated, we have $\boldsymbol{px} = \boldsymbol{p}\omega(a)$ for each $\boldsymbol{x} \in \phi(a, \boldsymbol{p})$. Hence, property (i) follows from Proposition F.11. We have already shown property (ii) in Theorem 3.7 (a). To prove (iii), let $(\boldsymbol{p}_n)$ be a sequence in $\mathring{\Delta}$ converging to $\boldsymbol{p} = (p^t)$ with $p^s = 0$ for some s. It suffices to show that

$$\inf \left\{ \sum_{t=1}^{\ell} x^t \,\middle|\, \boldsymbol{x} = (x^t) \in \phi(a, \boldsymbol{p}_n) \right\} \to \infty \text{ as } n \to \infty$$

for every $a \in A \backslash A_{\boldsymbol{p}}$, because $\lambda(A_{\boldsymbol{p}}) = 0$. Suppose not. Then there exists a bounded set B in $\mathbb{R}^\ell$ such that $\phi(a, \boldsymbol{p}_n) \cap B \neq \emptyset$ for infinitely many ns. Take $\boldsymbol{x}_n \in \phi(a, \boldsymbol{p}_n) \cap B$ for each n. We can take a converging subsequence, still denoted by $\{\boldsymbol{x}_n\}$ with $\boldsymbol{x}_n \to \boldsymbol{x}$. In the proof of Theorem 3.7 (b), we showed that the relation $\phi(a, \cdot)$ is closed at $\boldsymbol{p}$, because $a \notin A_{\boldsymbol{p}}$. Hence, $\boldsymbol{x} \in \phi(a, \boldsymbol{p})$. By the weak desirability condition (WD), there exists $\boldsymbol{y} \in X$ such that $y^s > x^s$, $y^t \leq x^t$ for $t \neq s$, and $\boldsymbol{x} \prec_a \boldsymbol{y}$. Because $p^s = 0$, $\boldsymbol{py} \leq \boldsymbol{p}\omega(a)$. A contradiction. □

3.6 On the Law of Demand

In this section, we present a remarkable observation on the law of demand, due to Hildenbrand [95]. Throughout this section, we assume that the consumer's income $w \in \mathbb{R}_+$ is independent of the market price vector $\boldsymbol{p} \in \mathbb{R}^{\ell}_{+}$, hence the budget set and the demand set of a consumer are given by

$$\beta(\boldsymbol{p}, w) = \{\boldsymbol{x} \in X |\, \boldsymbol{px} \leq w\},$$
$$\phi(\boldsymbol{p}, w) = \{\boldsymbol{x} \in X |\, \boldsymbol{px} \leq w, \boldsymbol{x} \succsim \boldsymbol{z} \text{ whenever } \boldsymbol{pz} \leq w\},$$

respectively. Suppose that the distribution of income w is given by the *density function* $\rho(w)$ over the unit interval $[0, 1]$. The market demand function $\Phi(\boldsymbol{p}) = (\Phi^t(\boldsymbol{p}))$ is then defined by

$$\Phi(\boldsymbol{p}) = \int_0^1 \phi(\boldsymbol{p}, w)\rho(w)dw.$$

The *law of demand* means that the market demand function for each commodity t is decreasing in the price of the commodity t when the prices of the other commodities $s \neq t$ remain constant

$$\frac{\partial \Phi^t(\boldsymbol{p})}{\partial p^t} \leq 0.$$

Hildenbrand [95] found that when the density function $\rho(w)$ is decreasing, then the law of demand holds. This fact is not obvious when one sees the behavior of the individual demand function $\phi(\boldsymbol{p}, w)$, which is described by the Slutzky equation (Theorem 2.3)

$$\frac{\partial}{\partial p^t}\phi^t(\boldsymbol{p}, w) = \frac{\partial}{\partial p^t}\phi^t_{\boldsymbol{x}}(\boldsymbol{p}) - \phi(\boldsymbol{p}, w)\frac{\partial}{\partial w}\phi^t(\boldsymbol{p}, w).$$

By Theorem 2.2 (b), we know that the *substitution effect* is negative,

$$\frac{\partial}{\partial p^t}\phi^t_{\boldsymbol{x}}(\boldsymbol{p}) \leq 0;$$

however, the sign of the *income effect* is undetermined,

$$\phi(\boldsymbol{p}, w)\frac{\partial}{\partial w}\phi^t(\boldsymbol{p}, w) \gtreqless 0.$$

Remarkably, however, we can determine the sign of the total income effect and it is negative, because

$$
\begin{aligned}
&-\int_0^1 \phi(\boldsymbol{p}, w)\frac{\partial}{\partial w}\phi^t(\boldsymbol{p}, w)\rho(w)dw \\
&= -(1/2)\int_0^1 \frac{\partial}{\partial w}(\phi^t(\boldsymbol{p}, w))^2\rho(w)dw \\
&= (1/2)\int_0^1 (\phi^t(\boldsymbol{p}, w))^2\rho'(w)dw \leq 0,
\end{aligned}
$$

under the assumptions that $\rho'(w) \leq 0$, $\rho(1) = 0$, and $\phi^t(\boldsymbol{p}, 0) = 0$.

This simple observation can be generalized as follows. We say that the market demand function $\Phi(\boldsymbol{p})$ is *monotone* if for any price vectors $\boldsymbol{p}$ and $\boldsymbol{q}$,

$$(\boldsymbol{q} - \boldsymbol{p})(\Phi(\boldsymbol{q}) - \Phi(\boldsymbol{p})) \leq 0.$$

It is said to be *strictly monotone* if for any price vectors $\boldsymbol{p}$ and $\boldsymbol{q}$ with $\boldsymbol{q} \neq \boldsymbol{p}$,

$$(\boldsymbol{q} - \boldsymbol{p})(\Phi(\boldsymbol{q}) - \Phi(\boldsymbol{p})) < 0.$$

The monotonicity of Φ clearly implies that the law of demand holds for all markets, $t = 1 \ldots \ell$.

When the individual demand function $\phi(a, \cdot)$ is continuously differentiable and the density function $\rho(w)$ of the wealth level w is continuous, then the mean demand function $\Phi(\boldsymbol{p}) = \int_0^1 \phi(\boldsymbol{p}, w)\rho(w)dw$ is continuously differentiable and it is easily verified by the Taylor expansion that the mean demand $\Phi(\boldsymbol{p})$ is strictly monotone if its Jacobian matrix is negative definite, or

$$\boldsymbol{v}\partial\Phi(\boldsymbol{p})\boldsymbol{v} = \sum_{t=1}^{\ell}\sum_{s=1}^{\ell} v^t v^s \partial_s \Phi^t(\boldsymbol{p}) < 0$$

for every $\boldsymbol{v} = (v^t) \neq \boldsymbol{0}$. The following theorem is due to [95].

Theorem 3.9. *Suppose that the individual demand function $\phi(\mathbf{p}, w)$ is continuously differentiable in $\mathbf{p}$ and the substitution matrix*

$S_{\boldsymbol{x}}(\boldsymbol{p}) = (\partial_s \phi^t_{\boldsymbol{x}}(\boldsymbol{p}))$ is of the rank $\ell - 1$. If the density $\rho(w)$ is continuous and decreasing, then the function $\Phi(\boldsymbol{p})$ is negative definite.

Proof. Let $S_{\boldsymbol{x}}(\boldsymbol{p}) = (\partial_s \phi^t_{\boldsymbol{x}}(\boldsymbol{p}))$ be the substitution matrix (see Theorem 2.2), and let $A(\boldsymbol{p}, w)$ be the matrix defined by $A(\boldsymbol{p}, w) = (a^s_t) = (\phi^s(\boldsymbol{p}, w)\partial_w \phi^t(\boldsymbol{p}, w))$, $s, t = \ldots \ell$. Then the Slutzky equation can be written as

$$\partial\phi(\boldsymbol{p}, w) = S_{\boldsymbol{x}}(\boldsymbol{p}) - A(\boldsymbol{p}, w).$$

Integrating over w, we have

$$\partial\Phi(\boldsymbol{p}) = \int_0^1 (S_{\boldsymbol{x}}(\boldsymbol{p}) - A(\boldsymbol{p}, w))\rho(w)dw.$$

By Theorem 2.2 (b), the substitution matrix is negative semi-definite. Because we assumed that $\operatorname{rank} S_{\boldsymbol{x}}(\boldsymbol{p}) = \ell - 1$ and $\boldsymbol{p}S_{\boldsymbol{x}}(\boldsymbol{p}) = \boldsymbol{0}$ by Proposition 2.8, it follows that

$$\boldsymbol{v}S_{\boldsymbol{x}}(\boldsymbol{p})\boldsymbol{v} < 0$$

for every $\boldsymbol{v} \in \mathbb{R}^\ell$, which is not collinear with $\boldsymbol{p}$, or which is not of the form $\boldsymbol{v} = \tau\boldsymbol{p}$, $\tau \neq 0$. Hence, the matrix $\int_0^1 S_{\boldsymbol{x}}(\boldsymbol{p})\rho(w)dw$ is also negative definite for all $\boldsymbol{v}$, which is not collinear with $\boldsymbol{p}$. Therefore, it remains to show that

$$\boldsymbol{v}\left(\int_0^b A(\boldsymbol{p}, w)\rho(w)dw\right)\boldsymbol{v} \geq 0$$

for every $\boldsymbol{v} \in \mathbb{R}^\ell$ with strict inequality for $\boldsymbol{p} = \tau\boldsymbol{v}$, $\tau \neq \boldsymbol{0}$. Now

$$\begin{aligned}
\boldsymbol{v}\left(\int_0^b A(\boldsymbol{p}, w)\rho(w)dw\right)\boldsymbol{v} &= \sum_{s=1}^{\ell}\sum_{t=1}^{\ell} v^s v^t \int_0^b \phi^s(\boldsymbol{p}, w)\partial_w \phi^t(\boldsymbol{p}, w)\rho(w)dw \\
&= \int_0^b \left(\sum_{s=1}^{\ell} v^s \phi^s(\boldsymbol{p}, w)\right)\left(\sum_{t=1}^{\ell} v^t \partial_w \phi^t(\boldsymbol{p}, w)\right)\rho(w)dw \\
&= \int_0^b (\partial_w \phi(\boldsymbol{p}, w)\boldsymbol{v})(\phi(\boldsymbol{p}, w)\boldsymbol{v})\rho(w)dw \\
&= (1/2)\int_0^b \partial_w \left(\phi(\boldsymbol{p}, w)\boldsymbol{v}\right)^2 \rho(w)dw.
\end{aligned}$$

Because ρ is decreasing, it follows from the intermediate value theorem of the elementary calculus that there exists $0 \leq \xi \leq b$ such that

$$\begin{aligned}\int_0^b \partial_w(\phi(\boldsymbol{p},w)\boldsymbol{v})^2\rho(w)dw &= \rho(0)\int_0^\xi \partial_w(\phi(\boldsymbol{p},w)\boldsymbol{v})^2 dw \\ &= \rho(0)(\phi(\boldsymbol{p},\xi)\boldsymbol{v})^2 \geq 0.\end{aligned}$$

If $\boldsymbol{v} = \tau\boldsymbol{p}$, then $\phi(\boldsymbol{p},w)\boldsymbol{v} = \tau w$. Hence,

$$\boldsymbol{v}\left(\int_0^b A(\boldsymbol{p},w)\rho(w)dw\right)\boldsymbol{v} = (1/2)\int_0^b \partial_w(\tau w)^2\rho(w)dw > 0.$$

Therefore, $\Phi(\boldsymbol{p})$ is negative definite. □

3.7 Notes

Section 3.2: A basic reference is Hildenbrand [94], Chapter 1. We also recommend Hildenbrand and Kirman [96] as a readable introductory textbook.

Section 3.3: The idea of the proof of the equilibrium existence theorem that truncates the consumption set and invokes Fatou's lemma in ℓ-dimension is originally due to Aumann [17] and Schmeidler [225]. We owe Yamazaki [265] for the proof of Theorem 3.3. The proof of Theorem 3.5 is due to [94]. For the case that the consumption sets can be different among consumers, see Yamazaki [266].

Section 3.4: Theorem 3.6 is originally due to Aumann [15]. For the proof, we follow Hildenbrand [94]. Chapter 3 of his book presents a formulation of the limit theorem of the core in terms of the sequences of economies. For the core equivalence theorem for the non-convex economies, see Khan and Yamazaki [122].

Vind [259] gave an alternative formulation of a continuum (large) economy. Let $(A, \mathcal{A})$ be a measurable space of consumers and $X \subset \mathbb{R}^\ell$ be a common consumption set. An allocation ξ is defined as a *vector measure* (σ-countable set function with values on $\mathbb{R}^\ell$) $\xi : \mathcal{A} \to X$, $C \mapsto \xi(C)$. Preference $\prec_C \subset X \times X$ is also defined for each coalition C

as follows. First, he postulated a function $\mathscr{S} : X \times X \to \mathcal{A}$. The interpretation of $\mathscr{S}(\xi, \zeta) \in \mathcal{A}$ is the set of the consumers who prefer ζ to ξ. Then we define $\xi \prec_C \zeta$ if and only if $C \subset \mathscr{S}(\xi, \zeta)$. The core and competitive equilibrium are defined naturally and he proved their equivalence under reasonable assumptions. References [7, 8] proved the existence of competitive allocations and their equivalence with core on a non-atomic Boolean algebra, and Reference [52] discussed the conditions under which Aumann–Hildenbrand approach and Vind approach are equivalent.

Section 3.5: The idea of dispersed endowment is originally due to Mas-Colell [146] and was established by Yamazaki [265]. Theorems 3.7 and 3.8 and Lemma 3.2 are due to [265]. In the following, we discuss the measurability of $\mathcal{P}_{wd}$ following Yamazaki [267].

Let $(\Omega, \mathcal{M})$ be a measurable space, and let λ be a (finite) measure on Ω. Let $(\Omega, \mathcal{M}_\lambda, \bar{\lambda})$ be the completion of $(\Omega, \mathcal{M}, \lambda)$ (see Appendix G). Define $\mathcal{M}_u = \cap\{\mathcal{M}_\lambda | \ \lambda$ is a finite measure on $(\Omega, \mathcal{M})\}$. The measurable space $(\Omega, \mathcal{M}_u)$ is called the universal completion of $(\Omega, \mathcal{M})$, and each set in $\mathcal{M}_u$ is called *universally measurable.*

For a topological space X, let $(X, \mathcal{B}_u(X))$ be the universal completion of the measurable space $(X, \mathcal{B}(X))$. Then we can prove the following theorem.

Theorem 3.10. *The set $\mathcal{P}_{wd}$ is universally measurable.*

To prove this theorem, we need some preparations. Let Ω be a set. The concept of the sequences of subsets of Ω is generalized as follows. Let $\mathcal{N}_f$ be the set of all finite sequences of positive integers, $\mathcal{N}_f = \{(n_0 \dots n_k) | \ n_j \in \mathbb{N}, 0 \leq j \leq k, k \in \mathbb{N}\}$. Consider a map from $\mathcal{N}_f$ to 2^Ω, which maps a finite sequence to a subset of Ω, $(n_0 \dots n_k) \mapsto B_{(n_0 \dots n_k)} \subset \Omega$. For the image $\{B_{(n_0 \dots n_k)}\}$ of this map, we can associate the set $B = \cup_{\boldsymbol{n}} \cap_{k \geq 0} B_{(n_0 \dots n_k)}$, where the union is taken over all infinite sequences $\boldsymbol{n} = (n_0 \dots n_k, n_{k+1} \dots)$ of non-negative integers. We call this the *Suslin operation.* Note that the countable unions and intersections are examples of the Suslin operation. Let $\mathcal{S}$ be a family of sets. We denote by $\mathcal{S}_s$ the family of all sets obtained from the sets of $\mathcal{S}$ by the Suslin operations. We call every element S of $\mathcal{S}_s$ a *Suslin set*. Then we have

Math. Theorem 1. *Let* $(\Omega, \mathcal{M})$ *be a measurable space. Then* $\mathcal{M}_s \subset \mathcal{M}_u$.

Let X be a topological space. A subset A of X is called an *analytic set* if there exists a complete separable metric space C and a continuous map $f : C \to X$ such that $A = f(C)$. Let $\mathcal{A}(X)$ be the set of all analytic sets of X. Then it follows that

Math. Theorem 2. *If* X *is a complete and separable metric space, then* $\mathcal{B}(X) \subset \mathcal{A}(X)$, *or a Borel set is an analytic set.*

Math. Theorem 3. *Let* $\mathcal{F}(X)$ *be the family of all closed subsets of a complete separable metric space* X. *Then* $\mathcal{A}(X) = \mathcal{F}(X)_s$.

We know that the inverse image of a Borel set by a measurable map is a Borel set. However, it may not be the case that the image of a Borel set by a measurable map is also a Borel set. For the analytic sets, however, the following proposition holds.

Math. Theorem 4. *Let* X, Y *be complete and separable metric spaces, and* $f : (X, \mathcal{B}(X)) \to (Y, \mathcal{B}(Y))$ *be a Borel measurable map. Then*

(*i*) *for every* $B \in \mathcal{A}(Y)$, $f^{-1}(B) \in \mathcal{A}(X)$,

(*ii*) *for every* $C \in \mathcal{A}(X)$, $f(C) \in \mathcal{A}(Y)$.

Finally, we have the following theorem:

Math. Theorem 5. *Let* $(\Omega, \mathcal{M})$ *be a measurable space,* X *a complete and separable metric space. Let* $pr_\Omega : \Omega \times X \to \Omega$ *be a projection map. Then for every* $C \in (\Omega \times X)_s$, *it follows that* $pr_\Omega(C) \in \mathcal{M}_s$.

For details of the Suslin operation and related concepts and results, see [59, 187, 213]. In [187], however, the Suslin operation is called the *operation (A)*.

Then we can prove the Theorem as follows. For given $t(=1\ldots\ell)$ and $\boldsymbol{x}=(x^t)\in\mathbb{R}^\ell$, define the subsets of $\mathbb{R}^\ell$ by

$$\mathscr{Q}^t(\boldsymbol{x})=\{\boldsymbol{z}\in\mathbb{R}^\ell|\ z^t>x^t, z^s\leq x^s \text{ for } s\neq t\}.$$

Then setting for each $t=1\ldots\ell$,

$$\mathcal{Q}^t=\{(X,\prec)\in\mathcal{P}|. \text{ There exists } \boldsymbol{x}\in X \text{ such that for each } \boldsymbol{z}\in\mathscr{Q}^t(\boldsymbol{x})\cap X, \boldsymbol{x}\succsim\boldsymbol{z}\},$$

we have $\mathcal{P}_{wd}=\mathcal{P}\backslash\cup_{t=1}^{\ell}\mathcal{Q}^t$. Therefore, it suffices to show that each $\mathcal{Q}^t$ is universally measurable. Fix a t and for each non-negative integer $i\in\mathbb{N}$, define

$$\mathscr{Q}_i^t(\boldsymbol{x})=\left\{\boldsymbol{z}\in\mathbb{R}^\ell\middle|\ z^t\geq x^t+\frac{1}{i+1},\ z^s\leq x^s \text{ for } s\neq t\right\},$$

$$\mathscr{H}_i=\{(X,\prec,\boldsymbol{x})\in\mathcal{P}\times\mathbb{R}^\ell|\ \mathscr{Q}_i^t(\boldsymbol{x})\cap X\neq\emptyset\}.$$

They induce the correspondence $\varphi_i:\mathscr{H}_i\to\mathbb{R}^\ell$ defined by $\varphi_i(X,\prec,\boldsymbol{x})=\mathscr{Q}_i^t\cap X$. The correspondences $(X,\prec,\boldsymbol{x})\to X$ and $(X,\prec,\boldsymbol{x})\to\mathscr{Q}_i^t(\boldsymbol{x})$ are closed, so is the correspondence φ_i, hence it is measurable. By Proposition F.9, there exists a countable family of measurable mappings $\{f_{ij}\}_{j\in\mathbb{N}}$ of $\mathscr{H}_i\to\mathbb{R}^\ell$ such that

$$\varphi_i(X,\prec,\boldsymbol{x})=\overline{\{f_{ij}(X,\prec,\boldsymbol{x})|\ j\in\mathbb{N}\}}.$$

Then defining for $i,j\in\mathbb{N}$,

$$\mathcal{P}_{ij}=\{(X,\prec)\in\mathcal{P}|\ \boldsymbol{x}\succsim f_{ij}(X,\prec,\boldsymbol{x}) \text{ for some } \boldsymbol{x}\in X\},$$

we have $\mathcal{Q}^t=\cap_{i=0}^{\infty}\cap_{j=0}^{\infty}\mathcal{P}_{ij}$.

Indeed, it is obvious that $\mathcal{Q}^t\subset\cap_{i=0}^{\infty}\cap_{j=0}^{\infty}\mathcal{P}_{ij}$. Suppose that there exists $(X,\prec)$ such that $(X,\prec)\in\cap_{i=0}^{\infty}\cap_{j=0}^{\infty}\mathcal{P}_{ij}$ and $(X,\prec)\notin\mathcal{Q}^t$. Then for every $\boldsymbol{x}\in X$, there exists $\boldsymbol{z}\in\mathscr{Q}_i^t(\boldsymbol{x})\cap X$ such that $\boldsymbol{x}\prec\boldsymbol{z}$. Then we can take some $\epsilon>0$ such that $\boldsymbol{z}\in\mathscr{Q}_i^t(\boldsymbol{x})\cap X$ and $\boldsymbol{x}\prec B(\boldsymbol{z},\epsilon)\cap X$. Because

$$\mathscr{Q}_i^t(\boldsymbol{x})\cap X=\overline{\{f_{ij}(X,\prec,\boldsymbol{x})|\ j\in\mathbb{N}\}},$$

it follows that $\boldsymbol{x}\prec f_{ij}(X,\prec,\boldsymbol{x})$ for some j, contradicting that $(X,\prec)\in\mathcal{P}_{ij}$ for all i,j. Therefore, the proof is complete if we show that $\mathcal{P}_{ij}$

is universally measurable. Define a relation $F_{ij} : \mathcal{P} \to \mathbb{R}^\ell$ by

$$F_{ij} = \{(X, \prec, \boldsymbol{x}) \in \mathcal{P} \times \mathbb{R}^\ell | \ \boldsymbol{x} \succsim f_{ij}(X, \prec, \boldsymbol{x})\},$$

and define a measurable mapping $g_{ij} : \mathscr{H}_i \to \mathcal{P} \times \mathbb{R}^\ell \times \mathbb{R}^\ell$ by

$$g_{ij}(X, \prec, \boldsymbol{x}) = (X, \prec, f_{ij}(X, \prec, \boldsymbol{x}), \boldsymbol{x}).$$

Obviously, $F_{ij} = g_{ij}^{-1}(\{(X, \prec, \boldsymbol{y}, \boldsymbol{x}) | \ \boldsymbol{x} \succsim \boldsymbol{y}\})$. Because the set $\{(X, \prec, \boldsymbol{y}, \boldsymbol{x}) | \ \boldsymbol{x} \succsim \boldsymbol{y}\}$ is closed, F_{ij} is a Borel set in $\mathcal{P} \times \mathbb{R}^\ell$. Because $\mathcal{P}_{ij} = \{(X, \prec) \in \mathcal{P} | \ F_{ij} \neq \emptyset\}$, it follows from Math. Theorems 1, 2, 3, and 5 that the set $\mathcal{P}_{ij}$ is a $\mathcal{B}(\mathcal{P})$-Suslin set; hence, it is a universally measurable set. □

Section 3.6: For Theorem 3.9, we entirely follow Hildenbrand [95].

Chapter 4

Economies with Infinitely Many Commodities

4.1 Markets with Infinitely Many Goods

In this chapter, we discuss the markets where infinitely many commodities are traded. As noted in Section 2.1, a commodity is distinguished from another based on its physical characteristics, location, and the time when it is available to traders. If any of these properties are indexed by infinitely many parameters, the commodity space will naturally have infinite dimensions.

Let us start by considering the available time period for commodities. As we quoted in Section 1.7, Debreu [51] stated "there are conceptual difficulties in postulating a predetermined instant beyond which all economic activity either ceases or is outside the scope of analysis." Therefore, a natural candidate for the commodity space to handle this situation is a sequence space for a discrete time model or a function space when the model is in continuous time. In the following, we discuss the case of discrete time models, and thus we work with the spaces of sequences. The commodity vector is represented by a sequence $\boldsymbol{x} = (x^0, x^1 \dots x^t \dots)$. The tth coordinate x^t of $\boldsymbol{x}$ denotes the x^t amount of goods available to the market on the date (period) t. All of the contracts are performed "now" or on the initial date 0. Hence, for the price vector $\boldsymbol{p}$, which is also given by a sequence $\boldsymbol{p} = (p^0, p^1 \dots p^t \dots)$, the tth coordinate p^t of $\boldsymbol{p}$

denotes the p^t amount of money for one unit of the good available in period t. It should be noted that the money is paid "now" and not at the date t.

To describe this type of market structure, the most natural space appears to be $\mathbb{R}^\infty$, which is the space of all sequences endowed with the topology of pointwise convergence or the product topology. It seems natural that the commodity space contains all the streams of commodities because we do not have any particular *a priori* reason to exclude some commodity stream from the analysis. However, the space $\mathbb{R}^\infty$ is technically not readily manageable. This space is excessively large because the dual space of $\mathbb{R}^\infty$ is the space comprising the vectors with all but finite number for the 0 coordinates. Thus, if we require that all the commodity bundles $\boldsymbol{x}$ should have finite value $\boldsymbol{px} < \infty$, then we have to restrict the price vectors $\boldsymbol{p} = (p^t)$ within the class of vectors where only a finite number of p^t are non-zero. If we allow price vectors with infinitely many non-zero coordinates, the market value of some of the commodity vectors will be infinite, $\boldsymbol{px} = \pm\infty$. Clearly, in this type of model, it would be technically difficult to handle the budget constraints of the consumers and/or the profit maximization for the firms.

Bewley [27] proposed the space ℓ^∞, which is the space of all bounded sequences as the commodity space. The space ℓ^∞ is a Banach space with the supremum norm. The dual space of ℓ^∞ is the space *ba*, which is the set of finitely additive set functions on $\mathbb{N}$ with the total variation norm (Appendix G). The space *ba* is not a sequence space. However, from an economic viewpoint, it is more natural to consider that the price vector is an element of the subspace *ca* of *ba*, which is the space of countably additive set functions on $\mathbb{N}$. The space *ca* is identified with the space ℓ^1, which is also a Banach space with the norm

$$\|\boldsymbol{p}\| = \sum_{t=0}^{\infty} |p^t|, \ \boldsymbol{p} = (p^t) \in \ell^1,$$

where we set $p^t = p(\{t\})$. Appendix G presents the mathematical properties of these spaces and related results. In Section 4.2, we define and prove the existence of the competitive equilibrium for an exchange economy with the commodity space ℓ^∞.

Another case where the commodity space is an infinite dimensional vector space is the model of *commodity differentiation*. In this case, the commodities are described by specifying their *characteristics*, which are points in an *a priori* defined compact metric space K. In particular, a commodity bundle is represented by a distribution (finite measure) on K. For example, let $K = [0, A] \times [0, P]$, where $A > 0$ and $P > 0$ are sufficiently large numbers, and the commodities under consideration are all foods indexed by their vitamin A and protein contents. Then, one unit of carrot is represented by the *Dirac measure*:

$$\delta_c(B) = \begin{cases} 1 & \text{if } c \in B, \\ 0 & \text{otherwise,} \end{cases}$$

for every Borel set $B \subset K$, where the point $c \in K$ representing the carrot will have a large first coordinate, and similarly, steak is represented by δ_s, where $s \in K$ has a large second coordinate.

More sophisticated (and delicious?) types of "food" such as an expensive French dinner can be represented by a more complicated measure, such as $\sum_{t_i \in K} x_i \delta_{t_i}$, where each $t_i \in K$ represents a particular food material used in French cuisine. Every Borel measure $\boldsymbol{x} \in ca(K)$ is a weak* limit of linear combinations of the Dirac measures (Proposition G.12), or $\sum_{t_i \in K} x_i \delta_{t_i} \to \boldsymbol{x}$, so it is natural to postulate that the "food" in general is defined by a Borel measure $\boldsymbol{x}$ on the set of characteristics K. Clearly, this example shows why the measures are more appropriate than the (measurable) functions on K for describing commodity differentiation situations. In general, the choice problem is not how much of each commodity characteristic to buy, but instead, the problem comprises which commodity bundle to buy.

Let $ca(K)$ be the set of countably additive set functions (or measure) on K. Then, the dual space of $ca(K)$ is $C(K)$, which is the set of continuous (real-valued) functions on K. Therefore, in the models with the commodity space $ca(K)$, the price vector is a continuous function on K, and for a commodity vector $\boldsymbol{x} \in ca(K)$ and a price vector $\boldsymbol{p} = p(t) \in C(K)$, the market value of $\boldsymbol{x}$ evaluated by $\boldsymbol{p}$ is given by the inner product

$$\boldsymbol{px} = \int_K p(t)d\boldsymbol{x}.$$

We discuss the model of commodity differentiation for an exchange economy in Section 4.3. The proof of the existence of the competitive equilibria for these models with infinite dimensional commodity spaces is obtained by reducing them to the proof for the classical (finite) economies discussed in Chapter 2. At least two alternative proof methods are suitable.

First, we consider the projection of the original economies $\mathcal{E}^\infty$ onto ℓ-dimensional subeconomies $\mathcal{E}^\ell$, $\ell = 1, 2 \ldots$, i.e., the commodity space L_ℓ of the economy $\mathcal{E}^\ell$ is the linear subspace of the commodity space L of $\mathcal{E}^\infty$, which is spanned by (at least) ℓ vectors $\boldsymbol{x}_1 \ldots \boldsymbol{x}_\ell$ in L. The preferences of $\mathcal{E}^\ell$ are obtained by the restrictions on those in $\mathcal{E}^\infty$ to L_ℓ. The endowment vectors $\omega_{a,\ell}$ are defined in an appropriate manner such that $\omega_{a,\ell} \to \omega_a$ as $\ell \to \infty$, where ω_a is the endowment vector of $\mathcal{E}^\infty$, $a = 1 \ldots m$. Intuitively, we obtain a sequence of finite dimensional economies $\mathcal{E}^\ell$, which converge to the infinite dimensional economy $\mathcal{E}^\infty$. Based on the results given in the previous chapters, for every ℓ, a competitive equilibrium $(\boldsymbol{p}_\ell, \boldsymbol{x}_{a,\ell})$ exists for the economy $\mathcal{E}^\ell$. Under suitable conditions, a limit point $(\boldsymbol{p}, \boldsymbol{x}_a)$ exists for the sequence $(\boldsymbol{p}_\ell, \boldsymbol{x}_{a,\ell})$ such that $(\boldsymbol{p}, \boldsymbol{x}_a)$ is a competitive equilibrium of the original economy $\mathcal{E}^\infty$. We apply this method in Sections 4.2 and 4.3.

The second proof is based on the Negishi-type proof, which exploits the fundamental theorems of welfare economics. This method operates on the utility possibility frontier of m consumers by constructing a fixed point mapping onto the $m-1$-dimensional simplex. Therefore, this proof works well for economies with a finite number of consumers, even if the commodity space is infinite dimensional. In Section 4.4, we prove following Mas-Colell [150] the existence of the competitive equilibria for an exchange economy by employing the Negishi-type proof. We assume that the commodity space is a topological vector lattice, which is a topological vector space with the lattice structure (see Appendix G). An important concept called *proper* preferences will be defied. We will also see that the lattice structure of the space is relevant to the existence of the competitive equilibrium. This is interesting because we would not know this when working only within the finite dimensional models.

Finally, in Section 4.5, we discuss the differentiability of the consumer demand on the infinite dimensional commodity spaces. We present a result given by Araujo [5], which states that if a demand function is differentiable, then the commodity space must have the Hilbert space structure. Therefore, if the commodity space is assumed to be ℓ^p, $1 \leq p \leq +\infty$, and the demand function is smooth, then p is equal to 2. This result shows that the mathematical properties of the demand function will impose a strong condition on the underlying commodity space, which is also interesting because we would not recognize this phenomenon when working only within the finite dimensional setup.

4.2 An Exchange Economy with Infinite Horizon

As explained in the previous section, the commodity space of the infinite time horizon economy is given by the space of all bounded sequences ℓ^∞. The price vector of this economy is an element of ℓ^1, which is the space of all countably additive set functions or summable sequences.

As usual, m consumers exist in the economy indexed by $a = 1 \dots m$. Throughout this chapter, and thus in the present section, the consumption set X_a is assumed to be the non-negative orthant (positive cone) of the commodity space, or

$$X_a = \ell^\infty_+ = \{\boldsymbol{x} = (x^t) \in \ell^\infty | \, x^t \geq 0, \; t \in \mathbb{N}\}, \quad a = 1 \dots m.$$

The consumer a has a preference relation $\prec_a \subset X_a \times X_a$ that satisfies (NTR) as well as the basic postulates comprising (IR), (TR), and (CT). The (CT) assumption on the infinite dimensional commodity space is more intricate than that on the finite dimensional space because there are several topologies on this space. We assume that the preferences are continuous with respect to the *Mackey topology*,

(CT_∞): for the preference relation $\prec_a$ is open in $\ell^\infty_+ \times \ell^\infty_+$ in the $\tau(\ell^\infty, \ell^1)$ topology.

We note that the assumption is stronger when the topology is weaker. We show that the relevant topology in the proof of the existence theorem is the *weak* topology* $\sigma(\ell^\infty, \ell^1)$, which is weaker than the Mackey topology. The Mackey continuity is sufficient because both topologies coincide on a convex set of ℓ^∞. Moreover, on a bounded set of ℓ^∞, the Mackey topology coincides with the *product topology* (Proposition G.5). We shall also assume the monotonicity (MT) of preferences which brings to us non-negative equilibrium price vectors.

Each consumer has a non-negative vector $\omega_a \geq \mathbf{0}$ as the initial endowment, or we assume the survival condition from the outset. Later, we also require an assumption regarding the positivity of total endowment on ℓ^∞ as follows.

(PTE_∞): a positive number $\gamma > 0$ exists such that $\sum_{a=1}^m \omega_a^t \geq \gamma$ for all $t \in \mathbb{N}$.

The (PTE_∞) condition is simply written as $\sum_{a=1}^m \omega_a \gg \mathbf{0}$.

The $2m$-tuple $(\precsim_a, \omega_a)_{a=1}^m$ is called an *infinite time horizon economy*, or simply an economy and denoted by $\mathcal{E}^\infty$. An m-tuple of consumption vectors $(\boldsymbol{x}_1 \cdots \boldsymbol{x}_m)$ such that $\boldsymbol{x}_a \in X_a = \ell^\infty_+$ is called an allocation. The definition of the competitive equilibrium is the same as that in Definition 2.3 in Chapter 2 but the commodity space is replaced by ℓ^∞.

Let $(\boldsymbol{x}_a)_{a=1}^m$ be a feasible allocation. The concept of Pareto optimality is defined for $\mathcal{E}^\infty$ in exactly the same manner as that in the case of the classical exchange economies (see Section 2.5). Then, we have the following.

Proposition 4.1. *Let $(\boldsymbol{p}, \boldsymbol{x}_1 \dots \boldsymbol{x}_m) \in \ell^1_+ \times \prod_{a=1}^m X_a$ be a competitive equilibrium of an economy $\mathcal{E}^\infty$ that satisfies the local non-satiation (on ℓ^∞),*

(LNS_∞): for every $\boldsymbol{x} \in X_a$ and every neighborhood U of $\boldsymbol{x}$ in the $\sigma(\ell^\infty, \ell^1)$-topology, a vector $\boldsymbol{z} \in U$ exists such that $\boldsymbol{x} \prec_a \boldsymbol{z}$, $a = 1 \dots m$.

Then, the allocation $(\boldsymbol{x}_1 \cdots \boldsymbol{x}_m)$ is Pareto-optimal,

We note that in the definition of the LNS_∞, weak* neighborhoods are sufficient. The proof of Proposition 4.1 is the same as that of Proposition 2.10 in Chapter 2. The fundamental existence theorem for the economy $\mathcal{E}^\infty$ now reads as follows.

Theorem 4.1. *Suppose that an economy $\mathcal{E}^\infty = (\precsim_a, \omega_a)$ satisfies the basic postulates, (NTR), (CV), and (MT) for every $a = 1 \dots m$, and the (PTE_∞) assumption. Then, a competitive equilibrium $(\boldsymbol{p}, \boldsymbol{x}_a)$ exists for $\mathcal{E}^\infty$.*

The idea of the proof involves approximating the economy $\mathcal{E}^\infty$ based on finite dimensional economies $\mathcal{E}^T$. We define the set of sequences with coordinates that are 0 after T as

$$R^T = \{\boldsymbol{x} = (x^t) \mid 0 = x^{T+1} = x^{T+2} = \cdots\} \subset L, \quad T = 0, 1, \dots$$

The set R^T is naturally identified with $\mathbb{R}^{T+1}$. By (MT), $X_a = \ell^\infty_+$. Let

$$X_a^T = X_a \cap R^T \approx \mathbb{R}^{T+1}_+, \quad a = 1 \dots m,$$

and we have a preference relation $\prec_a^T$ on X_a^T, which is the restriction of $\prec_a$ on the truncated consumption set X_a^T, $a = 1 \dots m$. Let $\omega_a(T) \in R^T \approx \mathbb{R}^{T+1}$ be the truncated initial endowment vector, which is defined by

$$\omega_a^t(T) = \begin{cases} \omega_a^t & \text{for } 0 \le t \le T, \\ 0 & \text{for } t \ge T+1. \end{cases}$$

Then, we have obtained the T-period economy: $\mathcal{E}^T = ((X_a^T, \prec_a^T), \omega_a(T))_{a=1}^m$. The following lemma is straightforward.

Lemma 4.1. *Under the assumptions of Theorem 4.1, a competitive equilibrium $(\boldsymbol{p}(T), \boldsymbol{x}_a(T))$ exists for the T-period economy $\mathcal{E}^T$, $T = 0, 1 \ldots$*

Proof. It is clear that the truncated T-period preference relation $\prec_a^T$ satisfies (CT) (in the usual topology on $\mathbb{R}^{T+1}$), (MT), and (CV). The strong positive endowment assumption (PTE_∞) implies (PTE) for $\mathcal{E}^T$. Hence, from Theorem 2.5 in Chapter 2, it follows that a competitive equilibrium $(\boldsymbol{p}(T), \boldsymbol{x}_a(T))$ exists for $\mathcal{E}^T$. □

Let $(\boldsymbol{p}(T), \boldsymbol{x}_a(T)$ be an equilibrium for $\mathcal{E}^T$. We note that (MT) implies that $\boldsymbol{p}(T) \gg \mathbf{0}$, and thus $\boldsymbol{p}(T) \neq \mathbf{0}$.

We can normalize the equilibrium price vector $\boldsymbol{p}(T) = (p^t(T))$ as $\|\boldsymbol{p}(T)\| = \sum_{t=0}^T p^t(T) = 1$. We identify the vectors in $\mathbb{R}^{T+1}$ and *ba*, and denote $\pi(T) = (\boldsymbol{p}(T), 0, 0, \ldots)$ for notational simplicity. Then, $\pi(T) \in \ell^1 \subset ba$ and $\pi(T) \geq \mathbf{0}$ for all $T \in \mathbb{N}$. Hence, $\|\pi(T)\| = \sum_{t=1}^T p^t(T) = 1$, where the norm denotes the ℓ^1-norm on the sequence space. Similarly, we define $\boldsymbol{x}_a(T) = (\boldsymbol{x}_a(T), 0, 0 \ldots)$, $a = 1 \ldots m$. Because $(\boldsymbol{x}_a(T))_{a=1}^m$ is a feasible allocation, then we have

$$\mathbf{0} \leq \boldsymbol{x}_a(T) \leq \sum_{b=1}^m \omega_b(T), \quad a = 1 \ldots m,$$

hence

$$\|\boldsymbol{x}_a(T)\| \leq \left\| \sum_{b=1}^m \omega_b(T) \right\| \leq \left\| \sum_{b=1}^m \omega_b \right\|, \quad a = 1 \ldots m,$$

where the norm is the sup-norm on the space ℓ^∞.

Then, by Alaoglu's theorem (Theorem G.3), the norm bounded sets are weakly compact in the weak* topologies, or the weak* closures of the bounded sets are compact in the weak* topologies. Hence, by Propositions B.1 and B.4, converging subnets $\pi_\kappa \to \pi \in ba$ and $\boldsymbol{x}_{a,\kappa} \to \boldsymbol{x}_a \in \ell^\infty$ exist for the sequences $\{\pi(T)\}$ and $\{\boldsymbol{x}_a(T)\}$, respectively, where we denote $\pi_\kappa = \pi(T_\kappa)$ and $\boldsymbol{x}_{a,\kappa} = \boldsymbol{x}_a(T_\kappa)$ for simplicity. The positive orthant of ℓ^∞ is weak* closed, so it follows that $\boldsymbol{x}_a \in X_a = \ell^\infty_+$, $a = 1 \ldots m$. Similarly, we have $\pi \geq \mathbf{0}$. Let $\mathbf{1} = (1, 1, \ldots) \in \ell^\infty$. Because $\pi(T)\mathbf{1} = 1$ for all $T \in \mathbb{N}$, then it follows that $\pi\mathbf{1} = 1$ in the limit. Hence, $\pi \neq \mathbf{0}$.

We have almost completed the proof.

Lemma 4.2. *The vector $\pi \in ba$ and the allocation $(\boldsymbol{x}_a)_{a=1}^m$ satisfy the following conditions:*

(BA-1) $\boldsymbol{x}_a \succsim_a \boldsymbol{z}$ *whenever* $\boldsymbol{z} \in \ell^\infty_+$ *and* $\pi\boldsymbol{z} < \pi\omega_a$, $a = 1 \ldots m$,
(E-2) $\sum_{a=1}^m \boldsymbol{x}_a \leq \sum_{a=1}^m \omega_a$.

Proof. If $\pi\omega_a = 0$, the condition (BA-1) holds trivially. Since $\sum_{a=1}^m \omega_a \ggg \mathbf{0}$, we have $\pi\omega_a > 0$ for some a. Suppose that condition (BA-1) does not hold for such an a. Then a vector $\boldsymbol{z} \in X_a$ exists such that $\boldsymbol{x}_a \prec_a \boldsymbol{z}$ and $\pi\boldsymbol{z} < \pi\omega_a$. Let $\boldsymbol{z}(T) = (z^0, z^1 \ldots z^T, 0, 0 \ldots)$ be the projection of $\boldsymbol{z}$ onto $\mathbb{R}^{T+1}$. Given $\boldsymbol{q} = (q^t) \in \ell^1$, we have $\boldsymbol{q}\boldsymbol{z}(T) = \sum_{t=0}^T q^t z^t \to \sum_{t=0}^\infty q^t z^t = \boldsymbol{q}\boldsymbol{z}$ as $T \to +\infty$. Hence, $\boldsymbol{z}(T) \to \boldsymbol{z}$ in the $\sigma(\ell^\infty, \ell^1)$-topology by Proposition G.2. Because $\pi \geq \mathbf{0}$ and $\boldsymbol{z}(T) \leq \boldsymbol{z}$ for all T, then $\pi\boldsymbol{z}(T) \leq \pi\boldsymbol{z} < \pi\omega_a$ for all T. Fix a $T_0 \in \mathbb{N}$ such that $\boldsymbol{x}_a \prec_a \boldsymbol{z}(T_0)$. Because the set $\{\boldsymbol{x} \in \ell^\infty_+ | \, \boldsymbol{x} \succsim_a \boldsymbol{z}\}$ is convex and Mackey closed, then it is closed in the weak* topology by Theorem G.2, and thus its complement $\{\boldsymbol{x} \in \ell^\infty_+ | \, \boldsymbol{x} \prec_a \boldsymbol{z}\}$ is weak* open. Therefore, we can take the neighborhoods $Q \subset ba$ and $U \subset \ell^\infty$ for π and $\boldsymbol{x}_a$, respectively, in the weak* topologies, which satisfy $\pi'\boldsymbol{z}(T_0) < \pi'\omega_a$ and $\boldsymbol{x} \prec_a \boldsymbol{z}(T_0)$ for all $\pi' \in Q$ and for all $\boldsymbol{x} \in U$. Because $\pi_\kappa \to \pi$ and $\boldsymbol{x}_{a,\kappa} \to \boldsymbol{x}_a$, then a $\hat{\kappa}$ with $T_{\hat{\kappa}} > T_0$ exists such that $\pi_{\hat{\kappa}} \in Q$ and $\boldsymbol{x}_{a,\hat{\kappa}} \in U$. In addition,

$$\pi_{\hat{\kappa}}\boldsymbol{z}(T_0) < \pi_{\hat{\kappa}}\omega_a = \sum_{t=0}^{T_{\hat{\kappa}}} p^t_{\hat{\kappa}}\omega^t_a = \pi_{\hat{\kappa}}\omega_{a,\hat{\kappa}},$$

and $\boldsymbol{x}_{a,\hat{\kappa}} \prec_a \boldsymbol{z}(T_0)$, which contradicts the fact that $\boldsymbol{x}_{a,\hat{\kappa}}$ is an equilibrium consumption bundle of the consumer a in the economy $\mathcal{E}^{T_{\hat{\kappa}}}$. This establishes the condition (BA-1). Because $\sum_{a=1}^m \boldsymbol{x}_a(T) \leq \sum_{a=1}^m \omega_a(T)$ for all T, then it follows that $\sum_{a=1}^m \boldsymbol{x}_a \leq \sum_{a=1}^m \omega_a$ in the limit. Hence, we have proved condition (E-2). □

In the proof given above, we employed Theorem G.2 as in the same way as Bewley [27], which deals with continuous time models. In fact, when we work with discrete time models, such as that in Lemma 4.2, this is not necessary because the sequence $\boldsymbol{z}(T)$ converges to $\boldsymbol{z}$ in the Mackey topology by Proposition G.5 (see the proof of Theorem 4.2). The existence of an equilibrium price vector in ℓ^1 can be established by the following lemma.

Lemma 4.3. *A vector* $\boldsymbol{p} \in \ell^1_+$ *exists such that* $(\boldsymbol{p}, \boldsymbol{x}_a)_{a=1}^m$ *satisfies conditions* (E-1) *and* (E-2) *in Definition* 2.3 *in Chapter* 2.

Proof. We have already established condition (E-2) in the previous lemma. Because $\pi \geq \mathbf{0}$, then from the Yosida–Hewitt theorem (Theorem G.4), it follows that we can write $\pi = \pi_c + \pi_p$, where $\pi_c \geq \mathbf{0}$ and $\pi_p \geq \mathbf{0}$ are the countably additive part and the purely finitely additive part of π, respectively.

For a given $a = 1 \dots m$, we take $\boldsymbol{z} = (z^t) \in X_a = \ell^\infty_+$ with $\boldsymbol{x}_a \prec_a \boldsymbol{z}$ and let $\boldsymbol{z}(T) = (z^0, z^1 \dots z^T, 0, 0 \dots)$ be the projection of $\boldsymbol{z}$ to $\mathbb{R}^{T+1}$. Then, as in the proof of Lemma 4.2, we can show that $\boldsymbol{x}_a \prec_a \boldsymbol{z}(T)$ for a sufficiently large T. By Lemma 4.2, we have $\pi \boldsymbol{z}(T) \geq \pi \omega_a$. Because π_p is purely finitely additive, then $\pi_p(\{0, 1, \dots, T\}) = 0$, and thus, from this and $\pi_c \geq \mathbf{0}$, it follows that $\pi \boldsymbol{z}(T) = (\pi_c + \pi_p)\boldsymbol{z}(T) = \pi_c \boldsymbol{z}(T) \leq \pi_c \boldsymbol{z}$ because $\boldsymbol{z}(T) \leq \boldsymbol{z}$. In addition, $\pi_p \geq \mathbf{0}$ and $\omega_a \geq \mathbf{0}$ imply that $\pi \omega_a = (\pi_c + \pi_p)\omega_a \geq \pi_c \omega_a$, so we have $\pi_c \boldsymbol{z} \geq \pi_c \omega_a$. If we denote $\pi_c(\{t\}) = p^t$, we obtain a vector $\boldsymbol{p} = (p^t)$ in ℓ^1 that satisfies

$$\boldsymbol{x}_a \prec_a \boldsymbol{z} \text{ implies } \boldsymbol{p}\boldsymbol{z} \geq \boldsymbol{p}\omega_a, \; a = 1 \dots m.$$

By (MT), we can take $\boldsymbol{z} \in X_a = \ell^\infty$, which is arbitrarily close to $\boldsymbol{x}_a$ such that $\boldsymbol{x}_a \prec_a \boldsymbol{z}$. Thus, the relation given above implies that $\boldsymbol{p}\boldsymbol{x}_a \geq \boldsymbol{p}\omega_a$, $a = 1 \dots m$. Moreover, because $\sum_{a=1}^m (\boldsymbol{x}_a - \omega_a) \leq \mathbf{0}$ by Lemma 4.2, then we have $0 \leq \sum_{a=1}^m \boldsymbol{p}(\boldsymbol{x}_a - \omega_a) \leq 0$. Therefore, we have $\boldsymbol{p}(\boldsymbol{x}_a - \omega_a) = 0$ for all a, and thus, $\boldsymbol{p}\boldsymbol{x}_a = \boldsymbol{p}\omega_a$, $a = 1 \dots m$.

Because $\sum_{a=1}^m \omega_a \gg \mathbf{0}$, then a consumer a exists with $\boldsymbol{p}\omega_a > 0$. As usual, for such an a, we can show that

$$\boldsymbol{x}_a \prec_a \boldsymbol{z} \text{ implies } \boldsymbol{p}\boldsymbol{z} > \boldsymbol{p}\omega_a, \quad a = 1 \dots m$$

and $\boldsymbol{p} \gg \mathbf{0}$. Then, condition (E-1) holds for a with positive income. When $\boldsymbol{p}\omega_a = 0$, then $\{\boldsymbol{x} | \, \boldsymbol{p}\boldsymbol{x} \le \boldsymbol{p}\omega_a\} = \{\mathbf{0}\}$, and thus, (E-1) holds trivially. □

We recall that a feasible allocation $(\boldsymbol{x}_a)_{a=1}^m$ is said to be an equilibrium relative to the price vector $\boldsymbol{p}$ if and only if for all $\boldsymbol{z} \in X_a$,

$$\boldsymbol{x}_a \prec_a \boldsymbol{z} \text{ implies that } \boldsymbol{p}\boldsymbol{x}_a < \boldsymbol{p}\boldsymbol{z}, \quad a = 1 \dots m$$

(see Definition 2.4 in Chapter 2). The second fundamental theorem of welfare economics, which is the converse of Proposition 4.1, still holds as it does for the classical exchange economies (Theorem 2.6). The key mathematical theorem employed was the separating hyperplane theorem (Theorem A.1), and it was extended to the infinite dimensional vector spaces, which is known as the Hahn–Banach theorem (Theorems E.2 and G.6), with a cost that the separated convex set must have an interior point.

A mathematical advantage of the space ℓ^∞ is that this interiority condition is easily obtained. For instance, the positive orthant of ℓ^∞ has an interior point in the norm topology, i.e., $\mathbf{1} = (1, 1 \dots)$. This is in contrast to other Banach spaces, such as ℓ^p for $p \neq \infty$ or $ca(K)$. In these spaces, we must impose additional assumptions on the preferences (and production sets) to prove the existence theorem and/or the second welfare theorem (see the subsequent sections).

Theorem 4.2. *Let $\mathcal{E}^\infty$ be an infinite time horizon exchange economy. For every consumer $a \in A$, we assume that the preference relation satisfies (NTR), (CV) and (MT). Then, every Pareto-optimal allocation is an equilibrium relative to some price vector $\boldsymbol{p} > \mathbf{0}$.*

Proof. Let $(\boldsymbol{x}_a)_{a=1}^m$ be a Pareto-optimal allocation. Then, the condition (E-2) is satisfied because it is a feasible allocation. For each consumer a, we define the strictly preferred set $P_a(\boldsymbol{x})$ by

$$P_a(\boldsymbol{x}) = \{\boldsymbol{z} \in X_a | \, \boldsymbol{x} \prec_a \boldsymbol{z}\}.$$

The set $P_a(\boldsymbol{x})$ is non-empty for all $\boldsymbol{x} \in X_a$ because $\boldsymbol{x} + \mathbf{1} \in P_a(\boldsymbol{x})$ by the (MT) assumption, where $\mathbf{1} = (1, 1 \dots)$. Similar to the proof of

Theorem 2.6 in Chapter 2, we can show that the set $P_a(\boldsymbol{x})$ is convex for all $\boldsymbol{x} \in X_a$. Moreover, the set $P_a(\boldsymbol{x})$ has an interior point $\boldsymbol{x}+\mathbf{1}$ because the ball $B(\boldsymbol{x}+\mathbf{1}, 1/2)$ centered at $\boldsymbol{x}+\mathbf{1}$ with a radius $1/2$ is contained in $P_a(\boldsymbol{x})$.

Let $Q = \sum_{a=1}^m P_a(\boldsymbol{x}_a)$. Then, Q is convex because it is the sum of the convex sets and $\mathring{Q} \neq \emptyset$. The allocation $(\boldsymbol{x}_a)_{a=1}^m$ is Pareto-optimal, so $\sum_{a=1}^m \omega_a \notin Q$. Hence, we can apply Theorem E.2 and obtain a vector $\pi \in ba$ with $\pi \neq \mathbf{0}$ and

$$\pi \sum_{a=1}^{m} \omega_a \leq \pi \sum_{a=1}^{m} \boldsymbol{z}_a$$

for every $\boldsymbol{z}_a \in P_a(\boldsymbol{x}_a)$, $a = 1 \dots m$. By (MT), it follows that $\pi \geq \mathbf{0}$. Fix an a. For $b \neq a$, we can take $\boldsymbol{z}_b$ as being arbitrarily close to $\boldsymbol{x}_b$ in the norm topology. Hence,

$$\pi \sum_{a=1}^{m} \omega_a \leq \pi \sum_{b \neq a} \boldsymbol{x}_b + \pi \boldsymbol{z}_a \leq \pi \sum_{a=1}^{m} \omega_a - \pi \boldsymbol{x}_a + \pi \boldsymbol{z}_a,$$

and thus, we obtain

$$\pi \boldsymbol{x}_a \leq \pi \boldsymbol{z}_a \text{ for every } \boldsymbol{z}_a \in P_a(\boldsymbol{x}_a), \quad a = 1 \dots m.$$

As in the same way of the proof of Lemma 4.3, we can show

$$\boldsymbol{p}\boldsymbol{x}_a < \boldsymbol{p}\boldsymbol{z} \text{ whenever } \boldsymbol{x}_a \prec_a \boldsymbol{z}, \quad a = 1 \dots m,$$

where $\boldsymbol{p} = (p^t) = (\pi(\{t\}))$, which shows that the optimal allocation $(\boldsymbol{x}_a)$ is an equilibrium relative to $\boldsymbol{p}$. This proves the theorem. □

Finally, we give a result regarding the continuity of the preferences on ℓ^∞. In applications of infinite time horizon economies, the preferences of the consumers are sometimes given in a time-separable form:

$$u(\boldsymbol{x}) = \sum_{t=0}^{\infty} v(x^t, t) \quad \text{for } \boldsymbol{x} = (x^t) \in X = \ell_+^\infty.$$

The utility function of the discounted sum for a stationary one-period utility

$$u(\boldsymbol{x}) = \sum_{t=0}^{\infty} \rho^t v(x^t) \quad \text{for } \boldsymbol{x} = (x^t) \in X = \ell_+^\infty, \quad 0 < \rho < 1$$

is often discussed as a special case. The following theorem shows that the time-separable utility functions are continuous in the Mackey topology under very general assumptions.

Theorem 4.3. *Suppose that the one-period utility function*

$$v : \mathbb{R}_+ \times \mathbb{N} \to \mathbb{R}, \ (x, t) \mapsto v(x, t)$$

satisfies the following: (a) for each $t \in \mathbb{N}$, $v(\cdot, t)$ is a continuous, non-decreasing, concave function with $v(0, t) = 0$; (b) for each $x \in \mathbb{R}_+$, $\sum_{t=0}^{\infty} v(x, t) < \infty$. Then, the utility function

$$u(\boldsymbol{x}) = \sum_{t=0}^{\infty} v(x^t, t) \quad \textit{for } \boldsymbol{x} = (x^t) \in X = \ell_+^\infty$$

is continuous with respect to the $\tau(\ell^\infty, \ell^1)$-topology.

Proof. First, we show that if a net $(\boldsymbol{x}_\kappa)$ of non-negative sequences, or $\boldsymbol{x}_\kappa \geq \boldsymbol{0}$ for all κ, converges to $\boldsymbol{x}$ in the $\tau(\ell^\infty, \ell^1)$-topology, then $|\boldsymbol{x}_\kappa - \boldsymbol{x}| \to \boldsymbol{0}$ in the $\sigma(\ell^\infty, \ell^1)$-topology, where we set $\boldsymbol{x}_+ = (x_+^t)$ and $\boldsymbol{x}_- = (x_-^t)$ as $x_+^t = \max\{0, x^t\}$ and $x_-^t = \max\{0, -x^t\}$, respectively, $|\boldsymbol{x}| = \boldsymbol{x}_+ + \boldsymbol{x}_-$.

To demonstrate this, we first show that a net $\{\boldsymbol{z}_\kappa\}$ of non-negative vectors that converges to a vector $\boldsymbol{z}_* \geq \boldsymbol{0}$ in the $\tau(\ell^\infty, \ell^1)$-topology is bounded in the ℓ^∞ norm. For each κ, we can take $\boldsymbol{p}_\kappa \in S_+ \equiv \{\boldsymbol{p} \in \ell^1 |\, \boldsymbol{p} \geq \boldsymbol{0}, \|\boldsymbol{p}\| = \boldsymbol{p}\boldsymbol{1} = 1\}$ such that $\|\boldsymbol{z}_\kappa\| = \sup\{|\boldsymbol{p}\boldsymbol{z}_\kappa|\, |\, \boldsymbol{p} \in \ell^1, \ \|\boldsymbol{p}\| = 1\} = \boldsymbol{p}_\kappa \boldsymbol{z}_\kappa$, because $\boldsymbol{z}_\kappa \geq \boldsymbol{0}$ and the set S_+ is $\sigma(\ell^\infty, \ell^1)$-compact. Let $\boldsymbol{p}_*$ be a limit point of the net $\{\boldsymbol{p}_\kappa\} \subset S_+$. The evaluation map $\boldsymbol{p}\boldsymbol{z}$ is $\sigma(\ell^1, \ell^\infty) \times \tau(\ell^\infty, \ell^1)$ jointly continuous by Proposition G.6, so it follows that $\|\boldsymbol{z}_\kappa\| = \boldsymbol{p}_\kappa \boldsymbol{z}_\kappa \to \boldsymbol{p}_* \boldsymbol{z}_*$. Hence, the set $\{\boldsymbol{z}_\kappa\}$ is bounded. Let $\boldsymbol{x}_\kappa \to \boldsymbol{x}$ be in the $\tau(\ell^\infty, \ell^1)$-topology with $\boldsymbol{x}, \boldsymbol{x}_\kappa \geq \boldsymbol{0}$ for all κ. Then, $\|(\boldsymbol{x}_\kappa - \boldsymbol{x})_+\| \leq \|\boldsymbol{x}_\kappa - \boldsymbol{x}\| \leq \|\boldsymbol{x}_\kappa\| + \|\boldsymbol{x}\|$, and thus the net $(\boldsymbol{x}_\kappa - \boldsymbol{x})_+$ is norm bounded and converges to $\boldsymbol{0}$ in a pointwise manner.

Hence, by Proposition G.5, $(\boldsymbol{x}_\kappa - \boldsymbol{x})_+ \to \mathbf{0}$ in the $\sigma(\ell^\infty, \ell^1)$-topology, which is also the case for $(\boldsymbol{x}_\kappa - \boldsymbol{x})_-$ and $|\boldsymbol{x}_\kappa - \boldsymbol{x}|$.

Suppose that $\{\boldsymbol{x}_\kappa\}$ is a net of non-negative vectors in ℓ^∞ that converge to $\boldsymbol{x}$ in the Mackey topology $\tau(\ell^\infty, \ell^1)$. Let $\tilde{\boldsymbol{x}}_\kappa = (\tilde{x}^t_\kappa)$ and $\hat{\boldsymbol{x}}_\kappa = (\hat{x}^t_\kappa)$ be defined by

$$\tilde{x}^t_\kappa = \max\{0, x^t - |x^t_\kappa - x^t|\},$$
$$\hat{x}^t_\kappa = \max\{0, x^t + |x^t_\kappa - x^t|\},$$

respectively. Then, $\tilde{\boldsymbol{x}}_\kappa \leq \boldsymbol{x} \leq \hat{\boldsymbol{x}}_\kappa$ and according to the proof in the previous paragraph, both $\tilde{\boldsymbol{x}}_\kappa$ and $\hat{\boldsymbol{x}}_\kappa$ converge to $\boldsymbol{x}$ in the $\sigma(\ell^\infty, \ell^1)$-topology. $v(x,t)$ is non-decreasing by the assumption, so we have

$$|v(x^t, t) - v(x^t_\kappa, t)| \leq v(\hat{x}^t_\kappa, t) - v(x^t, t) + v(x^t, t) - v(\tilde{x}^t_\kappa, t).$$

Therefore, it is sufficient to prove that

Case 1: $\boldsymbol{x}_\kappa \leq \boldsymbol{x}$ for all κ and $\sum_{t=0}^\infty \left(v(x^t, t) - v(x^t_\kappa, t)\right) \to 0$,
Case 2: $\boldsymbol{x}_\kappa \geq \boldsymbol{x}$ for all κ and $\sum_{t=0}^\infty \left(v(x^t_\kappa, t) - v(x^t, t)\right) \to 0$.

First, we give the proof for Case 1 where we assume that $\boldsymbol{x}_\kappa \leq \boldsymbol{x}$ for all κ. For all $\boldsymbol{z} = (z^t) \in \ell^\infty$, the utility function $u(\boldsymbol{z})$ is summable and well defined because $v(z^t, t) \leq v(\|\boldsymbol{z}\|, t)$.

Let $\epsilon > 0$ and for each $\tau \geq 0$, let $P(\tau) = \{t \in \mathbb{N} \mid x^t \geq \tau\}$. When $\tau_1 \geq \tau_2 \geq \cdots \to 0$, we have $P(\tau_1) \subset P(\tau_2) \subset \cdots \subset P(0) = \mathbb{N}$. Moreover, we have $\sum_{t \in \mathbb{N}} v(\tau, t) \to 0$ as $\tau \to 0$. Therefore, a $\tau > 0$ exists such that $\sum_{t \notin P(\tau)} v(x^t, t) < \epsilon$. Because $v(x^t, t)$ is concave, then

$$\frac{v(x^t, t) - v(x^t_\kappa, t)}{x^t - x^t_\kappa} \leq \frac{v(x^t, t)}{x^t} \leq \frac{v(x^t, t)}{\tau},$$

for all $t \in P(\tau)$, and thus,

$$\begin{aligned}\sum_{t \in \mathbb{N}} \left(v(x^t, t) - v(x^t_\kappa, t)\right) &\leq (1/\tau) \sum_{t \in P(\tau)} v(x^t, t)(x^t - x^t_\kappa) \\ &\quad + \sum_{t \notin P(\tau)} \left(v(x^t, t) - v(x^t_\kappa, t)\right) \\ &\leq (1/\tau) \sum_{t \in \mathbb{N}} v(x^t, t)(x^t - x^t_\kappa) + \sum_{t \notin P(\tau)} v(x^t, t).\end{aligned}$$

$\boldsymbol{x}_\kappa \to \boldsymbol{x}$ in the $\sigma(\ell^\infty, \ell^1)$-topology, so we have $\sum_{t\in\mathbb{N}} v(x^t, t)(x^t - x^t_\kappa) < \tau\epsilon$ for a sufficiently large κ. Hence, $u(\boldsymbol{x}) - u(\boldsymbol{x}_\kappa) \leq \epsilon + \epsilon$ for a sufficiently large κ, and thus, we have proved Case 1.

Now, suppose that $\boldsymbol{x} \leq \boldsymbol{x}_\kappa$ for all κ. Let $\epsilon > 0$. Then, we can choose $\tau > 0$ such that $\sum_{t=0}^{\infty} v(\tau, t) < \epsilon$. If $\tau \leq x^t$, then $v(x_\kappa, t) - v(x^t, t) \leq \frac{1}{\tau} v(x^t, t)(x^t_\kappa - x^t)$. If $x^t < \tau \leq x^t_\kappa$, then we have

$$\begin{aligned} v(x^t_\kappa, t) - v(x^t, t) &\leq v(x^t_\kappa, t) - v(x^t, t) + (v(\tau, t) - v(x^t, t)) \\ &\leq (1/\tau) v(\tau, t)(x^t_\kappa - x^t) + v(\tau, t) \end{aligned}$$

by the concavity and the monotonicity of $v(\cdot, t)$.

Finally, if $x^t_\kappa < \tau$, then it follows that $v(x^t_\kappa, t) - v(x^t, t) \leq v(\tau, t)$. Therefore, in all cases, $v(x^t_\kappa, t) - v(x^t, t) \leq (1/\tau) v(x^t, t)(x^t_\kappa - x^t) + (1/\tau) v(\tau, t)(x^t_\kappa - x^t) + v(\tau, t)$. $\boldsymbol{x}_\kappa \to \boldsymbol{x}$ in the $\sigma(\ell^\infty, \ell^1)$-topology, so we can take κ_0 such that for all $\kappa \geq \kappa_0$,

$$\sum_{t=0}^{\infty} v(x^t, t)(x^t_\kappa - x^t) < \tau\epsilon, \sum_{t=0}^{\infty} v(\tau, t)(x^t_\kappa - x^t) < \tau\epsilon.$$

Therefore, if $\kappa \geq \kappa_0$, then we obtain

$$u(\boldsymbol{x}_\kappa) - u(\boldsymbol{x}) < (1/\tau)\tau\epsilon + (1/\tau)\tau\epsilon + \epsilon = 3\epsilon.$$

This verifies Case 2 and the proof of Theorem 4.3 is established. □

4.3 An Exchange Economy with Differentiated Commodities

Let (K, d) be a compact metric space, or K is a compact space and d is a metric on K. The intended economic meaning of the set K is that it should represent the set of *commodity characteristics*, i.e., an element $t \in K$ is interpreted as a complete description of all the economically relevant characteristics of commodities.

Example 4.1 (Hotelling [98]). $K = [0, L]$, an interval with the length L. Every $t \in K$ represents a location.

Example 4.2 (Lancaster [127] and Rosen [215]). In this example, K is not compact; $K = \mathbb{R}^\ell_+$. Each $\boldsymbol{t} = (t^i) \in K$ represents a characteristics profile, containing t^i units of the characteristic

$i(= 1 \ldots \ell)$. In Lancaster [127], characteristics profiles $\boldsymbol{t}$ and $\boldsymbol{s}$ are considered to represent (different amounts of) the same commodity bundle if and only if $\tau \boldsymbol{t} = \boldsymbol{s}$ for some $\tau > 0$. In Rosen [215], each characteristics profile $\boldsymbol{t}$ is identified with a commodity bundle including it, and (possibly nonlinear) price functional $p(\boldsymbol{t})$ of the characteristics profile $\boldsymbol{t}$ which is not directly observable in the market is called the *hedonic price.*

Example 4.3 (Jones [103]). $K = [0, A] \times [0, P]$ $A, P > 0$. Every $\boldsymbol{t} = (v, p) \in K$ represents a food, which is characterized by its vitamin A and protein contents.

In this section, the commodity vector or the commodity bundle is defined as a Borel signed measure on K. In particular, we denote the collection of all Borel measurable subsets of K as $\mathcal{B}(K)$, where the commodity vector $\boldsymbol{x}$ is a bounded countably additive set function on the measurable space $(K, \mathcal{B}(K))$. Then, the economic interpretation of $\boldsymbol{x}$ is that for every $B \in \mathcal{B}(K)$, $\boldsymbol{x}(B)$ is the total amount of the commodity with its characteristics in B. Let $ca(K)$ be the set of all Borel signed measures on K (see Appendix F). In the classical examples 4.1 and 4.2, the commodities are considered to be some functions on K rather than measures. We will come back to this classical idea of differentiated commodities in Section 7.4.

For $\boldsymbol{x} \in ca(K)$, $\boldsymbol{x} \geq \boldsymbol{0}$ means that $\boldsymbol{x}(B) \geq 0$ for all $B \in \mathcal{B}(K)$. $\boldsymbol{x} > \boldsymbol{0}$ means that $\boldsymbol{x} \geq \boldsymbol{0}$ and $\boldsymbol{x} \neq \boldsymbol{0}$. Let $ca_+(K) = \{\boldsymbol{x} \in ca(K) |\ \boldsymbol{x} \geq \boldsymbol{0}\}$ be the positive orthant of $ca(K)$. $ca_+(K)$ is the set of all Borel measures on $(K, \mathcal{B}(K))$, or $ca_+(K) = \mathscr{M}(K)$. The *support* of $\boldsymbol{x} \in \mathscr{M}(K)$ is the smallest closed subset of K with full $\boldsymbol{x}$-measure and it is denoted by $support(\boldsymbol{x})$; see Proposition F.2.

In this market, the natural candidate of a price vector is a continuous function $p(t)$ on K. Let $C(K)$ be the set of all continuous functions on K. The space $C(K)$ is a Banach space with the norm $\|\boldsymbol{p}\| = \sup\{|p(t)|\,|\ t \in K\}$ (see Appendix G for details). The market value of a commodity vector $\boldsymbol{x} \in ca(K)$ evaluated by the price system $\boldsymbol{p} = p(t) \in C(K)$ is defined by

$$\boldsymbol{p}\boldsymbol{x} = \int_K p(t) d\boldsymbol{x}.$$

This "inner product" shows that the dual space of $C(K)$ is equal to $ca(K)$ (Proposition G.10). Then, by definition, the norm of $\boldsymbol{x} \in \mathscr{M}(K)$ is given by $\|\boldsymbol{x}\| = \sup\{|\boldsymbol{p}\boldsymbol{x}|| \boldsymbol{p} \in C(K),\ \|\boldsymbol{p}\| = 1\}$. We note that for $\boldsymbol{x} \geq \boldsymbol{0}$, $\|\boldsymbol{x}\| = \boldsymbol{1}\boldsymbol{x} = \boldsymbol{x}(K)$, where $\boldsymbol{1}(t) = 1$ for all $t \in K$.

We can consider an alternative topology on $ca(K)$, i.e., the weak* topology, which is weaker than the norm topology, and it is the topology of the pointwise convergence on $C(K)$. Hence, a net $\{\boldsymbol{x}_\kappa\}$ in $ca(K)$ converges to $\boldsymbol{x}$ in the *weak* topology* if and only if

$$\int_K p(t)d\boldsymbol{x}_\kappa \to \int_K p(t)d\boldsymbol{x} \text{ for every } \boldsymbol{p} = p(t) \in C(K).$$

It is well known that the norm bounded subsets of $ca(K)$ are compact and metrizable in the weak* topology (Proposition G.11). Throughout this section, we consider the weak* topology unless stated otherwise. A mathematical advantage of the weak* topology is that the set of finite linear combinations of the Dirac measures is dense in $ca(K)$ (Proposition G.12). A mathematical disadvantage for economics is that the norm interior of the positive orthant $ca_+(K)$ is empty.

m consumers exist in the economy indexed by $a = 1 \dots m$ or $a \in A = \{1 \dots m\}$. The consumption set X_a of consumer a is assumed to be the positive (non-negative) orthant of the commodity space $ca(K)$,

$$X_a = ca_+(K) = \mathscr{M}(K), \quad a = 1 \dots m.$$

The preference relation $\prec_a \subset X_a \times X_a$ is always assumed to satisfy the basic postulates as well as (NTR), (CV), and (MT). As stated above, the continuity of the preference is assumed with respect to the weak* topology:

($CT_{\text{weak}*}$): the set $\prec = \{(\boldsymbol{x}, \boldsymbol{y}) \in X_a \times X_a | \ \boldsymbol{x} \prec_a \boldsymbol{y}\}$ is open in $X_a \times X_a$ in the weak* topology.

In the previous section, we showed that the non-negative orthant of the space ℓ^∞ has an interior point $\boldsymbol{1} = (1, 1, \dots)$ in the norm topology and vector $\boldsymbol{1}$ played a crucial role in the proofs of the existence theorem and the second welfare theorem. However, the norm interior of $ca_+(K)$ is empty, so we need an additional assumption called the assumption of *bounded marginal rate of substitution* to prove the existence of the competitive equilibrium on the space $ca(K)$.

(BRS): for all sequences $t_n, s_n \in K$, $a_n, b_n > 0$, and $\boldsymbol{x}_n \in \mathscr{M}(K)$ such that $\lim_n t_n = \lim_n s_n$, $\lim_n \boldsymbol{x}_n = \boldsymbol{x}$ and $\lim_n a_n/b_n > 1$, an N exists such that $\boldsymbol{x}_N + a_N\delta_{t_N} \succsim_a \boldsymbol{x}_N + b_N\delta_{t_N}$, where the limit of $\{\boldsymbol{x}_n\}$ is taken with respect to the weak* topology, and δ_t is the Dirac measure.

The (BRS) assumption states that if t and s are sufficiently close in K, then the consumer a is willing to accept any trade of t and s, where the "terms" are strongly greater than 1. In Proposition 4.2, we show that if the preference relation is represented by a smooth utility function and its second derivatives are uniformly bounded, then the BRS assumption is satisfied.

The consumer a has the initial endowment vector $\omega_a \in \mathscr{M}(K)$. We consider an exchange economy, so all of the characteristics should be available initially in the market (*adequate endowment*):

(AE): $\omega_a \geq \mathbf{0}$ for all $a = 1 \dots m$ and *support* $(\sum_{a=1}^m \omega_a) = K$.

For the definition of *support*$(\boldsymbol{x})$ for $\boldsymbol{x} \in \mathscr{M}(K)$, see Proposition F.2. A $2m$-tuple of the preferences and initial endowment vectors $(\prec_a, \omega_a)_{a=1}^m$ is called an *economy with differentiated commodities*, which is denoted by $\mathcal{E}_{\mathscr{M}}$. The definition of the competitive equilibrium for $\mathcal{E}_{\mathscr{M}}$ should be sufficiently clear. The equilibrium existence theorem for the economy $\mathcal{E}_{\mathscr{M}}$ is now as follows.

Theorem 4.4. *Let $\mathcal{E}_{\mathscr{M}} = (\prec_a, \omega_a)$ be an economy such that the preference relations with the basic postulates satisfy (NTR), (CV), (MT), and (BRS) for every $a = 1 \dots m$. Furthermore, the (AE) assumption is satisfied. Then, a competitive equilibrium $(\mathbf{p}, \boldsymbol{x}_a)$ exists for $\mathcal{E}_{\mathscr{M}}$.*

Proof. A compact metric space is separable (Proposition B.11), so a countable dense subset $\{t_1, t_2 \dots\}$ of K exists. Let $K^\ell = \{t_1 \dots t_\ell\}$ and $LS(K^\ell)$ be the linear subspace spanned by $\delta_{t_1} \dots \delta_{t_\ell}$. We now define the ℓ-dimensional subeconomy $\mathcal{E}^\ell$ as follows.

First, let $X_a^\ell = X_a \cap LS(K^\ell)$ and $\prec_a^\ell = \prec_a \cap (X_a^\ell \times X_a^\ell)$. For each ℓ, take disjoint measurable sets M_k^ℓ, $k = 1 \dots \ell$ with $\cup_{k=1}^\ell M_k^\ell = K$, $t_k \in M_k^\ell$ for all k, and $B(t_k, r_k) \subset M_k^\ell$ for some $r_k > 0$, and finally, $\sup\{\text{diam} M_k^\ell\} \to 0$ as $\ell \to \infty$, where $\text{diam} M_k^\ell = \sup\{d(\boldsymbol{x}, \boldsymbol{y}) | \ \boldsymbol{x}, \boldsymbol{y} \in M_k^\ell\}$. We set $\omega_a^\ell = \sum_{k=1}^\ell \omega_a(M_k^\ell)\delta_{t_k}$. Then, it is easy to see that $\omega_a^\ell \in LS(K^\ell)$, $\sum_{a=1}^m \omega_a^\ell \gg \mathbf{0}$ for all ℓ, and $\omega_a^\ell \to \omega_a$ in the weak* topology, $a = 1 \dots m$. The $3m$-tuple $(X_a^\ell, \prec_a^\ell, \omega_a^\ell)$ is called the ℓ-dimensional subeconomy and it is denoted by $\mathcal{E}^\ell$.

Lemma 4.4. *A competitive equilibrium* $(\boldsymbol{p}^\ell, \boldsymbol{x}_a^\ell)$ *exists for* $\mathcal{E}^\ell$.

Proof. The preference relation $\prec_a^\ell$ satisfies (CT), (MT), and (CV) on the ℓ-dimensional space $LS(K^\ell) \approx \mathbb{R}^\ell$, and (PTE) is satisfied. Then, by Theorem 2.5 in Chapter 2, a competitive equilibrium $(\boldsymbol{p}^\ell, \boldsymbol{x}_a^\ell)$ exists for $\mathcal{E}^\ell$. □

Without loss of generality, we can normalize the price vector $\boldsymbol{p}^\ell = p^\ell(t)$ as $\|\boldsymbol{p}^\ell\| = \sup\{|p^\ell(t)| \mid t \in K\} = 1$. We recall that a sequence $(K^\ell, \boldsymbol{p}^\ell)$ is said to be equicontinuous if for all $\epsilon > 0$, a $\delta > 0$ exists such that for all ℓ and $t, s \in K^\ell$ with $d(t, s) < \delta$, it follows that $|p^\ell(t) - p^\ell(s)| < \epsilon$ (see Appendix G). We have the following.

Lemma 4.5. *The sequence* $(K^\ell, \boldsymbol{p}^\ell)$ *is equicontinuous.*

Proof. We assumed that the opposite holds. Then, the sequences $t_{\ell_k}, s_{\ell_k} \in K_{\ell_k}$ exist such that $t_{\ell_k} \to t$, $s_{\ell_k} \to t$ and for some $r > 0$, $p^{\ell_k}(s_{\ell_k}) > p^{\ell_k}(t_{\ell_k}) + r$. Hence, $\lim p^{\ell_k}(s_{\ell_k})/p^{\ell_k}(t_{\ell_k}) \geq 1 + r'$ for some $r' > 0$. In the following, we drop the subscript k for ℓ and denote the sequences as $t_\ell, s_\ell \in K^\ell$.

By the monotonicity of the preferences, the feasibility condition for $(\boldsymbol{x}_a^\ell)$ holds with the equality. Hence, $\sum_{a=1}^m \boldsymbol{x}_a^\ell(\{s_\ell\}) = \sum_{a=1}^m \omega_a^\ell(\{s_\ell\}) > 0$, so an a exists such that $\boldsymbol{x}_a^\ell(\{s_\ell\}) > 0$ for infinitely many ℓs. Take $1 < \gamma < 1 + r'$ and define

$$\boldsymbol{z}_a^\ell = \boldsymbol{x}_a^\ell - \boldsymbol{x}_a^\ell(\{s_\ell\})\delta_{s_\ell} + \gamma \boldsymbol{x}_a^\ell(\{s_\ell\})\delta_{t_\ell}.$$

Then, we have $\boldsymbol{p}^\ell \boldsymbol{z}_a^\ell < \boldsymbol{p}^\ell \boldsymbol{x}_a^\ell$ and $\boldsymbol{z}_a^\ell$ is in the a's budget set for infinitely many ℓs. By the (BRS) assumption, for this ℓ, we have

$\boldsymbol{z}_a^\ell \succsim_a^\ell \boldsymbol{x}_a^\ell$. By (MT), we can define $\tilde{\boldsymbol{z}}_a^\ell$ such that $\boldsymbol{p}\tilde{\boldsymbol{z}}_a^\ell < \boldsymbol{p}^\ell \boldsymbol{x}_a^\ell$ and $\boldsymbol{z}_a^\ell \prec_a \tilde{\boldsymbol{z}}_a^\ell$, and thus by (NTR), we have $\boldsymbol{x}_a^\ell \prec_a \tilde{\boldsymbol{z}}_a^\ell$, which is a contradiction. Therefore, the proof is complete. □

By applying Proposition G.14, we have a price vector $\boldsymbol{p} = p(t) \in C(K)$ with

$$(K^\ell, \boldsymbol{p}^\ell) \to (K, \boldsymbol{p}).$$

Recall that $(K^\ell, \boldsymbol{p}^\ell) \to (K, \boldsymbol{p})$ means that $K^\ell \to K$ in the closed convergence and for all sequences $t_\ell \in K^\ell$ with $t_\ell \to t$, $p^\ell(t_\ell) \to p(t)$. Then, by Proposition G.13, $\boldsymbol{p}^\ell \omega_a^\ell \to \boldsymbol{p}\omega_a$ and $\boldsymbol{p}^\ell \boldsymbol{x}_a^\ell \to \boldsymbol{p}\boldsymbol{x}_a$ for all a. Because $\sum_{a=1}^m \boldsymbol{x}_a^\ell \leq \sum_{a=1}^m \omega_a^\ell$ for all ℓ, then $\sum_{a=1}^m \boldsymbol{x}_a \leq \sum_{a=1}^m \omega_a^\ell$ in the limit. Hence, condition (E-2) is satisfied. We claim that for each a, $\boldsymbol{x}_a$ is maximal for $\prec_a$ in the budget set. For some t, $p(t) = 1$. Then, by the (AE) assumption, $\boldsymbol{p}\omega_a > 0$ for some a. If a vector $\boldsymbol{z} \in X_a$ exists with $\boldsymbol{p}\boldsymbol{z} \leq \boldsymbol{p}\omega_a$ and $\boldsymbol{x}_a \prec_a \boldsymbol{z}$, then by the $(CT_{\text{weak}*})$ assumption, we can take $\boldsymbol{z}' \in X_a$ such that $\boldsymbol{p}\boldsymbol{z}' < \boldsymbol{p}\omega_a$ and $\boldsymbol{x}_a \prec_a \boldsymbol{z}'$. Hence, for a sufficiently large ℓ, $\boldsymbol{p}^\ell \boldsymbol{z}^\ell < \boldsymbol{p}^\ell \omega_a^\ell$ and $\boldsymbol{x}_a^\ell \prec_a^\ell \boldsymbol{z}^\ell$, where $\boldsymbol{z}^\ell = \sum_{k=1}^\ell \boldsymbol{z}'(M_k^\ell)\delta_{t_k}$ which is a contradiction. By the (MT) assumption, $p(t) > 0$ for all $t \in K$, and thus, $\{\boldsymbol{x} \in X_a \mid \boldsymbol{p}\boldsymbol{x} = 0\} = \{\boldsymbol{0}\}$. Therefore,

$$\boldsymbol{p}\boldsymbol{x}_a \leq \boldsymbol{p}\omega_a \text{ and if } \boldsymbol{p}\boldsymbol{z} \leq \boldsymbol{p}\omega_a, \text{ then } \boldsymbol{x}_a \succsim_a \boldsymbol{z}, \; a = 1 \ldots m.$$

This completes the proof. □

We now make some remarks regarding the (BRS) assumption. Suppose that the preference relation $\prec$ is represented by a utility function $u : \mathscr{M}(K) \to \mathbb{R}$ and consider the Gateaux derivative, which is defined by

$$D_t u(\boldsymbol{x}) = \lim_{h \to 0} \frac{u(\boldsymbol{x} + h\delta_t) - u(\boldsymbol{x})}{h}$$

if the limit exists, where it is understood that $h \to 0^+$ if $\boldsymbol{x}(\{t\}) = 0$. $D_t u(\boldsymbol{x})$ is called the Gateaux derivative or the directional derivative of u at $\boldsymbol{x}$ in the direction of δ_t (see Appendix E). Let $R(\boldsymbol{x}, t, h)$ be the remainder term after approximating $u(\boldsymbol{x} + h\delta_t)$ by the first term in a Taylor series,

$$R(\boldsymbol{x}, t, h) = u(\boldsymbol{x} + h\delta_t) - (u(\boldsymbol{x}) + hD_t u(\boldsymbol{x})).$$

If the second derivative exists and it is uniformly bounded, then $(1/h)R(\boldsymbol{x}, h, t) \to 0$ as $h \to 0$. Now, we consider sequences that are described according to the (BRS) assumption. If $D_t u(\boldsymbol{x})$ exists for all $\boldsymbol{x} \in \mathscr{M}(K)$ and $t \in K$,

$$u(\boldsymbol{x}_\ell + a_\ell \delta_{s_\ell}) \approx u(\boldsymbol{x}_\ell) + a_\ell D_{t_\ell} u(\boldsymbol{x}_\ell),$$
$$u(\boldsymbol{x}_\ell + b_\ell \delta_{t_\ell}) \approx u(\boldsymbol{x}_\ell) + b_\ell D_{s_\ell} u(\boldsymbol{x}_\ell).$$

Hence, $u(\boldsymbol{x}_\ell + a_\ell \delta_{t_\ell}) \geq u(\boldsymbol{x}_\ell + b_\ell \delta_{s_\ell})$ if $a_\ell / b_\ell \geq D_{s_\ell} u(\boldsymbol{x}_\ell)/D_{t_\ell} u(\boldsymbol{x}_\ell)$ provided that $D_t u(\boldsymbol{x})$ is well behaved, e.g., $D_t u(\boldsymbol{x}) > 0$ and $D_t u(\boldsymbol{x})$ is jointly continuous with respect to $(t, \boldsymbol{x})$. This discussion explains the reason for the (BRS) assumption. In summary, we have obtained the following.

Proposition 4.2. *If the utility function $u(\boldsymbol{x})$ representing the preference $\prec$ is of class (Gâeaux) C^2 such that*

(a) $D_t u(\boldsymbol{x}) > 0$ *and* $D_t u(\boldsymbol{x})$ *is jointly continuous with respect to* $(t, \boldsymbol{x})$,

(b) *there exists a number* $b > 0$ *with* $\|D_t^2 u(\boldsymbol{x})\| \leq b$ *for all* $\boldsymbol{x} \in \mathscr{M}(K)$,

then $\prec$ satisfies the (BRS) assumption.

The first welfare theorem still holds for the economy $\mathcal{E}_{\mathscr{M}}$. However, the second welfare theorem is more intricate because the norm interior of $\mathscr{M}(K)$ is empty. We discuss this more generally in Section 4.4.

4.4 An Exchange Economy on a Topological Vector Lattice

In this section, we discuss a market with a mathematically general commodity space, which contains the spaces ℓ^p and $L^p([c, d])$ $(1 \leq p \leq +\infty)$ as special cases. This mathematical generalization makes the structure and logic of the equilibrium existence theorems and the welfare theorems more transparent.

The commodity space considered in this section is assumed to be a locally convex topological vector lattice (see Appendix G). A topological vector space L with *order relation* $\leq$ is denoted by $(L, \leq)$. The ordered linear space or the linear space with an order relation $\leq$ is called a *vector lattice* (*Riesz space*) if and only if for any $\boldsymbol{x}, \boldsymbol{y} \in L$, elements of L denoted by $\boldsymbol{x} \vee \boldsymbol{y}$ (and $\boldsymbol{x} \wedge \boldsymbol{y}$) exist such that (i) $\boldsymbol{x} \leq \boldsymbol{x} \vee \boldsymbol{y}$ and $\boldsymbol{y} \leq \boldsymbol{x} \vee \boldsymbol{y}$ and (ii) $\boldsymbol{x} \leq \boldsymbol{z}$ and $\boldsymbol{y} \leq \boldsymbol{z}$ imply $\boldsymbol{x} \vee \boldsymbol{y} \leq \boldsymbol{z}$ ((i') $\boldsymbol{x} \wedge \boldsymbol{y} \leq \boldsymbol{x}$ and $\boldsymbol{x} \wedge \boldsymbol{y} \leq \boldsymbol{y}$ and (ii') $\boldsymbol{z} \leq \boldsymbol{x}$ and $\boldsymbol{z} \leq \boldsymbol{y}$ imply $\boldsymbol{z} \leq \boldsymbol{x} \wedge \boldsymbol{y}$). Then, for any $\boldsymbol{x} \in L$, we can define the *positive part* $\boldsymbol{x}_+ = \boldsymbol{x} \vee \mathbf{0}$, the *negative part* $\boldsymbol{x}_- = (-\boldsymbol{x}) \vee \mathbf{0}$, and the *absolute value* $|\boldsymbol{x}| = \boldsymbol{x}_+ + \boldsymbol{x}_-$.

A map $f : L \to M$ is *uniformly continuous* if for every neighborhood $W \subset M$ of $\mathbf{0}$, there exists a neighborhood $V \subset L$ of $\mathbf{0}$ with $f(\boldsymbol{x}) - f(\boldsymbol{y}) \in W$ whenever $\boldsymbol{x} - \boldsymbol{y} \in V$. A vector lattice is called a topological vector lattice if it is a topological vector space and the lattice operations $\vee$ and $\wedge$ are uniformly continuous.

An element of the dual space L^* or a continuous linear functional on L is called a price vector or a *price functional.*

m consumers $a = 1 \dots m$ exist. The consumer a has the common closed consumption set $X_a = L_+$, the preference relation $\prec_a \subset X_a \times X_a$ that satisfies the basic postulates, (NTR), (CV) and (MT), and the initial endowment vector $\omega_a \in L_+$. The (CT) assumption in this section is stated as the continuity in the topology equipped with L. The $2m$-tuple $\mathcal{E} = (\prec_a, \omega_a)$ is called an (*exchange*) *economy*.

The first problem in this setting was identified in Section 4.3 where the commodity space is $ca(K)$, although we did not comment on the problem explicitly. We simply claimed that the (BRS) assumption is needed and proved the theorem with it. Example 4.4 given by Mas-Colell shows that in the space $ca(K)$, the non-zero supporting price vector, or the vector $\boldsymbol{p} \geq \mathbf{0}$ with $\boldsymbol{p} \neq \mathbf{0}$ such that $\boldsymbol{x} \succsim_a \boldsymbol{z}$ implies that $\boldsymbol{p}\boldsymbol{x} \geq \boldsymbol{p}\boldsymbol{z}$, may cease to exist.

Example 4.4. Let $\hat{K} = \mathbb{N} \cup \{\infty\}$ be the one-point compactification of $\mathbb{N}$ (Theorem B.2). The commodity space is assumed to be $L = ca(\hat{K})$. For $\boldsymbol{x} \in L$ and $t \in \hat{K}$, let $x^t = \boldsymbol{x}(\{t\})$ for $0 \leq t \leq \infty$.

For every $t \in \hat{K}$, define a function $u_t : \mathbb{R}_+ \to \mathbb{R}_+$ by

$$u_t(s) = \begin{cases} 2^t s & \text{for } s \leq 2^{-2t}, \\ 2^{-t} - 2^{-2t} + s & \text{otherwise,} \end{cases}$$

see Figure 4.1. The preference relation $\prec$ on the consumption set L_+ is then defined by the following utility function:

$$U(\boldsymbol{x}) = \sum_{t=0}^{\infty} u_t(x^t) \quad \text{for } \boldsymbol{x} = (x^t) \in L_+.$$

It is easy to see that $U(\boldsymbol{x})$ satisfies (MT) and the continuity $(CT_{\text{weak*}})$. Now, let $\omega = (\omega^t) \in L_+$ be defined by $\omega^t = 2^{-2t}$ for $t < \infty$ and $\omega^\infty = 1$. Then, we can show that a non-zero continuous linear functional $\boldsymbol{p}$ does not exist such that $\boldsymbol{pz} \geq \boldsymbol{p}\omega$ whenever $\boldsymbol{z} \succsim \omega$.

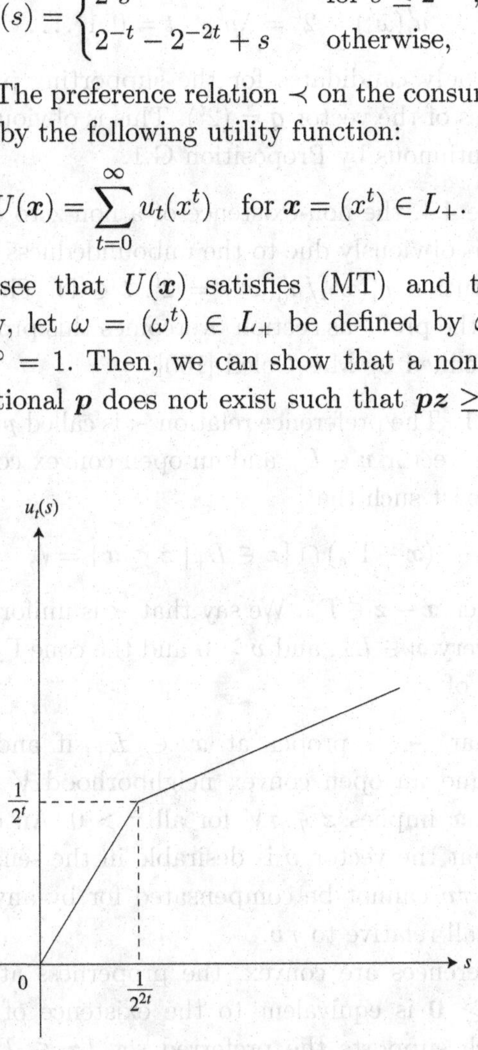

Figure 4.1. Graph of u_t.

Indeed, let $\boldsymbol{p}$ be this functional. For $\boldsymbol{x} \geq \mathbf{0}$, we have $\omega + \boldsymbol{x} \succsim \omega$. Hence, $\boldsymbol{px} \geq 0$, or $\boldsymbol{p}$ is a positive linear functional. For $t \in \hat{K}$, we denote $p^t = \boldsymbol{pe}_t$, where $\boldsymbol{e}_t = (e_t^s) = (\delta_t^s)$ and $\delta_t^s = 1$ for $t = s$; and $\delta_t^s = 0$ otherwise. From the first-order condition of the utility maximization, we have

$$u_t'(\omega^t) = 2^t = \lambda p^t, \quad t = 0, 1, \ldots,$$

and thus the only candidates for the supporting price vectors are scalar multiples of the vector $\boldsymbol{q} = (2^t)$. This is obviously unbounded, so it is not continuous by Proposition G.1.

In Example 4.4, the non-existence of a non-zero continuous supporting price is obviously due to the unboundedness of the marginal rate of substitution $u_t'(\omega^t)/u_0'(\omega^0) = 2^t$, $t \in \mathbb{N}$. The (BRS) condition stated in the previous section overcomes this problem. The next concept was defined by Mas-Colell [150].

Definition 4.1. The preference relation $\prec$ is called *proper* at $\boldsymbol{x} \in L_+$ if and only if a vector $\boldsymbol{v} \in L_+$ and an open convex cone $\Gamma_{\boldsymbol{x}}$ at $\mathbf{0} \in L$ containing $\boldsymbol{v}$ exist such that

$$(\boldsymbol{x} - \Gamma_{\boldsymbol{x}}) \cap \{\boldsymbol{z} \in L_+ | \, \boldsymbol{z} \succsim \boldsymbol{x}\} = \emptyset,$$

or if $\boldsymbol{z} \succsim \boldsymbol{x}$, then $\boldsymbol{x} - \boldsymbol{z} \notin \Gamma_{\boldsymbol{x}}$. We say that $\prec$ is uniformly proper if it is proper at every $\boldsymbol{x} \in L_+$, and $\boldsymbol{v} \geq \mathbf{0}$ and the cone $\Gamma_{\boldsymbol{x}}$ can be chosen independently of $\boldsymbol{x}$.

In particular, $\prec$ is proper at $\boldsymbol{x} \in L_+$ if and only if there exist $\boldsymbol{v} \geq \mathbf{0}$ and an open convex neighborhood V of $\mathbf{0}$ such that $\boldsymbol{x} - \tau\boldsymbol{v} + \boldsymbol{z} \succsim \boldsymbol{x}$ implies $\boldsymbol{z} \notin \tau V$ for all $\tau > 0$. An economic interpretation is that the vector $\boldsymbol{v}$ is desirable in the sense that the loss of an amount $\tau\boldsymbol{v}$ cannot be compensated for by any vector that is excessively small relative to $\tau\boldsymbol{v}$.

When preferences are convex, the properness at $\boldsymbol{x} \in L_+$ with respect to $\boldsymbol{v} \geq \mathbf{0}$ is equivalent to the existence of a price vector $\boldsymbol{p} \in L^*$, which supports the preferred set $\{\boldsymbol{z} \in L_+ | \, \boldsymbol{z} \succsim \boldsymbol{x}\}$ at $\boldsymbol{x}$ with the additional property that $\boldsymbol{pv} > 0$. Indeed, if the supporting price $\boldsymbol{p}$ exists, then we can take the properness cone as

$\Gamma_{\boldsymbol{x}} = \{\boldsymbol{z} \in L | \, \boldsymbol{pz} > 0\}$. Conversely, if $\prec$ is proper at $\boldsymbol{x}$ with respect to $\boldsymbol{v}$, then $\{\boldsymbol{z} \in L_+ | \, \boldsymbol{z} \succsim \boldsymbol{x}\}$ and $\boldsymbol{x} - \Gamma_{\boldsymbol{x}}$ are disjoint convex sets and $\text{int}(\boldsymbol{x} - \Gamma_{\boldsymbol{x}}) \neq \emptyset$, and thus the Hahn–Banach separation theorem (Theorem G.6) provides a continuous linear functional $\boldsymbol{p} \in L^*$ that separates them, or $\boldsymbol{p}(\boldsymbol{x} - \boldsymbol{y}) \leq \boldsymbol{pz}$ for each $\boldsymbol{y} \in \Gamma_{\boldsymbol{x}}$ and $\boldsymbol{z} \succsim \boldsymbol{x}$. Because $\Gamma_{\boldsymbol{x}}$ is an open cone at $\mathbf{0}$ and contains $\boldsymbol{v}$, then it follows that $\boldsymbol{py} > 0$ for each $\boldsymbol{y} \in \Gamma_{\boldsymbol{x}}$, and thus $\boldsymbol{pv} > 0$ and $\boldsymbol{pz} \geq \boldsymbol{px}$ for $\boldsymbol{z} \succsim \boldsymbol{x}$.

It is easy to show that if the positive cone L_+ of the commodity space L has a non-empty interior such as the spaces ℓ^∞ or $L^\infty[0,1]$ with the sup-norm, then the assumption (MT) implies the properness. In Section 4.2, we could have also shown that the conditions (MT), (CV), and the (Mackey) continuity (CT_∞) yield the properness at strictly positive vectors. This is actually a reason for the success of Bewley [27] in his proof of the existence of competitive equilibria on $L^\infty[0,1]$.

Let $\mathcal{F}$ be the set of feasible allocations such that

$$\mathcal{F} = \left\{ (\boldsymbol{x}_a) \in L_+^m \,\middle|\, \sum_{a=1}^{m} \boldsymbol{x}_a \leq \sum_{a=1}^{m} \omega_a \right\}.$$

A feasible allocation $(\boldsymbol{x}_a) \in \mathcal{F}$ is said to be *weakly Pareto-optimal* if no other feasible allocation $(\boldsymbol{z}_a) \in \mathcal{F}$ such that $\boldsymbol{x}_a \prec_a \boldsymbol{z}_a$ exists for all $a = 1 \dots m$. We can show that the first fundamental theorem of welfare economics still holds without any modification of Proposition 2.10 in Chapter 2. However, the second welfare theorem is technically more intricate on the infinite dimensional commodity space. To prove this, we recall the concepts of competitive equilibrium $(\boldsymbol{p}, \boldsymbol{x}_a) \in L_+^* \times L_+^m$ and quasi-equilibrium $(\boldsymbol{x}_a) \in L_+^m$ relative to the price vector $\boldsymbol{p} \in L_+^*$, which we defined as Definitions 2.4 and 2.5 in Chapter 2, respectively. The properness condition ensures the individual supportability. We may expect that (weakly) optimal feasible allocations can be supported by some price vector, i.e., the second welfare theorem holds by virtue of the properness. However, Example 4.5 given by Jones shows that this conjecture is incorrect.

Example 4.5. Let $L = L^{\infty}[0,1]$ be the set of essentially bounded functions on $[0,1]$ and $P = C^1[0,1]$ is the set of all continuously differentiable functions on $[0,1]$. (A function is called continuously differentiable if it has the first derivative which is continuous.) We consider the dual pair (L, P) and the $\sigma(L,P)$-topology on L. This topology is described by the net as $\boldsymbol{x}_\kappa = x_\kappa(t) \to \boldsymbol{x} = x(t)$ if and only if $\int_0^1 x_\kappa(t)p(t)dt \to \int_0^1 x(t)p(t)dt$ for every continuously differentiable function $p(t) \in C^1[0,1]$ (see Appendix G). Two consumers exist in the economy, $a = 1, 2$. Consumer 1 has the following utility function:

$$u_1(\boldsymbol{x}) = \int_0^1 (1-t)x(t)dt \text{ for every } \boldsymbol{x} = x(t) \in L_+,$$

and consumer 2's utility function is given as follows:

$$u_2(\boldsymbol{x}) = \int_0^1 tx(t)dt \text{ for every } \boldsymbol{x} = x(t) \in L_+.$$

The utility functions are linear and continuous, so they are proper. Let $\omega(t) = 1$ for all t. It is easy to verify that the feasible allocation $(\boldsymbol{x}_1, \boldsymbol{x}_2) = (x_1(t), x_2(t))$, which is defined by $x_1(t) = 1$ if $t \leq 1/2$, $x_1(t) = 0$ if $t > 1/2$ and $x_2(t) = 0$ if $t \leq 1/2$, $x_2(t) = 1$ if $t > 1/2$, is weakly optimal. However, it cannot be supported by any continuous linear functional on L because the only candidate for the supporting price is given by $p(t) = 1 - t$ if $t \leq 1/2$, $p(t) = t$ if $t > 1/2$, which is obviously not differentiable at $t = 1/2$, and thus, it is not continuous as a linear functional on L.

The topology considered in the example given above is quite artificial. Indeed, we give the topology on L as the dual pairing of $C^1[0,1]$, which does not make L a topological vector lattice. The key property of the second fundamental theorem is that the commodity space is a locally solid Riesz space which is equivalent to the uniform continuity of the lattice operations [2, Theorem 8.41, p. 334]. In this section, we are mainly concerned with a quasi-equilibrium rather than a competitive equilibrium because we prove the existence theorem via the Negishi-type proof, which utilizes a version of the second fundamental theorem which supports the former.

Theorem 4.5. *Let $\mathcal{E} = (\prec_a, \omega_a)$ be an exchange economy on a topological vector lattice L. For every consumer $a = 1 \ldots m$, we suppose that the preference relations with the basic postulates satisfy (NTR), (CV), (MT), and the uniform properness. Then, every weakly Pareto-optimal allocation is a quasi-equilibrium relative to some price vector $\boldsymbol{p} \in L_+^*$ such that $\boldsymbol{p}\sum_{a=1}^m \boldsymbol{v}_a = 1$, where $\boldsymbol{v}_a \in L_+$ is a vector, as given in Definition 4.1.*

Proof. Let an allocation $(\boldsymbol{x}_a)_{a=1}^m$ be weakly Pareto-optimal and set

$$Z = \left\{ \sum_{a=1}^{m} (\boldsymbol{z}_a - \boldsymbol{x}_a) \,\middle|\, \boldsymbol{z}_a \succsim_a \boldsymbol{x}_a,\ a = 1 \ldots m \right\}.$$

Clearly, the set Z is convex. Let $V_a \subset L$ be a neighborhood of $\boldsymbol{0}$ such that $\boldsymbol{x} - \tau \boldsymbol{v}_a + \boldsymbol{z} \succsim_a \boldsymbol{x}$ implies that $\boldsymbol{z} \notin \tau V_a$ for all $\tau > 0$, where $\boldsymbol{v}_a$ is a vector of the properness (Definition 4.1). Let $V = \cap_{a=1}^m V_a$ and $\boldsymbol{v} = \sum_{a=1}^m \boldsymbol{v}_a$. Without loss of generality, we can assume that V is convex, $-V = V$, and it is solid, i.e., $|\boldsymbol{u}| \leq |\boldsymbol{v}|$ and $\boldsymbol{v} \in V$ imply that $\boldsymbol{u} \in V$ by the uniform continuity of the lattice operations; see [2, Theorem 8.41, p. 334]. Let Γ be the open convex cone spanned by $\boldsymbol{v} + V$. Then, we claim that $Z \cap (-\Gamma) = \emptyset$.

Suppose $Z \cap (-\Gamma) \neq \emptyset$. Then, $\boldsymbol{z}_a \succsim_a \boldsymbol{x}_a$ denoted by $\boldsymbol{z} = \sum_{a=1}^m \boldsymbol{z}_a$ and $\boldsymbol{x} = \sum_{a=1}^m \boldsymbol{x}_a$ exist such that we have $\boldsymbol{z} - \boldsymbol{x} \in -\Gamma$, or $\boldsymbol{z} - (\boldsymbol{x} - \tau \boldsymbol{v}) \in \tau V$ for some $\tau > 0$. Set $\boldsymbol{y} = \boldsymbol{x} - \tau \boldsymbol{v}$. Because $\boldsymbol{z} \geq \boldsymbol{0}$, then we have $\boldsymbol{y} - \boldsymbol{z} \leq \boldsymbol{y} \leq \boldsymbol{x}$, and thus, $(\boldsymbol{y} - \boldsymbol{z})_+ = (\boldsymbol{y} - \boldsymbol{z}) \vee \boldsymbol{0} \leq \boldsymbol{x}$. Because $\boldsymbol{y} - \boldsymbol{z} = (\boldsymbol{y} - \boldsymbol{z})_+ - (\boldsymbol{y} - \boldsymbol{z})_-$, then it follows that

$$\boldsymbol{z} = (\boldsymbol{y} - \boldsymbol{x}) + \boldsymbol{x} - (\boldsymbol{y} - \boldsymbol{z})_+ + (\boldsymbol{y} - \boldsymbol{z})_- \geq -\tau \boldsymbol{v} + (\boldsymbol{y} - \boldsymbol{z})_-,$$

or $(\boldsymbol{y} - \boldsymbol{z})_- \leq \boldsymbol{z} + \tau \boldsymbol{v}$.

Because $\boldsymbol{z} + \tau \boldsymbol{v} = \sum_{a=1}^m (\boldsymbol{z}_a + \tau \boldsymbol{v}_a)$, Theorem G.5 implies that we can write $(\boldsymbol{y} - \boldsymbol{z})_- = \sum_{a=1}^m \boldsymbol{w}_a$, where $\boldsymbol{0} \leq \boldsymbol{w}_a \leq \boldsymbol{z}_a + \tau \boldsymbol{v}_a$ for each a. Define $\boldsymbol{z}'_a = \boldsymbol{z}_a + \tau \boldsymbol{v}_a - \boldsymbol{w}_a \geq \boldsymbol{0}$. Suppose that $\boldsymbol{z}_a \succsim_a \boldsymbol{z}'_a$ for some a. The preference is proper, so this implies that $\boldsymbol{w}_a \notin \tau V$. In addition, $\boldsymbol{0} \leq \boldsymbol{w}_a \leq (\boldsymbol{y} - \boldsymbol{z})_- \leq |\boldsymbol{y} - \boldsymbol{z}|$ and $\boldsymbol{y} - \boldsymbol{z} \in \tau V$, and thus, $\boldsymbol{w}_a \in \tau V$, which is a contradiction. Thus, $Z \cap (-\Gamma) = \emptyset$, or $\boldsymbol{0} \notin Z + \Gamma$.

$Z + \Gamma$ is convex and an open set, and $\mathbf{0} \notin Z + \Gamma$, so we can apply the Hahn–Banach separation theorem (Theorem G.6), and a non-zero, continuous linear functional $\boldsymbol{p}$ exists such that $\boldsymbol{py} > 0$ for all $\boldsymbol{y} \in Z + \Gamma$. Γ is a cone and $\mathbf{0} \in Z$, so we find that $\boldsymbol{pz} \geq 0$ for all $\boldsymbol{z} \in Z$ and $\boldsymbol{py} > 0$ for all $\boldsymbol{y} \in \Gamma$. In particular, $\boldsymbol{pv} > 0$, so without loss of generality, we can put $\boldsymbol{pv} = 1$. If $\boldsymbol{z} \in V$, then $\boldsymbol{v} - \boldsymbol{z} \in \Gamma$ and so $\boldsymbol{pz} \leq \boldsymbol{pv} = 1$. The same argument applies to $-\boldsymbol{z}$, so we obtain $|\boldsymbol{pz}| \leq 1$. Finally, let $\boldsymbol{z} \succsim_a \boldsymbol{x}_a$. Then, $\boldsymbol{z} - \boldsymbol{x}_a \in Z$, and thus $\boldsymbol{pz} \geq \boldsymbol{px}_a$, which shows that $(\boldsymbol{p}, \boldsymbol{x}_a)$ is a quasi-equilibrium relative to $\boldsymbol{p}$. The positivity of $\boldsymbol{p} \geq \mathbf{0}$ follows from the assumption (MT) of $\prec_a$. □

We can assume $\boldsymbol{v}_a = \omega_a$ for all a in order to obtain a desirable condition $\boldsymbol{p} \sum_{a=1}^{m} \omega_a > 0$. Example 4.6 given by Araujo shows that the utility possibility frontier does not need to be closed.

Example 4.6. The commodity space L is assumed to be ℓ^∞, which is the space of the bounded sequences with the supremum norm (see Section 4.2). Suppose that the total endowment $\omega = (1, 1 \ldots) = \mathbf{1}$. Two consumers exist. Consumer 1 has the utility function

$$u_1(\boldsymbol{x}) = \liminf_{t \to \infty} x^t \text{ for } \boldsymbol{x} = (x^t) \in \ell_+^\infty,$$

and consumer 2's utility function is given by

$$u_2(\boldsymbol{x}) = \sum_{t=0}^{\infty} \frac{x^t}{2^{t+1}} \text{ for } \boldsymbol{x} = (x^t) \in \ell_+^\infty.$$

It is easy to show that the utility possibility frontier

$$U(\mathcal{F}) = \{(u_1(\boldsymbol{x}_1), u_2(\boldsymbol{x}_2)) |\ \boldsymbol{x}_1 + \boldsymbol{x}_2 \leq \omega,\ \boldsymbol{x}_1, \boldsymbol{x}_2 \geq \mathbf{0}\}$$

has the form depicted in Figure 4.2. In the figure, the point $(0, 1)$ is included in $U(\mathcal{F})$.

We exclude this type of situation according to the next assumption of closedness of the utility possibility frontier:

(CL): Let $(\boldsymbol{x}_a^k) \in \mathcal{F}$ be a sequence of feasible allocations such that $j > k$ implies that $\boldsymbol{x}_a^j \succsim_a \boldsymbol{x}_a^k$ for all $a = 1 \ldots m$. Then, a

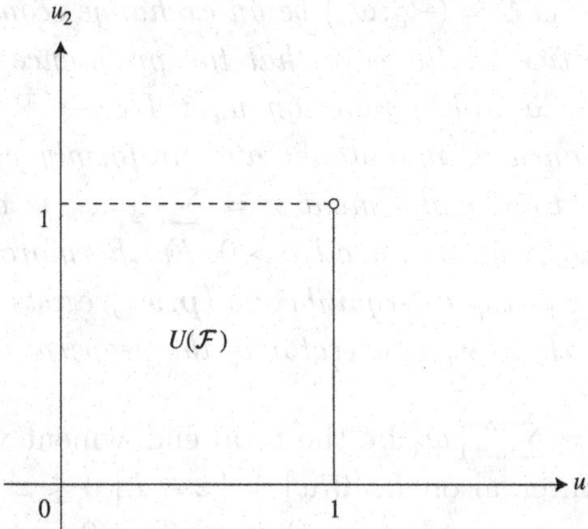

Figure 4.2. Example 4.6.

feasible allocation $(\boldsymbol{x}_a) \in \mathcal{F}$ exists such that $\boldsymbol{x}_a \succsim_a \boldsymbol{x}_a^k$ for all $k \in \mathbb{N}$ and all $a = 1 \dots m$.

It is easy to show that the (CL) assumption is equivalent to the closedness of $U(\mathcal{F})$, but which conditions on the individual preferences yield the (CL) condition? Let the sequences $(\boldsymbol{x}_a^k)$ be as defined by (CL). Then, $Z^k = \{(\boldsymbol{z}_a) \in \mathcal{F} |\ \boldsymbol{z}_a \succsim_a \boldsymbol{x}_a^k$ for all $a\}$ constitutes a decreasing sequence of closed subsets of $\mathcal{F}$. We need to find the conditions such that $\cap_{k=0}^{\infty} Z_k \neq \emptyset$. Clearly, this intersection is non-empty if the sets $\{\boldsymbol{z} \in L_+ |\ \boldsymbol{z} \succsim_a \boldsymbol{x}_a^k\}$ are closed for all a and the feasible allocation set $\mathcal{F}$ is compact, since $Z^k = \cap_{b=1}^{m} \{(\boldsymbol{z}_a) \in \mathcal{F} |\ \boldsymbol{z}_b \succsim_b \boldsymbol{x}_b^k\}$. When $L = ca(K)$, we assumed that the sets $\{\boldsymbol{z} \in L_+ |\ \boldsymbol{z} \succsim_a \boldsymbol{x}_a^k\}$ are closed in the weak* topology and we ensured the compactness of $\mathcal{F}$ in the weak* topology by Alaoglu's theorem (Theorem G.3). When $L = \ell^\infty$, we assumed that the sets $\{\boldsymbol{x} \in L_+ |\ \boldsymbol{x} \succsim_a \boldsymbol{x}_a\}$ are closed in the Mackey topology, but the Mackey topology is equal to the weak* topology on the convex sets by Theorem G.2. This is exactly how we obtained the (CL) condition in Sections 4.2 and 4.3. Using these methods, we can prove the existence theorem on a topological vector lattice. In the following theorem, the preferences are represented by utility functions (this assumption is actually unnecessary; see Section 4.6).

Theorem 4.6. *Let $\mathcal{E} = (\prec_a, \omega_a)$ be an exchange economy on a topological vector lattice L. Suppose that the preference of consumer a is represented by a utility function $u_a : L_+ \to \mathbb{R}$, which is continuous, quasi-concave, monotone, and uniformly proper. We also assume that the total endowment $\omega = \sum_{a=1}^m \omega_a$ is desirable in the sense that $u_a(\tau\omega) > u_a(\mathbf{0})$ for all $\tau > 0$. Furthermore, the (CL) condition holds. Then, a quasi-equilibrium $(\boldsymbol{p}, \boldsymbol{x}_a)$ exists for $\mathcal{E}$ such that $\boldsymbol{p}\sum_{a=1}^m \boldsymbol{v}_a = 1$, where $\boldsymbol{v}_a$ is a vector of the properness condition.*

Proof. Let $\omega = \sum_{a=1}^m \omega_a$ be the total endowment vector and consider the order interval on L, $[\mathbf{0}, \omega] = \{\boldsymbol{z} \in L | \, \mathbf{0} \leq \boldsymbol{z} \leq \omega\}$. Without loss of generality, we can assume that $u_a(\mathbf{0}) = 0$ and $u_a(\omega) = 1$, $a = 1 \dots m$. Define a map $U : [\mathbf{0}, \omega]^m \to [0,1]^m$ by $U((\boldsymbol{x}_a)) = (u_a(\boldsymbol{x}_a))$, and $h : \Delta \to \mathbb{R}$ by $h(\boldsymbol{r}) = \sup\{\tau \in \mathbb{R} | \, \tau\boldsymbol{r} \in U(\mathcal{F})\}$, where as usual, $\Delta = \{\boldsymbol{r} = (r^a) \in \mathbb{R}_+^m | \, \sum_{a=1}^m r^a = 1\}$ is the $(m-1)$-simplex. Then, we have the following.

Lemma 4.6. *$h : \Delta \to [0, m]$ is a well-defined continuous function with $h(\boldsymbol{r}) > 0$ for all $\boldsymbol{r} \in \Delta$.*

Proof. Because $\omega \geq \mathbf{0}$ is desirable, it follows that $0 = u_a(\mathbf{0}) < u_a(\tau\omega)$ for all $\tau > 0$, $a = 1 \dots m$. Therefore, we have $h(\boldsymbol{r}) > 0$ for all $\boldsymbol{r} \in \Delta$. Let $\boldsymbol{r}_n \to \boldsymbol{r} \in \Delta$ and $\tau\boldsymbol{r} = U(\boldsymbol{x})$ for some $\boldsymbol{x} \in \mathcal{F}$ and $\tau > 0$. Take $0 < \upsilon < \tau$. Without loss of generality, we can assume that $\upsilon r_n^a < \tau r_n^a$ for all n and for all a with $r^a > 0$. For a with $r^a = 0$, let $u_a(\gamma_n^a\omega) = \upsilon v_n^a$. Then, $\gamma_n^a \to 0$ for a with $r^a = 0$, and thus, $\gamma_n = \sum_{r^a=0} \gamma_n^a \to 0$. Define an allocation $\boldsymbol{y}^n = (\boldsymbol{y}_a^n)$ by $\boldsymbol{y}_a^n = \gamma_n^a\omega$ if $r^a = 0$, and $\boldsymbol{y}_a^n = (1 - \gamma_n)\boldsymbol{x}_a$ otherwise. Then,

$$\sum_{a=1}^m \boldsymbol{y}_a^n = \sum_{r^a=0} \boldsymbol{y}_a^n + \sum_{r^a>0} \boldsymbol{y}_a^n = \gamma_n\omega + (1-\gamma_n)\sum_{r^a>0} \boldsymbol{x}_a$$
$$\leq \gamma_n\omega + (1-\gamma_n)\sum_{a=1}^m \boldsymbol{x}_a \leq \omega,$$

or $(\boldsymbol{y}_a^n) \in \mathcal{F}$. For a sufficiently large n, we have $u_a(\boldsymbol{y}_a^n) \geq vr_n^a$ for $a = 1 \dots m$. Take $\mu_a \in [0,1]$ such that $u_a(\mu_a \boldsymbol{y}_a^n) = vr_n^a$ for $a = 1 \dots m$. Then, $\sum_{a=1}^m \mu_a \boldsymbol{y}_a^n \leq \sum_{a=1}^m \boldsymbol{y}_a^n \leq \omega$, or we obtain a feasible allocation and its utility is precisely $v\boldsymbol{r}_n$. Because $v < \tau$ is arbitrary, then we conclude that $h(\boldsymbol{r}) \leq \liminf_{n\to\infty} h(\boldsymbol{r}_n)$. Similarly, we can show that $\limsup_{n\to\infty} h(\boldsymbol{r}_n) \leq h(\boldsymbol{r})$. □

By the (CL) assumption, for any $\boldsymbol{r} \in \Delta$ we can find a feasible allocation $\boldsymbol{x}(\boldsymbol{r}) = (\boldsymbol{x}_a(\boldsymbol{r})) \in \mathcal{F}$ such that $U(\boldsymbol{x}(\boldsymbol{r})) = h(\boldsymbol{r})\boldsymbol{r}$. By the assumption (MT), it follows that $\sum_{a=1}^m \boldsymbol{x}_a(\boldsymbol{r}) = \omega$. Let $V = \cap_{a=1}^m V_a$ and $\boldsymbol{v} = \sum_{a=1}^m \boldsymbol{v}_a$ as in the proof of Theorem 4.5 by which we can define

$$P(\boldsymbol{r}) = \left\{ \boldsymbol{p} \in L_+^* \,\middle|\, \boldsymbol{p}\boldsymbol{x}_a(\boldsymbol{r}) \leq \boldsymbol{p}\boldsymbol{x} \text{ whenever } \boldsymbol{x} \succsim_a \boldsymbol{x}_a(\boldsymbol{r}), a = 1 \dots m, \right.$$
$$\left. \boldsymbol{p}\boldsymbol{v} = 1, |\boldsymbol{p}\boldsymbol{z}| \leq 1 \text{ for all } \boldsymbol{z} \in V \right\}.$$

Theorem 4.5 implies that $P(\boldsymbol{r}) \neq \emptyset$ for all $\boldsymbol{r} \in \Delta$. We define

$$\boldsymbol{\sigma}(\boldsymbol{r}) = \{(\boldsymbol{p}(\omega_1 - \boldsymbol{x}_1(\boldsymbol{r})) \cdots \boldsymbol{p}(\omega_m - \boldsymbol{x}_m(\boldsymbol{r}))) \in \mathbb{R}^m |\, \boldsymbol{p} \in P(\boldsymbol{r})\}.$$

If $\boldsymbol{s} = (s^a) \in \boldsymbol{\sigma}(\boldsymbol{r})$, then $\sum_{a=1}^m s^a = 0$, and $\boldsymbol{0} \in \boldsymbol{\sigma}(\boldsymbol{r})$ if and only if $(\boldsymbol{p}, \boldsymbol{x}(\boldsymbol{r}))$ is a quasi-equilibrium, where $\boldsymbol{p} \in P(\boldsymbol{r})$.

Lemma 4.7. *The correspondence* $\boldsymbol{\sigma}$: $\Delta \to \mathbb{R}^m$ *is upper hemicontinuous.*

Proof. Let $\boldsymbol{r}_n \to \boldsymbol{r} \in \Delta$, $\boldsymbol{s}_n \in \boldsymbol{\sigma}(\boldsymbol{r}_n)$ for all $n \in \mathbb{N}$ and $\boldsymbol{s}_n \to \boldsymbol{s}$. We want to show that $\boldsymbol{s} \in \boldsymbol{\sigma}(\boldsymbol{r})$. Let $\boldsymbol{p}_n \in P(\boldsymbol{r}_n)$ be such that $\boldsymbol{s}_n = (s_n^a) = (\boldsymbol{p}_n(\omega_a - \boldsymbol{x}_a(\boldsymbol{r}_n)))$. Because $|\boldsymbol{p}_n\boldsymbol{z}| \leq 1$ for all $\boldsymbol{z} \in V$, then we can assume that the sequence $(\boldsymbol{p}_n)$ has a weak* limit $\boldsymbol{p}$ by Alaoglu's theorem (Theorem G.3) or $\boldsymbol{p}_n\boldsymbol{x} \to \boldsymbol{p}\boldsymbol{x}$ for all $\boldsymbol{x} \in L$. Because $\boldsymbol{p}_n\boldsymbol{v} = 1$ for all n, then it follows that $\boldsymbol{p}\boldsymbol{v} = 1$. Let $\boldsymbol{z} \succsim_a \boldsymbol{x}_a(\boldsymbol{r})$. Then, by (MT), a $\boldsymbol{z}'$ exists that is arbitrarily close to $\boldsymbol{z}$ and $\boldsymbol{z} \prec_a \boldsymbol{z}'$. Hence, $\boldsymbol{x}_a(\boldsymbol{r}_n) \prec_a \boldsymbol{z}'$ for all sufficiently large n. Therefore, $\boldsymbol{p}_n\boldsymbol{z}' \geq \boldsymbol{p}_n\boldsymbol{x}_a(\boldsymbol{r}_n) = \boldsymbol{p}_n\omega_a - s_n^a$ for all sufficiently large n. By taking the limit, we have $\boldsymbol{p}\boldsymbol{z}' \geq \boldsymbol{p}\omega_a - s^a = \boldsymbol{p}\boldsymbol{x}_a(\boldsymbol{r})$. $\boldsymbol{z}'$ is arbitrarily close to $\boldsymbol{z}$, so we conclude that

$$\boldsymbol{pz} \geq \boldsymbol{p}\omega_a - s^a (= \boldsymbol{px}_a(\boldsymbol{r})) \text{ whenever } \boldsymbol{z} \succsim_a \boldsymbol{x}_a(\boldsymbol{r}),\ a = 1 \dots m.$$

Indeed, if $\boldsymbol{px}_a(\boldsymbol{r}) = 0$, then $\boldsymbol{pz} \geq 0$ holds trivially. Suppose that $\boldsymbol{z} \succsim_a \boldsymbol{x}_a(\boldsymbol{r})$ and $\boldsymbol{pz} < \boldsymbol{px}_a(\boldsymbol{r})$. Then, by (MT), we can take a $\boldsymbol{z}'$ that is sufficiently close to $\boldsymbol{z}$ such that $\boldsymbol{z} \prec_a \boldsymbol{z}'$ and $\boldsymbol{pz}' < \boldsymbol{px}_a(\boldsymbol{r})$, which is a contradiction.

By setting $\boldsymbol{z} = \boldsymbol{x}_a(\boldsymbol{r})$, we have $\boldsymbol{px}_a(\boldsymbol{r}) \geq \boldsymbol{p}\omega_a - s^a$ for all a. Because $\sum_{a=1}^m \boldsymbol{x}_a(\boldsymbol{r}) = \sum_{a=1}^m \omega_a$, then it follows that $\boldsymbol{px}_a(\boldsymbol{r}) = \boldsymbol{p}\omega_a - s^a$, which proves $\boldsymbol{s} \in \boldsymbol{\sigma}(\boldsymbol{r})$.

The last step of the proof is now standard. For each $\boldsymbol{r}$, an $\epsilon > 0$ exists such that $\epsilon(\omega_a - \boldsymbol{x}_a(\boldsymbol{r})) \in V$, and thus $|\boldsymbol{p}(\omega_a - \boldsymbol{x}_a(\boldsymbol{r}))| \leq 1/\epsilon$ for all $\boldsymbol{p} \in P(\boldsymbol{r})$. $\boldsymbol{\sigma}(\boldsymbol{r})$ is upper hemicontinuous, so it is compact-valued. Hence by Proposition D.3, $\boldsymbol{\sigma}(\Delta)$ is a compact subset of $\mathbb{R}^m$, or we can define

$$S = \left\{ \boldsymbol{s} = (s^a) \,\middle|\, \sum_{a=1}^m s^a = 0, \sum_{a=1}^m |s^a| \leq \bar{v} \right\}$$

for some $\bar{v} > 0$ such that $\boldsymbol{\sigma}(\Delta) \subset S$. Define a function $\boldsymbol{\rho} : \Delta \times S \to \Delta$, $\boldsymbol{\rho}(\boldsymbol{r}, \boldsymbol{s}) = (\rho^a(\boldsymbol{r}, \boldsymbol{s}))$ by

$$\rho^a(\boldsymbol{r}, \boldsymbol{s}) = \frac{\max\{0, r^a + s^a\}}{\sum_{a=1}^m \max\{0, r^a + s^a\}}, \quad a = 1 \dots m.$$

Because $\sum_{a=1}^m \max\{0, r^a + s^a\} \geq \sum_{a=1}^m (r^a + s^a) = 1 > 0$, the function $\boldsymbol{\rho}(\boldsymbol{r}, \boldsymbol{s})$ is continuous. Finally, we define a correspondence

$$\Phi : \Delta \times S \to \Delta \times S,\ \Phi(\boldsymbol{r}, \boldsymbol{s}) = (\boldsymbol{\rho}(\boldsymbol{r}, \boldsymbol{s}), \boldsymbol{\sigma}(\boldsymbol{r})).$$

Because $\boldsymbol{\sigma}(\boldsymbol{r})$ is convex for each $\boldsymbol{r}$, then $\Phi(\boldsymbol{r}, \boldsymbol{s})$ is an upper hemicontinuous, compact, and convex-valued correspondence from a compact and convex set to itself. Therefore, by Kakutani's fixed point theorem (Theorem D.1), a fixed point exists such that $(\boldsymbol{r}^*, \boldsymbol{s}^*) \in \Phi(\boldsymbol{r}^*, \boldsymbol{s}^*)$, or $\boldsymbol{r}^* = \boldsymbol{\rho}(\boldsymbol{r}^*, \boldsymbol{s}^*)$ and $\boldsymbol{s}^* \in \boldsymbol{\sigma}(\boldsymbol{r}^*)$. For a such that $r^{a*} = 0$, we have $s^{a*} \geq 0$. For $u_a(\boldsymbol{x}_a(\boldsymbol{r}^*)) = 0$, or $\boldsymbol{0} \succsim_a \boldsymbol{x}_a(\boldsymbol{r}^*)$, it is implied that $0 \geq \boldsymbol{px}_a(\boldsymbol{r}^*) \geq 0$ for every $\boldsymbol{p} \in P(\boldsymbol{r}^*)$. Therefore, we have

$\max\{0, r^{a*} + s^{a*}\} = r^{a*} + s^{a*}$ for all $a = 1 \ldots m$, and thus,

$$r^{a*} = \rho^a(\boldsymbol{r}^*, \boldsymbol{s}^*) = \frac{r^{a*} + s^{a*}}{\sum_{a=1}^{m}(r^{a*} + s^{a*})} = r^{a*} + s^{a*}, \quad a = 1 \ldots m,$$

or $s^{a*} = 0$, $a = 1 \ldots m$, which proves that $\mathbf{0} \in \boldsymbol{\sigma}(\boldsymbol{r}^*)$. Take $\boldsymbol{p}^* \in P(\boldsymbol{r}^*)$. Then, $(\boldsymbol{p}^*, \boldsymbol{x}_a(\boldsymbol{r}^*))$ is a quasi-equilibrium. This completes the proof. □

4.5 On the Differentiability of Demand

Let the commodity space L be a Banach space and the consumption set X is a convex subset of L. To perform differential calculus on X, we assume that X is an open subset of L in the norm topology. This assumption implies that X must contain negative vectors because the norm interior of the positive orthant of some Banach spaces is empty. We also assume that the consumer's preference is represented by a utility function on X, $\boldsymbol{x} \mapsto u(\boldsymbol{x})$, which belongs to the class C^2. From Proposition E.8, it follows that for each $\boldsymbol{x} \in X$, a linear functional $Du(\boldsymbol{x}) \in L^* = \mathscr{L}(L, \mathbb{R})$ and a bilinear form $D^2u(\boldsymbol{x}) \in \mathscr{L}^2(L, \mathbb{R})$ exist such that for every $\boldsymbol{h} \in L$,

$$u(\boldsymbol{x} + \boldsymbol{h}) = u(\boldsymbol{x}) + Du(\boldsymbol{x})\boldsymbol{h} + (1/2)D^2u(\boldsymbol{x})(\boldsymbol{h}, \boldsymbol{h}) + o(\|\boldsymbol{h}\|^3),$$

where $o(\|\boldsymbol{h}\|^3)/\|\boldsymbol{h}\|^3 \to 0$ as $\|\boldsymbol{h}\| \to 0$. In the following, we denote the value $D^2u(\boldsymbol{x})(\boldsymbol{h}, \boldsymbol{k})$ of the bilinear form by $\boldsymbol{h}D^2u(\boldsymbol{x})\boldsymbol{k}$ (recall that the bilinear form $D^2u(\boldsymbol{x})$ is a matrix when $L = \mathbb{R}^\ell$). By Proposition E.6, the map $D^2u(\boldsymbol{x})$ is *symmetric*:

$$\boldsymbol{h}D^2u(\boldsymbol{x})\boldsymbol{k} = \boldsymbol{k}D^2u(\boldsymbol{x})\boldsymbol{h} \text{ for all } \boldsymbol{h}, \boldsymbol{k} \in L.$$

In some cases, we consider $\boldsymbol{h}D^2u(\boldsymbol{x})(\cdot)$ as an element of L^*. We impose the condition that the utility function is differentiably concave and non-degenerate, which we call the *differentiably and regularly concave* condition, as follows:

(DRC): for every $\boldsymbol{x} \in X$ and every $\boldsymbol{h} \in L$,

(i) $\boldsymbol{h}D^2u(\boldsymbol{x})\boldsymbol{h} \leq 0$,

(ii) there exists an $\boldsymbol{h}_0 \in L$ with $\boldsymbol{h}_0 D^2 u(\boldsymbol{x}) \neq \mathbf{0}$, and $\boldsymbol{h} D^2 u(\boldsymbol{x}) \boldsymbol{h} < 0$ holds whenever $\boldsymbol{h} D^2 u(\boldsymbol{x}) \neq \mathbf{0}$.

Let $\boldsymbol{p} \in L_+^*$ and $w > 0$. In this section, we are concerned with the differentiability of the individual demand,

$$\phi(\boldsymbol{p}, w) = \{\boldsymbol{x} \in X \mid \boldsymbol{p}\boldsymbol{x} \leq w, \text{ and } u(\boldsymbol{z}) \leq u(\boldsymbol{x}) \text{ whenever } \boldsymbol{p}\boldsymbol{z} \leq w\}.$$

The main result is Theorem 4.7 given by Araujo [5], which states that the demand function $\phi(\boldsymbol{p}, w)$ belongs to the class C^1 only if the commodity space L is isomorphic to a Hilbert space. Proposition 4.3 is an immediate consequence of the Kuhn–Tucker theorem (Theorem E.3) and it may be of independent interest.

Proposition 4.3. *$\hat{\boldsymbol{x}} \in X$ satisfies $\hat{\boldsymbol{x}} = \phi(\hat{\boldsymbol{p}}, \hat{w})$ if and only if an $\lambda \geq 0$ exists such that $Du(\hat{\boldsymbol{x}}) = \lambda \hat{\boldsymbol{p}}$ and $\hat{\boldsymbol{p}}\hat{\boldsymbol{x}} = \hat{w}$.*

The main result in this section now reads as follows:

Theorem 4.7. *Suppose that $\phi(\hat{\boldsymbol{p}}, \hat{w}) \neq \emptyset$ with $\hat{w} > 0$ and a neighborhood $V \subset L^*$ of $\hat{\boldsymbol{p}}$ and $\epsilon > 0$ with $\epsilon < \hat{w}$ exist such that $\phi(\cdot, \cdot)$ is non-empty and it belongs to the class C^1 on $V \times (\hat{w} - \epsilon, \hat{w} + \epsilon)$. Then, the commodity space L is isomorphic to a Hilbert space H.*

We need Lemma 4.8 to prove the theorem.

Lemma 4.8. *Let L be a Banach space. Suppose that an isomorphism $T : L \to L^*$ exists such that*

(i) *$T(\boldsymbol{x})(\boldsymbol{y}) = T(\boldsymbol{y})(\boldsymbol{x})$ for all $\boldsymbol{x}, \boldsymbol{y} \in L$ and*
(ii) *$T(\boldsymbol{x})(\boldsymbol{x}) \leq 0$ for all $\boldsymbol{x} \in L$ and with strict inequality for $\boldsymbol{x} \neq \mathbf{0}$.*

Then, L is isomorphic to a Hilbert space with the inner product $\langle \boldsymbol{x}, \boldsymbol{y} \rangle = -T(\boldsymbol{x})(\boldsymbol{y})$. Moreover, the norm induced from this inner product is equivalent to the original norm on L.

Proof. Clearly, L is a pre-Hilbert space with the inner product $\langle \boldsymbol{x}, \boldsymbol{y} \rangle = -T(\boldsymbol{x})(\boldsymbol{y})$. Next, we prove that L is complete with respect to the reduced norm. Let H be the completion of L with respect to the norm $\|\boldsymbol{x}\| = \sqrt{-T(\boldsymbol{x})(\boldsymbol{x})}$ (Theorem G.1). Then, H is a Hilbert space with the inner product $\langle \boldsymbol{x}, \boldsymbol{y} \rangle$ for all $\boldsymbol{x}, \boldsymbol{y} \in H$, and we have an inclusion $J : L \to H$, which is continuous because of the continuity of T. Let $J^* : H^* = H \to L^*$ be the adjoint of J defined by $J^*(\boldsymbol{x})(\boldsymbol{y}) = J(\boldsymbol{y})(\boldsymbol{x})$ or $\langle J^*(\boldsymbol{x}), \boldsymbol{y} \rangle = \langle \boldsymbol{x}, J(\boldsymbol{y}) \rangle$ for all $\boldsymbol{x}, \boldsymbol{y} \in H$ (see Appendix E). We show that the following diagram commutes:

$$\begin{array}{ccc} L & \xrightarrow{\ J\ } & H \\ {\scriptstyle -T}\Big\downarrow & & \Big\downarrow \\ L^* & \xleftarrow{\ J^*\ } & H^* = H \end{array}$$

Indeed, for every $\boldsymbol{x}, \boldsymbol{y} \in L$, we have

$$\begin{aligned} J^*(J(\boldsymbol{x}))(\boldsymbol{y}) &= J(\boldsymbol{y})(J(\boldsymbol{x})) = \langle J(\boldsymbol{x}), J(\boldsymbol{y}) \rangle \\ &= -T(\boldsymbol{x})(\boldsymbol{y}), \quad \text{or } J^* \circ J(\boldsymbol{x}) = -T(\boldsymbol{x}). \end{aligned}$$

By the assumption, T is a bijection, or one-to-one and onto. Thus, it follows that J^* is onto. J has a dense range and $\mathscr{N}(J^*) = \mathscr{R}(J)^\perp$ by Proposition E.3, so it follows that $\mathscr{N}(J^*) = \{\boldsymbol{0}\}$, or J is one-to-one. Therefore J^* is a bijection, and thus, J is also onto because $J(L) = -(J^*)^{-1} \circ T(L) = H$. From the open mapping theorem (Theorem E.1), it follows that J is an isomorphism. □

We now consider the proof of Theorem 4.7. Define a mapping

$$\psi : X \times \mathbb{R}_{++} \to L^* \times \mathbb{R},\ (\boldsymbol{x}, \lambda) \mapsto (\lambda^{-1} Du(\boldsymbol{x}), \lambda^{-1} Du(\boldsymbol{x})\boldsymbol{x}).$$

Then, ψ belongs to the class C^1. For $w > 0$, we put

$$\lambda(\boldsymbol{p}, w) = \frac{Du(\phi(\boldsymbol{p}, w))\phi(\boldsymbol{p}, w)}{w}$$

and consider the "demand" defined by

$$\xi(\boldsymbol{p}, w) = (\phi(\boldsymbol{p}, w), \lambda(\boldsymbol{p}, w)).$$

By Proposition 4.3, ψ is locally the right and left inverse of ξ around $(\hat{\boldsymbol{x}}, \hat{\lambda}) = \xi(\hat{\boldsymbol{p}}, \hat{w})$ and $(\hat{\boldsymbol{p}}, \hat{w})$, respectively. The demand function $\phi(\boldsymbol{p}, w)$ is also assumed to belong to the class C^1 on $V \times (\hat{w} - \epsilon, \hat{w} + \epsilon)$, so it follows that $\xi(\boldsymbol{p}, w)$ belongs to the class C^1 on $V \times (\hat{w} - \epsilon, \hat{w} + \epsilon)$ and $D\psi(\hat{\boldsymbol{x}}, \hat{\lambda})$ is an isomorphism between $L \times \mathbb{R}$ and $L^* \times \mathbb{R}$.

Consider the map $\eta : X\times\mathbb{R}\times L^*\times\mathbb{R}_{++} \to L^*\times\mathbb{R}$ defined by

$$(\boldsymbol{x}, \lambda, \boldsymbol{p}, w) \mapsto (Du(\boldsymbol{x}) - \lambda\boldsymbol{p}, \boldsymbol{p}\boldsymbol{x} - w).$$

Then, η belongs to the class C^1 and we have

$$D\eta(\hat{\boldsymbol{x}}, \hat{\lambda}, \hat{\boldsymbol{p}}, \hat{w}) = \begin{pmatrix} D^2u(\hat{\boldsymbol{x}}) & -\hat{\boldsymbol{p}} & -\hat{\lambda} & 0 \\ \hat{\boldsymbol{p}} & 0 & \hat{\boldsymbol{x}} & -1 \end{pmatrix}.$$

Hence, from

$$D_{\boldsymbol{p},w}\eta(\hat{\boldsymbol{x}}, \hat{\lambda}, \hat{\boldsymbol{p}}, \hat{w}) = \begin{pmatrix} -\hat{\lambda} & 0 \\ \hat{\boldsymbol{x}} & -1 \end{pmatrix},$$

it follows that $D_{\boldsymbol{p},w}\eta(\hat{\boldsymbol{x}}, \hat{\lambda}, \hat{\boldsymbol{p}}, \hat{w})$ is an isomorphism. Moreover, by the implicit function theorem (Theorem E.5), we have

$$D\psi(\hat{\boldsymbol{x}}, \hat{\lambda}) = -\left(D_{\boldsymbol{p},w}\eta(\hat{\boldsymbol{x}}, \hat{\lambda}, \hat{\boldsymbol{p}}, \hat{w})\right)^{-1} D_{\boldsymbol{x},\lambda}\eta(\hat{\boldsymbol{x}}, \hat{\lambda}, \hat{\boldsymbol{p}}, \hat{w}).$$

Therefore, $D_{\boldsymbol{x},\lambda}\eta(\hat{\boldsymbol{x}}, \hat{\lambda}, \hat{\boldsymbol{p}}, \hat{w})$ is also an isomorphism. From

$$D_{\boldsymbol{x},\lambda}\eta(\hat{\boldsymbol{x}}, \hat{\lambda}, \hat{\boldsymbol{p}}, \hat{w}) = \begin{pmatrix} D^2u(\hat{\boldsymbol{x}}) & -\hat{\boldsymbol{p}} \\ \hat{\boldsymbol{p}} & 0 \end{pmatrix} = \begin{pmatrix} D^2u(\hat{\boldsymbol{x}}) & -\hat{\lambda}^{-1}Du(\hat{\boldsymbol{x}}) \\ \hat{\lambda}^{-1}Du(\hat{\boldsymbol{x}}) & 0 \end{pmatrix},$$

it follows that $D^2u(\hat{\boldsymbol{x}})$ satisfies $\boldsymbol{h}D^2u(\hat{\boldsymbol{x}})\boldsymbol{h} < 0$ if $Du(\hat{\boldsymbol{x}})\boldsymbol{h} = \boldsymbol{0}$ with $\boldsymbol{h} \neq \boldsymbol{0}$. It should be noted that we have used here the DRC condition, where

$$\boldsymbol{h}D^2u(\hat{\boldsymbol{x}})(\cdot) \neq \boldsymbol{0} \text{ implies that } \boldsymbol{h}D^2u(\hat{\boldsymbol{x}})\boldsymbol{h} < 0.$$

$L_0 = \{\boldsymbol{h} \in L|\ Du(\hat{\boldsymbol{x}})\boldsymbol{h} = 0\}$ is a Banach space of codimension 1, so L will be a Hilbert space if we can show that L_0 is a Hilbert space. Therefore, if we can show that $D^2u(\hat{\boldsymbol{x}})$ is an isomorphism from L_0 onto itself, then the proof is complete by Lemma 4.8. $D_{\boldsymbol{x},\lambda}\eta(\hat{\boldsymbol{x}}, \hat{\lambda}, \hat{\boldsymbol{p}}, \hat{w})$ is an isomorphism from $L\times\mathbb{R}$ onto $L^*\times\mathbb{R}$, so $D^2u(\hat{\boldsymbol{x}})$ is one-to-one and continuous as a restriction of the one-to-one and continuous map.

We show that $D^2u(\hat{\boldsymbol{x}})$ is also onto. $D_{\boldsymbol{x},\lambda}\eta(\hat{\boldsymbol{x}}, \hat{\lambda}, \hat{\boldsymbol{p}}, \hat{w})$ is onto, so given $(\boldsymbol{p}, 0)$ in $L_0^*\times\mathbb{R}$, then $(\boldsymbol{h}, \varrho) \in L\times\mathbb{R}$ exists such that $\boldsymbol{h}D^2u(\hat{\boldsymbol{x}}) - (\varrho/\lambda)Du(\hat{\boldsymbol{x}}) = \boldsymbol{p}$ and $(1/\lambda)Du(\hat{\boldsymbol{x}})\boldsymbol{h} = 0$. By the second equation, we have $\boldsymbol{h} \in L_0$. For $\boldsymbol{k} \in L_0$, we have $Du(\hat{\boldsymbol{x}})\boldsymbol{k} = 0$, and thus, $\boldsymbol{h}D^2u(\hat{\boldsymbol{x}})\boldsymbol{k} = \boldsymbol{pk}$, which means that

$$\boldsymbol{h}D^2u(\hat{\boldsymbol{x}}) = \boldsymbol{p} \text{ on } L_0,$$

and we have proved that $D^2u(\hat{\boldsymbol{x}})$ is onto. □

By Lemma 4.8, the Hilbert norm $\left(-\boldsymbol{x}D^2u(\hat{\boldsymbol{x}})\boldsymbol{x}\right)^{1/2}$ is equivalent to the original norm. Hence, an $\epsilon > 0$ exists such that

$$\boldsymbol{x}D^2u(\hat{\boldsymbol{x}})\boldsymbol{x} \leq -\epsilon\|\boldsymbol{x}\|^2.$$

Conversely, Araujo [5] showed that this inequality is sufficient for the demand function $\phi(\boldsymbol{p}, w)$ comprising C^1 at $(\hat{\boldsymbol{p}}, \hat{w})$, where $\hat{\boldsymbol{x}} = \phi(\hat{\boldsymbol{p}}, \hat{w})$. Then, it is easy to show that the norm $\left(-\boldsymbol{x}D^2u(\hat{\boldsymbol{x}})\boldsymbol{x}\right)^{1/2}$ is a Hilbert norm on L_0, which is equivalent to the original norm. Therefore, we can say that the existence of the C^1 demand function at $(\hat{\boldsymbol{p}}, \hat{w})$ is essentially equivalent to the map onto $L \times L$ defined by $\langle \boldsymbol{x}, \boldsymbol{y} \rangle = -\boldsymbol{x}D^2u(\hat{\boldsymbol{x}})\boldsymbol{y}$ being a Hilbert inner product on L_0 and, thus, on L.

4.6 Notes

Section 4.1: Peleg and Yaari [188] worked with the commodity space $\mathbb{R}^\infty$. To prove the existence of an equilibrium, they applied the limit theorem of the core according to Debreu and Scarf (Theorem 2.7 in Chapter 2). Their technique and results were developed by Aliprantis *et al.* [3].

Section 4.2: Theorems 4.1 and 4.3 were given by Bewley [27]. He proved these theorems more generally for a commodity space of $L^\infty(\Omega, \mathcal{A}, \mu)$, which is the set of essentially bounded measurable functions on the (probably non-atomic) measure space $(\Omega, \mathcal{A}, \mu)$, and thus, his theorems include the case of continuous time models. Theorem 4.2 is essentially that given by Debreu [50].

Section 4.3: An equilibrium model of the commodity differentiation based on the commodity space $ca(K)$ was initiated by

Mas-Colell [145]. To avoid the problem due to the unbounded marginal rate of substitution indicated in Example 4.4, he introduced indivisible commodities into the model. Consequently, his proof is very complicated. Remarkably, he also proved the core equivalence theorem in this model. The expositions in this section are based on Jones [103], who defined the BRS condition and simplified Mas-Colell's proof.

Section 4.4: Mas-Colell [150] developed the BRS condition to the concept of proper preferences, as explained in the text. However, it should be noted that the space $ca(K)$ with the weak* topology is not a topological vector lattice. For the sequence $\boldsymbol{x}^n = \delta_{1/n} - \delta_0$ in $ca([0,1])$, $\boldsymbol{x}^n \to \mathbf{0}$ but $\boldsymbol{x}^n_+ = \delta_{1/n} \to \delta_0 \neq \mathbf{0}$ in the weak* topology. Mas-Colell [150] also proved the following:

Proposition. Let $\prec$ be a continuous preference relation on the order interval $[\boldsymbol{a}, \boldsymbol{b}] \subset L$ for some fixed vectors $\boldsymbol{a}, \boldsymbol{b} \in L$. Suppose that the preference relation is weakly monotone, i.e., $\boldsymbol{x} \geq \boldsymbol{y}$ implies that $\boldsymbol{x} \succsim \boldsymbol{y}$. Then, a continuous function $u : [\boldsymbol{a}, \boldsymbol{b}] \to \mathbb{R}$ exists such that $\boldsymbol{x} \succsim \boldsymbol{y}$ if and only if $u(\boldsymbol{x}) \geq u(\boldsymbol{y})$.

Hence, the assumption in Theorem 4.6 that the preference relations are represented by utility functions is not necessary. There are so many articles written on this topic. We just cite [70, 104, 136, 204, 256, 268, 270, 271, 277]. The readers can consult a survey article [152] and references in it.

Section 4.5: Theorem 4.7 was proved by Araujo [5] and the exposition in this section entirely follows that given by Araujo. He also proved that if a demand function exists (i.e., well defined) on the dual space of a Banach space of the commodities, then the commodity space must be reflexive. Therefore, in the infinite dimensional setting, strong non-existence results for the demand functions are the rule, but the competitive equilibria themselves generally exist. These results seem to suggest that the concept of a "competitive equilibrium" is more solid and fundamental than that of a "demand" in general equilibrium theory.

Chapter 5

Large Infinite-Dimensional Economies

5.1 Aumann's Thesis

A *large economy* in the sense of Hildenbrand [94], which was discussed thoroughly in Chapter 3, means an economy with an atomless measure space of economic agents. It defines the economy by a measurable map $\mathcal{E} : A \to \mathcal{P} \times \Omega$, where $\mathcal{P}$ is the set of consumers' preferences and Ω is that of initial endowments. It assigns to each consumer his/her economic characteristics. Hence, it is also called the *individualized economy.* In Chapter 4, we examined economies with infinitely many commodities. This chapter integrates both the models and presents a large infinite-horizon economy and proves the existence of equilibrium and its equivalence with the core. In this section, we shall indicate some technical points that should be taken into account when constructing such a model.

Up to now, no direct proofs for an equilibrium existence theorem with the continuum of consumers and the infinite-dimensional commodity space have been available. We have to resort to some approximations. Recall that we have already used the same technique for infinite dimensional economies with finitely many consumers in the previous chapter. In [248] for instance, the proof of the existence of equilibrium $(\boldsymbol{p}, \xi)$ is carried out by approximating the large infinite-dimensional economy by large finite-dimensional subeconomies with equilibria $(\boldsymbol{p}_n, \xi_n)$. In the course of the approximation $\int_A \xi_n(a)d\lambda \to \int_A \xi(a)d\lambda$, we need the Fatou's lemma. On the finite-dimensional spaces, we have the limit function of the sequence of allocations which is obtained from the limit set of that sequence, namely that $\xi(a) \in L_s(\xi_n(a))$; see the proof of Theorem 3.3. The infinite-dimensional versions of Fatou's lemma, however, hold only "approximately", or they have only ensured that $\xi(a) \in \overline{co}L_s(\xi_n(a))$, where $\overline{co}S$ means the closed convex hull of a set S. Hence, the convex-valuedness of the demand correspondences themselves has been considered to be necessary. The previous studies for economic models with infinitely many commodities and consumers such as [6, 30, 123, 143, 177, 178, 185, 248] assumed that the preferences are convex. The convexity assumptions obviously weaken the impact of the Aumann's classical result, which revealed the "convexfying effect" of large numbers of the economic agents.

The first paper, probably by Rustichini and Yannelis [218], tackled the equilibrium existence problem for a market model on an infinite-dimensional commodity space without the convexity of preferences. They concluded that in order to obtain any Fatou's lemma in infinite-dimensional spaces, one has to have "many more agents than commodities". According to a footnote of Mertens ([163], p. 189), this "many more agents than commodities" thesis seemingly at first was addressed by Aumann in the context of the core-equivalence theorem; hence, we call it *Aumann's thesis*.

In the Rustichini–Yannelis paper, the Aumann's thesis was stated as follows: Let $(A, \mathcal{A}, \lambda)$ be a finite measure space (of consumers) and $\mathcal{A}_E = \{A \cap E | \ A \in \mathcal{A}\}$ the sub-σ algebra of $\mathcal{A}$ restricted to $E \in \mathcal{A}$. We denote the restriction of λ to $\mathcal{A}_E$ by λ_E. Recall that for any (real) vector space, an algebraic Hamel basis exists. The cardinality of any Hamel basis of a vector space L is the same, and we denote it $dim(L)$. Rustichini and Yannelis proposed the next condition which is their version of the Aumann's thesis.

$$\text{(RY): For any } E \in \mathcal{A} \text{ with } \lambda(E) > 0,\ \dim(L_E^\infty(\lambda)) > \dim(L),$$

where $L_E^\infty(\lambda)$ is the space of essentially bounded functions on E and L is the commodity space. Note that this condition involves the spaces of the both consumers and the commodities.

Podczeck criticized this condition that "one may wish to interpret an atomless measure space as an idealization of a large but finite number of them. From this point of view, it is preferable to keep a measure space of agents 'small' [190, p. 386]". We admit the Podczeck's criticism to be reasonable. We would like to, however, point out that he had to pay the cost for requiring the extrinsic and an artificial mathematical structure of the measure space. He took a quotient space $\tilde{A} = A/\sim$, where we set $a \sim a'$ if and only if $(\succsim_a, \omega(a)) = (\succsim_{a'}, \omega(a'))$. Then $\tilde{A}$ represents the set of the equivalent classes of the consumers with the same characteristic. In other words, $\tilde{A} = \{[a] = \mathcal{E}^{-1}(\succsim, \omega) | \, (\succsim, \omega) \in \mathcal{P} \times \Omega\}$. He then postulates that the population measure λ is decomposed into a family of $\{\lambda_{[a]}\}$ and each measure $\lambda_{[a]}$ is concentrated on the set $[a] = \mathcal{E}^{-1}(\succsim, \omega)$ of consumers in A associated with the type $(\succsim, \omega)$. The essential part of Podczeck's version of the thesis is that every $\lambda_{[a]}$ is an atomless probability measure; see also Martin-da-Rocha [142].

Some theorists including Podczeck himself, however, have recognized that the Podczeck's condition contains a correct path to the true statement of the Aumann's thesis. This is accomplished by considering *Maharam types*, instead of simply the types of consumers. The Maharam type of a measure space $(A, \mathcal{A}, \lambda)$ is the smallest cardinal of generating subalgebras of $\mathcal{A}$ and it indicates in some sense the "size" of $\mathcal{A}$. For instance, if a (finite) measure space $(A, \mathcal{A}, \lambda)$ is atomless, its Maharam types of submeasure spaces $(E, \mathcal{A}_E, \lambda_E)$ are infinite for any $E \subset A$ with $\lambda(E) > 0$.

We propose that the Aumann's thesis is manifestly represented when we set a *saturated* or *super-atomless*[1] measure space of consumers (Definition F.1 of Appendix F). A measure space $(A, \mathcal{A}, \lambda)$ is saturated if its Maharam types of submeasure spaces $(E, \mathcal{A}_E, \lambda_E)$ are uncountable for any $E \subset A$ with $\lambda(E) > 0$; see Appendix H. Hence, the saturated measure space strengthens the atomless measure space (hence called super-atomless). We will give some historical remarks related with it in Section 5.5.

[1]The name "super-atomless" was coined by Podczeck [193].

As Podczeck [193] pointed out, the saturated measure space itself does not necessarily have an extraordinarily large cardinality. This can be explained by the non-trivial Loeb measure spaces (e.g., [111]), which are important examples of the saturated measure spaces. It is possible for some non-trivial Loeb measure spaces to have cardinality of the continuum, and hence, it can be identified with the unit interval on the real line. Keisler and Sun [111] showed that for any atomless Loeb space $(A, \mathcal{A}, \lambda)$, Z any Polish space, and $\mathcal{E} : A \to Z$ any $\mathcal{A}$-$\mathcal{B}(Z)$-measurable mapping, the set $\mathcal{E}^{-1}(z)$ has cardinality of greater than or equal to the continuum for almost all z (where "for almost all" means in the image measure on Z of $\mathcal{E}$). This captures essentially the Podczeck version of thesis when $Z = \mathcal{P} \times \Omega$. The Aumann's thesis is now embodied intrinsically in the measure space of consumers, rather than a condition imposed on it from outside; it is realized naturally in our models.

Since we shall work with the commodity space ℓ^∞ with the weak* topology, it is appropriate to define the resource condition in terms of the Gelfand integral (Appendix G). Therefore, we need the Fatou's lemma for the Gelfand integrable maps. This has indeed been obtained by [83, 118], who proved an exact version of the Fatou's lemma for Gelfand integrable maps on a saturated measure space. The statement of their version of the lemma is "exact" in the sense of the finite-dimensional version, or $L_s(\int_A \xi_n(a)d\lambda) \subset \int_A L_s(\xi_n(a))d\lambda$; hence, we can discard the convexity of preferences. Moreover, an exact version of the Lyapunov-type convexity theorem for the Gelfand integrable correspondences on a saturated measure space was established by [193, 242]. We shall apply their theorem and prove a core equivalence theorem for our large infinite-horizon economy in Section 5.4.

There is an alternative formulation of a large economy, which defines the economy as a probability measure μ on the set of agents' characteristics $\mathcal{P} \times \Omega$ and is called the *distributionalized economy*. Then the competitive equilibrium of this economy is also defined as a price vector $\boldsymbol{p} \in \ell^1$ and a probability measure ν on $X \times \mathcal{P} \times \Omega$, where the set $X \subset \ell^\infty$ is a consumption set, which is assumed to be identical among all consumers. These definitions of the economy and the competitive equilibrium on it were first proposed by Hart and Kohlberg [92] and applied to the model on the space $ca(K)$ by [39, 102, 145] (the model of the commodity differentiation). A technical reason of those authors to adopt the distributionalized form is that the procedure of finite approximation can be conducted in terms of not maps but measures; hence, the Fatou's lemma or the Lyapunov's convexity theorem are not necessary. Although a natural formulation for an economy with the continuum of consumers, the distributionalized economy and its equilibria are defined only distributionally, all of the information on an individual level will be lost. If we want to obtain the individual information, we have to set up the individualized economy $\mathcal{E} : A \to \mathcal{P} \times \Omega$.

The saturated measure space also casts the light on the relation between the individualized and the distributionalized form of economies and their equilibria. When the space of the characteristics $\mathcal{P} \times \Omega$ is a complete separable metric space, the distributional economy μ has a representation (Theorem F.7). Let a distributional economy μ and its equilibrium ν be given. An important question is the following; for each representation $\mathcal{E}$ of μ, do we have an equilibrium allocation map ξ such that $(\xi, \mathcal{E})$ represents ν? Generally, the answer is no. As stated above, all of the information of $\mathcal{E}$ are lost from μ. However, we will obtain the affirmative answer when the measure space is saturated (Theorem 5.2). Hence, the distributional equilibrium ν does not loose any individual information when one works with the saturated economies; the individualized and the distributionalized equilibria are equivalent in a strong sense for the saturated measure spaces.

5.2 A Large Infinite-Horizon Economy

Let $(A, \mathcal{A}, \lambda)$ be a complete probability measure space of consumers. The commodity space of the economy in this chapter is set to be ℓ^∞. Let $\bar{\xi} > 0$ be a given positive number. We will assume that the consumption set X of each consumer is the set of non-negative vectors whose coordinates after ℓ are bounded by $\bar{\xi}$,

$$X = \{\xi = (\xi^t) \in \ell^\infty_+ |\ \xi^t \leq \bar{\xi} \text{ for } t > \ell\}.$$

Let Z be a cube in ℓ^∞ of length $\bar{\xi}$,

$$Z = \{\xi = (\xi^t) \in \ell^\infty |\ \mathbf{0} \leq \xi^t \leq \bar{\xi}\mathbf{1}\ \},$$

where $\mathbf{1} = (1, 1 \dots)$. Then we can write $X = \mathbb{R}^{\ell+1}_+ \times Z \subset \ell^\infty$ (recall that we start from $t = 0$). Of course, $\bar{\xi} > 0$ is intended to be a very large number. We will denote the commodity vector $\xi = (\boldsymbol{x}, \boldsymbol{z}) \in \mathbb{R}^{\ell+1} \times Z$. Correspondingly, we shall write a price vector $\pi \in \ell^1$ as $\pi = (\boldsymbol{p}, \boldsymbol{q})$, where $\boldsymbol{p} \in \mathbb{R}^{\ell+1}$ and $\boldsymbol{q} \in \ell^1$.

A Gelfand integrable map $f : A \to X$, $f(a) = (\boldsymbol{x}(a), \boldsymbol{z}(a))$ is called an allocation which means that the mean allocation $\int_A f(a) d\lambda$ satisfies $\pi \int_A f(a) d\lambda = \int_A \pi f(a) d\lambda$ for every $\pi \in \ell^1$; see Appendix G for Gelfand integral. We can write the total (mean) allocation $\int_A f(a) d\lambda$ by

$$\int_A f(a) d\lambda = \left(\int_A \boldsymbol{x}(a) d\lambda, \int_A \boldsymbol{z}(a) d\lambda \right),$$

where $\int_A \boldsymbol{x}(a) d\lambda$ is the usual Lebesgue integral for the map $\boldsymbol{x} : A \to \mathbb{R}^{\ell+1}$ and $\int_A \boldsymbol{z}(a) d\lambda)$ is the Gelfand integral for the map $\boldsymbol{z} : A \to Z$.

As usual, a preference $\prec$ is a strict preference relation on X with the basic postulates. As seen below, the continuity assumption in this chapter is stated in terms of the weak* topology.

(CT_∞) (Weak* continuity): the preference relation $\prec_a$ is open in $X \times X$ with respect to the $\sigma(\ell^\infty, \ell^1)$-topology.

In this chapter, we also assume that $\prec$ satisfies the monotonicity,

(MT): For each $\xi \in X$ and $\eta \in X$, if $\xi < \eta$ then $\xi \prec \eta$.

We restated the (MT) assumption here, since the consumption set X defined above is different from the positive orthant of the commodity space as in the usual case. It follows from Theorem G.3 that Z is compact and metrizable in the $\sigma(\ell^\infty, \ell^1)$-topology; hence, X is a locally compact metric space. Therefore, $\mathcal{F}(X \times X)$ is a compact metric space with respect to the topology of closed convergence τ_c by Theorem D.6, so that it is complete and separable.

Let $\mathcal{P}_{mo} \subset \mathcal{F}(X \times X)$ be the collection of all allowed preference relations satisfying the assumptions stated above. Since the monotonicity will be assumed throughout this chapter, below we will denote $\mathcal{P}_{mo}$ simply as $\mathcal{P}$. Since X is a locally compact, complete, and separable metric space, we can prove as in Theorem 3.2 that $\mathcal{P}$ is a G_δ set, hence a Borel measurable set.

An endowment vector is an element of ℓ^∞. We denote the set of all endowment vectors by Ω and assume that it is of the form

$$\Omega = \{\omega = (\omega^t) \in \ell^\infty |\, \mathbf{0} \leq \omega \leq \varpi \mathbf{1}\},$$

for some $\varpi > 0$. We assume that $0 < \varpi < \bar{\xi}$. The set Ω is also a compact metric space by the same reason as the space Z. We also denote $\omega = (\boldsymbol{e}, \boldsymbol{f})$, $\boldsymbol{e} \in \mathbb{R}^{\ell+1}_+$, $\boldsymbol{f} \in Z$. As usual, the economy $\mathcal{E}$ is a measurable map of A to $\mathcal{P} \times \Omega$, $\mathcal{E}(a) = (\prec_a, \omega(a))$. Here, the endowment map $\omega : A \to \Omega$ is Gelfand integrable by Theorem G.8. An Allocation $f : A \to X$ is called feasible if $\int_A f(a) d\lambda \leq \int_A \omega(a) d\lambda$. The definition of the competitive equilibrium should be clear enough. In this section, we shall assume

(PE) (Positive endowment): $\boldsymbol{e}(a) > \mathbf{0}$ a.e., and $\int_\Omega \omega d\mu_\Omega \ggg \mathbf{0}$.

The first main result of this chapter reads

Theorem 5.1. *Let $\mathcal{E} : A \to \mathcal{P} \times \Omega$ be an economy which satisfies the assumptions stated above and the measure space $(A, \mathcal{A}, \lambda)$ is saturated. Then there exist a price vector $\pi \in \ell^1$ with $\pi > \mathbf{0}$ and a feasible allocation $\xi : A \to X$ such that (π, ξ) is a competitive equilibrium for $\mathcal{E}$.*

Proof. The basic idea is the same as that of Theorem 4.1. For each $n \in \mathbb{N}$, let $R^n = \{\xi = (\xi^t) \in \ell^\infty |\, \xi = (\xi^0, \xi^1 \ldots \xi^n, 0, 0 \ldots)\} \approx \mathbb{R}^{n+1}$. We define

$$X^n = X \cap R^n,\ \precsim^n = \precsim \cap (X^n \times X^n),\ \mathcal{P}^n = \mathcal{P} \cap 2^{X^n \times X^n},$$
$$\text{and } \Omega^n = \Omega \cap R^n.$$

For every $\omega = (\omega^0, \omega^1 \dots \omega^n, \omega^{n+1} \dots) \in \Omega$, we denote

$$\omega_n = (\omega^0, \omega^1 \dots \omega^n, 0, 0 \dots) \in \Omega^n,$$

the canonical projection of ω. They induce finite-dimensional economies $\mathcal{E}^n : A \to \mathcal{P}^n \times \Omega^n$ defined by $\mathcal{E}^n(a) = (\succsim_a^n, \omega_n(a))$, $n = 1, 2 \dots$. We then have the following lemma

Lemma 5.1. $\mathcal{E}^n(a) \to \mathcal{E}(a)$ *a.e.*

Proof. We show that $X^n \times X^n \to X \times X$ in the topology of closed convergence τ_c. It is clear that $L_i(X^n \times X^n) \subset L_s(X^n \times X^n) \subset X \times X$. Therefore, it suffices to show that $X \times X \subset L_i(X^n \times X^n)$. Let $(\xi, \eta) = ((\xi^t), (\eta^t)) \in X \times X$, and set $\xi_n = (\xi^1 \dots \xi^n, 0, 0 \dots)$ and similarly η_n for η. Then $(\xi_n, \eta_n) \in X^n \times X^n$ for all n and $(\xi_n, \eta_n) \to (\xi, \eta)$. Hence, $(\xi, \eta) \in L_i(X^n \times X^n)$. Then it follows that $\succsim^n = \succsim \cap (X^n \times X^n) \to \succsim$.

Obviously, one obtains $\omega_n \to \omega$ in the $\sigma(\ell^\infty, \ell^1)$-topology. Consequently, we have $\mathcal{E}^n(a) \to \mathcal{E}(a)$ a.e., on A. □

We also have the following lemma.

Lemma 5.2. *For each n, there exists a quasi-competitive equilibrium for the economy $\mathcal{E}^n$, or a price–allocation pair $(\pi_n, \xi_n(a))$ which satisfies*

(Q-1n) $\pi_n \xi_n(a) \le \pi_n \omega_n(a)$ *and* $\xi_n(a) \succsim_a \eta$ *whenever* $\pi_n \eta \le \pi_n \omega_n(a)$ *and* $\pi_n \omega_n(a) > 0$ *a.e.*,

(E-2n) $\int_A \xi_n(a) d\lambda \le \int_A \omega_n(a) d\lambda$.

Proof. The proof of this lemma is essentially contained in that of Theorem 3.3. Since we do not assume (MI) for each consumer, we only obtain quasi-equilibrium rather than full competitive equilibrium. □

Note that we don't invoke Theorem 3.4, since the consumption set is not the positive orthant. Since $\omega_n(a) \to \omega(a)$ a.e., we have

$$\int_A \omega_n(a) d\lambda \to \int_A \omega(a) d\lambda$$

by Theorem G.9. Without loss of generality, we can assume that

$$\pi_n \mathbf{1} = \sum_{s=0}^{\ell} p_n^s + \sum_{t=n-\ell-1}^{n} q_n^t = 1$$

for all $n(> \ell+1)$, where $\pi_n = ((p_n^s), (q_n^t))$ and $\mathbf{1} = (1, 1 \dots)$. Here, we have identified $\pi_n \in \mathbb{R}_+^{n+1}$ with a vector in ℓ_+^1 which is also denoted by π_n as $\pi_n = (\pi_n, 0, 0 \dots)$.

We denote $\xi_n(a) = (\boldsymbol{x}_n(a), \boldsymbol{z}_n(a)) \in X^n$. The finite-dimensional Fatou's lemma (Theorem F.13) implies that there exists a measurable map $\boldsymbol{x} : A \to \mathbb{R}_+^{\ell+1}$ such that $\boldsymbol{x}(a) \in L_s(\boldsymbol{x}_n(a))$ a.e., in A and $\int_A \boldsymbol{x}(a)d\lambda \leq \lim_{n\to\infty} \int_A \boldsymbol{x}_n(a)d\lambda \leq \int_A \boldsymbol{e}(a)d\lambda$. By Theorem G.10, we also have a Gelfand integrable function $\boldsymbol{z} : A \to Z$ such that $\boldsymbol{z}(a) \in L_s(\boldsymbol{z}_n(a))$ a.e., and $\int_A \boldsymbol{z}(a)d\lambda \leq \int_A \boldsymbol{f}(a)d\lambda$. Let $\xi(a) = (\boldsymbol{x}(a), \boldsymbol{z}(a))$. Then we have obtained that

$$\int_A \xi(a)d\lambda \leq \int_A \omega(a)d\lambda.$$

Since the set $\Delta = \{\pi \in ba_+ | \, \|\pi\| = \pi\mathbf{1} = 1\}$ is weak* compact by the Alaoglu's theorem (Theorem G.3), we have a subnet $(\pi_{n(\kappa)}, \xi_{n(\kappa)}(a))$ with

$$(\pi_{n(\kappa)}, \xi_{n(\kappa)}(a)) \to (\hat{\pi}, \xi(a)) \text{ a.e., in the weak* topology,}$$

where $\hat{\pi} \in ba_+$ with $\hat{\pi}\mathbf{1} = 1$.

Define the set $\mathscr{P} = \{a \in A | \; \hat{\pi}\omega(a) > 0\}$. Since $\pi_n \in \ell^1$ for all n, $\pi_{n(\kappa)} \int_A \omega(a)d\lambda = \int_A \pi_{n(\kappa)}\omega(a)d\lambda \to \int_A \hat{\pi}\omega(a)d\lambda > 0$ by the assumption (PE) and $\hat{\pi}\mathbf{1} = 1$, we obtain that $\lambda(\mathscr{P}) > 0$. The essence of the proof is contained in the following Lemma.

Lemma 5.3. *$\xi(a) \prec_a \eta$ implies that $\hat{\pi}\omega(a) < \hat{\pi}\eta$ a.e., on $\mathscr{P}$.*

Proof. If the lemma was false, there exists $\eta(a) = (\eta^t(a)) \in X$ such that $\hat{\pi}\eta(a) \leq \hat{\pi}\omega(a)$ and $\xi(a) \prec_a \eta(a)$ on a subset of $\mathscr{P}$ with λ-positive measure. Let $\mathscr{Q}_n = \{a \in A | \; a \text{ does not satisfy (Q-1n)}\}$ and $\mathscr{R} = \{a \in A | \; \xi(a) \notin L_s(\xi_n(a))\}$. Set $\mathscr{Q} = \cup_{n=1}^{\infty} \mathscr{Q}_n$. Since $\mathscr{Q} \cup \mathscr{R}$ is of measure 0, $\mathscr{P} \backslash (\mathscr{Q} \cup \mathscr{R})$ is non-empty. Let $a \in \mathscr{P} \backslash (\mathscr{Q} \cup \mathscr{R})$. We can assume without loss of generality that $\hat{\pi}\eta(a) < \hat{\pi}\omega(a)$ and

$\xi(a) \prec_a \eta(a)$. Let $\eta_n(a) = (\eta^1(a) \dots \eta^n(a), 0, 0 \dots)$ be the projection of $\eta(a)$ to X^n. Since $\eta_n(a) \to \eta(a)$, we have for sufficiently large n_0 that $\hat{\pi}\eta_{n_0}(a) \leq \hat{\pi}\eta(a) < \hat{\pi}\omega(a)$ and $\xi(a) \prec_a \eta_{n_0}(a)$.

Since $(\pi_{n(\kappa)}, \xi_{n(\kappa)}) \to (\hat{\pi}, \xi(a))$, there is a κ_1 with $n(\kappa_1) \equiv n_1 \geq n_0$ such that $0 \leq \pi_{n_1}\eta_{n_0}(a) < \pi_{n_1}\omega(a) = \pi_{n_1}\omega_{n_1}(a)$, and $\xi_{n_1}(a) \prec_a \eta_{n_0}(a)$, or $\xi_{n_1}(a) \prec_a^{n_1} \eta_{n_0}(a)$. This contradicts the fact that $(\pi_{n_1}, \xi_{n_1}(a))$ is a quasi-equilibrium for $\mathcal{E}^{n_1}$. □

Let $\hat{\pi} = \pi + \pi_p$ be the Yosida–Hewitt decomposition where $\pi \equiv (\boldsymbol{p}, \boldsymbol{q}) \in \mathbb{R}_+^{\ell+1} \times \ell_+^1$ is the countably additive part and π_p is the purely finitely additive part. As usual, we can show the budget condition $\pi\xi(a) = \pi\omega(a)$ a.e. By Lemma 5.3 and (MT), we obtain $\boldsymbol{p} \gg \boldsymbol{0}$; hence, $\lambda(\mathscr{P}) = 1$ by (PE), and we conclude

$$\xi(a) \prec_a \eta \text{ implies } \pi\eta > \pi\omega_a, a.e.$$

This completes the proof. □

5.3 Large Distributionalized Economies

Let L be a topological vector lattice (Appendix G) and $X \subset L_+$ be a common consumption set which is assumed to be complete, separable and metrizable, and let Ω be a compact subset of L_+. Let $\mathcal{P} \subset \mathcal{F}(X \times X)$ be the set of allowed preferences which are also assumed to be a complete and separable metric space in the topology of closed convergence τ_c. A probability measure μ on $\mathcal{P} \times \Omega$ which is a distribution over the space of consumers' characteristics can be considered as another formulation of a large economy. In other words, for each Borel set $B \in \mathcal{B}(\mathcal{P} \times \Omega)$, $\mu(B)$ is the population ratio of consumers belonging to the set B.

Definition 5.1. A *distributionalized economy* is a probability measure μ on the measurable space $(\mathcal{P}\times\Omega, \mathcal{B}(\mathcal{P}\times\Omega))$.

A probability measure ν on $X \times \mathcal{P}\times\Omega$ represents a distribution of allocations. Let $\pi \in L^*$ be a price vector. The equilibrium concept of the distributionalized economy is defined as follows.

Definition 5.2. A pair (π, ν) of a price vector $\pi \in L^*\backslash\mathbf{0}$ and a probability measure ν on $X \times \mathcal{P} \times \Omega$ is a *distributionalized equilibrium* of the economy μ if and only if

(D-1) $\nu(\{(\xi, \succsim, \omega) \in X \times \mathcal{P} \times \Omega |\ \pi\xi \leq \pi\omega$, and $\xi \succsim \eta$ whenever $\pi\eta \leq \pi\omega\}) = 1$,
(D-2) $\int_X \iota(\xi) d\nu_X \leq \int_\Omega \iota(\omega) d\nu_\Omega$, where ι is the inclusion map,
(D-3) $\nu_{\mathcal{P}\times\Omega} = \mu$,
where ν_X is the marginal of ν on X and similarly for $\nu_{\mathcal{P}\times\Omega}$ and ν_Ω, respectively (Appendix F).

Since $\iota(\xi) = \xi$ and so on, the condition (D-2) can be written as $\int_X \xi d\nu_X \leq \int_\Omega \omega d\nu_\Omega$. A conceptual advantage of the distributionalized economies and equilibria is that the economies are described by smaller information than the individual form. From the economic point of view, what we are really interested in is the performance of the market itself rather than behaviors of individuals. For this, it is enough to know the distribution of consumers' characteristics, and we do not have to know who has which character, and who demands which commodities in equilibrium. In other words, even if the economy and its equilibrium are defined by the distributions ν and ν rather than the maps $\mathcal{E}$ and ϕ, respectively, almost nothing is lost from the point of view of economic theorists and/or policy makers.

We admit that when at least one of the numbers of agents or commodities is finite, the individual form is epistemologically powerful in the sense that we know everything of each individual in the economy at the equilibrium. However, we are now facing market models including infinite numbers of both the agents and the commodities. Those models are considered to conceptualize "huge" markets. When the market scale is very large in this sense, it will be generally hard to get all the information on the market; hence, it is usually advisable to see the market from a macroeconomic point of view. In this case, the distributionalized form seems to be more natural and appropriate.

Let $(A, \mathcal{A}, \lambda)$ be a measure space of consumers. If $(\pi, \xi(a))$ is a competitive equilibrium of an economy $\mathcal{E} : A \to \mathcal{P} \times \Omega$, it naturally yields a distributionalized equilibrium of the economy $\mu \equiv \lambda \circ \mathcal{E}^{-1}$; set $\nu = \mu \circ (\xi, \mathcal{E})^{-1}$. Then (π, ν) obviously satisfies the conditions (D-1), (D-2), and (D-3) of Definition 5.2. Hence, Theorems 3.3, 3.4, 3.8, and 5.1 imply the corresponding existence theorems for distributionalized equilibria. In Theorem 3.8, the endowment distribution is now defined to be dispersed if for every $\pi > \mathbf{0}$, the measure ν_π on $\mathbb{R}$ defined by $\mu_\pi(B) = \mu_\Omega(\{\omega \in \Omega | \ \pi\omega \in B\})$ is equal to 0 whenever $B = \{b\}$.

But not vice versa. Given a distributionalized equilibrium (π, ν) and a mapping $\mathcal{E} : (A, \mathcal{A}, \lambda) \to \mathcal{P} \times \Omega$ such that $\nu_{\mathcal{P}\times\Omega} = \lambda \circ \mathcal{E}^{-1}$, we do not necessarily have a measurable map $\xi : A \to X$ satisfying $\nu = (\xi, \mathcal{E})^{-1}$. However, we will be able to recover competitive equilibrium from distributionalized equilibrium when the measure space $(A, \mathcal{A}, \lambda)$ is *saturated.*

Theorem 5.2. *Given (π, ν) is an equilibrium of a distributionalized economy $\mu \in \mathscr{M}(\mathcal{P}\times\Omega)$, and an economy $\mathcal{E} : A \to \mathcal{P}\times\Omega$ such that $\nu_{\mathcal{P}\times\Omega} = \lambda\circ\mathcal{E}^{-1}$ where $(A, \mathcal{A}, \lambda)$ is saturated, there exists an allocation $\xi : A \to X$ such that $\lambda \circ (\xi, \mathcal{E})^{-1} = \nu$ and (π, ξ) is a competitive equilibrium of $\mathcal{E}$.*

Proof. First we prove the following lemma.

Lemma 5.4. *Let $(A, \mathcal{A}, \lambda)$ be an atomless measure space, $\mathcal{E} : A \to \mathcal{P} \times \Omega$ be a representation of a distributionalized economy μ and $\xi : A \to X$ a measurable mapping. Define $\nu = \lambda \circ (\xi, \mathcal{E})^{-1}$. Then ξ is an equilibrium allocation of $\mathcal{E}$ if and only if ν is an equilibrium distribution of μ.*

Proof. Suppose that ξ is an equilibrium allocation of $\mathcal{E}$. Then there exists a price vector $\pi(\neq \mathbf{0}) \in L^*$ with $\lambda(G) = 1$, and $\int_A \xi(a) d\lambda = \int_A \omega(a) d\lambda$, where

$$G = \{a \in A | \ \pi\xi(a) = \pi\omega(a) \text{ and } \xi(a) \succsim_a \eta \text{ whenever } \pi\eta \leq \pi\omega(a)\}.$$

Define the set $H \subset X \times \mathcal{P} \times \Omega$ by

$$H = \{(\xi, \succsim, \omega) \in X \times \mathcal{P} \times \Omega |\ \pi\xi = \pi\omega \text{ and } \xi \succsim \eta \text{ whenever } \pi\eta \leq \pi\omega\}.$$

Then $(\xi, \mathcal{E})(G) = H$; hence, $\nu(H) = \lambda \circ (\xi, \mathcal{E})^{-1}(H) = \lambda(G) = 1$, which proves the condition (D-1) of Definition 5.2. Since $\lambda \circ \xi^{-1} = \nu_X$ and $\lambda \circ \omega^{-1} = \nu_\Omega = \mu_\Omega$, the change of variable formula gives $\int_X \xi d\nu_X = \int_\Omega \omega d\mu_\Omega$. Hence, the condition (D-2) is verified. Finally, the condition (D-3) follows from $\nu_{\mathcal{P}\times\Omega} = \lambda \circ \mathcal{E}^{-1} = \mu$. The converse is also proved in a similar way. □

Now let (π, ν) be an equilibrium of a distributionalized economy $\mu \in \mathscr{M}(\mathcal{P}\times\Omega)$, and let an economy $\mathcal{E} : A \to \mathcal{P}\times\Omega$ such that $\nu_{\mathcal{P}\times\Omega} = \lambda \circ \mathcal{E}^{-1}$ be given. By the saturation property (see Definition F.1), there exists a measurable map $\xi : A \to X$ such that $\lambda \circ (\xi, \mathcal{E})^{-1} = \nu$. By Lemma 5.4, the map ξ is an equilibrium allocation and (π, ξ) is a competitive equilibrium of the economy $\mathcal{E}$. □

5.4 A Core Equivalence Theorem

In this section, we shall establish the core–equilibrium equivalence theorem for the economy described in Section 5.2. Let $X = \mathbb{R}_+^{\ell+1} \times Z \subset \ell^\infty$ and Ω be the common consumption set and the space of initial endowment vectors defined in that section, respectively. The space of preferences $\mathcal{P}$ is also defined as before, which is a G_δ subset of $\mathcal{F}(X\times X)$. Recall that $\mathscr{W}(\mathcal{E})$ and $\mathscr{C}(\mathcal{E})$ are the sets of equilibrium allocations and core allocations, respectively. We then prove the following theorem:

Theorem 5.3. *Let $\mathcal{E} : A \to \mathcal{P} \times \Omega$ be an economy which satisfies the assumptions stated in Section* 5.2 *and the measure space $(A, \mathcal{A}, \lambda)$ is saturated. Then $\mathscr{W}(\mathcal{E}) = \mathscr{C}(\mathcal{E})$.*

Proof. The basic idea is the same as that of Theorem 3.6. $\mathscr{W}(\mathcal{E}) \subset \mathscr{C}(\mathcal{E})$ can be proved as in Proposition 3.4; hence, we skip it. We shall show $\mathscr{C}(\mathcal{E}) \subset \mathscr{W}(\mathcal{E})$. Let $f \in \mathscr{C}(\mathcal{E})$ and define $P : A \to \ell^\infty$ by $P(a) = \{\xi \in \ell^\infty |\ f(a) \prec_a (\xi + \omega(a))\}$ and $\Psi : A \to \ell^\infty$ by

$\Psi(a) = P(a) \cup \{\mathbf{0}\}$, respectively. It follows from (MT) and unboundedness of the first $\ell + 1$ goods that $P(a) \neq \emptyset$, a.e.

Lemma 5.5. *Graph*$(\Psi) \in \mathcal{B}(A \times \ell^\infty)$.

Proof. The set

$$G = \{(\prec, \xi, \eta) \in \mathcal{P} \times \ell^\infty \times \ell^\infty | \, \eta \prec \xi\}$$

is obviously a Borel set of $\mathcal{P} \times \ell^\infty \times \ell^\infty$. Defining a map $\phi : A \times \ell^\infty \to \mathcal{P} \times \ell^\infty \times \ell^\infty$ by $\phi(a, \xi) = (\prec_a, \xi + \omega(a), f(a))$, it follows that *Graph*$(P) = \phi^{-1}(G)$. Since the map ϕ is measurable by Proposition F.6 and Remark G.1, the graph of Ψ is measurable. □

Since $\Psi(a)$ is weak* measurable by Lemma 5.5 and Remark G.1, we can define the integral $\int_A \Psi(a) d\lambda$ by Theorem G.8. Since $\mathbf{0} \in \int_A \Psi(a) d\lambda$, it is non-empty and convex by Theorem G.11. We now show that $\int_A \Psi(a) d\lambda \cap \ell^\infty_- = \{\mathbf{0}\}$. Suppose not. Then there exists a function $h : A \to \ell^\infty$ such that $h(a) \in \Psi(a)$ a.e., and $\int_A h(a) d\lambda < \mathbf{0}$. Set $C = \{a \in A | \; h(a) \neq \mathbf{0}\}$. Obviously, $\lambda(C) > 0$. Define an integrable function $g : A \to \ell^\infty$ by

$$g(a) = h(a) + \omega(a) - \frac{\int_A h(a) d\lambda}{\lambda(C)}.$$

We can write $g(a) = (\boldsymbol{x}(a), \boldsymbol{z}(a)) = ((x^s(a)), (z^t(a))) \in \mathbb{R}^{\ell+1} \times \ell^\infty$, and let $\hat{\boldsymbol{z}}(a) = (\hat{z}^t(a))$ by $\hat{z}^t(a) = \inf\{z^t(a), \bar{\xi}\}$. Setting $\hat{g}(a) = (\boldsymbol{x}(a), \hat{\boldsymbol{z}}(a))$, it can be easily seen that $f(a) \prec_a \hat{g}(a)$ a.e., in C and $\int_C \hat{g}(a) d\lambda \leq \int_C \omega(a) d\lambda$, contradicting $f \in \mathscr{C}(\mathcal{E})$.

We can apply Theorem E.2 to the two disjoint convex sets $\int_A \Psi(a) d\lambda$ and $\ell^\infty_- \backslash \{\mathbf{0}\}$, and obtain a vector $\hat{\pi} \in ba$ with $\hat{\pi} > \mathbf{0}$ such that

$$0 \leq \hat{\pi}\zeta \text{ whenever } \zeta \in \int_A \Psi(a) d\lambda.$$

We shall show that

$$\hat{\pi}\eta \geq \hat{\pi}\omega(a) \text{ whenever } f(a) \prec_a \eta \text{ a.e.}$$

In order to do this, we first verify $\hat{\pi} f(a) = \hat{\pi}\omega(a)$ a.e. in A. Let $C \subset A$ be a measurable set with $\lambda(C) > 0$ and take $\epsilon > 0$ and $\boldsymbol{d} \in \mathbb{R}^{\ell+1}_+ \backslash \mathbf{0}$.

We denote $\omega(a) = (\boldsymbol{e}(a), \boldsymbol{f}(a)) \in \mathbb{R}_+^{\ell+1} \times Z$ and define $k : A \to X$ by

$$k(a) = \begin{cases} f(a) - \omega(a) + \epsilon(\boldsymbol{d}, \boldsymbol{0}) & \text{for } a \in C, \\ 0 & \text{for } a \notin C. \end{cases}$$

Then it follows from (MT) that $k(a) \in \Psi(a)$, hence

$$\hat{\pi}\left(\int_C f(a) + \epsilon\lambda(C)(\boldsymbol{d}, \boldsymbol{0}) - \int_C \omega(a) d\lambda\right) \geq 0,$$

and rearranging this, we obtain $\int_C \hat{\pi} f(a) d\lambda \geq \int_C \hat{\pi}\omega(a) d\lambda - \epsilon\lambda(C)\hat{\pi}(\boldsymbol{d}, \boldsymbol{0})$; hence, $\int_C \hat{\pi} f(a) d\lambda \geq \int_C \hat{\pi}\omega(a) d\lambda$ for any $C \subset A$, since $\epsilon > 0$ is arbitrary. We then conclude $\hat{\pi} f(a) \geq \hat{\pi}\omega(a)$ a.e. It follows from $\int_A f(a) d\lambda \leq \int_A \omega(a) d\lambda$ that $\hat{\pi} f(a) = \hat{\pi}\omega(a)$ a.e., as desired. Replacing $f(a)$ in the definition of $k(a)$ by η with $f(a) \prec_a \eta$ for $a \in C$, we obtain $\int_C \hat{\pi}\eta d\lambda \geq \int_C \hat{\pi}\omega(a) d\lambda$. Since $C \subset A$ is arbitrary, we conclude $\hat{\pi}\eta \geq \hat{\pi}\omega(a)$ for every η such that $f(a) \prec_a \eta$ a.e. in A.

The remaining part of the proof proceeds in exactly the same way as in the last part of the proof of Theorem 5.1. Let $\hat{\pi} = \pi_c + \pi_p$ be the Yosida–Hewitt decomposition where $\pi_c \equiv (\boldsymbol{p}, \boldsymbol{q}) \in \mathbb{R}_+^{\ell+1} \times \ell_+^1$ is the countably additive part and π_p is purely finitely additive part. The budget conditions $\pi_c f(a) = \pi_c \omega(a)$ a.e. can be shown as usual. The above condition and (MT) imply $\boldsymbol{p} \gg \boldsymbol{0}$; hence, $\pi_c \omega(a) > 0$ a.e. by (PE). We then conclude $f(a) \prec_a \eta$ implies $\pi_c \eta > \pi_c \omega_a$, a.e. This completes the proof. □

5.5 Notes

Section 5.1: In what follows, we will give some historical remarks on the studies of saturated measure spaces and their applications to game theory and equilibrium theory. The saturated measure spaces have been grown out from the study of the structures of measure algebras (Appendix H). Following classical isomorphism theorem is due to [37, 87]. For an example of modern textbook-level expositions, see [216, p. 399].

Isomorphism Theorem. Every separable measure algebra associated with an atomless probability space is isomorphic to the measure algebra associated with the Lebesgue space.

Maharam [137, 138] extended this classical result and established the fundamental structure theorem of measure algebras ([69, Theorem 3B.6], see also [72, 332B]).

Structure Theorem. Every measure algebra associated with an atomless probability measure space is isomorphic to the (finite or countably infinite) convex combination of the algebras of the product spaces $[0,1]^{m_i}$ of the unit interval [0, 1] with Lebesgue measure, $i = 1, 2 \ldots.$

In the structure theorem, $\mathbf{m_i}$ are arbitrary infinite cardinals and the set of cardinals $\{\mathbf{m_1}, \mathbf{m_2}, \ldots\}$ is unique. The set is called the *Maharam spectrum* of the algebra. The measure algebra for which the spectrum is a singleton $\{\mathbf{m}\}$ is called *homogeneous* and the cardinal $\mathbf{m}$ is the *Maharam type* of the algebra. Therefore, the structure theorem asserts that the homogeneous and atomless (probability) measure algebras are completely characterized as their isomorphic classes by the Maharam types.

Note that the isomorphism theorem corresponds to the case of a homogeneous algebra where $\mathbf{m}$ is countable. Since a separable algebra has a countable and dense subset, every subalgebra is countably generated, or it is homogeneous with its Maharam type of the countable cardinal. The measure algebra associated with the Lebesgue space is also homogeneous and the Maharam type is countable ([72, 331X, p.130]). Hence, by the structure theorem both the algebras belong to the same isomorphism class of the measure algebras with the Maharam type of the countable (infinite) cardinal (represented by $[0,1]^{\mathbb{N}}$). This is nothing but the statement of the isomorphism theorem.

Based on the Maharam's work, Hoover and Keisler [97] defined the saturated measure space (they called it the $\aleph_1$ atomless measure space) and proved its equivalence with the saturation property (Definition F.1) in their study of the stochastic process (the existence of strong solutions for stochastic integral equations). They also observed that the atomless Loeb space introduced by Loeb [132] is a special case of the saturated measure space. By the subsequent works ([112, 134, 242]; among others), it has become apparent that results on Loeb spaces can be transferred in a straightforward manner to saturated measure spaces. Fajardo and Keisler [69] elaborated on the Maharam spectra of the saturated measure spaces. They also proved the next theorem [69, Theorem 3B.7].

Theorem. A probability space is saturated if and only if its Maharam spectrum is a set of uncountable cardinals.

The very comprehensive and systematic expositions for all these results and observations can be found in the monumental monograph of Fremlin [72].

Sun [241] and Keisler and Sun [112] applied saturated measure spaces for analysis of the distributions of correspondences. Let $(A, \mathcal{A}, \lambda)$ be a saturated measure space and F be a correspondence from A to a complete separable metric space X. The distribution of F denoted by $\mathcal{D}_F$ is the set of all distributions of the measurable selections of F, $\mathcal{D}_F = \{\lambda \circ f^{-1} |\ f$ is a measurable selection of $F\}$. The following theorem was proved by Sun [241] for the atomless Loeb space and by Keisler and Sun [112] for a general saturated space.

Theorem. For any correspondence F, $\mathcal{D}_F$ is convex.

They also proved that $\mathcal{D}_F$ is closed and compact (in the weak* topology of probability measures) if F is closed-valued and convex-valued, respectively. Moreover, the following is true. For a metric space Y, let G be a correspondence from $A \times Y$ to X with $G(a, y) \subset F(a)$ for a compact-valued correspondence F a.e., $G(\cdot, y)$ is measurable for all y and $G(a, \cdot)$ is upper hemicontinuous for almost all a. Then $\mathcal{H}(y) \equiv \mathcal{D}_{G(\cdot,y)}$ is upper hemicontinuous.

The following theorem which was also proved by [241] for atomless Loeb spaces and by [112] for saturated spaces roughly says that for any "randomized solution", there exists a corresponding "purified (non-random) solution" (see also [134, 194]).

Theorem. Let Φ be a measurable mapping from A to the space $\mathscr{M}(X)$ of probability measures on X. Then there exists a measurable mapping ϕ from A to X such that (a) for every Borel set B in X, $\lambda \circ \phi^{-1}(B) = \int_A \Phi(a)(B) d\lambda$, and (b) $\phi(a) \in support(\Phi(a))$ a.e.

Keisler and Sun [112] showed that all these results failed for every atomless probability space that is not saturated. Hence, the saturation is also necessary for each of these theorems to hold.

As seen in this book, the theory of integrations for maps and correspondences defined on a saturated measure space which take their values in Banach spaces is important in game theory and equilibrium theory. The study of these integrals started from Sun [241] for atomless Loeb spaces, and subsequently [193, 242] for general saturated spaces. They extended the classical results of Aumann [16] and Richter [205], which assert that for an integrably bounded correspondence from an atomless measure space to a finite-dimensional euclidean space its integral is compact and convex, to Banach space-valued correspondences.

Podczeck [193] in particular observed a remarkable fact that one can extend the Lebesgue space to a saturated measure space (without any use of the non-standard technique) by "enriching" the σ-algebra. Hence, for a saturated measure space, the space itself does not necessarily have an extraordinarily large cardinality. As Podczeck stressed, this has an important implication when a measure space of agents for game theory or economic theory is assumed to be saturated as considered in this chapter (this point has been already addressed in Section 5.1).

The results including the Fatou's lemmas of [115, 116, 118] were obtained exactly on this line of research. Studies of the Fatou's lemma have their own history and it is too long to be presented here. See the introductory sections of [83, 116, 118] and the references therein. Here, we only cite Sun [241] and Loeb-Sun [133] for the exact versions of the lemma, compared to approximate versions (see Section 5.1) of [22, 46, 248], among others.

The games with a continuum of players (the large games) were introduced by Schmeidler [226]. He proved the existence of strategic (individual) Nash equilibria when the reactions of players in the payoff functions are contained as the integral of the strategy profiles. On the other hand, Mas-Colell [148] defined a distributionalized form of the large game. The set of players' strategies (action set) X is assumed to be a compact metric space and the payoff function is a continuous function on $X \times \mathcal{M}(X)$, where a measure $\mu \in \mathcal{M}(X)$ means the distribution of actions taken by the players of the game. The *distributionalized game* $\mathcal{G}$ is defined to be a probability measure on the set of payoff functions $\mathcal{U}(X \times \mathcal{M}(X))$, and the Nash equilibrium is a probability measure ν on $X \times \mathcal{U}(X \times \mathcal{M}(X))$, an element of $\mathcal{M}(X \times \mathcal{U}(X \times \mathcal{M}(X)))$. Mas-Colell proved the existence of a Nash equilibrium distribution in a very simple and elegant manner.

We can now define the strategic *individualized game* to be a measurable map Υ from a measure space $(A, \mathcal{A}, \lambda)$ to $\mathcal{U}(X \times \mathcal{M}(X))$, $a \mapsto \Upsilon(a) = u(a; \xi, \mu)$. The strategy profile (and the Nash equilibrium) is defined to be a measurable map $\xi : A \to X$. Rath [200] gave a simple proof for the existence of Nash equilibria for a case that X is a compact subset of $\mathbb{R}^\ell$ and the second variable of $u(a, \cdot, \cdot)$ is not a measure μ but an integral $\int_A \xi(a)d\lambda$ (as in [226]).

In Section 5.1, we pointed out that the condition on a measure space postulated by Podczeck [190] contains essential ingredients of the saturation. Indeed, Podczeck [190] showed that his condition on the measure space implied the convexity of the integral (Lyapunov's theorem) and an exact version of Fatou's lemma. This observation was also verified by Noguchi [179] in which he proved the existence of Nash equilibria for a strategic large game $\Upsilon : A \to \mathcal{U}(X \times \mathcal{M}(X))$, where X is a compact metric space. In order to prove this, he showed that the "Podczeck condition" admits essentially the saturation property (Definition F.1); hence, we can go "back and forth" between the distributional equilibria and the strategic equilibria. More precisely, we have a distributional equilibrium $\nu \in \mathcal{M}(X \times \mathcal{U}(X \times \mathcal{M}(X)))$ for the distributional game $\mathcal{G} \equiv \lambda \circ \Upsilon^{-1}$, which has been proven to exist by [148], and apply the saturated property to obtain a strategic equilibrium $\xi : A \to X$ satisfying $\lambda \circ (\xi, \Upsilon)^{-1} = \nu$.

Fully armed with the concept of the saturated measure space, Carmona and Podczeck [38] showed the equivalence between the existence of distributional and strategic equilibria and discussed the results previously obtained by [121, 148, 194, 199, 226] in a systematic and unified manner.

Rath *et al.* [201] and Khan *et al.* [114] constructed examples of strategic games which do not have any Nash equilibria. These counterexamples culminated in the following fundamental theorem due to [112] which shows that the saturation is necessary and sufficient for the existence of Nash equilibria.

Theorem. Let $(A, \mathcal{A}, \lambda)$ be an atomless probability space, and X an uncountable compact metric space. Then $(A, \mathcal{A}, \lambda)$ is saturated if and only if every game $\Upsilon : A \to \mathcal{U}(X \times \mathscr{M}(X))$ has a Nash equilibrium.

Section 5.2: Obviously, Bewley [30] was the first to discuss the existence of equilibria for an economy with measure space of consumers and the commodity space ℓ^∞. He postulated a space of consumers as the set of characteristics or $A = \mathcal{P} \times \Omega$, and an allocation $f : \mathcal{P} \times \Omega \to X$ as a map which assigns each consumer's characteristic $(\succsim, \omega)$ to a commodity bundle which is demanded by the consumers with this characteristic. Preceding 5 years of [91, 92], his model is considered to be a prototype of the distributionalized economy discussed in Section 5.3: All those authors who studied the space ℓ^∞, namely [30, 178, 246–248], assumed that the consumption set is an identical and norm-bounded subset of the positive orthant. We relaxed the latter assumption slightly by allowing finitely many unbounded consumable goods in our consumption set. This generalization makes it possible to invoke the monotonicity of preferences and helps the proof of core equivalence theorem in Section 5.4. Note that this remark is not required for the models on $ca(K)$ such as [39, 102, 120, 145], since all consumers are assumed to have the entire non-negative orthant of $ca(K)$ as their common consumption set. Martin-da-Rocha [143] discussed a model on $ca(K)$ including a weak* closed, convex and comprehensive consumption set and production activities. On the other hand, in [123, 142, 177, 218] the consumption sets can depend upon the individuals and are assumed to be integrably bounded.

For the assumption on the initial endowments, however, we departed from most of those authors. Indeed, [30, 123, 142, 177, 178, 218, 247, 248] assumed that almost all consumers have their initial endowments in the (norm) interior of the consumption set. Obviously, such an assumption is very strong in the economies with a continuum of traders.[2] Lee [129] discussed a model with the saturated measure space of consumers and the same commodity space with [123, 142, 177, 218].

Section 5.3: As stated in Section 5.1, [102, 145] first discussed the distributionalized economies on the commodity space $ca(K)$; see also [39, 120]. Subsequently, [246, 247] applied this idea to the space ℓ^∞. Suzuki [249] discussed the competitive equilibrium for a distributionalized production economy with infinitely many indivisible commodities, and applied Theorem 5.2 to show the existence of equilibrium for an economy of the individual form. The power of the saturated spaces is particularly manifest in the models with indivisible commodities, since one cannot assume the convexity of preferences for such models. Noguchi and Zame [180] discussed the externalities in a distributionalized economy.

Section 5.4: The core equivalence theorems on infinite-dimensional spaces are discussed by [28, 84, 145, 163, 185, 192, 219, 246, 255] and others. The idea of the proof of Theorem 5.3 follows that of [94, 219]. However, in order to prove Theorem 5.3, the saturation is not required (cf. [192, 219]). Since the weak* closure of the Gelfand integral is weak* compact and convex (see [269]), the Hahn-Banach separation theorem can be still applied with minor modifications. However, if we drop the saturation, non-emptiness of the equilibrium hence the core can be no longer guaranteed, therefore relevance of such a generalization is questionable. Tourky and Yannelis [255] constructed a class of counter examples with non-separable infinite dimensional commodity spaces in which the existence or

[2] Precisely, Khan-Yannelis, Noguchi, Rustichini-Yannelis assumed that there exists $\eta \in X$ such that $\omega - \eta$ belongs to the (norm) interior of X.

the core equivalence theorems do not hold when aggregation means the Bochner integration. Suzuki [252] proved the existence and core equivalence theorems systematically for exchange economies on the commodity spaces of both ℓ^∞ and $ca(K)$ and discussed the relations between individualized and distributionalized equilibria. Finally a survey article [119] by Khan and Sagara will be helpful for readers who are interested in the topics of this chapter.

Part II
Theory of Production

Chapter 6

Competitive Production Economies

6.1 A Classical Competitive Economy

Herein, we introduce markets firms, thus making production decisions independent of consumers. We assume the number of producers or firms is finite, say n, with each firm indexed by $b(= 1 \dots n)$. In the following, we sometimes write $B = \{1 \dots n\}$. The *production technology* of firm b is represented by a closed set $Y_b \subset \mathbb{R}^\ell$. As usual, for $\boldsymbol{y} = (y^t) \in Y_b$, the commodity t is the input commodity when $y^t < 0$ and an output commodity when $y^t > 0$. We refer to Y_b as the *production set* of firm b. The production set Y_b is said to display *non-increasing returns to scale* if it is convex. As a special case, it exhibits *constant returns to scale* if it is a convex cone with vertex $\{\mathbf{0}\}$, or for every $\boldsymbol{y} \in Y_b$ and every $\tau \geq 0$, $\tau\boldsymbol{y} \in Y_b$. The set Y_b is said to exhibit *increasing returns* if for every $\boldsymbol{y} \in Y_b$ and every $\tau \geq 1$, $\tau\boldsymbol{y} \in Y_b$. Figures 6.1(a), 6.1(b), and 6.1(c) represent the decreasing, constant, and increasing returns to scale production sets, respectively.

Chapter 2 assumed that there are m consumers in the economy indexed by $a(= 1 \dots m)$ and the consumption set $X_a \subset \mathbb{R}^\ell$ of the consumer a is convex and bounded from below, $a = 1 \dots m$. As common, the consumer a is characterized by the preference relation $\prec_a \subset X_a \times X_a$, which satisfies the basic postulates (IR), (TR), (CT). In this section, we will sometimes assume (NTR) and (CV). The endowment vector of consumer a is denoted by $\omega_a \in \mathbb{R}^\ell$, $a = 1 \dots m$. Let $\boldsymbol{x}_a \in X_a$ be a consumption vector of a and $\boldsymbol{y}_b \in Y_b$ a production

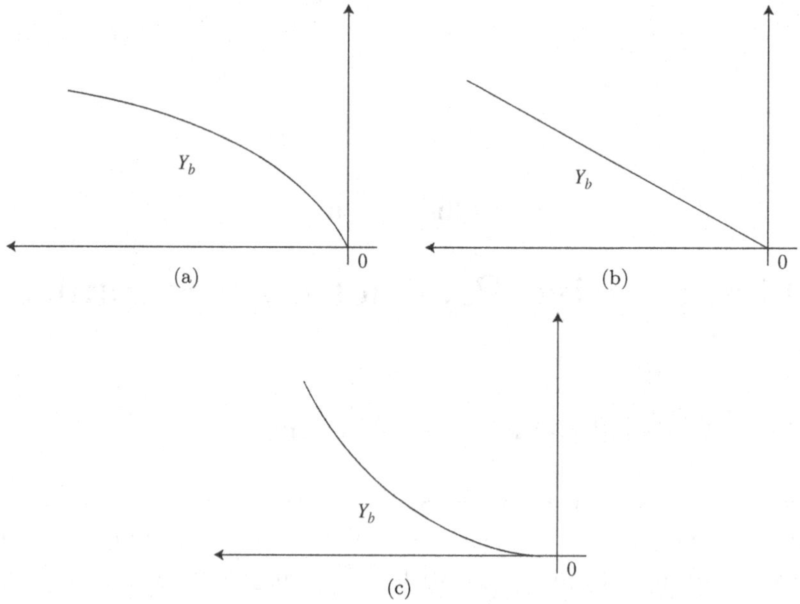

Figure 6.1. (a) Decreasing returns. (b) Constant returns. (c) Increasing returns.

vector of firm b. Then the $m+n$-tuple of vectors $(\boldsymbol{x}_a, \boldsymbol{y}_b)_{a \in A, b \in B}$ is called an *allocation*. The allocation $(\boldsymbol{x}_a, \boldsymbol{y}_b)$ is said to be *feasible* if and only if

$$\sum_{a=1}^{m} \boldsymbol{x}_a \leq \sum_{b=1}^{n} \boldsymbol{y}_b + \sum_{a=1}^{m} \omega_a.$$

It is called *exactly feasible* if the equality holds.

In this chapter, we consider private ownership economies in which the consumer a owns the firm b with a fixed share $\theta_{ab} \geq 0$, $a = 1 \ldots m$, $b = 1 \ldots n$. Hence,

$$\sum_{a=1}^{m} \theta_{ab} = 1$$

holds for each $b = 1 \ldots n$. The list $\mathcal{E}_Y = (X_a, \prec_a, \omega_a, \theta_{ab}, Y_b)_{a \in A, b \in B}$ is referred to as the *private ownership economy* or simply the *production economy*. The definition of the equilibrium of the production economy $\mathcal{E}_Y$ in which each firm behaves competitively is as follows:

Definition 6.1. An $m+n+1$-tuple $(\boldsymbol{p}, \boldsymbol{x}_a, \boldsymbol{y}_b)$ of an allocation $(\boldsymbol{x}_a, \boldsymbol{y}_b) \in \prod_{a=1}^m X_a \times \prod_{b=1}^n Y_b$ and a price vector $\boldsymbol{p} \in \mathbb{R}_+^\ell \backslash \{\boldsymbol{0}\}$ is said to consist of a *competitive equilibrium* or a *Walras equilibrium* if and only if

(E-1) $\boldsymbol{p}\boldsymbol{x}_a \leq \sum_{b=1}^n \theta_{ab}\boldsymbol{p}\boldsymbol{y}_b + \boldsymbol{p}\omega_a$ and
$\boldsymbol{x}_a \succsim_a \boldsymbol{z}$ whenever $\boldsymbol{p}\boldsymbol{z} \leq \sum_{b=1}^n \theta_{ab}\boldsymbol{p}\boldsymbol{y}_b + \boldsymbol{p}\omega_a$, $a = 1 \dots m$,
(E-2) $\boldsymbol{p}\boldsymbol{y}_b \geq \boldsymbol{p}\boldsymbol{y}$ for all $\boldsymbol{y} \in Y_b$, $b = 1 \dots n$,
(E-3) $\sum_{a=1}^m \boldsymbol{x}_a \leq \sum_{b=1}^n \boldsymbol{y}_b + \sum_{a=1}^m \omega_a$.

We now provide a few simple remarks. In condition (E-1), the income of consumer a is the market value of the initial endowment plus the total value of dividends $\theta_{ab}\boldsymbol{p}\boldsymbol{y}_b$, which are distributed from the firm $b = 1 \dots n$. The condition (E-2) describes the profit maximization of the firm, which maximizes its profit value taken from the market price system $\boldsymbol{p}$ as given. Hence, a firm in Definition 6.1 is a *competitive firm*. Obviously, this condition is not generally compatible with increasing returns to scale technologies. Generally speaking, the competitive equilibrium will prevail only in a market where each firm operates under the non-increasing returns to scale technology. A simple glance at actual markets obviously tells us the limitation of this equilibrium concept. In Chapter 8, we discuss a more realistic equilibrium concept in which the firms have a technology with non-trivial setup costs and behave monopolistically. Finally, condition (E-3) is a standard market condition that says the equilibrium allocation must be feasible.

The concept of Pareto optimality is defined in the same manner as that for exchange economies; a feasible allocation $(\boldsymbol{x}_a, \boldsymbol{y}_b) \in \prod_{a=1}^m X_a \times \prod_{b=1}^n Y_b$ is *Pareto-optimal* if and only if there exists no other feasible allocation $(\boldsymbol{x}'_a, \boldsymbol{y}'_b) \in \prod_{a=1}^m X_a \times \prod_{b=1}^n Y_b$ such that

$$\boldsymbol{x}'_a \succsim_a \boldsymbol{x}_a \text{ for all } a \in A, \text{ and } \boldsymbol{x}_a \prec \boldsymbol{x}'_a \text{ holds for at least one } a \in A.$$

The first and second welfare theorems hold for the competitive equilibrium of the private ownership economy.

Proposition 6.1. *Let* $(\boldsymbol{p}^*, \boldsymbol{x}_a^*, \boldsymbol{y}_b^*) \in \mathbb{R}_+^\ell \backslash \{\boldsymbol{0}\} \times \prod_{a=1}^m X_a \times \prod_{b=1}^n Y_b$ *be a competitive equilibrium of an economy* $\mathcal{E}_Y = (X_a, \prec_a, \omega_a, \theta_{ab},$

$Y_b)_{a\in A, b\in B}$, *which satisfies (LNS) for all* $a \in A$. *Then the allocation* $(\boldsymbol{x}_a^*, \boldsymbol{y}_b^*)$ *is Pareto-optimal.*

Proof. Suppose that $(\boldsymbol{x}_a^*, \boldsymbol{y}_b^*)$ is not Pareto-optimal. Then there exists a feasible allocation $(\boldsymbol{x}_a, \boldsymbol{y}_b)$ such that $\boldsymbol{x}_a \succsim_a \boldsymbol{x}_a^*$ for all $\in A$ and $\boldsymbol{x}_a^* \prec \boldsymbol{x}_a$ holds for at least one $a \in A$.

Then it follows that $\boldsymbol{p}^*\boldsymbol{x}_a \geq \sum_{b=1}^n \theta_{ab}\boldsymbol{p}^*\boldsymbol{y}_b^* + \boldsymbol{p}^*\omega_a$ for all $a \in A$. If not, there is an a such that $\boldsymbol{p}^*\boldsymbol{x}_a < \sum_{b=1}^n \theta_{ab}\boldsymbol{p}^*\boldsymbol{y}_b^* + \boldsymbol{p}^*\omega_a$. By (LNS), there exists a bundle $\boldsymbol{z} \in X_a$, which is close enough to $\boldsymbol{x}_a$ so that $\boldsymbol{p}^*\boldsymbol{z} < \sum_{b=1}^n \theta_{ab}\boldsymbol{p}^*\boldsymbol{y}_b^* + \boldsymbol{p}^*\omega_a$, and $\boldsymbol{x}_a \prec_a \boldsymbol{z}$, contradicting the condition (E-1). Furthermore, for a such that $\boldsymbol{x}_a^* \prec_a \boldsymbol{x}_a$, we have $\boldsymbol{p}^*\boldsymbol{x}_a > \sum_{b=1}^n \theta_{ab}\boldsymbol{p}^*\boldsymbol{y}_b + \boldsymbol{p}^*\omega_a$. Summing these inequalities over a, and using $\sum_{a=1}^m \theta_{ab} = 1$, we obtain $\sum_{a=1}^m \boldsymbol{p}^*\boldsymbol{x}_a > \sum_{b=1}^n \boldsymbol{p}^*\boldsymbol{y}_b + \sum_{a=1}^m \boldsymbol{p}^*\omega_a$. Conversely, as the allocation $(\boldsymbol{x}_a, \boldsymbol{y}_b)$ is feasible, we have $\sum_{a=1}^m \boldsymbol{x}_a \leq \sum_{b=1}^n \boldsymbol{y}_b + \sum_{a=1}^m \omega_a$; hence, $\sum_{a=1}^m \boldsymbol{p}^*\boldsymbol{x}_a \leq \sum_{b=1}^n \boldsymbol{p}^*\boldsymbol{y}_b + \sum_{a=1}^m \boldsymbol{p}^*\omega_a$, which is a contradiction. □

To state the second welfare theorem, we recall the following definition:

Definition 6.2. Suppose a price vector $\boldsymbol{p} \in \mathbb{R}_+^\ell \backslash \{\boldsymbol{0}\}$ is given. An allocation $(\boldsymbol{x}_a, \boldsymbol{y}_b) \in \prod_{a=1}^m X_a \times \prod_{b=1}^n Y_b$ is said to be an *equilibrium relative to the price vector* $\boldsymbol{p}$ if and only if

(R-1) $\boldsymbol{x}_a \prec_a \boldsymbol{z}$ implies that $\boldsymbol{p}\boldsymbol{x}_a < \boldsymbol{p}\boldsymbol{z}$ for all $a \in A$,
(E-2) $\boldsymbol{p}\boldsymbol{y}_b \geq \boldsymbol{p}\boldsymbol{y}$ for all $\boldsymbol{y} \in Y_b$, $b = 1 \ldots n$,
(E-3) $\sum_{a=1}^m \boldsymbol{x}_a \leq \sum_{b=1}^n \boldsymbol{y}_b + \sum_{a=1}^m \omega_a$.

As in the case of exchange economies, if the condition (R-1) is replaced by

(Q-1) $\boldsymbol{x}_a \prec_a \boldsymbol{z}$ implies that $\boldsymbol{p}\boldsymbol{x}_a \leq \boldsymbol{p}\boldsymbol{z}$ for all $a \in A$,

the allocation $(\boldsymbol{x}_a, \boldsymbol{y}_b)$ is said to be a *quasi-equilibrium relative to the price vector* $\boldsymbol{p}$. The second fundamental theorem of welfare economics reads as follows:

Theorem 6.1. *For every consumer* $a \in A$, *suppose that the preference relation satisfies (NTR), (CV) and (MT) and for all* $b \in B$,

(*YCV*) (*Convex production sets*): *the production set Y_b is convex.*

Then, every Pareto-optimal allocation is a quasi-equilibrium.

Proof. Let $(\boldsymbol{x}_a, \boldsymbol{y}_b)$ be a Pareto-optimal allocation. Given this is a feasible allocation, the condition (E-3) is met. For each consumer a, define the strictly preferred set $P_a(\boldsymbol{x})$ by

$$P_a(\boldsymbol{x}) = \{\boldsymbol{z} \in X_a |\ \boldsymbol{x} \prec_a \boldsymbol{z}\}.$$

The set $P_a(\boldsymbol{x})$ is non-empty for all $\boldsymbol{x} \in X_a$ by the assumption (MT). As in the proof of Theorem 2.6 in Chapter 2, we can show that the set $P_a(\boldsymbol{x})$ is convex for all $\boldsymbol{x} \in X_a$.

Let $Q = \sum_{a=1}^m P_a(\boldsymbol{x}_a) - \sum_{b=1}^n Y_b$. Then, Q is convex as the sum of the convex sets. Given the allocation $(\boldsymbol{x}_a, \boldsymbol{y}_b)$ is Pareto-optimal, $\sum_{a=1}^n \omega_a \notin Q$. If not, there exists a vector $\boldsymbol{z} \in Q$ or $\boldsymbol{z} = \sum_a \boldsymbol{x}'_a - \sum_b \boldsymbol{y}'_b = \sum_{a=1}^m \omega_a$ satisfying $\boldsymbol{x}'_a \in P_a(\boldsymbol{x}_a)$, $\boldsymbol{y}'_b \in Y_b$, $a = 1 \dots m$, $b = 1 \dots n$, a contradiction. Hence, we can apply the separation theorem (Theorem A.1) in Appendix A and obtain a vector $\boldsymbol{p} \in \mathbb{R}^\ell$ with $\boldsymbol{p} \neq \boldsymbol{0}$ and

$$\boldsymbol{p} \sum_{a=1}^m \omega_a \leq \boldsymbol{p} \sum_{a=1}^m \boldsymbol{z}$$

for every $\boldsymbol{z} \in Q$, or for every $\boldsymbol{z}$ of the form $\boldsymbol{z} = \sum_{a=1}^m \boldsymbol{x}'_a - \sum_{b=1}^n \boldsymbol{y}'_b$, $\boldsymbol{x}'_a \in P_a(\boldsymbol{x}_a)$, $\boldsymbol{y}'_b \in Y_b$, $a = 1 \dots m$, $b = 1 \dots n$. By (MT), it follows that $\boldsymbol{p} \geq \boldsymbol{0}$. For $c \neq a$, we can take $\boldsymbol{x}'_c$ arbitrarily close to $\boldsymbol{x}_c$. Hence, in the limit, setting $\boldsymbol{y}'_b = \boldsymbol{y}_b$, $b = 1 \dots n$, we have

$$\begin{aligned}
\boldsymbol{p} \sum_{a=1}^m \omega_a &\leq \boldsymbol{p} \sum_{a=1}^m \boldsymbol{x}'_a - \boldsymbol{p} \sum_{b=1}^n \boldsymbol{y}_b \leq \boldsymbol{p} \sum_{c \neq a} \boldsymbol{x}_c + \boldsymbol{p}\boldsymbol{x}'_a - \boldsymbol{p} \sum_{b=1}^n \boldsymbol{y}_b \\
&\leq \boldsymbol{p} \sum_{c=1}^m \boldsymbol{x}_c - \boldsymbol{p}\boldsymbol{x}_a + \boldsymbol{p}\boldsymbol{x}'_a + \boldsymbol{p} \sum_{a=1}^m \omega_a - \boldsymbol{p} \sum_{a=1}^m \boldsymbol{x}_a \\
&= \boldsymbol{p} \sum_{a=1}^m \omega_a - \boldsymbol{p}\boldsymbol{x}_a + \boldsymbol{p}\boldsymbol{x}'_a.
\end{aligned}$$

and from this, we obtain

$$\boldsymbol{p}\boldsymbol{x}_a \leq \boldsymbol{p}\boldsymbol{x}'_a \text{ for every } \boldsymbol{x}'_a \in P_a(\boldsymbol{x}_a),\ a = 1 \dots m.$$

Therefore, the condition (Q-1) is met. Next, we set $\boldsymbol{x}'_a = \boldsymbol{x}_a$ for all a and $\boldsymbol{y}'_c = \boldsymbol{y}_c$ for $c \neq b$ and obtain

$$\begin{aligned}
\boldsymbol{p}\sum_{a=1}^{m} \omega_a &\leq \boldsymbol{p}\sum_{a=1}^{m} \boldsymbol{x}_a - \boldsymbol{p}\sum_{c \neq b} \boldsymbol{y}_c - \boldsymbol{p}\boldsymbol{y}'_b \\
&\leq \boldsymbol{p}\sum_{a=1}^{m} \boldsymbol{x}_a - \boldsymbol{p}\sum_{c=1}^{n} \boldsymbol{y}_c + \boldsymbol{p}\boldsymbol{y}_b - \boldsymbol{p}\boldsymbol{y}'_b \\
&\leq \boldsymbol{p}\sum_{b=1}^{n} \boldsymbol{y}_b + \boldsymbol{p}\sum_{a=1}^{m} \omega_a - \boldsymbol{p}\sum_{c=1}^{m} \boldsymbol{y}_c + \boldsymbol{p}\boldsymbol{y}_b - \boldsymbol{p}\boldsymbol{y}'_b.
\end{aligned}$$

Therefore, we have

$$\boldsymbol{p}\boldsymbol{y}'_b \leq \boldsymbol{p}\boldsymbol{y}_b \text{ for every } \boldsymbol{y}'_b \in Y_b,\ b = 1 \dots n.$$

This proves the condition (E-2) and the theorem is established. □

As in the case of exchange economies, if the situation is that $\boldsymbol{p}\boldsymbol{x}_a = \inf \boldsymbol{p}X_a$ is excluded, then we have an equilibrium rather than a quasi-equilibrium.

Corollary 6.1. *Under the assumptions of Theorem 6.1, if the Pareto-optimal allocation $(\boldsymbol{x}_a, \boldsymbol{y}_b)$ satisfies $\boldsymbol{p}\boldsymbol{x}_a > \inf \boldsymbol{p}X_a$, then it is an equilibrium relative to $\mathbf{p}$.*

The fundamental existence theorem of the competitive equilibrium now reads as follows:

Theorem 6.2. *Suppose that an economy $\mathcal{E}_Y = (X_a, \prec_a, \omega_a, \theta_{ab}, Y_b)_{a \in A, b \in B}$ satisfies the assumptions (NTR), (CV) and (MI) for every $a \in A$, and for every $b \in B$, Y_b satisfies the assumptions (YCV) and*

(NP) (Possibility of no production): $\mathbf{0} \in Y_b$,

(FD) (Free disposal) $Y_b + \mathbb{R}^\ell_- \subset Y_b$,
and the total production set satisfies

(*BTP*) (*Bounded total production*): *The set* $\{(\boldsymbol{y}_b) \in \prod_{b=1}^n Y_b | \ \boldsymbol{z} \leq \sum_{b=1}^n \boldsymbol{y}_b\}$ *is bounded for all* $\boldsymbol{z} \in \mathbb{R}^\ell$.

Then there exists a competitive equilibrium $(\boldsymbol{p}^*, \boldsymbol{x}_a^*, \boldsymbol{y}_b^*)$ *for* $\mathcal{E}_Y$.

Proof. The basic idea of the proof is the same as that of Theorem 2.4. Take a positive number $\bar{k}$ such that $\|\sum_{a=1}^m \omega_a\| < \bar{k}$, and for each integer $k \geq \bar{k}$, set $\hat{K} = \{\boldsymbol{x} \in \mathbb{R}^\ell | \ \|\boldsymbol{x}\| \leq k\mathbf{1}\}$, where $\mathbf{1} = (1 \ldots 1)$. As before, for each integer $k \geq \bar{k}$, we define the restricted consumption and production sets,

$$\hat{X}_a(k) = X_a \cap \hat{K}, \quad a = 1 \ldots m,$$
$$\hat{Y}_b(k) = Y_b \cap \hat{K}, \quad b = 1 \ldots n,$$

respectively. The sets $\hat{X}_a(k)$ and $\hat{Y}_b(k)$ are non-empty, compact, and convex subsets of $\mathbb{R}^\ell$ for all $a = 1 \ldots m$ and $b = 1 \ldots n$. For each b, we define the (restricted) supply correspondence $\hat{\psi}_b$

$$\hat{\psi}_b : \mathbb{R}_+^\ell \to \hat{Y}_b(k), \ \boldsymbol{p} \mapsto \hat{\psi}_b(\boldsymbol{p}) = \{\boldsymbol{y} \in \hat{Y}_b(k) | \ \boldsymbol{p}\boldsymbol{y} \geq \boldsymbol{p}\hat{Y}_b(k)\}$$

and the (restricted) profit function by

$$\hat{\pi}_b : \mathbb{R}_+^\ell \to \mathbb{R}_+, \ \boldsymbol{p} \mapsto \boldsymbol{p}\hat{\psi}_b(\boldsymbol{p}),$$

respectively. Clearly, for all $\tau > 0$, we have $\hat{\psi}_b(\tau\boldsymbol{p}) = \hat{\psi}_b(\boldsymbol{p})$ and $\hat{\pi}_b(\tau\boldsymbol{p}) = \tau\hat{\pi}_b(\boldsymbol{p})$, $b = 1 \ldots n$. Therefore, without loss of generality, we can restrict the domain $\mathbb{R}_+^\ell$ to Δ, the unit simplex. Moreover, given $\mathbf{0} \in \hat{Y}_b(k)$, we have $\hat{\pi}_b(\boldsymbol{p}) \geq 0$ for all $\boldsymbol{p} \in \Delta$. As $\hat{Y}_b(k)$ is compact, it follows from Berge's maximum theorem (Theorem D.2) that the correspondence $\hat{\psi}_b$ is upper hemicontinuous and the function $\hat{\pi}_b$ is continuous. It follows from the assumption (YCV) that $\hat{\psi}_b$ is convex-valued.

We now consider the budget correspondence $\hat{\beta}_a : \Delta \to \hat{X}_a(k)$, which is defined by

$$\hat{\beta}_a(\boldsymbol{p}) = \left\{ \boldsymbol{x} \in \hat{X}_a(k) \,\middle|\, \boldsymbol{p}\boldsymbol{x} \leq \sum_{b=1}^n \theta_{ab}\hat{\pi}_b(\boldsymbol{p}) + \boldsymbol{p}\omega_a \right\}, \quad a = 1 \ldots m,$$

and the demand correspondence $\hat{\phi}_a : \Delta \to \hat{X}_a(k)$ defined by

$$\hat{\phi}_a(\boldsymbol{p}) = \{\boldsymbol{x} \in \hat{\beta}_a(\boldsymbol{p}) |\ \boldsymbol{x} \succsim_a \boldsymbol{z} \text{ for all } \boldsymbol{z} \in \hat{\beta}_a(\boldsymbol{p})\},\ a = 1 \ldots m.$$

By assumption of minimum income (MI) and $\hat{\pi}_b(\boldsymbol{p}) \geq 0$, the budget set $\hat{\beta}_a(\boldsymbol{p})$ is non-empty for all $\boldsymbol{p} \in \Delta$. Obviously, it is a compact and convex subset of $\mathbb{R}^\ell$. Given $\hat{\pi}_b(\boldsymbol{p})$ is continuous, we can easily show that the correspondence $\hat{\beta}_a$ is upper hemicontinuous. As in the proof of Proposition 2.2 in Section 2.3, we can prove that $\hat{\beta}_a$ is lower hemicontinuous; hence, it is continuous. Then similar to the proof of Propositions 2.1 and 2.3 in Chapter 2, the correspondence $\hat{\phi}_a(\boldsymbol{p})$ is non-empty, compact-valued, and upper hemicontinuous. By the assumption (CV), it is also convex-valued. Then, we define the aggregate excess demand correspondence,

$$\hat{\zeta} : \Delta \to \mathbb{R}^\ell, \hat{\zeta}(\boldsymbol{p}) = \sum_{a=1}^{m} \hat{\phi}_a(\boldsymbol{p}) - \sum_{b=1}^{n} \hat{\psi}_b(\boldsymbol{p}) - \sum_{a=1}^{m} \omega_a.$$

Let $\hat{Z}(k) = \sum_{a=1}^m \hat{X}_a(k) - \sum_{b=1}^n \hat{Y}_b(k) - \sum_{a=1}^m \omega_a$. Then, the set $\hat{Z}(k)$ is also non-empty, compact, and convex, and $\zeta(\boldsymbol{p}) \subset \hat{Z}(k)$ for every $\boldsymbol{p} \in \Delta$. The correspondence $\hat{\zeta}$ is upper hemicontinuous and convex-valued and satisfies $\boldsymbol{p}\zeta(\boldsymbol{p}) \leq 0$ by the budget condition of each a. Applying the Gale–Nikaido lemma (Theorem D.3), we obtain $\boldsymbol{p}(k) \in \Delta$, such that

$$\hat{\zeta}(\boldsymbol{p}(k)) = \sum_{a=1}^{m} \hat{\phi}_a(\boldsymbol{p}(k)) - \sum_{b=1}^{n} \hat{\psi}_b(\boldsymbol{p}(k)) - \sum_{a=1}^{m} \omega_a \leq \boldsymbol{0}.$$

Take $\boldsymbol{x}_a(k) \in \hat{\phi}_a(\boldsymbol{p}(k))$, $a = 1 \ldots m$ and $\boldsymbol{y}_b(k) \in \hat{\psi}_b(\boldsymbol{p}(k))$, $b = 1 \ldots n$. We then have

(E-1_k) $\boldsymbol{p}(k)\boldsymbol{x}_a(k) \leq \sum_{b=1}^n \theta_{ab}\boldsymbol{p}(k)\boldsymbol{y}_b(k) + \boldsymbol{p}(k)\omega_a$ and $\boldsymbol{x}_a(k) \succsim_a \boldsymbol{z}$ whenever $\boldsymbol{p}(k)\boldsymbol{z} \leq \sum_{b=1}^n \theta_{ab}\boldsymbol{p}(k)\boldsymbol{y}_b(k) + \boldsymbol{p}(k)\omega_a$, $a = 1 \ldots m$,
(E-2_k) $\boldsymbol{p}(k)\boldsymbol{y}_b(k) \geq \boldsymbol{p}(k)\boldsymbol{y}$ for every $\boldsymbol{y} \in \hat{Y}_b(k)$, $b = 1 \ldots n$,
(E-3_k) $\sum_{a=1}^m \boldsymbol{x}_a(k) \leq \sum_{b=1}^n \boldsymbol{y}_b(k) + \sum_{a=1}^m \omega_a$.

As Δ is compact, we can assume that $\boldsymbol{p}(k) \to \boldsymbol{p}^* \in \Delta$. By the assumption (BTP) and the consumption sets being bounded from below, such that the sequences $(\boldsymbol{y}_b(k))$ are bounded and the production sets are closed, we may assume that $\boldsymbol{y}_b(k) \to \boldsymbol{y}_b^* \in Y_b$, $b = 1 \dots n$. Consequently, the sequences $(\boldsymbol{x}_a(k))$ are also bounded, and we can assume that $\boldsymbol{x}_a(k) \to \boldsymbol{x}_a^* \in X_a$.

We assert that $(\boldsymbol{p}^*, \boldsymbol{x}_a^*, \boldsymbol{y}_b^*)$ is a desired competitive equilibrium. Passing to the limit in the above market conditions (E-3_k), we obtain $\sum_{a=1}^m \boldsymbol{x}_a^* \leq \sum_{b=1}^n \boldsymbol{y}_b^* + \sum_{a=1}^m \omega_a$. Next, we shall show that for all $b = 1 \dots n$, $\boldsymbol{p}^*\boldsymbol{y}_b^* \geq \boldsymbol{p}^*\boldsymbol{y}$ for every $\boldsymbol{y} \in Y_b$. Suppose to the contrary that for some b, there exists $\boldsymbol{y} \in Y_b$ such that $\boldsymbol{p}^*\boldsymbol{y}_b < \boldsymbol{p}^*\boldsymbol{y}$. For k that are sufficiently large, $\boldsymbol{y} \in \hat{Y}_b(k)$. Given $\boldsymbol{p}(k) \to \boldsymbol{p}^*$ and $\boldsymbol{y}_b(k) \to \boldsymbol{y}_b^*$, it follows that $\boldsymbol{p}(k)\boldsymbol{y}_b(k) < \boldsymbol{p}(k)\boldsymbol{y}$, contradicting the condition (E-2_k). As $\boldsymbol{p}(k)\boldsymbol{x}_a(k) \leq \sum_{b=1}^n \theta_{ab}\boldsymbol{p}(k)\boldsymbol{y}_b(k) + \boldsymbol{p}(k)\omega_a$, passing to the limit, we have

$$\boldsymbol{p}^*\boldsymbol{x}_a^* \leq \sum_{b=1}^n \theta_{ab}\boldsymbol{p}^*\boldsymbol{y}_b^* + \boldsymbol{p}^*\omega_a, \quad a = 1 \dots m.$$

Suppose there exists $\boldsymbol{z} \in X_a$ such that $\boldsymbol{p}^*\boldsymbol{z} \leq \sum_{b=1}^n \theta_{ab}\boldsymbol{p}^*\boldsymbol{y}_b^* + \boldsymbol{p}^*\omega_a$ and $\boldsymbol{x}_a^* \prec_a \boldsymbol{z}$. For k that are sufficiently large, we have $\boldsymbol{z} \in \hat{X}_a(k)$. By (CT) and (MI), we can take $\boldsymbol{z}' \in \hat{X}_a(k)$, which is close enough to $\boldsymbol{z}$ such that $\boldsymbol{x}_a^* \prec_a \boldsymbol{z}'$ and $\boldsymbol{p}^*\boldsymbol{z}' < \sum_{b=1}^n \theta_{ab}\boldsymbol{p}^*\boldsymbol{y}_b^* + \boldsymbol{p}^*\omega_a$. Hence, for k that are sufficiently large, we have $\boldsymbol{x}_a(k) \prec_a \boldsymbol{z}'$ and $\boldsymbol{p}(k)\boldsymbol{z}' < \sum_{b=1}^n \theta_{ab}\boldsymbol{p}(k)\boldsymbol{y}_b(k) + \boldsymbol{p}(k)\omega_a$, contradicting the condition (E-1_k). Therefore, the condition (E-1) is also met. □

Note that the above proof shows that the assumption (BTP) is replaced by

(BTP'): the set $\{(\boldsymbol{y}_b) \in \prod_{b=1}^n Y_b | \ \boldsymbol{b} - \sum_{a=1}^m \omega_a \leq \sum_{b=1}^n \boldsymbol{y}_b\}$ is bounded,

where $\boldsymbol{b} = (b^t)$ is defined by $b^t = \min\{b_a^t | a = 1 \dots m\}$ and recall that $\boldsymbol{b}_a = (b_a^t)$ is the lower bound of the consumption set of consumer a.

6.2 A Production Economy with Externalities

In this section, we present a competitive market that includes firms with production externalities. Theorem 6.3 proved below is used in Chapter 7. We distinguish between the input and output vectors and denote them as $\boldsymbol{z}$ and $\boldsymbol{y}$, respectively. Hence, the net output is $\boldsymbol{y}-\boldsymbol{z}$. Note that both $\boldsymbol{y}$ and $\boldsymbol{z}$ are non-negative vectors. Let a production function $F : \mathbb{R}^\ell_+ \times \mathbb{R}^\ell_+ \to \mathbb{R}^\ell_+$ be given by $(\boldsymbol{z}, \varsigma) \mapsto \boldsymbol{y} = F(\boldsymbol{z}, \varsigma)$, where $\varsigma \in \mathbb{R}^\ell_+$ is a parameter representing the externalities. The function $F(\boldsymbol{z}, \varsigma)$ is assumed to be continuous, concave in $\boldsymbol{z}$ and $\boldsymbol{0} = F(\boldsymbol{0}, \varsigma)$ for every ς, and it is called a *technology function with externalities*, as distinguished from the (standard) production function.

In the following, suppose that there exist m consumers in the economy indexed by $a = 1 \dots m$. As usual, the consumer a has a preference–consumption set pair $(X_a, \prec_a)$, there is an initial endowment vector $\omega_a \in \mathbb{R}^\ell$, and the firm share is $\theta_{ab} \geq 0$ with $\sum_{a=1}^m \theta_{ab} = 1$. As in Section 6.1, we assume that $\precsim_a$ satisfies the basic postulates, (NTR) and (CV). The consumption set X_a is assumed a closed and convex subset of $\mathbb{R}^\ell$ bounded from below.

We also assume that there exist n firms in the economy indexed by $b = 1 \dots n$. The production technology of the firm b is described by the technology function with externalities $F_b(\boldsymbol{z}_b, \varsigma_b)$, $b = 1 \dots n$. In the following, we assume that for every $b = 1 \dots n$, the parameter ς_b of the firm b depends on the input variables of all firms $(\boldsymbol{z}_1 \dots \boldsymbol{z}_n)$ and

(YCT): the function $\varsigma_b : \mathbb{R}^{n\ell}_+ \to \mathbb{R}^\ell_+$ $(\boldsymbol{z}_1 \dots \boldsymbol{z}_n) \mapsto \varsigma_b(\boldsymbol{z}_1 \dots \boldsymbol{z}_n)$ is continuous.

Consider an economy $(\precsim_a, \omega_a, \theta_{ab}, F_b(\boldsymbol{z}_b, \varsigma_b))$, $a = 1 \dots m$, $b = 1 \dots n$. We assume the firms take the parameter ς_b as given when it maximizes profit. ς_b is determined endogenously by the input level of each firm in equilibrium. The equilibrium concept for the economy $\mathcal{E}_\varsigma = (\precsim_a, \omega_a, \theta_{ab}, F_b)$ is the *competitive equilibrium with the externalities*.

Definition 6.3. An $m+2n+1$-tuple $(\boldsymbol{p}, \boldsymbol{x}_a, \boldsymbol{y}_b, \boldsymbol{z}_b)$ is said to consist of a competitive equilibrium (with externalities) if and only if

(EX-1) $\boldsymbol{p}\boldsymbol{x}_a \leq \sum_{b=1}^{n} \theta_{ab}\boldsymbol{p}(\boldsymbol{y}_b - \boldsymbol{z}_b) + \boldsymbol{p}\omega_a$ and $\boldsymbol{x}_a \succsim_a \boldsymbol{x}'$ whenever $\boldsymbol{p}\boldsymbol{x}' \leq \sum_{b=1}^{n} \theta_{ab}\boldsymbol{p}(\boldsymbol{y}_b - \boldsymbol{z}_b) + \boldsymbol{p}\omega_a$, $a = 1 \ldots m$,

(EX-2) $\boldsymbol{p}(\boldsymbol{y} - \boldsymbol{z}) \leq \boldsymbol{p}(\boldsymbol{y}_b - \boldsymbol{z}_b)$ for all $\boldsymbol{y} \leq F_b(\boldsymbol{z}, \varsigma_b(\boldsymbol{z}_1 \ldots \boldsymbol{z}_n))$, $b = 1 \ldots n$,

(EX-3) $\sum_{a=1}^{m} \boldsymbol{x}_a \leq \sum_{b=1}^{n} (\boldsymbol{y}_b - \boldsymbol{z}_b) + \sum_{a=1}^{m} \omega_a$.

Note that the condition (EX-2) makes sense through the assumption of concavity of $F_b(\cdot, \varsigma_b)$ for every ς_b. The following theorem plays a significant role in Chapter 7.

Theorem 6.3. *Suppose that an economy* $\mathcal{E}_\varsigma = (X_a, \prec_a, \omega_a, \theta_{ab}, F_b)_{a \in A, b \in B}$ *satisfies the assumptions of Theorem 6.2 for every* $a = 1 \ldots m$. *We also assume the condition (YCT) for all* $b = 1 \ldots n$ *and that the set of feasible allocations*

$$\mathcal{F} = \left\{ (\boldsymbol{x}_a, \boldsymbol{y}_b, \boldsymbol{z}_b) \in \prod_{a=1}^{m} X_a \times \mathbb{R}_+^{2\ell} \,\middle|\, \sum_{a=1}^{m} \boldsymbol{x}_a \leq \sum_{b=1}^{n} (\boldsymbol{y}_b - \boldsymbol{z}_b) + \sum_{a=1}^{m} \omega_a, \right.$$
$$\left. \boldsymbol{y}_b \leq F_b(\boldsymbol{z}_b, \varsigma_b(\bar{\boldsymbol{z}}_c)) \right\}$$

is bounded, where we denote $\varsigma_b(\boldsymbol{z}_1 \ldots \boldsymbol{z}_n)$ *by* $\varsigma_b(\bar{\boldsymbol{z}}_c)$. *Then, there exists a competitive equilibrium* $(\boldsymbol{p}^*, \boldsymbol{x}_a^*, \boldsymbol{y}_b^*, \boldsymbol{z}_b^*)$ *for* $\mathcal{E}_\varsigma$.

Proof. We construct an $m+n+1$-person game $\mathcal{G}_\varsigma$ with the strategy space $\mathbb{R}^{2\ell}$ and apply Theorem 2.8 from Chapter 2. As the set $\mathcal{F}$ is bounded, we can take a compact and convex set $\hat{K} \subset \mathbb{R}^\ell$ such that if $(\boldsymbol{x}_a, \boldsymbol{y}_b, \boldsymbol{z}_b) \in \mathcal{F}$, then $\boldsymbol{x}_a, \boldsymbol{y}_b, \boldsymbol{z}_b \in int\hat{K}$ for all $a \in A$ and all $b \in B$. Define

$$\hat{X}_a = \{X_a \cap \hat{K}\} \times \{\boldsymbol{0}\} \subset \mathbb{R}^\ell \times \mathbb{R}^\ell,\ a = 1 \ldots m,$$

$$\hat{\Delta} = \Delta \times \{\boldsymbol{0}\},\ \text{where } \Delta = \left\{ (p^t) \in \mathbb{R}_+^\ell \,\middle|\, \sum_{t=1}^{\ell} p^t = 1 \right\}.$$

The first m players are described as follows. The player $a(= 1 \ldots m)$ has the choice set $\hat{X}_a$, the constraint correspondence

$\mathcal{C}_a : \prod_{c=1}^m \hat{X}_c \times \hat{K}^{2n} \times \hat{\Delta} \to \hat{X}_a$, which is defined by

$$\mathcal{C}_a((\boldsymbol{x}_a, \boldsymbol{0}), (\boldsymbol{y}_b, \boldsymbol{z}_b), (\boldsymbol{p}, \boldsymbol{0})) = \left\{ (\boldsymbol{x}, \boldsymbol{0}) \in \hat{X}_a \,\middle|\, \boldsymbol{p}\boldsymbol{x} \leq \sum_{b=1}^{n} \theta_{ab} \max\{\boldsymbol{p}(\boldsymbol{y}_b - \boldsymbol{z}_b), 0\} + \boldsymbol{p}\omega_a \right\},$$

and a has the preference correspondence $P_a : \prod_{c=1}^m \hat{X}_c \times \hat{K}^{2n} \times \hat{\Delta} \to \hat{X}_a$, which is defined by

$$P_a((\boldsymbol{x}_a, \boldsymbol{0}), (\boldsymbol{y}_b, \boldsymbol{z}_b), (\boldsymbol{p}, \boldsymbol{0})) = \{(\boldsymbol{x}, \boldsymbol{0}) \in \hat{X}_a |\ \boldsymbol{x}_a \prec_a \boldsymbol{x}\}.$$

By Proposition 2.2 in Chapter 2, the correspondence $\mathcal{C}_a$ is a non-empty, compact, and convex-valued continuous correspondence. In the proof of Theorem 2.6 in Chapter 2, we showed that the set $P_a((\boldsymbol{x}_a, \boldsymbol{0}), (\boldsymbol{y}_b, \boldsymbol{z}_b), (\boldsymbol{p}, \boldsymbol{0}))$ is convex, and as the preference $\prec_a$ is irreflexive, we have $\boldsymbol{x}_a \notin P_a((\boldsymbol{x}_a, \boldsymbol{0}), (\boldsymbol{y}_b, \boldsymbol{z}_b), (\boldsymbol{p}, \boldsymbol{0}))$. The open graph of P_a comes from the continuity of $\prec_a$.

The players $b(= 1 \ldots n)$ are described as follows. The choice set of player b is $\hat{K} \times \hat{K} = \hat{K}^2$ and displays the constraint correspondence $\mathcal{C}_b : \prod_{a=1}^m \hat{X}_a \times \hat{K}^{2n} \times \hat{\Delta} \to \hat{K}^2$, which is defined by

$$\mathcal{C}_b((\boldsymbol{x}_a, \boldsymbol{0}), (\boldsymbol{y}_b, \boldsymbol{z}_b), (\boldsymbol{p}, \boldsymbol{0})) = \left\{ (\boldsymbol{y}, \boldsymbol{z}) \in \hat{K}^2 \,\middle|\, \boldsymbol{y} \leq F_b\left(\boldsymbol{z}, \varsigma_b(\bar{\boldsymbol{z}}_c)\right) \right\},$$

and the preference correspondence $P_b : \prod_{a=1}^m \hat{X}_a \times \hat{K}^{2n} \times \hat{\Delta} \to \hat{K}^2$, which is defined by

$$P_b((\boldsymbol{x}_a, \boldsymbol{0}), (\boldsymbol{y}_b, \boldsymbol{z}_b), (\boldsymbol{p}, \boldsymbol{0})) = \{(\boldsymbol{y}, \boldsymbol{z}) \in \hat{K}^2 |\ \boldsymbol{p}(\boldsymbol{y} - \boldsymbol{z}) > \boldsymbol{p}(\boldsymbol{y}_b - \boldsymbol{z}_b)\}.$$

The correspondence $\mathcal{C}_b$ is continuous. Indeed, let

$$((\boldsymbol{x}_a(k), \boldsymbol{0}), (\boldsymbol{y}_b(k), \boldsymbol{z}_b(k)), (\boldsymbol{p}(k), \boldsymbol{0})) \in \prod_{a=1}^{m} \hat{X}_a \times \hat{K}^{2n} \times \hat{\Delta}, \quad k = 0, 1, \ldots$$

be a sequence that converges to $((\boldsymbol{x}_a, \boldsymbol{0}), (\boldsymbol{y}_b, \boldsymbol{z}_b), (\boldsymbol{p}, \boldsymbol{0}))$, and let $(\boldsymbol{y}(k), \boldsymbol{z}(k))$ be a sequence in $\hat{K}^2$ such that $\boldsymbol{y}(k) \leq F_b(\boldsymbol{z}(k), \varsigma_b(\bar{\boldsymbol{z}}_c(k)))$ for all k, and $(\boldsymbol{y}(k), \boldsymbol{z}(k)) \to (\boldsymbol{y}, \boldsymbol{z})$. As $\hat{K}$ is closed and the function F_b is continuous, we have $(\boldsymbol{y}, \boldsymbol{z}) \in \hat{K}^{2n}$ and $\boldsymbol{y} \leq F_b(\boldsymbol{z}, \varsigma_b(\bar{\boldsymbol{z}}_c))$.

This shows that $\mathcal{C}_b$ is closed. Given $\hat{K}$ is compact, it is upper hemicontinuous.

Next, let the sequence $((\boldsymbol{x}_a(k), \mathbf{0}), (\boldsymbol{y}_b(k), \boldsymbol{z}_b(k)), (\boldsymbol{p}(k), \mathbf{0}))$ be a sequence in $\prod_{a=1}^m \hat{X}_a \times \hat{K}^{2n} \times \hat{\Delta}$, which converges to $((\boldsymbol{x}_a, \mathbf{0}), (\boldsymbol{y}_b, \boldsymbol{z}_b), (\boldsymbol{p}, \mathbf{0}))$, and let $(\boldsymbol{y}, \boldsymbol{z}) \in \hat{K}^2$ be a point with $\boldsymbol{y} \leq F_b(\boldsymbol{z}, \varsigma_b(\bar{\boldsymbol{z}}_c))$. Define a sequence $(\boldsymbol{y}(k), \boldsymbol{z}(k)) \in \hat{K}^2$ in the following way. For each k, set $\boldsymbol{z}(k) = \boldsymbol{z}$. For $t(= 1 \ldots \ell)$ such that $y^t = F^t(\boldsymbol{z}, \varsigma_b(\bar{\boldsymbol{z}}_c))$, set $y^t(k) = F^t(\boldsymbol{z}(k), \varsigma_b(\bar{\boldsymbol{z}}_c(k)))$, $k = 0, 1 \ldots$. For t such that $y^t < F^t(\boldsymbol{z}, \varsigma_b(\bar{\boldsymbol{z}}_c))$, set $y^t(k) = y^t$, $k = 0, 1 \ldots$. Then $(\boldsymbol{y}(k), \boldsymbol{z}(k)) \to (\boldsymbol{y}, \boldsymbol{z})$ and

$$\boldsymbol{y}(k) \leq F_b(\boldsymbol{z}(k), \varsigma_b(\bar{\boldsymbol{z}}_c(k)))$$

for all k sufficiently large. Therefore, $\mathcal{C}_b$ is lower hemicontinuous, hence continuous. It follows from $\mathbf{0} = F(\mathbf{0}, \cdot)$ that $\mathcal{C}_b$ is non-empty, or

$$(\mathbf{0}, \mathbf{0}) \in \mathcal{C}_b((\boldsymbol{x}_a(k), \mathbf{0}), (\boldsymbol{y}_b(k), \boldsymbol{z}_b(k)), (\boldsymbol{p}(k), \mathbf{0})),$$

and convex by the assumption of concavity of $F_b(\cdot, \varsigma_b)$. Obviously, P_b has an open graph and is convex-valued and

$$(\boldsymbol{y}_b, \boldsymbol{z}_b) \notin P_b((\boldsymbol{x}_a, \mathbf{0}), (\boldsymbol{y}_b, \boldsymbol{z}_b), (\boldsymbol{p}, \mathbf{0})).$$

The last player, known as the market player, has the choice set $\hat{\Delta}$ and a constraint correspondence

$$\mathcal{C}((\boldsymbol{x}_a, \mathbf{0}), (\boldsymbol{y}_b, \boldsymbol{z}_b), (\boldsymbol{p}, \mathbf{0})) = \hat{\Delta},$$

which is a natural projection and the preference correspondence $P : \prod_{a=1}^m \hat{X}_a \times \hat{K}^{2n} \times \hat{\Delta} \to \hat{\Delta}$, which is defined by

$$\begin{aligned} &P((\boldsymbol{x}_a, \mathbf{0}), (\boldsymbol{y}_b, \boldsymbol{z}_b), (\boldsymbol{p}, \mathbf{0})) \\ &\quad = \left\{ (\boldsymbol{q}, \mathbf{0})) \in \hat{\Delta} \,\middle|\, \boldsymbol{q}\left(\sum_{a=1}^m \boldsymbol{x}_a - \sum_{b=1}^n (\boldsymbol{y}_b - \boldsymbol{z}_b) - \sum_{a=1}^m \omega_a \right) \right. \\ &\quad\quad \left. > \boldsymbol{p}\left(\sum_{a=1}^m \boldsymbol{x}_a - \sum_{b=1}^n (\boldsymbol{y}_b - \boldsymbol{z}_b) - \sum_{a=1}^m \omega_a \right) \right\}. \end{aligned}$$

It is clear that the correspondence $\mathcal{C}$ is non-empty and convex-valued and continuous, and the correspondence P has an open graph and is convex-valued and satisfies $(\boldsymbol{p}, \mathbf{0}) \notin P((\boldsymbol{x}_a, \mathbf{0}), (\boldsymbol{y}_b, \boldsymbol{z}_b), (\boldsymbol{p}, \mathbf{0}))$.

We can apply Theorem 2.8 from Chapter 2 and obtain a Nash equilibrium $((\boldsymbol{x}_a, \mathbf{0}), (\boldsymbol{y}_b, \boldsymbol{z}_b), (\boldsymbol{p}, \mathbf{0}))$ of the game $\mathcal{G}_\varsigma$. We show that it is a competitive equilibrium of the economy $\mathcal{E}_\varsigma$. The (EX-2) is met immediately. $\boldsymbol{p}(\boldsymbol{y}_b - \boldsymbol{z}_b) \geq 0$ by the assumption $\mathbf{0} = F(\mathbf{0}, \varsigma_b(\bar{\boldsymbol{z}}_c))$; hence, we have $\boldsymbol{p}\boldsymbol{x}_a \leq \sum_{b=1}^{n} \theta_{ab}\boldsymbol{p}(\boldsymbol{y}_b - \boldsymbol{z}_b) + \boldsymbol{p}\omega_a$, $a = 1 \dots m$. Now the condition (EX-1) follows immediately. Then, it follows from $\sum_{a=1}^{m} \theta_{ab} = 1$ that

$$\boldsymbol{p}\left(\sum_{a=1}^{m} \boldsymbol{x}_a - \sum_{b=1}^{n} (\boldsymbol{y}_b - \boldsymbol{z}_b) - \sum_{a=1}^{m} \omega_a\right) \leq 0.$$

Given the market player maximizes the value of the excess demand at $\boldsymbol{p}$, we obtain

$$\boldsymbol{q}\left(\sum_{a=1}^{m} \boldsymbol{x}_a - \sum_{b=1}^{n} (\boldsymbol{y}_b - \boldsymbol{z}_b) - \sum_{a=1}^{m} \omega_a\right) \leq 0$$

for all $\boldsymbol{q} \in \Delta$. Taking $\boldsymbol{q} = \boldsymbol{e}_k = (\delta_k^t)$, where $\delta_k^t = 1$ for $t = k$ and $\delta_k^t = 0$ for $t \neq k$, we conclude that

$$\sum_{a=1}^{m} \boldsymbol{x}_a \leq \sum_{b=1}^{n} (\boldsymbol{y}_b - \boldsymbol{z}_b) + \sum_{a=1}^{m} \omega_a.$$

Hence, the condition (EX-3) is met. Therefore, $(\boldsymbol{p}, \boldsymbol{x}_a, \boldsymbol{y}_b, \boldsymbol{z}_b)$ is a competitive equilibrium for the economy $(\hat{X}_a, \omega_a, \prec_a |_{\hat{K}}, F_b|_{\hat{K}})$.

Take compact and convex sets $\hat{K}_n \subset \mathbb{R}^\ell$ such that $\hat{K} \subset \hat{K}_0 \cdots \subset \hat{K}_n \subset \dots$ and $\cup_{n=0}^{\infty} \hat{K}_n = \mathbb{R}^\ell$. Repeating the above discussion for the game $\mathcal{G}_n$, which is obtained from $\mathcal{G}_\varsigma$ by replacing $\hat{K}$ by $\hat{K}_n$, we get a sequence of Nash or competitive equilibria $(\boldsymbol{p}(n), \boldsymbol{x}_a(n), \boldsymbol{y}_b(n), \boldsymbol{z}_b(n))$, $n \in \mathbb{N}$. In the same manner as the proof of Theorem 6.2, we show that the limit $(\boldsymbol{p}^*, \boldsymbol{x}_a^*, \boldsymbol{y}_b^*, \boldsymbol{z}_b^*)$ is a competitive equilibrium of the economy $\mathcal{E}_\varsigma$. □

6.3 Coalition Production Economies

In this section, we introduce a description of economies with a complete and atomless measure space $(A, \mathcal{A}, \lambda)$ of consumers and production activities. This formulation does not presuppose any firms or producers from the outset, rather it assumes that each coalition $C \in \mathcal{A}$ has its own production capability given by a production set $\boldsymbol{Y}(C)$. The production set $\boldsymbol{Y}(C)$ would be determined by purely technological as well as institutional conditions of the society. Let $\mathcal{E} : A \to \mathcal{P} \times \Omega$ be an atomless exchange economy considered in Chapter 3. Then, we reach the following definition.

Definition 6.4. A *coalition production economy* is a pair $(\mathcal{E}, \boldsymbol{Y})$ of an exchange economy $\mathcal{E} : A \to \mathcal{P} \times \Omega$ and a *production correspondence* $\boldsymbol{Y} : \mathcal{A} \to \mathbb{R}^\ell$, which associates with each coalition C a production set $\boldsymbol{Y}(C)$. An element $\boldsymbol{y} \in \boldsymbol{Y}(C)$ is called a *production plan* of the coalition C.

In coalition production economies, the input vectors referring to the various types of labor are significant; therefore, the consumption set X_a of an agent $a \in A$ is typically different from the positive orthant of the commodity space. However, we assume that all consumers have a common consumption set $X \subset \mathbb{R}^\ell$ which is closed and convex and has a lower bound $\boldsymbol{b} \in \mathbb{R}^\ell$. Obviously, a production plan attainable by a coalition is also possible for bigger coalitions. Hence, we can naturally assume that the correspondence $\boldsymbol{Y}$ is *additive*, or it satisfies

$$\boldsymbol{Y}(B) + \boldsymbol{Y}(C) = \boldsymbol{Y}(B \cup C)$$

for any disjoint measurable sets $B, C \in \mathcal{A}$.

A production correspondence $\boldsymbol{Y} : \mathcal{A} \to \mathbb{R}^\ell$ is called *countably additive* if it satisfies

$$\boldsymbol{Y}(\cup_{k=1}^{\infty} C_k) = \sum_{k=1}^{\infty} \boldsymbol{Y}(C_k)$$

for every sequence (C_k) of pairwise disjoint elements of $\mathcal{A}$. Consider the production activity of a coalition C under a prevailing market price $\boldsymbol{p}$. The profit earned by C in this situation is defined by $\Pi(C, \boldsymbol{p}) = \sup\{\boldsymbol{py}|\ \boldsymbol{y} \in \boldsymbol{Y}(C)\}$, which would be distributed among the members of the coalition. However, for the countably additive production correspondence, the profit for each $a \in C$ is unambiguously determined. Indeed, when $\boldsymbol{Y}(C)$ is countably additive, it is easy to see that the set function $\Pi(C, \boldsymbol{p})$ is countably additive and absolutely continuous with respect to λ, namely that $\lambda(C) = 0$ implies $\Pi(C, \boldsymbol{p}) = 0$. Then by the Radon–Nikodym theorem (Theorem F.9), there exists a measurable function $\pi(\cdot, \boldsymbol{p}) : A \to \mathbb{R}$ such that

$$\Pi(C, \boldsymbol{p}) = \int_C \pi(a, \boldsymbol{p}) d\lambda,$$

for every $C \in \mathcal{A}$.

An *allocation* of the coalition production economy is an integrable map $\boldsymbol{x} : A \to X_a$. It is feasible if

$$\int_A \boldsymbol{x}(a) d\lambda \in \boldsymbol{Y}(A) + \int_A \omega(a) d\lambda.$$

If an allocation $\boldsymbol{x}$ is feasible, then there exists a production vector $\boldsymbol{y} \in \boldsymbol{Y}(A)$ such that $\int_A \boldsymbol{x}(a) d\lambda = \boldsymbol{y} + \int_A \omega(a) d\lambda$. The vector $\boldsymbol{y}$ is a *production plan* of the economy $(\mathcal{E}, \boldsymbol{Y})$. The definition of the competitive equilibrium should be clear.

Definition 6.5. A triple $(\boldsymbol{p}, \boldsymbol{x}(a), \boldsymbol{y})$ of a price vector, an allocation, and a production plan is called a *competitive equilibrium* of a coalition production economy $(\mathcal{E}, \boldsymbol{Y})$ if the following conditions hold.

(Cl-1) $\boldsymbol{px}(a) \leq \pi(a, \boldsymbol{p}) + \boldsymbol{p}\omega(a)$ and $\boldsymbol{x}(a) \succsim_a \boldsymbol{z}$ whenever $\boldsymbol{z} \leq \pi(a, \boldsymbol{p}) + \boldsymbol{p}\omega(a)$ a.e.,
(Cl-2) $\boldsymbol{py} = \sup\{\boldsymbol{pz}|\ \boldsymbol{z} \in \boldsymbol{Y}(A)\}$,
(Cl-3) $\int_A \boldsymbol{x}(a) d\lambda = \boldsymbol{y} + \int_A \omega(a) d\lambda$.

As usual, the triple $(\boldsymbol{p}, \boldsymbol{x}(a), \boldsymbol{y})$ is referred to as a quasi-equilibrium if the condition (Cl-1) in Definition 6.7 is replaced by

(Ql-1) $\boldsymbol{px}(a) \leq \pi(a,\boldsymbol{p}) + \boldsymbol{p}\omega(a)$ and $\boldsymbol{x}(a) \succsim_a \boldsymbol{z}$ whenever $\boldsymbol{z} < \pi(a,\boldsymbol{p}) + \boldsymbol{p}\omega(a)$ a.e.

For later use, we consider the simplest production correspondence

$$\boldsymbol{Y}(C) = \begin{cases} T & \text{for } C \in \mathcal{A} \text{ with } \lambda(C) > 0, \\ \{\boldsymbol{0}\} & \text{for } C \in \mathcal{A} \text{ with } \lambda(C) = 0, \end{cases}$$

and the set $T \subset \mathbb{R}^\ell$ is a convex cone with the vertex at the origin. Note that this production correspondence is not countably additive, but the total production set T being a cone implies that $\Pi(C,\boldsymbol{p}) = 0$ for all $C \in \mathcal{A}$, hence the condition (Cl-1) is unambiguously defined. We can denote this economy simply as $\mathcal{E}_T$. We have the following theorem:

Theorem 6.4. *Suppose that for economy $\mathcal{E}_T$, the exchange economy $\mathcal{E}$ satisfies the assumptions of Theorem 3.3 and T satisfies the free disposal (FD) and*

(NFP) (No free production): $T \cap \mathbb{R}^\ell_+ = \{\boldsymbol{0}\}$,
(BTP') : the set $\{\boldsymbol{z} \in T |\, \boldsymbol{b} \leq \boldsymbol{z} + \int_A \omega(a) d\lambda\}$ *is bounded,*

where $\boldsymbol{b} = (b^t)$ *is a lower bound of the consumption sets of all consumers (see above).*

Then there exists a quasi-equilibrium $(\boldsymbol{p}, \boldsymbol{x}(a), \boldsymbol{y})$ *for* $\mathcal{E}_T$.

Proof. The proof will proceed along the same lines as that of Theorem 3.3; hence, we state it briefly. Let $M > 0$ be a non-negative number defined by $M = \int_A \omega^1(a) d\lambda + \cdots \int_A \omega^\ell(a) d\lambda$ and define the set $A_k \subset A$ by $A_k = \{a \in A |\ \omega(a) \leq k(M \ldots M)\}$. Then the set A_k is measurable and by the definition of M, we have $A_1 \neq \emptyset$. Because $A_1 \subset A_2 \subset \cdots \subset A_k \subset \ldots$, we have $A_k \neq \emptyset$ for all $k \geq 1$. Define the sub-σ-field $\mathcal{A}_k$ of $\mathcal{A}$ by $\mathcal{A}_k = \{C \cap A_k |\ C \in \mathcal{A}\}$, and let λ_k be the restriction of λ to $(A_k, \mathcal{A}_k)$. Because the measure space $(A, \mathcal{A}, \lambda)$ is atomless, so too are the measure spaces $(A_k, \mathcal{A}_k, \lambda_k)$, $k = 1, 2, \ldots$ Now, for each positive integer k, let $X_k = \{\boldsymbol{x} \in X |\ \boldsymbol{x} \leq k(M \ldots M)\}$, and for each $a \in A_k$, define the restricted budget sets $\beta_k(a, \boldsymbol{p}) = \{\boldsymbol{x} \in X_k |\ \boldsymbol{px} \leq \boldsymbol{p}\omega(a)\}$ and quasi-demand sets,

$$\tilde{\phi}(a,\boldsymbol{p}) = \begin{cases} \{\boldsymbol{x} \in \beta_k(a,\boldsymbol{p})|\ \boldsymbol{x} \succsim_a \boldsymbol{z} \\ \quad \text{whenever } \boldsymbol{z} \in \beta_k(a,\boldsymbol{p})\} & \text{if } \inf \boldsymbol{p}X_k < \boldsymbol{p}\omega_a, \\ \{\boldsymbol{x} \in X_a|\ \boldsymbol{p}\boldsymbol{x} = \inf \boldsymbol{p}X_k\} & \text{if } \inf \boldsymbol{p}X_k \geq \boldsymbol{p}\omega_a, \end{cases}$$

respectively. Finally, the restricted quasi-excess mean demand is defined by

$$\tilde{\zeta}_k(\boldsymbol{p}) = \int_{A_k} \phi_k(a,\boldsymbol{p})d\lambda - \int_{A_k} \omega(a)d\lambda.$$

As in the proof of Theorem 3.4, we can show that the correspondence $\tilde{\zeta}_k$ is compact and convex-valued, bounded from below and upper hemicontinuous. As T is a closed convex cone with the vertex at $\mathbf{0}$, $\pi(a,\boldsymbol{p}) = 0$ for $\boldsymbol{p}$ in the polar set $\mathscr{P}_T$ of T defined by

$$\mathscr{P}_T = \{\boldsymbol{p} \in \mathbb{R}^\ell \backslash \{\mathbf{0}\}|\ \boldsymbol{p}T \leq 0\}.$$

By (FD), $\mathscr{P}_T \subset \mathbb{R}^\ell_+$, we can assume that $\mathscr{P}_T \subset \Delta$ and $\boldsymbol{p}\tilde{\zeta}_k(\boldsymbol{p}) \leq 0$ for every $\boldsymbol{p} \in \mathscr{P}_T$. Given $\tilde{\phi}(a,\boldsymbol{p}) \subset X_k$, we can also assume without loss of generality that the range of the correspondence $\tilde{\zeta}_k$ is a compact and convex subset $\tilde{K}$ of $\mathbb{R}^\ell$. Then $\tilde{\zeta}_k : \mathscr{P}_T \to \tilde{K}$ satisfies the conditions of the Gale–Nikaido lemma (Theorem D.3) and obtains a price vector $\boldsymbol{p}_k$ and an allocation $\boldsymbol{x}_k(a) \in \tilde{\phi}_k(a,\boldsymbol{p}_k)$ such that $\tilde{\zeta}_k(\boldsymbol{p}_k) \in \mathscr{P}(\mathscr{P}_T) = T$ by Proposition A.2. Setting $\tilde{\zeta}_k(\boldsymbol{p}_k) = \boldsymbol{y}_k$, we obtain a sequence of a quasi-equilibria $(\boldsymbol{p}_k, \boldsymbol{x}_k(a), \boldsymbol{y}_k)$ of economies $(\mathcal{E}_k, T)$, where $\mathcal{E}_k$ is an exchange economy obtained from an obvious restriction of $\mathcal{E}$ on $(A_k, \mathcal{A}_k, \lambda_k)$ and X_k. Define a sequence of measurable functions $\{\boldsymbol{z}_k(a)\}_{n\in\mathbb{N}}$ of A to $\mathbb{R}^\ell$ by

$$\boldsymbol{z}_k(a) = \begin{cases} \boldsymbol{x}_k(a) & \text{if } a \in A_k, \\ \omega(a) & \text{otherwise}, \end{cases}$$

we obtain as in Theorem 3.3 an integrable function $\boldsymbol{x}(\cdot)$ of A to $\mathbb{R}^\ell$ such that $\boldsymbol{x}(a)$ is a cluster point of the sequence $\{\boldsymbol{z}_k(a)\}$ and $\int_A \boldsymbol{x}(a)d\lambda \leq \int_A \omega(a)d\lambda$. Without loss of generality, we can assume that the sequence $\{\boldsymbol{p}_k\}$ is convergent, say, $\lim_{k\to\infty} \boldsymbol{p}_k = \boldsymbol{p} \in \Delta$. By the assumption (BTP'), we can also assume $\boldsymbol{y}_k \to \boldsymbol{y} \in T$ given T

is closed. It is easy to verify that $(\boldsymbol{p}, \boldsymbol{x}(a), \boldsymbol{y})$ is a quasi-equilibrium of $\mathcal{E}_T$. □

The core of the production economy $(\mathcal{E}, \boldsymbol{Y})$ is also defined in the standard manner.

Definition 6.6. An allocation $f : A \to X_a$ a.e. in A is *blocked* by a coalition $C \in \mathcal{A}$ if there exists an allocation $g : A \to X_a$ a.e. in A such that

$$f(a) \prec_a g(a) \text{ for almost all } a \in C$$

and

$$\int_C g(a)d\lambda \in \boldsymbol{Y}(C) + \int_C \omega(a)d\lambda.$$

The set of feasible allocations not blocked by any non-null coalition is called the *core* and denoted by $\mathscr{C}(\mathcal{E}, \boldsymbol{Y})$.

Suppose that a countably additive production correspondence $\boldsymbol{Y}$ satisfies that $\boldsymbol{Y}(C)$ is a convex set for every $C \in \mathcal{A}$, and $\boldsymbol{Y}(A) = \{0\}$ whenever $\lambda(A) = 0$. Then by Theorem F.15, there exists a convex-valued correspondence of $Y : A \to \mathbb{R}^\ell$ such that (i) $\int_A Y(a)d\lambda \subset \boldsymbol{Y}(A)$ and $\overline{\int_A Y(a)d\lambda} = \overline{\boldsymbol{Y}(A)}$ for every $A \in \mathcal{A}$, (ii) the correspondence Y is measurable in the sense that

$$\{a \in A|\, Y(a) \cap F \neq \emptyset\} \in \mathcal{A} \text{ for every closed subset } F \text{ of } \mathbb{R}^\ell.$$

Under the assumption that $\int_A Y(a)d\lambda$ and $\boldsymbol{Y}(A)$ are closed for every $A \in \mathcal{A}$, the production set $\boldsymbol{Y}(C)$ for a coalition C in this case is decomposed into the individual correspondences $Y(a)$ for $a \in C$. Then, by Proposition F.11, $\pi(a, \boldsymbol{p}) = \sup\{\boldsymbol{p}\boldsymbol{y}|\ \boldsymbol{y} \in Y(a)\}$ and $\boldsymbol{y}(a) \in Y(a)$ is called a production plan of the individual $a \in A$. Then Definition 6.5 can be restated as follows:

Definition 6.7. A triple $(\boldsymbol{p}, \boldsymbol{x}(a), \boldsymbol{y}(a))$ consisting of a price vector, an allocation, and a production plan is a competitive equilibrium of a production economy $(\mathcal{E}, \boldsymbol{Y})$ if the following conditions hold.

(Id-1) $\boldsymbol{p}\boldsymbol{x}(a) \leq \pi(a,\boldsymbol{p}) + \boldsymbol{p}\omega(a)$ and $\boldsymbol{x}(a) \succsim_a \boldsymbol{z}$ whenever $\boldsymbol{z} \leq \pi(a,\boldsymbol{p}) + \boldsymbol{p}\omega(a)$ a.e.,
(Id-2) $\boldsymbol{p}\boldsymbol{y}(a) = \pi(a,\boldsymbol{p})$ a.e.,
(Id-3) $\int_A \boldsymbol{x}(a)d\lambda = \int_A \boldsymbol{y}(a)d\lambda + \int_A \omega(a)d\lambda$.

In Section 6.4, we discuss the competitive equilibria of a production economy of this type where each individual $a \in A$ has its own production technology and a non-convex consumption set.

6.4 Non-Convex Production Economies

Let $(A, \mathcal{A}, \lambda)$ be a complete and atomless measure space of consumers. In this section, we assume that each consumer $a \in A$ has its own production technology set $Y(a) \subset \mathbb{R}^\ell$ from the outset. For simplicity, we assume that the production sets $Y(a)$ are compact and convex subsets of $\mathbb{R}^\ell$ and satisfy no production condition (NP); $\mathbf{0} \in Y(a)$ for all a. Let $\mathcal{Y}$ be the family of all non-empty, compact, and convex subsets of $\mathbb{R}^\ell$ that contain $\mathbf{0}$. Given $\mathbb{R}^\ell$ is separable, the space $(\mathcal{K}, \delta)$ of all non-empty compact subsets of $\mathbb{R}^\ell$ with the Hausdorff distance metric δ is separable by Proposition D.11. Hence, so is the set $(\mathcal{Y}, \delta)$ by Proposition B.10. We assume that the correspondence

$$Y : (A, \mathcal{A}, \lambda) \to (\mathcal{Y}, \mathcal{B}(\mathcal{Y})),\ a \mapsto Y(a)$$

is (Borel) measurable. As explained in Section 6.3, for the economy $(\mathcal{E}, Y)$, each coalition has a possibility to access the production set $\int_C Y(a)d\lambda$, that is, the individual production correspondence $Y : A \to \mathcal{Y}$ on the set of agents A induces a coalition production correspondence on the set of coalitions $\mathcal{A}$,

$$\boldsymbol{Y} : \mathcal{A} \to \mathcal{Y},\ C \mapsto \boldsymbol{Y}(C) = \int_C Y(a)d\lambda.$$

We assume that the common consumption set $X \subset \mathbb{R}^\ell$ is closed and bounded from below. As usual, the consumer $a \in A$ has the preference $\prec_a$, and let $\mathcal{P}$ be the set of all preferences satisfying the basic postulates or (IR), (TR), and (CT). Then an exchange economy $\mathcal{E} : A \to \mathcal{P} \times \mathbb{R}^\ell$ is defined. We refer to the pair of measurable mappings $(\mathcal{E}, Y)$ as a *production economy*, or simply an *economy*. Every agent $a \in A$ is assumed to choose a production vector maximizing the profit $\boldsymbol{py}$ in its own production set $Y(a)$, given the price system $\boldsymbol{p}$. Therefore, we assume each agent in the economy behaves competitively, both as a consumer and as a producer. This seems to be natural in the equilibrium model of a large market.

In the following, we discuss the existence of equilibrium for an economy without the convexity assumption on the consumption set. We discussed this problem for an exchange economy and saw that the crucial condition is that the income distribution disperses. The income level of a consumer in exchange economies is just the market value of the initial endowment. For the production economy, however, it is the value of the endowment vector and the profit value earned by production activity. Therefore, even if the endowment distribution is dispersed in the sense of Definition 3.7 in Chapter 3, the distribution of profit may counteract the dispersed endowments, and the income level of the consumer could concentrate on some particular level, so that the equilibrium may cease to exist. The following example illustrates this situation.

Example 6.1. For each consumer $a \in [0, 1]$, the consumption set X and the utility function are the same as in Examples 3.2 and 3.3 in Chapter 3,

$$X = \{(x, y) \in \mathbb{R}^2 |\ x \geq 1,\ y \geq 0\} \cup \{(x, y) \in \mathbb{R}^2 |\ x \geq 0,\ y \geq 3\},$$
$$u(x, y) = x^2 + y^2,\ (x, y) \in X.$$

The endowment vector $\omega(a)$ is given as in Example 3.3,

$$\omega(a) = (1, a) \text{ for } a \in [0, 1].$$

Therefore, the endowment distribution is dispersed. Suppose that the consumer a has as its production possibility set

$$Y(a) = co\{(0, 0), (0, 1 - a)\},\ a \in [0, 1],$$

where coS as usual denotes the convex hull of the set S. Note that $Y(a)$ produces $1-a$ amount of the commodity y from nothing (free production). Hence, for every price system $(p, 1-p)$, $0 \leq p \leq 1$, the point $(0, 1-a)$ maximizes the profit in $Y(a)$. Therefore, the budget equation of the consumer $a \in [0,1]$ is now given by

$$px + (1-p)y = p + (1-p)a + (1-p)(1-a) = 1,$$

which does not depend on a, and the situation comes back to that of Example 3.2 in Chapter 3. The mean excess demand function was calculated there, namely

$$\zeta(p) = \begin{cases} \left(\frac{1}{p} - 1, -1\right) & \text{for } 0 \leq p < 2/3, \\ \left(-1, \frac{1}{p-1} - 1\right) & \text{for } 2/3 \leq p \leq 1, \end{cases}$$

noting that $\omega(a) + g(a) = (1, a) + (0, 1-a) = (1,1)$ for every $a \in A$. This economy has a dispersed endowment distribution, but does not have any equilibria.

Example 6.1 shows that we must impose some additional conditions on the distribution of production sets to obtain the existence of equilibria. The conditions of the following theorem are the simplest among those possible.

Theorem 6.5. *Let $(\mathcal{E}, Y)$ be a production economy such that the exchange economy $\mathcal{E} : A \to \mathcal{P} \times \mathbb{R}^\ell$ satisfies that $\mathcal{E}(a) \subset \mathcal{P}_{wd,lns} \times X$ a.e., and the endowment distribution is dispersed. Suppose that the production correspondence $Y : A \to \mathbb{R}^\ell$ is simple, or there exists a finite partition of A, $\{A_1 \dots A_n\}$ with $A_i \in \mathcal{A}$, $A_i \cap A_j = \emptyset$ and production sets $\{Y_1 \dots Y_n\} \subset \mathcal{Y}$ such that $Y(a) = Y_i$ on A_i, $i = 1 \dots n$. Then we have*

(a) *for every $\mathbf{p} \in \mathring{\Delta}$, there exists a λ-null set $A_p \subset A$ such that for all $a \in A \backslash A_p$, the individual demand correspondence $\phi(a, \mathbf{p})$ of $\mathring{\Delta}$ to $\mathbb{R}^\ell$ is upper hemicontinuous at $\mathbf{p}$,*
(b) *the mean demand correspondence $\Phi(\mathbf{p}) : \mathring{\Delta} \to \mathbb{R}^\ell$, $\Phi(\mathbf{p}) = \int_A \phi(a, \mathbf{p}) d\lambda$ is compact, convex-valued, and upper hemicontinuous.*

Proof. Let $\boldsymbol{p} \neq \mathbf{0}$ in $\mathbb{R}^\ell_+$ be given. Let the sets $HB_{\boldsymbol{p}}(\boldsymbol{x}, \delta)$, $NC_{\boldsymbol{p}}$, and $CW_{\boldsymbol{p}}$ be defined as in Theorem 3.7,

$$HB_{\boldsymbol{p}}(\boldsymbol{x}, \delta) = \{\boldsymbol{z} \in \mathbb{R}^\ell | \ \|\boldsymbol{x} - \boldsymbol{z}\| < \delta, \ \boldsymbol{pz} < \boldsymbol{px}\},$$

$$NC_{\boldsymbol{p}} = \{\boldsymbol{x} \in X | \ HB_{\boldsymbol{p}}(\boldsymbol{x}, \delta) \cap X = \emptyset \text{ for some } \delta > 0\},$$

$$CW_{\boldsymbol{p}} = \{w \in \mathbb{R} | \ w = \boldsymbol{px} \text{ for some } \boldsymbol{x} \in NC_{\boldsymbol{p}}\},$$

respectively. We now define

$$A_{\boldsymbol{p}} = \{a \in A | \ \boldsymbol{p}\omega(a) + \max \boldsymbol{p}Y(a) \in CW_{\boldsymbol{p}}\}.$$

By Lemma 3.2, the set $CW_{\boldsymbol{p}}$ is a countable set, $\{w_1, w_2, \dots\}$. Given the range of the map $Y : A \to \mathcal{Y}$ is a finite set $\{Y_1 \dots Y_n\}$, it follows that

$$\begin{aligned}\lambda(A_{\boldsymbol{p}}) &= \lambda(\{a \in A | \ \boldsymbol{p}\omega(a) + \max \boldsymbol{p}Y(a) \in CW_{\boldsymbol{p}}\}) \\ &\leq \sum_{j=1}^{\infty} \sum_{k=1}^{n} \lambda(\{a \in A | \ \boldsymbol{p}\omega(a) = w_j - \max \boldsymbol{p}Y_k\}) = 0,\end{aligned}$$

given the endowment distribution is dispersed. As in the proof of Theorem 3.7 (a), we can show that the relation $\phi(a, \boldsymbol{p})$ is closed at $\boldsymbol{p} \in \mathbb{R}^\ell_+$, $\boldsymbol{p} \neq \mathbf{0}$ for each $a \in A \backslash A_{\boldsymbol{p}}$.

Let $\boldsymbol{p} \in \mathring{\Delta}$ be given. Given $\omega(a) \in X$ a.e. and $Y(a)$ is a compact subset of $\mathbb{R}^\ell$ with $\mathbf{0} \in Y(a)$, the budget set

$$\beta(a, \boldsymbol{p}) = \{\boldsymbol{x} \in X | \ \boldsymbol{px} \leq \boldsymbol{p}\omega(a) + \max \boldsymbol{p}Y(a)\}$$

is non-empty and compact. Hence, by Proposition 2.1, we have $\phi(a, \boldsymbol{p}) \neq \emptyset$. Let V be a compact neighborhood of $\boldsymbol{p}$ with strictly positive vectors, and $Y_1 \dots Y_n$ be the values of the simple correspondence $Y : A \to \mathcal{Y}$. Given the consumption set X is bounded from below and each Y_i is compact, we can take an integrable function $h(a)$ such that $\|\phi(a, \boldsymbol{q})\| \leq h(a)$ for all $\boldsymbol{q} \in V$ as in the proof of Theorem 3.7 (a). Since $\phi(a, \cdot)$ is closed at $\boldsymbol{p} \in V$, it is upper hemicontinuous. Part (b) can be proved in exactly the same way as Theorem 3.7(b). □

Once we have obtained the upper hemicontinuity of the mean demand correspondence $\Phi(\boldsymbol{p}) = \int_A \phi(a, \boldsymbol{p})d\lambda$ thanks to the regularizing effect of the aggregation, the existence of equilibrium can be immediately deduced.

Theorem 6.6. *Let $(\mathcal{E}, Y) : A \to \mathcal{P} \times X \times \mathcal{Y}$ be a production economy such that $\mathcal{E}(a) \subset \mathcal{P}_{wd,lns} \times X$ a.e. Suppose that the endowment distribution is dispersed and the production correspondence Y is simple. Then there exists a competitive equilibrium for $(\mathcal{E}, Y)$.*

Proof. Define the mean supply correspondence $\Psi(\boldsymbol{p})$ of $\mathring{\Delta}$ to $\mathbb{R}^\ell$ by

$$\Psi(\boldsymbol{p}) = \left\{ \boldsymbol{y} \in \int_A Y(a)d\lambda \,\middle|\, \boldsymbol{p}\boldsymbol{y} = \max \boldsymbol{p} \int_A Y(a)d\lambda \right\}$$

and consider the mean excess demand correspondence from $\mathring{\Delta}$ to $\mathbb{R}^\ell$,

$$\zeta(\boldsymbol{p}) = \Phi(\boldsymbol{p}) - \Psi(\boldsymbol{p}) - \int_A \omega(a)d\lambda,$$

for every $\boldsymbol{p} \in \Delta$. We show that the correspondence ζ satisfies the conditions needed to apply Theorem D.4,

(i) for every $\boldsymbol{p} \in \mathring{\Delta}$, and $\boldsymbol{z} \in \zeta(\boldsymbol{p})$, $\boldsymbol{p}\boldsymbol{z} = 0$,
(ii) the correspondence $\zeta(\cdot)$ is compact and convex-valued, bounded from below and upper hemicontinuous,
(iii) for every sequence $\{\boldsymbol{p}_n\}$ in $\mathring{\Delta}$ converging to $\boldsymbol{p} \in \partial\Delta = \{\boldsymbol{p} = (p^t) \in \Delta |\, p^t = 0 \text{ for some } t\}$, it follows that

$$\inf \left\{ \sum_{t=1}^{\ell} z^t \,\middle|\, \boldsymbol{z} = (z^t) \in \zeta(\boldsymbol{p}) \right\} > 0 \text{ for } n \text{ large enough.}$$

Property (i) follows from Proposition F.11 and the assumption (LNS) of preferences. By Theorem 6.5 (b), the mean demand correspondence $\Phi(\boldsymbol{p})$ is upper hemicontinuous. Given the correspondence Y is compact-valued and simple, it is integrably bounded. By Lyapunov's theorem (Theorem F.12), $\int_A Y(a)d\lambda$ is a convex set. Hence, $\Psi(\boldsymbol{p})$ is compact, convex-valued, and upper hemicontinuous. Therefore, so is

$\zeta(\boldsymbol{p})$. Clearly, $\zeta(\boldsymbol{p})$ is bounded from below, given X is bounded from below and $\int_A Y(a)d\lambda$ is compact.

To prove (iii), let $\{\boldsymbol{p}_n\}$ be a sequence in $\mathring{\Delta}$ converging to $\boldsymbol{p} \in \partial\Delta$. Since $\int_A Y(a)d\lambda$ is compact, it suffices to show that

$$\inf \left\{ \sum_{t=1}^{\ell} x^t \,\middle|\, \boldsymbol{x} = (x^t) \in \phi(a, \boldsymbol{p}_n) \right\} \to +\infty \text{ as } n \to +\infty$$

for all $a \in A \backslash A_{\boldsymbol{p}}$. However, this can be shown in exactly the same way as the proof of Theorem 3.8. We can then apply Theorem D.4 and the proof is complete. □

The condition that the production correspondence Y is simple in Theorems 6.5 and 6.6 is very restrictive. However, we can apply these results to a generic analysis of the space of production economies. In Section 2.8 of Chapter 2, we examined the regularity of equilibria for exchange economies and observed that the regular economies are an open and dense subset of the space of all exchange economies. In the following, we pursue a similar analysis of the topological space of economies. Example 6.1 showed that the dispersed endowment distribution is not sufficient for the existence of equilibria. However, in Theorem 6.7, we see that the set of production economies with equilibria is a dense subset of the space of production economies with dispersed endowment distributions. Unfortunately, the set is not open, as Example 6.2 shows.

To present the results, we describe the topology endowed with the set of economies. We do so in terms of sequences. Let $(A_n, \mathcal{A}_n, \lambda_n)_{n\in\mathbb{N}}$ be a sequence of probability spaces. We say that a sequence of exchange economies

$$\mathcal{E}_n : (A_n, \mathcal{A}_n, \lambda_n) \to \mathcal{P} \times \mathbb{R}^\ell,\ a \mapsto (\prec_{a,n}, \omega_n(a))$$

converges to an exchange economy

$$\mathcal{E} : (A, \mathcal{A}, \lambda) \to \mathcal{P} \times \mathbb{R}^\ell,\ a \mapsto (\prec_a, \omega(a)),$$

if and only if

(i) the sequence $(\mathcal{E}_n)$ converges to $\mathcal{E}$ in distribution, or the sequence of measures $\mu_n \equiv \lambda_n \circ \mathcal{E}_n^{-1}$ converges weakly to the measure $\mu \equiv \lambda \circ \mathcal{E}^{-1}$,

(ii) the mean endowment converges, or $\int_{A_n} \omega_n(a) d\lambda_n \to \int_A \omega(a) d\lambda$.

Similarly, we say that a sequence of production economies

$$(\mathcal{E}_n, Y_n) : (A_n, \mathcal{A}_n, \lambda_n) \to \mathcal{P} \times \mathbb{R}^\ell \times \mathcal{Y},\ a \mapsto (\prec_{a,n}, \omega_n(a), Y_n(a))$$

converges to a production economy

$$(\mathcal{E}, Y) : (A, \mathcal{A}, \lambda) \to \mathcal{P} \times \mathbb{R}^\ell \times \mathcal{Y},\ a \mapsto (\prec_a, \omega(a), Y(a))$$

when the condition (i) is replaced by

$(i)'$ the sequence $(\mathcal{E}_n, Y_n)$ converges to $(\mathcal{E}, Y)$ in distribution, or the sequence of measures $\mu_n \equiv \lambda_n \circ (\mathcal{E}_n, Y_n)^{-1}$ converges weakly to the measure $\mu \equiv \lambda \circ (\mathcal{E}, Y)^{-1}$.

In the following, we discuss the case in which $(A_n, \mathcal{A}_n, \lambda_n) = (A, \mathcal{A}, \lambda)$ for all $n \in \mathbb{N}$. We now state the main result of this section.

Theorem 6.7. *Let $(\mathcal{E}, Y) : A \to \mathcal{P} \times X \times \mathcal{Y}$ be a production economy such that $\mathcal{E}(a) \subset \mathcal{P}_{wd,lns} \times X$ a.e. Suppose that the endowment distribution is dispersed. Then there exists a sequence of production correspondences $Y_n : A \to \mathcal{Y}$ such that*

(a) *for each n, the production economy $(\mathcal{E}, Y_n)$ has an equilibrium,*
(b) *the sequence $(\mathcal{E}, Y_n)$ converges to $(\mathcal{E}, Y)$ in distribution.*

Note that the statement (b) implies that the sequences of production economies converge to the production economy $(\mathcal{E}, Y)$ as the exchange economy (consumption sector) $\mathcal{E}$ does not change; hence, the mean endowments of the sequence economies are constant.

Proof of Theorem 6.7. By the theorem of Rådström (Theorem D.5 of Appendix D), the space $(\mathcal{K}_c, \delta)$ is embedded as a convex cone with the vertex at the origin to a real normed linear space L by an isomorphic embedding maps ι_1. Note that $\iota_1(\{\mathbf{0}\}) = \mathbf{0}$. According to the theorem of completion (Theorem G.1), there exists an isometric isomorphism ι_2 between L and a dense linear subspace of a Banach space $\tilde{L}$. Given $\mathcal{Y} \subset \mathcal{F}_0$ is a separable metric space, the set $\{Y(a) \in \mathcal{Y} | \ a \in A\}$ is separable by Proposition B.10; hence, the set $\{\iota_2 \circ \iota_1 \circ Y(a) \in \mathcal{Y} | \ a \in A\}$ is also a separable subset of $\tilde{L}$. As the maps ι_1 and ι_2 are both continuous and the correspondence $Y : A \to \mathcal{Y}$ is $\mathcal{A}$-measurable, it follows that for any continuous linear functional on $\tilde{L}$, the map $f \circ \iota_2 \circ \iota_1 \circ Y(\cdot)$ is $\mathcal{A}$-measurable. According to Theorem G.7, there exists a sequence $(\tilde{Y}_n)_{n \in \mathbb{N}}$ of simple functions of A into $\tilde{L}$ with $\|\tilde{Y}_n(a) - \iota_2 \circ \iota_1 \circ Y(a)\| \to 0$ a.e. For each n, we have $\{\tilde{Y}_n(a) \in \tilde{L} | \ a \in A\} \subset \{\iota_2 \circ \iota_1 \circ Y(a) \in \tilde{L} | \ a \in A\} \cup \{\mathbf{0}\}$. Therefore, we have a sequence of simple functions $(Y_n(\cdot))_{n \in \mathbb{N}}$ of A to $\mathcal{Y}$ with $\delta(Y_n(a), Y(a)) \to 0$ as $n \to \infty$, where δ is the Hausdorff distance. Let d be a metric on $\mathcal{P} \times \mathbb{R}^\ell \times \mathcal{Y}$ defined by

$$d((\prec, \omega, Y), (\prec', \omega', Y')) = \hat{\delta}(\prec, \prec') + \|\omega - \omega'\| + \delta(Y, Y'),$$

where $\hat{\delta}$ is the metric of the topology of a closed convergence τ_c. Then it follows that $d((\mathcal{E}(a), Y_n(a)), (\mathcal{E}(a), Y(a)) = \delta(Y_n(a), Y(a)) \to 0$ a.e. The map $\mathcal{E}$ is $\mathcal{B}(\mathcal{P} \times \mathbb{R}^\ell)$-measurable and the mappings Y, Y_n are $\mathcal{B}(\mathcal{Y})$-measurable. Hence, the mappings $(\mathcal{E}, Y)$, $(\mathcal{E}, Y_n)$ are $\mathcal{B}(\mathcal{P} \times \mathbb{R}^\ell) \times \mathcal{B}(\mathcal{Y}) = \mathcal{B}(\mathcal{P} \times \mathbb{R}^\ell \times \mathcal{Y})$-measurable by Propositions F.3 and F.6, given the sets $\mathcal{P} \times \mathbb{R}^\ell$ and $\mathcal{Y}$ are both separable metric spaces. Then by Theorems F.8 and F.5, we see that $(\mathcal{E}, Y_n) \to (\mathcal{E}, Y)$ in distribution, and for each n, the economy $(\mathcal{E}_n, Y)$ has an equilibrium according to Theorem 6.6. □

The following example shows that the set of production economies with equilibrium is not open.

Example 6.2. Consider the exchange economy in Example 3.2 of Chapter 3. For each consumer $a \in [0, 1]$, the consumption set X and the utility function are given as follows:

$$X = \{(x, y) \in \mathbb{R}^2 | \ x \geq 1, \ y \geq 0\} \cup \{(x, y) \in \mathbb{R}^2 | \ x \geq 0, \ y \geq 3\},$$
$$u(x, y) = x^2 + y^2, \ (x, y) \in X,$$

and the endowment vector $\omega(a)$ is

$$\omega(a) = (1, a) \text{ for } a \in [0, 1].$$

This exchange economy (with a dispersed endowment distribution) can be seen as a production economy if we give a production correspondence

$$Y(a) = \{(0, 0)\}, \ a \in [0, 1].$$

We know that the excess demand function for x-commodity is

$$\zeta(p) = \frac{1}{2p(1-p)}(10p^2 - 20p + 9),$$

where we normalized the price vector $\boldsymbol{p} = (p, 1-p)$. The equilibrium price of the commodity x is

$$10p^2 - 20p + 9 = 0, \ p = \frac{10 - \sqrt{10}}{10}.$$

We can approximate this "production" economy $(\mathcal{E}, Y)$ by a sequence of production economies without equilibrium $(\mathcal{E}, Y_n)$ in the following way. Note that this shows that $(\mathcal{E}, Y)$ is not an interior point of the space of production economies with equilibrium. Therefore, it is not an open subset of the economies with dispersed endowment distributions. Let $(\epsilon_n)_{n \in \mathbb{N}}$ be a sequence of positive numbers decreasing to 0. The demand function of the consumer $a \in [0, 1]$ has been already calculated as follows:

$$\phi(a, p) = (\phi^x(a, p), \phi^y(a, p)) = \begin{cases} \left(1 + \frac{(1-p)a}{p}, 0\right) & \text{for } 0 \le p < \frac{3-a}{4-a}, \\ \left(0, \frac{p}{1-p} + a\right) & \text{for } \frac{3-a}{4-a} \le p \le 1. \end{cases}$$

For a given p, we can identify consumer a who "jumps" at the price $p = (3-a)/(4-a)$, solving for a, $a = (3-4p)/(1-p)$. Then we calculate

$$\begin{aligned}
\zeta^x(p) &= \int_{\epsilon_n}^{1} (\phi^x(a,\boldsymbol{p}) - \omega^x(a))da \\
&= \frac{1-p}{p}\int_{\epsilon_n}^{3-4p/1-p} a da - \int_{3-4p/1-p}^{1} da \\
&= \frac{1-p}{2p}\left\{\left(\frac{3-4p}{1-p}\right)^2 - \epsilon_n^2\right\} - \left(1 - \frac{3-4p}{1-p}\right) \\
&= \frac{1}{2p(1-p)}((10-\epsilon_n^2)p^2 - 2(10-\epsilon_n^2)p + 9 - \epsilon_n^2).
\end{aligned}$$

Solving $(10-\epsilon_n^2)p^2 - 2(10-\epsilon_n^2)p + 9 - \epsilon_n^2 = 0$, we obtain

$$p = \frac{10-\epsilon_n^2 - \sqrt{10-\epsilon_n^2}}{10-\epsilon_n^2} \equiv p_n.$$

Note that $p_n \to p = \frac{10-\sqrt{10}}{10}$ as $n \to \infty$, given $\epsilon_n \to 0$. For each n, define the production correspondence $Y_n : A \to \mathcal{Y}$ by

$$Y(a) = \begin{cases} co\left\{(0,0), \left(0, \frac{3-4p_n}{1-p_n} - a\right)\right\} & \text{for } a \in [0,\epsilon_n], \\ \{(0,0)\} & \text{for } a \in (\epsilon_n, 1]. \end{cases}$$

Given $\epsilon_n \to 0$, it follows that $\delta(Y_n, Y) \to 0$ a.e. Hence, as in the proof of Theorem 6.7, we have from Theorems F.8 and F.5 that $(\mathcal{E}, Y_n) \to (\mathcal{E}, Y)$ in distribution. Furthermore, by the construction of Y_n, the excess demand for the x-commodity of the consumer $a \in [0,1]$ is

$$\phi^x(a,p) - \omega^x(a) = \begin{cases} \left(\frac{1-p}{p}\right)\left(\frac{3-4p_n}{1-p_n}\right) & \text{for } 0 \le p < p_n, \\ -1 & \text{for } p_n \le p \le 1. \end{cases}$$

Therefore, the mean demand correspondence for x given as

$$\begin{aligned}
\zeta^x(p) &= \int_0^1 (\phi^x(a,\boldsymbol{p}) - \omega^x(a))da \\
&= \int_0^{\epsilon_n} (\phi^x(a,\boldsymbol{p}) - \omega^x(a))da + \int_{\epsilon_n}^{1} (\phi^x(a,\boldsymbol{p}) - \omega^x(a))da
\end{aligned}$$

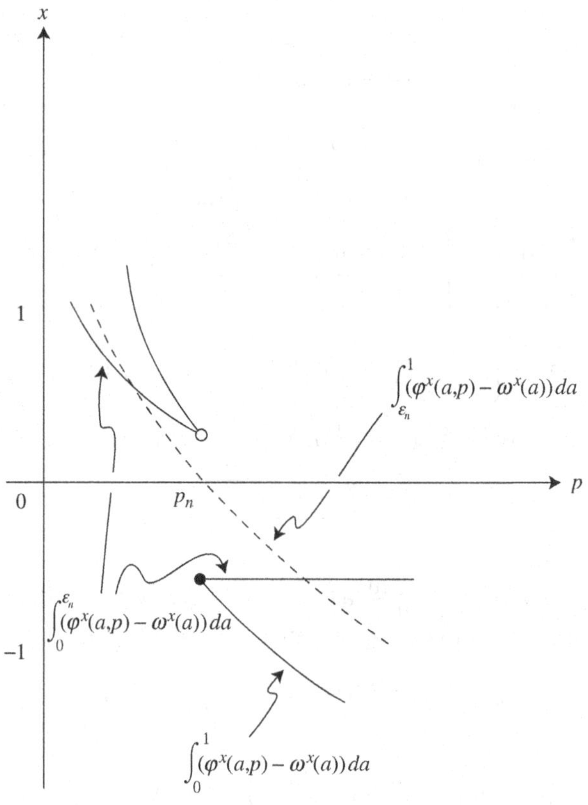

Figure 6.2. Example 6.2.

is discontinuous at $p = p_n$; hence, the economy $(\mathcal{E}, Y_n)$ does not have an equilibrium; see Figure 6.2.

6.5 Notes

Section 6.1: The contents of this section are standard. The most basic reference is Debreu [51, Chapters 5 and 6]. See also Arrow and Hahn [11, Chapters 3 and 5], McKenzie [160, Chapters 5 and 6], and Nikaido [174, Chapter 5]. Debreu proved the existence of competitive equilibrium assuming closure and convexity for the total production set Y [51, p.53] based on the idea of Uzawa; see [51, p. 88].

For definiteness, we have assumed that every individual production set is closed and convex. However, this dose not imply that the total production set is closed. If Y is closed and convex and the free disposability $\mathbb{R}^\ell_- \subset Y$ holds, one can show that the equilibrium allocation is exactly feasible without assuming the monotonicity of preferences; see [51, p. 86].

Section 6.2: Classic studies concerning externalities in consumption or production are [11, 155, 228, 229]. As shown, the result of Shafer and Sonnenschein [228] in particular is crucial for Theorem 6.3. See also Section 4 of McKenzie [159].

Section 6.3: The exposition in this section follows [94, Chapter 4]. See also [82]. Sondermann [234] allows the coalition production correspondence to be *superadditive*, or

$$\boldsymbol{Y}(B) + \boldsymbol{Y}(C) \subset \boldsymbol{Y}(B \cup C)$$

for any disjoint measurable sets $B, C \in \mathcal{A}$, which exhibits increasing returns to scale technology with respect to the size of coalitions. He showed that the profit functions were well defined in this setting and proved the existence of (quasi-) equilibria.

Section 6.4: All of the expositions in this section are due to Suzuki [243]. Greenberg *et al.* [82] discussed similar types of production economies. Shitovitz [232] formulated a model of oligopolistic markets with a continuum of traders including atoms.

Chapter 7

Theory of Increasing Returns

7.1 Marginal Cost Pricing Equilibria

In this section and in the following section, we consider the core and two distinct concepts of equilibria of a production economy where increasing returns prevail, or in other words, where we assume the production sets of the producers are not convex.

Consider the production economy described in Section 6.1 where there are ℓ commodities and $n+1$ firms in the economy, and each firm is indexed by $b(=1\ldots n+1)$. The production set of firm b is represented by a production set $Y_b \subset \mathbb{R}^\ell$. We assume that $Y_b \subset \mathbb{R}^\ell$ is closed and satisfies (YCV), (NP) and (FD), namely that it is convex, contains $\{\mathbf{0}\}$, and $Y_b - \mathbb{R}^\ell_+ \subset Y_b$ for $b = 1\ldots n$. The set of *(weakly) efficient production plans* is defined by

$$\partial Y_b = \left\{ \boldsymbol{y}_b \in Y_b \middle| \text{ There exists no } \boldsymbol{y} \in Y_b \text{ such that } \boldsymbol{y} \gg \boldsymbol{y}_b \right\}.$$

Let $\Gamma : \mathbb{R}^\ell \to \mathbb{R}$ be a production function that is continuously differentiable and $\Gamma(\mathbf{0}) = 0$. The $n+1$th firm's production set is defined as

$$Y_{n+1} = \{\boldsymbol{y} \in \mathbb{R}^\ell | \, \Gamma(\boldsymbol{y}) \leq 0\}.$$

As we do not assume $\Gamma(\boldsymbol{y})$ or Y_{n+1} to be convex, the technology of firm $n+1$ may exhibit increasing returns to scale. We denote the normalized gradient of $\Gamma(\boldsymbol{y})$ by

$$\partial\Gamma(\boldsymbol{y}) = \frac{D\Gamma(\boldsymbol{y})}{\sum_{t=1}^{\ell} \partial_t \Gamma(\boldsymbol{y})},$$

where $D\Gamma(\boldsymbol{y}) = (\partial_1\Gamma(\boldsymbol{y}) \dots \partial_\ell\Gamma(\boldsymbol{y}))$. We assume $D\Gamma(\boldsymbol{y}) > \boldsymbol{0}$, and hence; $\partial\Gamma(\boldsymbol{y})$ is well defined and Y_{n+1} satisfies (NP) and (FD). Mathematically, the (normalized) gradient $\partial\Gamma(\boldsymbol{y})$ is a normal vector to the hypersurface ∂Y_{n+1} at $\boldsymbol{y} \in \partial Y_{n+1}$. Obviously, the set of the weakly efficient production plans of firm $n+1$ is determined as the boundary of the production set, $\partial Y_{n+1} = \{\boldsymbol{y} \in \mathbb{R}^\ell | \, \Gamma(\boldsymbol{y}) = 0\}$.

Definition 7.1. A price vector $\boldsymbol{p} \in \Delta$ is called a *marginal cost price vector* if and only if $\boldsymbol{p} = \partial\Gamma(\boldsymbol{y})$ for some $\boldsymbol{y} \in \partial Y_{n+1}$.

In other words, the marginal cost price is a price vector that satisfies the first-order condition (FOC) for the local profit maximum or minimum of the $n + 1$th firm.

There exist m consumers in the economy indexed by $a(= 1 \dots m)$ with the consumption set $X_a \subset \mathbb{R}^\ell_+$ which is assumed to be convex, $a = 1 \dots m$. The consumer a has the preference relation $\prec_a \subset X_a \times X_a$, which satisfies the basic postulates (IR), (TR), (CT). We also assume (NTR) and (CV).

The total resource of the economy is provided by the aggregate endowment vector $\omega \gg \boldsymbol{0}$ and the income distribution is given in terms of a fixed structure of revenues $\theta_a > 0$ with $\sum_{a=1}^m \theta_a = 1$, so that consumer a's wealth level at price $\boldsymbol{p}$ and prevailing production plans $(\boldsymbol{y}_b)$ is defined by

$$w_a(\boldsymbol{p}, \boldsymbol{y}_b) = \theta_a \boldsymbol{p} \left(\sum_{b=1}^{n+1} \boldsymbol{y}_b + \omega \right).$$

A production economy or simply an economy $\mathcal{E}_\Gamma$ is defined by the list $(X_a, \prec_a, \theta_a, Y_b, \omega)$. The equilibrium of the economy $\mathcal{E}_\Gamma$ that includes a possible increasing-returns technology is defined as follows:

Definition 7.2. An $m + n + 2$-tuple $(\boldsymbol{p}, \boldsymbol{x}_a, \boldsymbol{y}_b)$ of an allocation $(\boldsymbol{x}_a, \boldsymbol{y}_b) \in \prod_{a=1}^m X_a \times \prod_{b=1}^{n+1} Y_b$ and a non-zero price vector $\boldsymbol{p} \in \mathbb{R}^\ell_+$

is said to consist of a *marginal cost pricing equilibrium* or an *MCP equilibrium* if, and only if,

(MC-1) $\boldsymbol{p}\boldsymbol{x}_a \leq w_a(\boldsymbol{p}, \boldsymbol{y}_b)$ and $\boldsymbol{x}_a \succsim_a \boldsymbol{z}$ whenever $\boldsymbol{p}\boldsymbol{z} \leq w_a(\boldsymbol{p}, \boldsymbol{y}_b)$, unless $\boldsymbol{p}(\sum_{b=1}^{n+1} \boldsymbol{y}_b + \omega) = 0$, $a = 1 \dots m$,
(MC-2) $\boldsymbol{p} = \partial\Gamma(\boldsymbol{y}_{n+1})$ and $\boldsymbol{p}\boldsymbol{y}_b \geq \boldsymbol{p}\boldsymbol{y}$ for all $\boldsymbol{y} \in Y_b$, $b = 1 \dots n$,
(MC-3) $\sum_{a=1}^{m} \boldsymbol{x}_a \leq \sum_{b=1}^{n+1} \boldsymbol{y}_b + \omega$.

The condition (MC-2) explains the title of this equilibrium concept. For notational convenience, we denote

$$ArgMax(\boldsymbol{p}Y_b) = \{\boldsymbol{y} \in Y_b |\ \boldsymbol{p}\boldsymbol{y} \geq \boldsymbol{p}\boldsymbol{z} \text{ for all } \boldsymbol{z} \in Y_b\}.$$

The main existence theorem reads as follows:

Theorem 7.1. *Suppose that an economy $\mathcal{E}_\Gamma = (X_a, \prec_a, \theta_a, Y_b, \omega)$ satisfies all of the assumptions detailed above, and*

(*PI*) (*Positive income*): *for all $\boldsymbol{y}_b \in ArgMax(\boldsymbol{p}Y_b)$ $(b = 1 \dots n)$, $\boldsymbol{y}_{n+1} \in \partial Y_{n+1}$ and $\boldsymbol{p} \in \partial\Gamma(\boldsymbol{y}_{n+1})$, it follows $\boldsymbol{p}(\sum_{b=1}^{n+1} \boldsymbol{y}_b + \omega) \geq 0$,*
(*YCL*) (*Closed total production set*): *$\sum_{b=1}^{n+1} Y_b \subset \mathbb{R}^\ell$ is closed,*
(*BTP*) (*Bounded total production*): *the set $\{(\boldsymbol{y}_b) \in \prod_{b=1}^{n+1} Y_b |\ \boldsymbol{z} \leq \sum_{b=1}^{n+1} \boldsymbol{y}_b\}$ is bounded for all $\boldsymbol{z} \in \mathbb{R}^\ell$.*

Then there exists a competitive equilibrium $(\boldsymbol{p}^*, \boldsymbol{x}_a^*, \boldsymbol{y}_b^*)$ for $\mathcal{E}_\Gamma$.

Remark 7.1. When the technology exhibits increasing returns, marginal cost pricing may yield negative profits, as displayed in Figure 7.1. Consequently, the wealth of consumers could be possibly negative. The assumption (PI) excludes this situation.

Proof of Theorem 7.1. Given the set of feasible allocations

$$\mathcal{F} = \left\{ (\boldsymbol{x}_a, \boldsymbol{y}_b) \in \prod_{a=1}^{m} X_a \times \prod_{b=1}^{n+1} Y_b \,\middle|\, \sum_{a=1}^{m} \boldsymbol{x}_a \leq \sum_{b=1}^{n+1} \boldsymbol{y}_b + \omega \right\}$$

is non-empty and bounded by the assumption (BTP), the feasible consumption and production sets, which are defined as the projections of $\mathcal{F}$ to X_a and Y_b, respectively, are bounded. Therefore, there

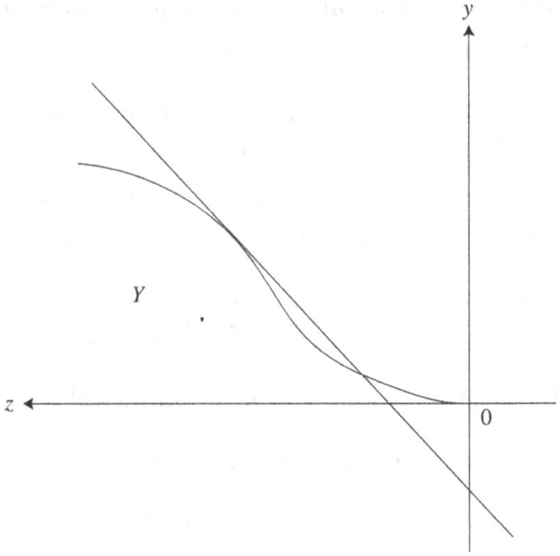

Figure 7.1. Negative profit.

exists a closed cube $\hat{K}$ in $\mathbb{R}^\ell$ with length $k > 0$, centered at the origin and containing in its interior the feasible consumption and production plans, $\boldsymbol{x}_a, \boldsymbol{y}_b \subset int\hat{K}$ for $(\boldsymbol{x}_a, \boldsymbol{y}_b) \in \mathcal{F}$. Then we define

$$\hat{X}_a = X_a \cap \hat{K}, \quad a = 1 \dots m,$$

$$\hat{Y}_b = Y_b \cap \hat{K}, \quad b = 1 \dots n,$$

$$\hat{Y}_{n+1} = (Y_{n+1} + (1 + nk)\mathbf{1} + \omega) \cap \mathbb{R}^\ell_+,$$

where $\mathbf{1} = (1 \dots 1)$. Let g denote the projection of points in $\mathbb{R}^\ell_+ \backslash \{0\}$ from the origin on the unit simplex Δ, and we define for each $\boldsymbol{q} \in \Delta$,

$$\tau(\boldsymbol{q}) = sup\{\tau > 0 |\ \tau\boldsymbol{q} \in \hat{Y}_{n+1}\},$$

$$h(\boldsymbol{q}) = \tau(\boldsymbol{q})\boldsymbol{q},$$

$$\partial\hat{Y}_{n+1} = \{h(\boldsymbol{q}) \in \hat{Y}_{n+1} |\ \boldsymbol{q} \in \Delta\}.$$

Obviously, $h(\boldsymbol{q})$ is weakly efficient up to $(1 + nk)\mathbf{1} + \omega$ or $h(\boldsymbol{q}) - (1 + nk)\mathbf{1} - \omega \in \partial Y_{n+1}$ for each $\boldsymbol{q} \in \Delta$, and if $\boldsymbol{y}$ is an attainable production plan, which means that $\boldsymbol{y}$ belongs to the projection of $\mathcal{F}$ to Y_{n+1}, then $\boldsymbol{y} + (1 + nk)\mathbf{1} + \omega \gg \mathbf{0}$; see Figure 7.2.

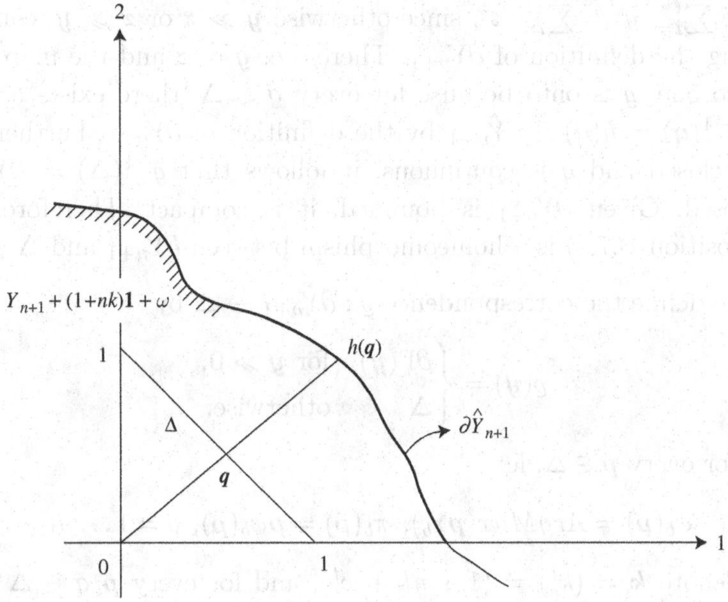

Figure 7.2. Lemma 7.1.

Lemma 7.1. *$g : \partial\hat{Y}_{n+1} \to \Delta$ is a homeomorphism between $\partial\hat{Y}_{n+1}$ and Δ which satisfies that $g(\boldsymbol{y}) \gg \mathbf{0}$ if and only if $\boldsymbol{y} \gg \mathbf{0}$.*

Proof. First note that the mappings $g : \partial\hat{Y}_{n+1} \to \Delta$ and $h : \Delta \to \partial\hat{Y}_{n+1}$ are the inverse map of each other, or $h \circ g$ and $g \circ h$ are identities on $\partial\hat{Y}_{n+1}$ and Δ, respectively. By definition, we can write $g : \partial\hat{Y}_{n+1} \to \Delta$ as

$$g(\boldsymbol{y}) = \left(\frac{1}{\sum_{t=1}^{\ell} y^t}\right)\boldsymbol{y} \quad \text{for } \boldsymbol{y} = (y^t) \in \partial\hat{Y}_{n+1}.$$

From this expression, it is immediate that $g(\boldsymbol{y}) \gg \mathbf{0}$ if and only if $\boldsymbol{y} \gg \mathbf{0}$, and g is continuous, since $\sum_{t=1}^{\ell} y^t > 0$ for every $\boldsymbol{y} \in \partial\hat{Y}_{n+1}$. Suppose that for $\boldsymbol{y} = (y^t) \in \partial\hat{Y}_{n+1}$ and $\boldsymbol{z} = (z^t) \in \partial\hat{Y}_{n+1}$, $g(\boldsymbol{y}) = g(\boldsymbol{z})$. Then we have

$$\boldsymbol{y} = \left(\frac{\sum_{t=1}^{\ell} y^t}{\sum_{t=1}^{\ell} z^t}\right)\boldsymbol{z},$$

hence $\sum_{t=1}^{\ell} y^t = \sum_{t=1}^{\ell} z^t$, since otherwise $\boldsymbol{y} \gg \boldsymbol{z}$ or $\boldsymbol{z} \gg \boldsymbol{y}$, contradicting the definition of $\partial\hat{Y}_{n+1}$. Therefore, $\boldsymbol{y} = \boldsymbol{z}$ and the map g is one to one. g is onto because for every $\boldsymbol{q} \in \Delta$, there exists a vector $g^{-1}(\boldsymbol{q}) = h(\boldsymbol{q}) \in \partial\hat{Y}_{n+1}$ by the definition of $\partial\hat{Y}_{n+1}$. Further, as Δ is closed and g is continuous, it follows that $g^{-1}(\Delta) = \partial\hat{Y}_{n+1}$ is closed. Given $\partial\hat{Y}_{n+1}$ is bounded, it is compact. Therefore, by Proposition B.7, g is a homeomorphism between $\partial\hat{Y}_{n+1}$ and Δ. $\square$

We define the correspondence $\varrho : \partial\hat{Y}_{n+1} \to \Delta$ by

$$\varrho(\boldsymbol{y}) = \begin{cases} \partial\Gamma(\boldsymbol{y}) & \text{for } \boldsymbol{y} \gg \boldsymbol{0}, \\ \Delta & \text{otherwise,} \end{cases}$$

and for every $\boldsymbol{p} \in \Delta$, let

$$\psi_b(\boldsymbol{p}) = ArgMax(\boldsymbol{p}\hat{Y}_b), \ \pi_b(\boldsymbol{p}) = \boldsymbol{p}\psi_b(\boldsymbol{p}), \ b = 1 \ldots n.$$

We denote $\hat{\boldsymbol{k}} = (\hat{k}^t) = (1 + nk + \omega^t)$, and for every $\boldsymbol{p}, \boldsymbol{q} \in \Delta$ and $h(\boldsymbol{q}) = g^{-1}(\boldsymbol{q})$ as in Lemma 7.1, define

$$\hat{w}_a(\boldsymbol{p}, \boldsymbol{q}) = \theta_a \left(\sum_{b=1}^{n} \pi_b(\boldsymbol{p}) + \boldsymbol{p}(h(\boldsymbol{q}) - \hat{\boldsymbol{k}}) + \boldsymbol{p}\omega \right),$$

$$\beta_a(\boldsymbol{p}, \boldsymbol{q}) = \{\boldsymbol{x} \in \hat{X}_a | \ \boldsymbol{p}\boldsymbol{x} \le \hat{w}_a(\boldsymbol{p}, \boldsymbol{q})\},$$

$$\phi_a(\boldsymbol{p}, \boldsymbol{q}) = \begin{cases} \{\boldsymbol{x} \in \beta_a(\boldsymbol{p}, \boldsymbol{q}) | \ \boldsymbol{x} \succsim_a \boldsymbol{z} \text{ for } \boldsymbol{z} \in \beta_a(\boldsymbol{p}, \boldsymbol{q})\} & \text{if } \hat{w}_a(\boldsymbol{p}, \boldsymbol{q}) > 0, \\ \beta_a(\boldsymbol{p}, \boldsymbol{q}) & \text{if } \hat{w}_a(\boldsymbol{p}, \boldsymbol{q}) = 0, \\ \{\boldsymbol{0}\} & \text{if } \hat{w}_a(\boldsymbol{p}, \boldsymbol{q}) < 0. \end{cases}$$

For notational simplicity, we denote $\sum_{a=1}^{m} \hat{X}_a \times \sum_{b=1}^{n} \hat{Y}_b$ by $\hat{X} \times \hat{Y}$ and define the following maps,

$$\rho : \Delta \to \Delta, \ \rho(\boldsymbol{q}) = \varrho \circ h(\boldsymbol{q}),$$

$$\sigma : \Delta \times \Delta \to \hat{X} \times \hat{Y}, \ \sigma(\boldsymbol{p}, \boldsymbol{q}) = \sum_{a=1}^{m} \phi_a(\boldsymbol{p}, \boldsymbol{q}) \times \sum_{b=1}^{n} \psi_b(\boldsymbol{p}),$$

$$\zeta : \hat{X} \times \hat{Y} \to \Delta, \ \zeta(\hat{\boldsymbol{x}}, \hat{\boldsymbol{y}}) = \left(\frac{\hat{x}^t - \hat{y}^t - \omega^t + \hat{k}^t}{\sum_{t=1}^{\ell} (\hat{x}^t - \hat{y}^t - \omega^t + \hat{k}^t)} \right),$$

where $\hat{\boldsymbol{x}} = (\hat{x}^t) = \sum_{a=1}^{m} \boldsymbol{x}_a = \sum_{a=1}^{m} (x_a^t)$ and similarly for $\hat{\boldsymbol{y}}$.

Note that given our truncation,

$$\hat{\boldsymbol{x}} - \hat{\boldsymbol{y}} - \omega + \hat{\boldsymbol{k}} \geq \mathbf{1} \gg \mathbf{0}.$$

We then consider a map Φ from $\Delta \times \Delta \times \hat{X} \times \hat{Y}$ to itself defined by

$$\Phi(\boldsymbol{p}, \boldsymbol{q}, \hat{\boldsymbol{x}}, \hat{\boldsymbol{y}}) = \rho(\boldsymbol{q}) \times \zeta(\hat{\boldsymbol{x}}, \hat{\boldsymbol{y}}) \times \sigma(\boldsymbol{p}, \boldsymbol{q}).$$

As all of the maps ρ, ζ, σ are compact and convex-valued, upper-hemicontinuous, so is the map ϕ; hence, it satisfies the conditions of Kakutani's fixed point theorem (Theorem D.1). Let $(\boldsymbol{p}^*, \boldsymbol{q}^*, \hat{\boldsymbol{x}}^*, \hat{\boldsymbol{y}}^*) = (\boldsymbol{p}^*, \boldsymbol{q}^*, \sum_{a=1}^m \boldsymbol{x}_a^*, \sum_{b=1}^n \boldsymbol{y}_b^*)$ be a fixed point.

Given $\hat{\boldsymbol{x}}^* - \hat{\boldsymbol{y}}^* - \omega + \hat{\boldsymbol{k}} \geq \mathbf{1} \gg \mathbf{0}$, $\boldsymbol{q}^* = \zeta(\hat{\boldsymbol{x}}^*, \hat{\boldsymbol{y}}^*) \gg \mathbf{0}$ so that $h(\boldsymbol{q}^*) \gg \mathbf{0}$ by Lemma 7.1. Hence, by the definition of ϱ, we obtain $\boldsymbol{p}^* = \partial\Gamma(\boldsymbol{y}_{n+1}^*)$. By the assumption (PI), we have $\hat{w}_a(\boldsymbol{p}^*, \boldsymbol{q}^*) \geq 0$ for all $a = 1 \ldots m$. It follows from the definition of ϕ_a that $\boldsymbol{x} \succsim_a \boldsymbol{z}$ for $\boldsymbol{z} \in \beta_a(\boldsymbol{p}, \boldsymbol{q})$. By the definition of ψ_b, we obviously have $\boldsymbol{p}^*\boldsymbol{y}_b^* \geq \boldsymbol{p}^*\boldsymbol{y}$ for all $\boldsymbol{y} \in Y_b, b = 1 \ldots n$. It remains to show that $\sum_{a=1}^m \boldsymbol{x}_a^* \leq \sum_{b=1}^n \boldsymbol{y}_b^* + \boldsymbol{y}_{n+1}^* + \omega$.

It follows from $\hat{w}_a(\boldsymbol{p}^*, \boldsymbol{q}^*) \geq 0$ for all $a = 1 \ldots m$ that

$$\boldsymbol{p}^*\hat{\boldsymbol{x}}^* \leq \sum_{a=1}^m \hat{w}_a(\boldsymbol{p}^*, \boldsymbol{q}^*) = \boldsymbol{p}^*\hat{\boldsymbol{y}}^* + \boldsymbol{p}^*\boldsymbol{y}_{n+1}^* + \boldsymbol{p}^*\omega.$$

Since $h(\boldsymbol{q}^*)$ is a scalar multiple of $\boldsymbol{q}^*$, we have $\hat{\boldsymbol{x}}^* - \hat{\boldsymbol{y}}^* - \omega + \hat{\boldsymbol{k}} = \kappa(\boldsymbol{y}_{n+1}^* + \hat{\boldsymbol{k}})$, where κ is a positive real number. Applying this to the above relation, we can obtain $\boldsymbol{p}^*(\boldsymbol{y}_{n+1}^* + \hat{\boldsymbol{k}}) \geq \kappa\boldsymbol{p}^*(\boldsymbol{y}_{n+1}^* + \hat{\boldsymbol{k}})$. Since $\boldsymbol{y}_{n+1}^* + \hat{\boldsymbol{k}} = h(\boldsymbol{q}^*) \gg \mathbf{0}$, we have $\boldsymbol{p}^*(\boldsymbol{y}_{n+1} + \hat{\boldsymbol{k}}) > 0$. Therefore, $\kappa \leq 1$ and the proof is complete. □

7.2 Core of an Economy with Increasing Returns

In this section, we consider the coalition production economy presented in Section 6.3 where the commodity space is $\mathbb{R}^{\ell+m}$ and $\mathcal{P}$ is the set of allowed preference relations that satisfy the basic postulates, and we assume that all consumers have the common consumption set $X = \mathbb{R}_+^{\ell+m}$. A typical commodity vector is denoted by

$\xi = (\boldsymbol{u}, \boldsymbol{x})$, $\eta = (\boldsymbol{v}, \boldsymbol{y})$ or $\zeta = (\boldsymbol{w}, \boldsymbol{z}) \in \mathbb{R}^{\ell+m}$, where $\boldsymbol{u}, \boldsymbol{v}, \boldsymbol{w} \in \mathbb{R}^{\ell}$ and $\boldsymbol{x}, \boldsymbol{y}, \boldsymbol{z} \in \mathbb{R}^m$. m-vectors $\boldsymbol{x} = (x^t)$, $\boldsymbol{y} = (y^t)$ and $\boldsymbol{z} = (z^t)$ are consumption vectors which are called *consumer vectors* (x^t, y^t and z^t, $t = 1 \dots m$, are *consumer commodities*) and produced from ℓ-vectors $\boldsymbol{u} = (u^s)$, $\boldsymbol{v} = (v^s)$ and $\boldsymbol{w} = (w^s)$ which are *producer vectors* (u^s, v^s and w^s, $s = 1 \dots \ell$, are *producer commodities*). We assume that the preference relations $\succsim \in \mathcal{P}$ satisfy (LNS) and

(CM) (Only consumer commodities matter): $(\boldsymbol{u}, \boldsymbol{x}) \sim (\boldsymbol{w}, \boldsymbol{z})$ if $x^t = z^t$ for $t = 1 \dots m$ for every $\boldsymbol{u}, \boldsymbol{w} \in \mathbb{R}^{\ell}_{+}$,

(MTC) (Monotonicity for consumer goods): $\boldsymbol{x} < \boldsymbol{z}$ implies $(\boldsymbol{u}, \boldsymbol{x}) \prec (\boldsymbol{w}, \boldsymbol{z})$ for every $\boldsymbol{u}, \boldsymbol{w} \in \mathbb{R}^{\ell}_{+}$.

Let $(A, \mathcal{A}, \lambda)$ be an atomless measure space of consumers, $\mathcal{E} : A \to (\mathcal{P}, X)$ is an exchange economy, and $\boldsymbol{Y} : \mathcal{A} \to \mathcal{Y}$ a production correspondence. We denote $\omega(a) = (\boldsymbol{e}(a), \boldsymbol{f}(a)) = ((e^s(a)), (f^t(a))) \in \mathbb{R}^{\ell}_{+} \times \mathbb{R}^m_{+}$ and assume

(PE) (Positive endowments): $\int_A \omega(a) d\lambda \gg \mathbf{0}$ and $\boldsymbol{f}(a) > \mathbf{0}$, a.e.

We consider the simplest case where the set of allowed production sets is a singleton $\mathcal{Y} = \{Y\}$; hence, the production correspondence is given by a constant

$$\boldsymbol{Y}(C) = \begin{cases} Y & \text{for } C \in \mathcal{A} \text{ with } \lambda(C) > 0, \\ \{\mathbf{0}\} & \text{for } C \in \mathcal{A} \text{ with } \lambda(C) = 0, \end{cases}$$

and the production set Y is a closed subset of $\mathbb{R}^{\ell+m}$ which satisfies (NFP) and (FD), and as already mentioned, we assume that the producer commodities are only used as inputs and are not producible.

Let Λ be a cone in which only the first ℓ components are restricted to be non-negative, $\Lambda = \{(\boldsymbol{v}, \boldsymbol{y}) \in \mathbb{R}^{\ell} \times \mathbb{R}^m | \ \boldsymbol{v} \geq \mathbf{0}\}$.

(NPC) (Non-productivity of producer commodities): $Y \subset -\Lambda$.

Scarf [224] introduced the concept of *distributive production sets*, which contain increasing returns to scale production sets.

Definition 7.3. Let Y be a set in $\mathbb{R}^{\ell+m}$ with $\mathbb{R}_-^{\ell+m}(= -\mathbb{R}_+^{\ell+m}) \subset Y \subset -\Lambda$. We say that Y is a *distributive set* if for any finite number of vectors $\xi_i \in Y$ and any non-negative $\tau_i \geq 0$, the vector $\xi = \sum_i \tau_i \xi_i$ is also in Y, if ξ satisfies the condition that $\xi_i - \xi \in \Lambda$ for all i.

In other words, a set Y is distributive if, and only if, every non-negative weighted sum of the vectors in Y will be in Y if it uses more of the producer commodities than any original plan. Note that if all commodities in the economy are consumer commodities or $\ell = 0$, then $\Lambda = \mathbb{R}^m$ and a distributive set is a convex cone with a vertex at the origin.

Example 7.1. Let $\ell = m = 1$ and assume that the first commodity is a producer commodity and the second a consumer commodity. Let

$$f : \mathbb{R}_+ \to \mathbb{R}_+, \; u \mapsto x = f(u),$$

be a production function that is continuous and monotonically increasing. Define the production set

$$Y = \{\xi = (-u, x) \in \mathbb{R}_- \times \mathbb{R}_+ |\; u \geq 0, \; x \leq f(u)\}$$

in which the consumer good is produced from the producer good using the production function f. It is obvious from Figure 7.3 that the set Y is distributive if, and only if, the function f has non-decreasing returns to scale, or $\tau f(u) \leq f(\tau u)$ for all $\tau \geq 1$.

In the general case where there are several commodities in each category, the distributive sets exhibit non-decreasing returns to scale. For if $\xi \in Y$ and $\tau \geq 1$, then the first ℓ components of the producer commodities in $\xi - \tau\xi$ will be non-negative, hence $\tau\xi \in Y$.

The following theorem, which is a version of the separating hyperplane theorem for the distributive sets, is useful for later discussions.

Proposition 7.1. *Let Y be a closed distributive set and $\xi \notin Y$. Then there exists a non-negative vector $\pi > \mathbf{0}$, such that $\pi\xi > 0$ and $\pi\zeta \leq 0$ for all $\zeta \in Y \cap \{\Lambda + \xi\}$. Moreover, if any of the first ℓ components of $\xi = (\boldsymbol{u}, \boldsymbol{x})$ are zero, or $u^s = 0$ for some $s = 1 \ldots \ell$, then we can take the vector $\pi = (\boldsymbol{p}, \boldsymbol{q}) = (p^s, q^t)$ as $p^s = 0$.*

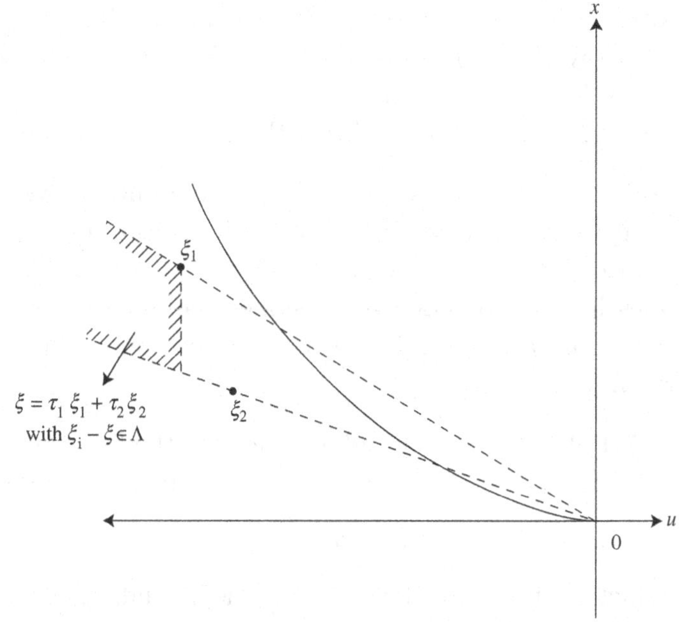

Figure 7.3. Distributive set.

Proof. We first prove the theorem in the case where $u^s < 0$ for $s = 1 \ldots \ell$. Define the set $S \subset \mathbb{R}^{\ell+m}$ by

$$S = \left\{ \sum_{i=1}^{k} \tau_i \xi_i \middle| \, \tau_i \geq 0, \ \xi_i \in Y, \text{ and } \xi_i - \xi \in \Lambda, \ k = 1, 2, \ldots \right\}.$$

S is the smallest convex cone with vertex $\mathbf{0}$, containing the points of $Y \cap (\Lambda + \xi)$. Note that the cone S satisfies the assumption (FD), as it is additive, or $\zeta_1, \zeta_2 \in S$ implies that $\zeta_1 + \zeta_2 \in S$, and contains $(-\delta \cdots - \delta)$ for δ that are sufficiently small $\delta > 0$. This implies that the closure of S does not contain ξ. Suppose on the contrary $\xi \in \bar{S}$. Then there exists a sequence $(\xi^n)_{n \in \mathbb{N}}$ such that $\xi^n \in S$ for all $n \in \mathbb{N}$ and $\xi^n \to \xi$. Using (FD), we can assume without loss of generality that $\xi^n \leq \xi$ for all n. However, we can write $\xi^n = \sum_i \tau_{i,n} \xi_i$ with $\xi_i \in Y$ and $\xi_i - \xi \in \Lambda$ and this implies that $\xi_i - \xi^n \in \Lambda$. By the definition of the distributive set, we see that $\xi^n \in Y$, and as Y is closed, it follows that $\xi \in Y$, a contradiction.

By the separating hyperplane theorem (Theorem A.1), there exists a non-zero vector π with $\pi\xi > 0$ and $\pi\zeta \leq 0$ for all $\zeta \in Y$ with $\zeta - \xi \in \Lambda$. Since $(-\delta \cdots - \delta) \in Y$ and satisfies $(-\delta \cdots - \delta) - \xi \in \Lambda$ for all sufficiently small $\delta > 0$, we see that $\pi \geq \mathbf{0}$; hence, $\pi > \mathbf{0}$.

Next, suppose that some of the producer commodities in $\xi = (\boldsymbol{u}, \boldsymbol{x})$ are zero. Then without loss of generality, we may assume that $u^1 = \ldots u^j = 0$ and $u^{j+1} < 0, \ldots u^\ell < 0$. Consider the set in $(\ell + m - j)$-dimensional space

$$\hat{Y} = \{(y^{j+1} \ldots y^{\ell+m}) | \, (0 \ldots 0, y^{j+1} \ldots y^{\ell+m}) \in Y\}.$$

It is a trivial matter to verify that $\hat{Y}$ is a distributive set with commodities $j+1 \ldots \ell$ being the producer commodities and that $(u^{j+1} \ldots u^\ell, x^1 \ldots x^m) \notin \hat{Y}$. Applying the previous argument, we obtain a vector $\hat{\pi} = (p^{j+1} \ldots p^\ell, q^1 \ldots q^m) \geq \mathbf{0}$ and $\pi \neq \mathbf{0}$ such that $\sum_{t=j+1}^{\ell} p^t u^t + \boldsymbol{qx} > 0$ and $\sum_{t=j+1}^{\ell} p^t v^t + \boldsymbol{qz} \leq 0$ for all $(v^{j+1} \ldots v^\ell, y^1 \ldots y^m) \in \hat{Y}$ with $v^s \geq u^s$ for $s = j+1 \ldots \ell$. The vector $(0 \ldots 0, p^{j+1} \ldots p^\ell, q^1 \ldots q^m)$ can be applied to Y, and this completes the proof. □

The situation of Proposition 7.1 is illustrated by Figure 7.4, where $\ell = 1$ and $m = 1$. Let Y be a distributive set. We call a production plan $\eta = (\boldsymbol{v}, \boldsymbol{y}) \in Y$ is *efficient* if there exists no $\zeta = (\boldsymbol{w}, \boldsymbol{z}) \in Y$ with $\zeta \geq \eta$ and $z^t > y^t$ for at least one $t = 1 \ldots m$. With a slight modification of the proof of Proposition 7.1, we prove the following proposition.

Proposition 7.2. *Let Y be a distributive set and $\eta = (\boldsymbol{v}, \boldsymbol{y}) \in Y$ be an efficient production plan. Then there exists a non-zero vector $\pi > \mathbf{0}$, such that $\pi\eta = 0$ and $\pi\zeta \leq 0$ for all $\zeta = (\boldsymbol{w}, \boldsymbol{z}) \in Y$ with $w^s \geq v^s$ for all $s = 1 \ldots \ell$.*

We now define the equilibrium concept in this section.

Definition 7.4. An allocation $(\xi(a), \eta)$ of a coalition production economy $(\mathcal{E}, \boldsymbol{Y})$ is a *social equilibrium* if there exists a price vector $\pi > \mathbf{0}$, such that

(S-1) $\pi\xi(a) \leq \pi\omega(a)$ and $\xi(a) \succsim_a \xi$ whenever $\pi\xi \leq \pi\omega(a)$, a.e.,

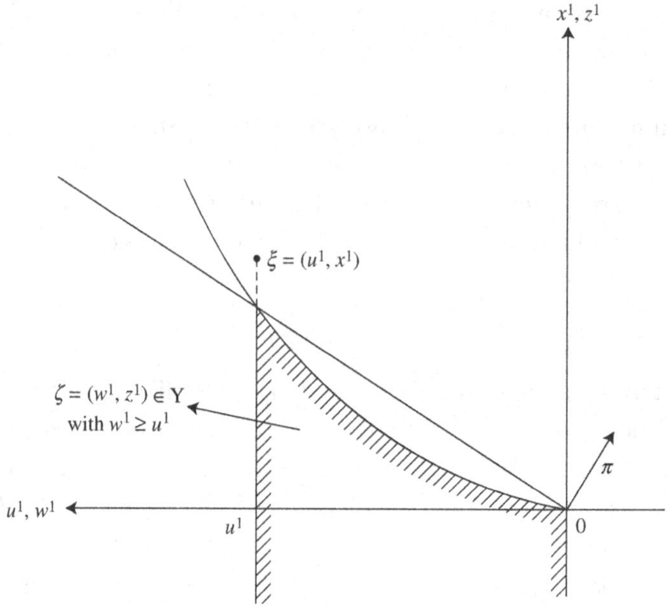

Figure 7.4. Proposition 7.1.

(S-2) $\pi\eta = 0$ and $\pi\zeta \leq 0$ for all $\zeta = (w^s, z^t) \in Y$ such that $w^s \geq -\int_A e^s(a)d\lambda$ for all $s = 1\ldots\ell$,

(S-3) $\int_A \xi(a)d\lambda \leq \eta + \int_A \omega(a)d\lambda$.

The condition (S-1) states that consumers maximize their utility under the budget constraint. Note that the income of consumer a is just the market value of the initial endowment vector, as in condition (S-2), the producers yield zero profits. At first glance, this is curious because these firms operate under increasing returns to scale technology, but we should note that the firms consider only those plans which use no more of the producer commodities than are available when they maximize their profits.

Remark 7.2. A significant property of the social equilibrium is that it is contained in the core, as Proposition 7.3 indicates. Although the core is certainly an important equilibrium concept, it seems rather less realistic as a descriptive concept for it imposes tremendously heavy information demands on economic agents when they form coalitions. The economic significance of the core seems to be more in its normative nature, namely that it demonstrates the social stability and the optimality. It is therefore desirable for a decentralized equilibrium concept to refine the core, or is at least contained in the core. The social equilibrium is exactly that which enjoys this property. We suggest that the social equilibrium deserves study for its own sake. This is because it is always contained in the core, hence it is unconditionally Pareto-optimal or efficient, which is on the same footing as the competitive equilibrium being always efficient, Moreover, it is in a "stable" state in the sense that any non-null coalitions do not block it.

Proposition 7.3. *Let $(\xi(a), \eta)$ be a social equilibrium of the coalition production economy $(\mathcal{E}, \boldsymbol{Y})$ that satisfies all of the assumptions stated earlier. Then the allocation $(\xi(a), \eta)$ is in the core.*

Proof. Given $\pi\eta = 0$ by condition (S-2), it follows from conditions (LNS), (S-1) and (S-3) that $\pi\xi(a) = \pi\omega(a)$ a.e., in A. Suppose, on the contrary, that the allocation $(\xi(a), \eta)$ is blocked by a non-null coalition $C \subset A$. Then, there exists a map $\zeta : C \to X$, such that $\int_C (\zeta(a) - \omega(a))d\lambda \equiv \hat{\eta} \in Y$ and $\xi(a) \prec_a \zeta(a)$ for almost all $a \in C$. Then by the condition (S-1),

$$\int_C \pi\zeta(a)d\lambda > \int_C \pi\xi(a)d\lambda \geq \int_C \pi\omega(a)d\lambda,$$

hence $\pi\hat{\eta} > 0$. But $\hat{\eta} \in Y$ and $\hat{\eta} \geq -\int_C \omega(a) \geq -\int_A \omega(a)$, which contradicts the condition (S-2). □

The following theorem, which is the main result of this section, asserts that the social equilibrium exists with the distributive production set; hence, by Proposition 7.3, the core is not empty, even under increasing returns to scale technologies.

Theorem 7.2. *Let $(\mathcal{E}, \boldsymbol{Y})$ be a coalition production economy that satisfies all of the assumptions stated above and (BTP') (cf. Theorem 6.4). Then there exists a social equilibrium $(\xi(a), \eta)$ for $(\mathcal{E}, \boldsymbol{Y})$.*

Proof. Let $\omega = \int_A \omega(a)d\lambda = ((e^s),(f^t))$ and define the set T by

$$T = \left\{ \sum_{i=1}^{k} \tau_i \zeta_i \middle| \tau_i \geq 0,\ \zeta_i \in Y,\ \text{and } \zeta_i + \omega \in \Lambda,\ k = 1, 2 \ldots \right\},$$

which is the smallest convex cone with its vertex at $\mathbf{0}$ containing $Y \cap (\Lambda - \omega)$. Consider the (private ownership) production economy $(\mathcal{E}, T)$. We show that there exists a competitive equilibrium $(\pi, \xi(a), \eta)$ for $(\mathcal{E}, T)$, and then $(\xi(a), \eta)$ is a social equilibrium for $(\mathcal{E}, \boldsymbol{Y})$.

In order to apply Theorem 6.4, it suffices to show that the set T satisfies (FD) and (NFP). Given T is a closed convex cone with the vertex at $\mathbf{0}$, in order to prove (FD), it is sufficient to show that

$$(-\delta^1 \cdots - \delta^\ell) \in T$$

for all sufficiently small $\delta^t > 0$, $t = 1 \ldots \ell$, and this is certainly true as $\omega \gg \mathbf{0}$ and Y has the free disposal property (FD).

In order to show (NFP) or $T \cap \mathbb{R}^\ell_+ = \{\mathbf{0}\}$, suppose that there exists $\zeta \in T \cap \mathbb{R}^\ell_+$ with $\zeta = ((w^s),(z^t)) \neq \mathbf{0}$. It follows from the assumption (NPC) that $w^s \leq 0$ for all $s = 1 \ldots \ell$. Hence, we have $\zeta = (0 \ldots 0, z^1 \ldots z^m)$. Consider the vector $\zeta' = (-\omega^1 \cdots - \omega^\ell, z^1 \ldots z^m)$. We assert that this latter vector must belong to Y, for if not, we can apply Proposition 7.1 and obtain a vector $\pi = (\boldsymbol{p}, \boldsymbol{q}) = ((p^s),(q^t)) > \mathbf{0}$, such that

$$\pi\zeta' = -\sum_{s=1}^{\ell} p^s \omega^s + \sum_{t=1}^{m} p^t z^t > 0,$$

$$\pi\eta \leq 0 \text{ for all } \eta = ((v^s),(y^t)) \in Y \text{ with } v^s \geq -\omega^s,\ s = 1 \ldots \ell.$$

Nevertheless, this implies that $\pi\eta \leq 0$ for all $\eta \in T$ and therefore $\pi\zeta' \leq 0$. This contradicts the previous inequality. Therefore,

$$(-\omega^1 \cdots - \omega^k, z^{t+1} \ldots z^\ell) \in Y.$$

Repeating the same argument with $\tau\zeta$ instead of ζ, one sees that

$$(-\omega^1 \cdots - \omega^\ell, \tau z^{t+1} \ldots \tau z^m) \in Y, \text{ for all } \tau > 0,$$

contradicting the assumption (BTP'). Therefore, by Theorem 6.4, there exists a quasi-equilibrium $(\pi, \xi(a), \eta)$ for the economy $(\mathcal{E}, T)$ which satisfies

$$\pi\xi(a) \leq \pi\omega(a) \text{ and } \xi(a) \succsim_a \xi \text{ whenever } \pi\xi < \pi\omega(a), \text{a.e.,}$$
$$\pi\eta = 0 \text{ and } \pi\zeta \leq 0 \text{ for all } \zeta \in T,$$
$$\int_A \xi(a) \leq \int_A \omega(a) + \eta.$$

Note that from the profit maximization over T we can see that

$$\pi\zeta \leq 0 \text{ for all } \zeta = ((w^s), (z^t)) \in Y, \text{ such that } w^s \geq -\int_A e^s(a) \text{ for } s = 1 \ldots \ell.$$

The assumption (MTC) implies $\boldsymbol{q} \gg \boldsymbol{0}$. Hence, by the assumption (PE) $\pi\omega(a) > 0$ a.e. Therefore, the quasi-equilibrium $(\pi, \xi(a), \eta)$ is a competitive equilibrium of the economy $(\mathcal{E}, T)$, so that $(\xi(a), \eta)$ is a social equilibrium of the economy $(\mathcal{E}, \boldsymbol{Y})$, if we can show that $\eta \in Y$.

Let $(\pi, \xi(a), \eta) = ((p^s, q^t), (u^s(a), x^t(a)), (v^s, y^t))$. First note that $\eta + \omega \geq \boldsymbol{0}$, given $\xi(a) \geq \boldsymbol{0}$ a.e. We now claim that if $v^s > -e^s$, then $p^s = 0$. To check this, suppose that $v^1 > -e^1$ and $p^1 > 0$. Then on a set of positive measures, $u^1(a) > 0$, as the feasibility condition for the first producer commodity holds with the equality, $\int_A u^1(a)d\lambda = e^1 + v^1$. Define $\hat{\xi}(a) = (0 \ldots 0, x^1(a) + \delta \ldots x^m(a) + \delta)$ for some $\delta > 0$ with $\delta \sum_{s=1}^{\ell} q^s \leq p^1 u^1(a)$. Then we have $\pi\hat{\xi}(a) \leq \pi\omega(a)$ and by the assumptions (CM) and (MTC), we see that $\xi(a) \prec_a \hat{\xi}(a)$, a contradiction. This means that if $v^s > -e^s$ for $1 \leq s \leq \ell$, we can reduce v^s without disturbing the preferences, maintaining $\pi\eta = 0$ and $\eta \in T$, given T has (FD). So, assume that $v^s = -e^s$ for $1 \leq s \leq \ell$. Now suppose that $\eta \notin Y$. Applying Proposition 7.1 again, we obtain a non-zero vector $\pi > \boldsymbol{0}$, such that $\pi\eta > 0$ and $\pi\zeta \leq 0$ for all $\zeta \in T$. As $\eta \in T$, this is a contradiction which establishes $\eta \in Y$. □

7.3 Economies with External Increasing Returns

In this section, we prove the existence of the competitive equilibria for an infinite time horizon economy with external increasing returns. To this end, while the space ℓ^∞ is mathematically nice, it is not adequate for presenting the competitive equilibrium of a production economy with external increasing returns. This is because external increasing returns make the economy grow without bound, as Young pointed out as early as 1928, so the commodity space of the bounded streams are too small. Consequently, we extend the space to include sequences growing without bound.

Let $\boldsymbol{g} = (g^t)$ be a vector of the maximal growth path of the economy, which we define more precisely later and set $\boldsymbol{\beta} = (\beta^t)$, $\beta^t = 1/g^t$. Then we can define the space $\ell^\infty_{\boldsymbol{\beta}}$ which is weighted by $\boldsymbol{\beta}$,

$$\ell^\infty_{\boldsymbol{\beta}} = \{\boldsymbol{x} = (x^t) | \sup_{t \geq 0} |\beta^t x^t| < \infty\}.$$

We can see that the feasible allocations are contained in the unit ball of $\ell^\infty_{\boldsymbol{\beta}}$, which is a nice, convex, and compact set (in the weak* topology; see Appendix G). The dual space of the space $\ell^\infty_{\boldsymbol{\beta}}$ is the weighted ba space of finitely additive set functions on $\mathbb{N}$, denoted by $ba_{\boldsymbol{\beta}^{-1}}$. Mathematically speaking, the weighted spaces $\ell^\infty_{\boldsymbol{\beta}}$ and $ba_{\boldsymbol{\beta}^{-1}}$ are isometrically isomorphic to the spaces ℓ^∞ and ba, respectively.

As we saw in Chapters 4 and 5, a mathematical difficulty arising from the infinite-dimensional commodity spaces is that the inner product $\boldsymbol{px}$ may not be jointly continuous with respect to the topology under consideration. That is to say, even if $\boldsymbol{p}_n \to \boldsymbol{p}$ and $\boldsymbol{x}_n \to \boldsymbol{x}$, they do not imply that $\boldsymbol{p}_n\boldsymbol{x}_n \to \boldsymbol{px}$. This problem is particularly serious when dealing with externalities in production; see Section 7.6. Fortunately, as we show, the structure, which is specific to external increasing returns, is sufficient to rescue the proof. In particular, the conditions that the producers earn zero profit because of the homogeneity of their production function and the periodic structure of the whole production process, $y^{t+1} = f_b(z^t, \varsigma^t)$, $t = 0, 1 \ldots$, are "just enough" to rescue the proof.

We first present the related concepts in a general form. Let L be an ordered vector space (Appendix G) that is to be understood as the commodity space. As usual, let $L_+ = \{\boldsymbol{x} \in L | \, \boldsymbol{x} \geq \boldsymbol{0}\}$ be the non-negative orthant of L. Consider the function

$$F : L_+ \times L_+ \to L_+, \ (\boldsymbol{z}, \varsigma) \mapsto \boldsymbol{y} = F(\boldsymbol{z}, \varsigma),$$

where $\boldsymbol{z} \in L_+$ is the input vector, $\boldsymbol{y} \in L_+$ output and $\varsigma \in L_+$ is a parameter. The function F is the technology function, as distinguished from the (standard) production function.

Definition 7.5. The technology function $F(\boldsymbol{z}, \varsigma)$ is said to exhibit *external increasing returns to scale* (or *social increasing returns to scale*) if $F(\boldsymbol{z}, \varsigma)$ is homogeneous of degree 1 in $\boldsymbol{z}$,

(Hom): $F(\tau \boldsymbol{z}, \varsigma) = \tau F(\boldsymbol{z}, \varsigma)$ for all $\tau \geq 0$ and all $(\boldsymbol{z}, \varsigma) \in L_+ \times L_+$,

and is monotonically increasing in ς,

(Mon): $\varsigma \leq \varsigma'$ implies $F(\boldsymbol{z}, \varsigma) \leq F(\boldsymbol{z}, \varsigma')$ for all $\boldsymbol{z} \in L_+$.

The firms are assumed to take the parameter ς as given when it maximizes profit, whereas ς is determined endogenously at the aggregate input level, $\varsigma = \sum \boldsymbol{z}$ in equilibrium, where the summation is over all firms in the industry, or sometimes in the economy as a whole, depending on the range of externalities. When the function F is of external returns to scale, we see that it exhibits "increasing returns" socially, for

$$\tau F(\boldsymbol{z}, \varsigma) = F(\tau \boldsymbol{z}, \varsigma) \leq F(\tau \boldsymbol{z}, \tau \varsigma) \text{ for all } \tau \geq 1.$$

In the following, the technology function is sometimes given by the (usual) production function with a parameter,

$$f : \mathbb{R}_+ \times \mathbb{R}_+ \to \mathbb{R}_+, \ (z, \varsigma) \mapsto y = f(z, \varsigma).$$

We then say that the production function f is also of external increasing returns if $f(z, \varsigma)$ is homogeneous of degree 1 in z and is monotonically increasing in ς.

Example 7.2 (Chipman [43]). There exist ℓ industries, each producing a single homogeneous commodity from a single factor, being labor z. Every firm in the industry $t(= 1 \dots \ell)$ has the same production function $y^t = \varsigma_t^{\varepsilon_t - 1} z_t$, where ε_t is a real number greater than 1, which is of external increasing returns to scale.

If there exist ν_t firms in the industry t, assuming that they operate under the same level of input, $\varsigma_t = \nu_t z_t$ will hold in equilibrium. Then the firm's objective production function is

$$y^t = \kappa_t (z_t)^{\varepsilon_t}, \ \kappa_t = (\nu_t)^{\varepsilon_t - 1}, \ t = 1 \dots \ell,$$

which exhibits increasing returns in the objective sense under the condition $\varepsilon_t > 1$. Chipman assumed that there are m consumers with the same utility function

$$u(x^1 \dots x^\ell) = \sum_{t=1}^{\ell} \rho_t \log x^t.$$

Note that the utility function does not depend on commodity 0, being labor. Hence, all consumers supply labor inelastically up to the amount they initially hold. Assume that all consumers have one unit of labor and a zero amount of other commodities, hence the initial endowment vector of all consumers is $(1, 0, \dots 0) \in \mathbb{R}^{\ell+1}$. The competitive equilibrium is a triple of vectors $(\boldsymbol{p}, \boldsymbol{x}, \boldsymbol{z}) = ((p^t), (x^t), (z_t)) \in \mathbb{R}_+^\ell \times \mathbb{R}_+^\ell \times \mathbb{R}_+^\ell$ which satisfies that

$$\begin{aligned}
&\boldsymbol{px} = w \text{ and } \rho_t = \lambda p^t x^t, \\
&\boldsymbol{py} = \boldsymbol{z}, \ \text{where } \boldsymbol{y} = (y^t) = (\varsigma_t^{\varepsilon_t - 1} z_t), \\
&\sum_{t=1}^{\ell} \nu_t z_t = m, \ \varsigma_t = \nu_t z_t,
\end{aligned}$$

where λ is the multiplier, and from the above equations, we have $w\lambda = \sum_{t=1}^{\ell} \rho_t$. In Section 7.4, we discuss the welfare property of the competitive equilibrium for a (generalized) Chipman model.

Example 7.3 (Romer [214]). For simplicity, we present a discrete time version of Romer's model, which is an optimal growth model of the infinite time horizon. At each time period $t+1 \geq 1$, the output commodity y^{t+1} is produced from the input commodity z^t available at the previous period $t \geq 0$ through the external increasing returns to scale technology,

$$y^{t+1} = f(z^t, \varsigma^t),\ t = 0, 1 \ldots,$$

and in equilibrium, $\varsigma^t = z^t$ for all t should hold. The representative consumer's preference is given by a time-separable utility function

$$U(\boldsymbol{x}) = \sum_{t=0}^{\infty} \rho^t u(x^t) \text{ for } \boldsymbol{x} = (x^t),$$

where $0 \leq \rho \leq 1$ is a discount factor. At the initial period $t = 0$, the consumer has the initial endowment $\omega^0 > 0$. The competitive equilibrium is a triple of vectors $(\boldsymbol{p}, \boldsymbol{x}, \boldsymbol{z})$ characterized as

$$\begin{aligned}
&\boldsymbol{px} \leq p^0\omega^0 \text{ and } U(\boldsymbol{x}) \geq U(\boldsymbol{x}') \text{ whenever } \boldsymbol{px}' \leq p^0\omega^0,\\
&p^{t+1} f(z^t, z^t) - p^t z^t = 0,\ t = 0, 1, \ldots,\\
&\boldsymbol{x} + \boldsymbol{z} \leq (\omega^0, y^1, y^2 \ldots),\ y^{t+1} = f(z^t, z^t),\ t = 0, 1 \ldots.
\end{aligned}$$

The main result of this section is to prove the existence of the competitive equilibrium for a (generalized) Romer model. Hence, in the following, we consider the infinite horizon economy.

Suppose that each firm $b (= 1 \ldots n)$ has a one-period production function with external increasing returns

$$f_b : \mathbb{R}_+ \times \mathbb{R}_+ \to \mathbb{R}_+,\ (z^t, \varsigma^t) \mapsto y^{t+1} = f_b(z^t, \varsigma^t),\ t = 0, 1, \ldots.$$

As the time coordinates t indicate, the firm uses the input commodity at date t and produces the output commodity at the next date $t+1$.

Let $\omega = \sum_{a=1}^{m} \omega_a = (\omega^t)$ be the total endowment vector, which is assumed to be strictly positive or to belong to $\mathbb{R}^{\infty}_{++} = \{\boldsymbol{x} = (x^t) \in$

$\mathbb{R}^\ell |\, x^t > 0, t = 0, 1, \ldots \}$. We define the path of pure capital accumulation (g^t) inductively by $g^0 = \omega^0$, and given g^{t-1}, set

$$g^t = \max \left\{ \sum_{b=1}^{n} f_b \left(z_b, \sum_{c=1}^{n} z_c \right) \middle| \sum_{c=1}^{n} z_c \leq g^{t-1} \right\} + \omega^t.$$

Set $\boldsymbol{\beta} = (\beta^t) = (1/g^t)$. Using the path (β^t), we define the commodity space L of the economy by

$$L = \ell^\infty_{\boldsymbol{\beta}} = \{ \boldsymbol{x} = (x^t) |\ \sup\{|\beta^t x^t|\, t \geq 0\} < +\infty \}.$$

The space $\ell^\infty_{\boldsymbol{\beta}}$ is the ℓ^∞ weighted by $\boldsymbol{\beta}$, or the order ideal generated by $\boldsymbol{\beta}$; see Appendix G for details. Conceptually, this market model is different from those presented so far. All of the previous models have been built from the commodity spaces given at the outset. Conversely in the present model, economic data ω and f_b are given at first, and then the commodity space $\ell^\infty_{\boldsymbol{\beta}}$ comes from them.

Given we assume that the total endowment vector is strictly positive, so is the path $\boldsymbol{\beta}$. Of course, our main concern is the case where $g^t \to +\infty$ as $t \to +\infty$. The dual space of $\ell^\infty_{\boldsymbol{\beta}}$ is the weighted *ba* space defined by

$$ba_{\boldsymbol{\beta}^{-1}} = \left\{ \pi : 2^{\mathbb{N}} \to \mathbb{R} \middle| \sup_{E \subset \mathbb{N}} \int_E \boldsymbol{\beta}^{-1} d|\pi| < +\infty,\ \pi(E \cup F) \right.$$
$$\left. = \pi(E) + \pi(F) \quad \text{whenever } E \cap F = \emptyset \right\},$$

and the subspace of countably additive set functions is identified with the weighted ℓ^1 space

$$\ell^1_{\boldsymbol{\beta}^{-1}} = \left\{ \boldsymbol{p} = (p^t) \middle| \sum_{t=0}^{\infty} |p^t/\beta^t| < +\infty \right\},$$

which is the mathematically natural candidate for the price space of the economy. However, by technical and economic reasons explained later, we search for the equilibrium price vector in the (standard) ℓ^1

space,

$$\ell^1 = \left\{ \boldsymbol{p} = (x^t) \,\middle|\, \sum_{t=0}^{\infty} |p^t| < +\infty \right\}.$$

Note that for $\boldsymbol{x} \in \ell^\infty_\beta$ and $\boldsymbol{p} \in \ell^1$, values of the bilinear form $\boldsymbol{px} = \sum_{t=0}^\infty p^t x^t$ could be $\pm\infty$, or even do not exist. Let $\ell^\infty_{\beta+}$ denote the positive orthant of ℓ^∞_β,

$$\ell^\infty_{\beta+} = \{\boldsymbol{x} \in \ell^\infty_\beta \,|\, \boldsymbol{x} \geq \boldsymbol{0}\}.$$

The technology function $F_b : \ell^\infty_{\beta+} \times \ell^\infty_{\beta+} \to \mathbb{R}^\infty$ of the firm b is defined by

$$F_b : ((z^t), (\varsigma^t)) \mapsto (y^t),\ y^0 = 0,\ y^{t+1} = f_b(z^t, \varsigma^t), t \geq 0.$$

Note that the range of the map F_b may not be contained in $\ell^\infty_{\beta+}$.

The consumer a has the consumption set $X_a = \ell^\infty_{\beta+}$, and the preference relation $\prec_a \subset \ell^\infty_{\beta+} \times \ell^\infty_{\beta+}$. Certainly, we have $\omega_a \in \ell^\infty_{\beta+}$. An allocation is denoted by $(\boldsymbol{x}_a, \boldsymbol{y}_b, \boldsymbol{z}_b) \in (\ell^\infty_{\beta+})^{m+2n}$. The list $(\ell^\infty_\beta, \prec_a, \omega_a, f_b)$ is called an *infinite-horizon economy with external increasing returns*, or simply an economy and denoted by $\mathcal{E}^\infty_\beta$. For definiteness, we state the definition of the competitive equilibrium.

Definition 7.6. An $m + 2n + 1$-tuple $(\boldsymbol{p}, \boldsymbol{x}_a, \boldsymbol{y}_b, \boldsymbol{z}_b) \in \ell^1_+ \backslash \{\boldsymbol{0}\} \times (\ell^\infty_\beta)^{m+2n}$ consisting of a price vector and an allocation is called a competitive equilibrium if, and only if,

(EX-1) $\boldsymbol{px}_a \leq \boldsymbol{p}\omega_a$ and $\boldsymbol{x}_a \succsim_a \boldsymbol{x}$ whenever $\boldsymbol{px} \leq \boldsymbol{p}\omega_a$, $a = 1 \ldots m$,

(EX-2) $\boldsymbol{y}_b \leq F_b\left(\boldsymbol{z}_b, \sum_{c=1}^n \boldsymbol{z}_c\right)$, and $\boldsymbol{p}(\boldsymbol{y} - \boldsymbol{z}) \leq \boldsymbol{p}(\boldsymbol{y}_b - \boldsymbol{z}_b) = 0$ for all $(\boldsymbol{y}, \boldsymbol{z})$ such that $\boldsymbol{y} \leq F_b\left(\boldsymbol{z}, \sum_{c=1}^n \boldsymbol{z}_c\right)$, $b = 1 \ldots n$,

(EX-3) $\sum_{a=1}^m \boldsymbol{x}_a \leq \sum_{b=1}^n (\boldsymbol{y}_b - \boldsymbol{z}_b) + \sum_{a=1}^m \omega_a$.

Remark 7.3. The economic meaning of all conditions should be clear enough. We shall assume that in Theorem 7.3 the endowment vector is contained in ℓ^∞_+. Hence, in condition (EX-1), we have $\boldsymbol{p}\omega_a < +\infty$ for $\boldsymbol{p} \in \ell^1_+$. In condition (EX-2), by the homogeneity in the first variable of F_b, the profit $\boldsymbol{p}(\boldsymbol{y}_b - \boldsymbol{z}_b) = 0$ in equilibrium. Hence, for all technologically feasible production vectors $(\boldsymbol{y}, \boldsymbol{z}) \in \ell^\infty_\beta \times \ell^\infty_\beta$, the value $\boldsymbol{p}(\boldsymbol{y} - \boldsymbol{z})$ exists, including $-\infty$.

Then our main theorem reads as follows:

Theorem 7.3. *Suppose that an economy $\mathcal{E}^\infty_\beta = (\ell^\infty_\beta, \prec_a, \omega_a, f_b)$ satisfies the basic postulates with*

$\left(CT^\infty_\beta\right)$ *(Continuity in Mackey topology): the preference relation $\prec_a \subset X_a \times X_a$ is open relative to $X_a \times X_a$ in the $\tau(\ell^\infty_\beta, \ell^1_{\beta-1})$-topology,*

(NTR), (CV), (MT) and the positive endowments,

$\left(PE^\infty_\beta\right)$: *$\omega_a \in \ell^\infty$ and $\omega_a \ggg \mathbf{0}$ for all $a = 1 \ldots m$.*

Suppose further that the (one-period) production function

$$f : \mathbb{R}_+ \times \mathbb{R}_+ \to \mathbb{R}_+,\ (z, \varsigma) \mapsto y = f_b(z, \varsigma)$$

is continuous (in the usual topology). Then, there exists a competitive equilibrium $(\boldsymbol{p}, \boldsymbol{x}_a, \boldsymbol{y}_b, \boldsymbol{z}_b)$ for $\mathcal{E}^\infty_\beta$.

Proof. The proof goes essentially as in that of Theorem 4.1. We first define the T-period truncated economy $\mathcal{E}^T_\beta$. Consider, as in the proof of Theorem 4.1, the set of sequences whose coordinates are 0 after T,

$$R^T = \{\boldsymbol{x} = (x^t) \in \ell^\infty_\beta \mid 0 = x^{T+1} = x^{T+2} = \cdots\} \subset L, \quad T = 0, 1, \ldots.$$

The set R^T is naturally identified with $\mathbb{R}^{T+1}$, $R^T \approx \mathbb{R}^{T+1}$. Let

$$X^T_a = X_a \cap R^T \approx \mathbb{R}^{T+1}_+, \quad a = 1 \ldots m,$$

and we have a preference relation $\prec^T_a$ on X^T_a, which is the restriction of $\prec_a$ on the truncated consumption set X^T_a, $a = 1 \ldots m$. Let $\omega_a(T) \in R^T \approx \mathbb{R}^{T+1}$ be the truncated initial endowment vector defined by $\omega_a(T) = (\omega^0_a, \omega^1_a \ldots \omega^T_a)$. By (PE^∞_β), $\omega_a(T) \gg \mathbf{0}$. The technology function F_b^T is also defined as the restriction of the map F_b to $\mathbb{R}_+^{T+1} \times \mathbb{R}_+^{T+1}$,

$$F_b^T : \mathbb{R}_+^{T+1} \times \mathbb{R}_+^{T+1} \to \mathbb{R}_+^{T+1}, \ (z^t, \varsigma^t) \mapsto y^{t+1} = f_b(z^t, \varsigma^t), \quad t = 0 \dots T,$$

where we set $y^0 = 0$. Then, we have obtained the T-period economy $\mathcal{E}_{\boldsymbol{\beta}}^T = (\prec_a^T, \omega_a(T), F_b^T)$. The competitive equilibrium for the economy $\mathcal{E}_{\boldsymbol{\beta}}^T$ is obviously defined as in Definition 6.3. The following lemma is easy.

Lemma 7.2. *Under the assumptions of Theorem 7.3, there is a competitive equilibrium* $(\boldsymbol{p}(T), \boldsymbol{x}_a(T), \boldsymbol{y}_b(T), \boldsymbol{z}_b(T))$ *for the T-period economy* $\mathcal{E}_{\boldsymbol{\beta}}^T$, $T = 0, 1 \dots$.

Proof. We show that the T-period economy $\mathcal{E}_{\boldsymbol{\beta}}^T$ satisfies the assumptions of Theorem 6.3. First we prove that the set of feasible allocations

$$\mathcal{F} = \left\{ (\boldsymbol{x}_a, \boldsymbol{y}_b, \boldsymbol{z}_b) \in (\mathbb{R}_+^{T+1})^m \times (\mathbb{R}_+^{T+1})^{2n} \,\middle|\, \sum_{a=1}^m \boldsymbol{x}_a \leq \sum_{b=1}^n (\boldsymbol{y}_b - \boldsymbol{z}_b) \right.$$
$$\left. + \sum_{a=1}^m \omega_a, \boldsymbol{y}_b \leq F\left(\boldsymbol{z}_b, \sum_{c=1}^n \boldsymbol{z}_c\right) \right\}$$

is bounded. Let $(\boldsymbol{x}_a, \boldsymbol{y}_b, \boldsymbol{z}_b)$ be a feasible allocation. Then we have

$$0 \leq z_b^s \leq g^s, \quad s = 0, \dots T, \ b = 1 \dots n.$$

Indeed, for $s = 0$, $0 \leq z_b^0 \leq \sum_{b=1}^n z_b^0 \leq \sum_{a=1}^m x_a^0 + \sum_{b=1}^n z_b^0 \leq \sum_{a=1}^m \omega_a^0 \leq g^0$. Given $0 \leq z_b^{s-1} \leq g^{s-1}$, it follows from the definition of $\boldsymbol{g} = (g^t)$ that

$$\begin{aligned} 0 \leq z_b^s &\leq \sum_{b=1}^n z_b^s \leq \sum_{a=1}^m x_a^s + \sum_{b=1}^n z_b^s \\ &\leq \sum_{b=1}^n f_b\left(z_b^{s-1}, \sum_{c=1}^n z_c^{s-1}\right) + \sum_{a=1}^m \omega_a^s \leq g^s. \end{aligned}$$

Hence, the mathematical induction is established. Similarly, we have

$$0 \leq x_a^s \leq g^s, \quad s = 0 \dots T, \ a = 1 \dots m,$$

$$0 \leq y_b^s \leq g^s, \quad s = 0 \dots T, \ b = 1 \dots n.$$

$\omega_a(T) \gg \mathbf{0}$ implies (MI) for every $a = 1 \dots m$. Setting $\varsigma_b(\boldsymbol{z}_1 \dots \boldsymbol{z}_n) = \sum_{c=1}^{n} \boldsymbol{z}_c$, the condition (YCT) is also satisfied. Hence, it follows from Theorem 6.3 that there is a competitive equilibrium $(\boldsymbol{p}(T), \boldsymbol{x}_a(T), \boldsymbol{y}_b(T), \boldsymbol{z}_b(T))$ for $\mathcal{E}_{\boldsymbol{\beta}}^T$. □

Let $(\boldsymbol{p}(T), \boldsymbol{x}_a(T), \boldsymbol{y}_b(T), \boldsymbol{z}(T))$ be an equilibrium for $\mathcal{E}_{\boldsymbol{\beta}}^T$. Note that the (MT) assumption implies that $\boldsymbol{p}(T) \gg \mathbf{0}$ for all $T \in \mathbb{N}$; hence, $\boldsymbol{p}(T) \neq \mathbf{0}$. Then we can normalize the equilibrium price vector $\boldsymbol{p}(T) = (p^t(T))$ as $\|\boldsymbol{p}(T)\| = \sum_{t=0}^{T} p^t(T) = 1$. Let $\pi(T) = (\boldsymbol{p}(T), 0, 0, \dots)$. Then $\pi(T) \in \ell^1 \subset ba$ and $\pi(T) \geq \mathbf{0}$ for all $T \in \mathbb{N}$. Hence, $\|\pi(T)\| = \sum_{t=1}^{T} p^t(T) = 1$, where the norm stands for ℓ^1-norm on the sequence space. Similarly, we write for simplicity $\boldsymbol{x}_a(T) = (\boldsymbol{x}_a(T), 0, 0 \dots)$, $a = 1 \dots m$, $\boldsymbol{y}_b(T) = (\boldsymbol{y}_b(T), 0, 0 \dots)$ and $\boldsymbol{z}_b(T) = (\boldsymbol{z}_b(T), 0, 0 \dots)$, $b = 1 \dots n$. Then, by Lemma 7.2, we have

$$\|\boldsymbol{x}_a(T)\|_{\boldsymbol{\beta}}, \|\boldsymbol{y}_b(T)\|_{\boldsymbol{\beta}}, \|\boldsymbol{z}_b(T)\|_{\boldsymbol{\beta}} \leq 1$$

for all T. Let $B = \{\boldsymbol{x} \in \ell_{\boldsymbol{\beta}}^{\infty} | \ \|\boldsymbol{x}\|_{\boldsymbol{\beta}} \leq 1\}$ and $B' = \{\pi \in ba | \ \|\pi\| \leq 1\}$. Then by Theorem G.3, $B \subset \ell_{\boldsymbol{\beta}}^{\infty}$ and $B' \subset ba$ are compact in the weak* topologies. Hence, by Propositions B.1 and B.4, we can take converging subnets $\pi_\kappa \to \pi \in ba$, $\boldsymbol{x}_{a,\kappa} \to \boldsymbol{x}_a \in \ell_{\boldsymbol{\beta}}^{\infty}$, $\boldsymbol{y}_{b,\kappa} \to \boldsymbol{y}_b \in \ell_{\boldsymbol{\beta}}^{\infty}$, $\boldsymbol{z}_{b,\kappa} \to \boldsymbol{z}_b \in \ell_{\boldsymbol{\beta}}^{\infty}$ of the sequences $\{\pi(T)\}$, $\{\boldsymbol{x}_a(T)\}$, $\{\boldsymbol{y}_b(T)\}$ and $\{\boldsymbol{z}_b(T)\}$, respectively, where we denote $\pi(T_\kappa) \equiv \pi_\kappa$, $\boldsymbol{x}_a(T_\kappa) \equiv \boldsymbol{x}_{a,\kappa}$ and so on for simplicity. Given the positive orthant of $\ell_{\boldsymbol{\beta}}^{\infty}$ is weak* closed, it follows that $\boldsymbol{x}_a, \boldsymbol{y}_b, \boldsymbol{z}_b \geq \mathbf{0}$, $a = 1 \dots m$, $b = 1 \dots n$. Similarly, we have $\pi \geq \mathbf{0}$. Let $\mathbf{1} = (1, 1, \dots) \in \ell^\infty$. As $\pi(T)\mathbf{1} = 1$ for all $T \in \mathbb{N}$, it follows that $\pi\mathbf{1} = 1$ in the limit. Hence, $\pi \neq \mathbf{0}$. We now claim,

Lemma 7.3. *The vector* $\pi \in ba$ *and the allocation* $(\boldsymbol{x}_a, \boldsymbol{y}_b, \boldsymbol{z}_b) = ((x_a^t), (y_b^t), (z_b^t))$ *satisfy the following conditions:*

(BA-1) $\boldsymbol{x}_a \succsim_a \boldsymbol{x}$ *whenever* $\pi\boldsymbol{x} < \pi\omega_a$, $a = 1 \dots m$,

(BA-2) $\boldsymbol{y}_b \leq F_b\left(\boldsymbol{z}_b, \sum_{c=1}^n \boldsymbol{z}_c\right)$ *and* $\pi(\{t+1\})y - \pi(\{t\})z \leq \pi(\{t+1\})y_b^{t+1} - \pi(\{t\})z_b^t = 0$ *whenever* $y \leq f_b\left(z, \sum_{c=1}^n z_c^t\right)$, $t \in \mathbb{N}$, $b = 1 \ldots n$,

(EX-3) $\sum_{a=1}^m \boldsymbol{x}_a \leq \sum_{b=1}^n (\boldsymbol{y}_b - \boldsymbol{z}_b) + \sum_{a=1}^m \omega_a$.

Proof. The condition (BA-1) is proved exactly in the same way as Lemma 4.2. We show the condition (BA-2). Given $\boldsymbol{y}_{b,\kappa}$ and $\boldsymbol{z}_{b,\kappa}$ converge in the weak* topology, they converge in the product topology. Since $y_{b,\kappa}^{t+1} \leq f_b(z_{b,\kappa}^t, \sum_{c=1}^n z_{c,\kappa}^t)$ for all κ, $y_b^{t+1} \leq f_b(z_b^t, \sum_{c=1}^n z_c^t)$ in the limit.

Let $y, z \geq 0$ satisfy that $y \leq f_b(z, \sum_{c=1}^n z_c^t)$ and set $\boldsymbol{y} = y\boldsymbol{e}_{t+1}$ and $\boldsymbol{z} = z\boldsymbol{e}_t$, where $\boldsymbol{e}_t = (0 \ldots 0, 1, 0 \ldots)$ is the tth elementary vector. We want to show $\pi(\boldsymbol{y} - \boldsymbol{z}) = \pi(\{t+1\})y - \pi(\{t\})z \leq 0$. Suppose $\pi(\{t+1\})y - \pi(\{t\})z > 0$. If $y = f_b(z, \sum_{c=1}^n z_c^t)$, then setting $y_\kappa = f_b(z, \sum_{c=1}^n z_{c,\kappa}^t)$, we have $\pi_\kappa(\{t+1\})y_\kappa - \pi_\kappa(\{t\})z > 0$ for a sufficiently large κ, a contradiction. If $y < f_b(z, \sum_{c=1}^n z_c^t)$, then $\pi_\kappa(\{t+1\})y - \pi_\kappa(\{t\})z > 0$ for a sufficiently large κ, which is also a contradiction. As $\pi_\kappa(\{t+1\})y_{b,\kappa}^{t+1} - \pi_\kappa(\{t\})z_{b,\kappa}^t = 0$ for all κ, we have $\pi(\{t+1\})y_b^{t+1} - \pi(\{t\})z_b^t = 0$ in the limit. Therefore, the condition (BA-2) is established. Finally, because $\sum_{a=1}^m \boldsymbol{x}_{a,\kappa} \leq \sum_{b=1}^n (\boldsymbol{y}_{b,\kappa} - \boldsymbol{z}_{b,\kappa}) + \sum_{a=1}^m \omega_{a,\kappa}$ for all κ, it follows that $\sum_{a=1}^m \boldsymbol{x}_a \leq \sum_{b=1}^n (\boldsymbol{y}_b - \boldsymbol{z}_b) + \sum_{a=1}^m \omega_a$ in the limit. Hence, the condition (EX-3) is also proved. □

The proof of Theorem 7.3 with an equilibrium price vector in ℓ^1 is established by the following lemma.

Lemma 7.4. *There exists a vector* $\boldsymbol{p} \in \ell_+^1$ *such that* $(\boldsymbol{p}, \boldsymbol{x}_a, \boldsymbol{y}_b, \boldsymbol{z}_b)$ *satisfies* (EX-1), (EX-2) *and* (EX-3) *of Definition* 7.6.

Proof. We have already established the condition (EX-3) in the previous lemma. Given $\pi \geq \boldsymbol{0}$, it follows from Theorem G.4 that we can write

$$\pi = \pi_c + \pi_p,$$

where $\pi_c \geq \boldsymbol{0}$ and $\pi_p \geq \boldsymbol{0}$ are the countably additive part and the purely finitely additive part of π, respectively. In exactly the same way as Lemma 4.3, we can show that

$$\boldsymbol{x}_a \prec_a \boldsymbol{x} \text{ implies that } \boldsymbol{p}\boldsymbol{x} > \boldsymbol{p}\omega_a,\ a = 1 \dots m,$$

where $\boldsymbol{p} = (p^t) = (\pi_c(\{t\}))$. Note that this implies that $\boldsymbol{p} \neq \mathbf{0}$. By the assumption (MT) of preferences, we can take $\boldsymbol{x} \in X_a = \ell^{\infty}_{\beta+}$, which is arbitrarily close to $\boldsymbol{x}_a$, such that $\boldsymbol{x}_a \prec_a \boldsymbol{x}$. Thus, the above relation implies that

$$\boldsymbol{p}\boldsymbol{x}_a \geq \boldsymbol{p}\omega_a, \quad a = 1 \dots m.$$

$\boldsymbol{p}\boldsymbol{x}_a$ is finite or $+\infty$ and $\boldsymbol{p}\omega_a$ is finite. Hence, $\sum_{a=1}^{m} \boldsymbol{p}(\boldsymbol{x}_a - \omega_a) \geq 0$ is finite or $+\infty$.

Take $(\boldsymbol{y}, \boldsymbol{z}) = ((y^t), (z^t)) \in \ell^{\infty}_{\beta+} \times \ell^{\infty}_{\beta+}$ such that $\boldsymbol{y} \leq F_b(\boldsymbol{z}, \sum_{c=1}^{n} \boldsymbol{z}_c)$. By Lemma 7.3, we have $p^{t+1}y^{t+1} - p^t z^t \leq 0$ for all $t \geq 0$. Setting $S(T) = \sum_{t=0}^{T}(p^{t+1}y^{t+1} - p^t z^t)$, it follows that $S(T)$ is monotonically decreasing, $0 \geq S(0) \geq S(1) \dots$. Hence, $S(+\infty) = \sum_{t=0}^{\infty}(p^{t+1}y^{t+1} - p^t z^t)$ exists, including $-\infty$. Therefore,

$$\boldsymbol{p}(\boldsymbol{y} - \boldsymbol{z}) \leq 0 \text{ whenever } \boldsymbol{y} \leq F_b\left(\boldsymbol{z}, \sum_{c=1}^{n} \boldsymbol{z}_c\right), \quad b = 1 \dots n.$$

Given $\boldsymbol{y}_b \leq F_b(\boldsymbol{z}_b, \sum_{c=1}^{n} \boldsymbol{z}_c)$ by Lemma 7.3, we have $\boldsymbol{p}(\boldsymbol{y}_b - \boldsymbol{z}_b) \leq 0$, $b = 1 \dots n$. As $\sum_{a=1}^{m}(\boldsymbol{x}_a - \omega_a) \leq \sum_{b=1}^{n}(\boldsymbol{y}_b - \boldsymbol{z}_b)$ by Lemma 7.3, we have

$$0 \leq \sum_{a=1}^{m} \boldsymbol{p}(\boldsymbol{x}_a - \omega_a) \leq \sum_{b=1}^{n} \boldsymbol{p}(\boldsymbol{y}_b - \boldsymbol{z}_b) \leq 0.$$

Consequently, we have $\sum_{a=1}^{m} \boldsymbol{p}(\boldsymbol{x}_a - \omega_a) = 0$, hence

$$\boldsymbol{p}\boldsymbol{x}_a = \boldsymbol{p}\omega_a,\ a = 1 \dots m.$$

This establishes the condition (EX-1). We also have

$$\boldsymbol{p}(\boldsymbol{y}_b - \boldsymbol{z}_b) = 0,\ b = 1 \dots n,$$

hence condition (EX-2) is also verified. This completes the proof. □

We took ℓ^1 rather than $\ell^1_{\beta-1}$ as the price space in Theorem 7.3. We need this to ensure that each consumer's income is positive in the condition (BA-1). We had $\pi(T)\omega_a(T) > 0$ for all T. What we indeed wanted is that $\pi\omega_a > 0$ in the limit. This was ensured by the (strong) positivity of the endowment ($PE^{\infty}_{\boldsymbol{\beta}}$) and the normalization $\|\pi(T)\| = \pi(T)\mathbf{1} = 1$ for all T. If we took $\ell^1_{\beta^{-1}}$ and normalized the price vector $\|\pi(T)\|_{\boldsymbol{\beta}^{-1}} = \pi(T)\boldsymbol{\beta}^{-1} = 1$, the assumption ($PE^{\infty}_{\boldsymbol{\beta}}$) has to be replaced by

$$\omega_a^t \geq \varepsilon g^t \text{ for some } \varepsilon > 0, \quad t = 0, 1 \ldots,$$

or that the endowment must grow at least at the same rate as the path of pure capital accumulation $\boldsymbol{\beta}^{-1} = (g^t)$, which is obviously unrealistic. Finally, we notice that the assumption ($PE^{\infty}_{\boldsymbol{\beta}}$) is weakened as

$$\omega_a \in \ell^{\infty},\ \omega_a \gg \mathbf{0} \text{ for all } a, \text{ and } \omega_a \ggg \mathbf{0} \text{ for some } a.$$

7.4 Welfare Analysis of External Increasing Returns

All of the results in this section will be obtained by direct computations. The expositions will be more intuitive and less mathematically rigorous than the other sections. We assume that there exist two categories of commodities, the homogeneous input commodity (labor) indexed by 0, and the differentiated consumption commodities, which are determined by their *characteristics* indexed by $t \in I$. The index set I is simply a subset of the real line $\mathbb{R}$, which is defined below. The amount of the commodity 0 is denoted by z, etc., and that of the commodity characteristic $t \in I$ by $x(t), y(t)$, and so on. In Section 4.3, *differentiated commodities* were represented by (signed) measures on the set of characteristics, I. In this section, we shall describe them instead as functions defined on I. Let us call the function a commodity bundle. Then it has a continuum of coordinates, or the commodity bundle is an infinite-dimensional vector. Let $\mathscr{F}(I)$ be the set of commodity bundles.

The economic interpretation of the differentiated commodities is the same as that of Section 4.3; when two values $t \in I$ and $s \in I$ are "close" (in the usual mathematical sense), so are the commodity characteristics t and s (in the economic sense). In other words, the characteristics t and s, while not exactly the same, are very similar. When each characteristic contained in the two commodity bundles with the same set of the characteristics are close, they are similar, or the commodities are "differentiated." Note that this interpretation is impossible for the usual finite-dimensional commodity bundles for which the coordinates are necessarily discrete.

Each commodity characteristic $t \in I$ is assumed to be produced from the input commodity 0 by a firm that is also for convenience indexed by t (hence, there exists a continuum of firms in the economy) using the production function that embodies external increasing returns,

$$y(t) = f_t(z_t, \varsigma_t).$$

We need to specify the range of the externality to extend this analysis. For simplicity, consider a finite number of measurable subsets $\{I_1 \dots I_n\}$ of $\mathbb{R}$ such that $\sup\{|t_i - s_i| \mid t_i, s_i \in I_i\} < \inf\{|t_i - s_j| \mid t_i \in I_i, s_j \in I_j\}$ for all $i, j = 1 \dots n$ with $i \neq j$. We set $I = \cup_{i=1}^n I_i$ and call I_i the *industry* i. The above condition implies that any characteristics in the same industry are closer than the characteristics in different industries.

It is important to distinguish the commodity vectors (bundles) from their characteristics. The industry i produces and supplies its own commodity bundles (hence, there are n distinct commodities in the economy) which are functions $\boldsymbol{x} = x(t)$ or members of an appropriate function space $\mathscr{F}(I_i)$ on I_i, and the value $x(t)$ is the amount of the characteristic $t \in I_i$ which is not traded among the consumers. These are then "intermediate goods" which constitute commodity i supplied by industry i. Of course, for each $t^* \in I$, we can consider the commodity bundle $\boldsymbol{x}_{t^*} = x(t; t^*)$ which contains only the characteristic t^* defined by $x(t; t^*) = 1$ for $t = t^*$ and $x(t; t^*) = 0$ otherwise. But this commodity would not have any economic significance, as its market values are always 0, as shown later. Functions on I_i can be naturally extended to I setting as 0 is outside I_i. Then, we can write $\mathscr{F}(I) = \mathscr{F}(I_1) \oplus \cdots \oplus \mathscr{F}(I_n)$ (direct sum).

There exist m consumers indexed by $a(= 1 \dots m)$. Consumer a's utility function takes the log-linear (Cobb–Douglas) form,

$$u_a(x_a(t)) = \int_I \rho_a(s) \log x_a(s) ds, \; a = 1 \dots m,$$

where for all $t \in I$, $\rho_a(t) \geq 0$ for each a and $\sum_{a=1}^m \rho_a(t) > 0$, and $\int_I \rho_a(s)ds = 1$, $a = 1 \dots m$. Note that u_a does not include the consumption of commodity 0. Hence, consumers do not demand commodity 0, but instead supply it inelastically for as long as it is owned. Precisely speaking, this utility function is not a function but a (nonlinear) *functional*; that is, a map assigning a real value to the function $x(\cdot)$.

We assume that consumer a is endowed with $\omega_a(> 0)$ units of commodity 0 as an initial endowment, but does not have an initial endowment for commodity t. Set $\omega = \sum_{a=1}^m \omega_a$. The market price of commodity 0 is denoted by w, and the price of commodity t by $p(t)$. The competitive equilibrium of this market is defined in the standard manner.

Definition 7.7. An $m+1$-tuple of the input and consumption bundles $(\hat{z}_t, \hat{x}_a(t))$, $a = 1 \dots m$, and prices $(\hat{w}, \hat{p}(t))$ is called the competitive equilibrium if and only if the following conditions are satisfied.

(DX-1) $\hat{x}_a(t)$ maximizes $u_a(\cdot)$ subject to $\int_I \hat{p}(s)x_a(s)ds \leq \hat{w}\omega_a$, $a = 1 \dots m$,

(DX-2) $\hat{p}(t)f_t(z, \int_{I_i} \hat{z}_s ds) - \hat{w}z \leq \hat{p}(t)f_t(\hat{z}_t, \int_{I_i} \hat{z}_s ds) - \hat{w}\hat{z}_t = 0$ for all $z \geq 0$ and $t \in I_i$ a.e., $i = 1 \dots n$,

(DX-3) $\sum_{a=1}^m \hat{x}_a(t) = f_t(\hat{z}_t, \int_{I_i} \hat{z}_s ds)$ for $t \in I_i$ a.e., $i = 1 \dots n$, and $\int_I \hat{z}_s ds = \omega$.

Remark 7.4. An important remark on the concept of prices in this economy is in order. Strictly speaking, given the price function $p(t)$ as in Definition 7.7, the *market price* is a (nonnegative) linear functional $\boldsymbol{p}$ on $\mathscr{F}(I)$ defined by $\boldsymbol{px} = \int_I p(s)x(s)ds$ for every $\boldsymbol{x} = x(t) \in \mathscr{F}(I)$. As the characteristic t is not traded by consumers, each value $p(t)$ of the functional is not directly observable to them, or $p(t)$ is the *hedonic price* in the sense of Rosen [215]. To see this, suppose that the price functional is strictly positive, $p(t) > 0$ for all $t \in I$. If $p(t^*)$ is observable in the market for some $t^* \in I$, it must be the market value of the commodity $\boldsymbol{x}_{t^*}$ defined above. However, we have $\boldsymbol{p}\boldsymbol{x}_{t^*} = \int_I p(s)x(s;t^*)ds = 0 \neq p(t^*)(> 0)$. This is exactly what we stated earlier. Here, the industry i is acting as a fictitious producer "selling" the consumption bundle $\boldsymbol{x} \in \mathscr{F}(I_a)$ as a "final output" at the price $\boldsymbol{p}$ and "buying" infinitely many characteristics $x(t) \in \mathbb{R}$ as "inputs" from the firm t at the prices $p(t)$.

In order to elucidate the functional calculus in an elementary way, the celebrated Dirac's delta function is used. For any $t \in I$, the "function" $\delta(t)$ on $\mathbb{R}$ is defined as[1]

$$\delta(t) = \begin{cases} 0 & \text{for } t \neq 0, \\ +\infty & \text{for } t = 0, \end{cases}$$

and assumed to satisfy $\int_{\mathbb{R}} \delta(s)ds = 1$. From this and the definition, we obtain that $\int_I g(s)\delta(t-s)ds = g(t)$ for any function $g(s)$ ([64, p.59]). Let $f(x)$ be a differentiable function. We define

$$\frac{df(x(s))}{dx(t)} \equiv \lim_{h \to 0} \frac{f(x(s) + h\delta(t-s)) - f(x(s))}{h} = f'(x(s))\delta(t-s).$$

This is a fundamental mathematical formula used throughout the section. Then we can differentiate $\int_I f(x(s))ds$ with respect to $x(t)$,

$$\begin{aligned} \frac{d}{dx(t)} \int_I f(x(s))ds &= \int_I \frac{df(x(s))}{dx(t)} ds \\ &= \int_I f'(x(s))\delta(t-s)ds = f'(x(t)). \end{aligned}$$

The following example, although economically simple, illustrates the mathematical structure of the problem and a technique of the delta function.

[1]The delta function $\delta(t)$ should not be confused with the Dirac measure δ_t.

Example 7.4. Consider a consumer with the utility function

$$u(x(t)) = \int_I \rho(s)\log x(s)ds$$

such that $\rho(t) > 0$ for all $t \in I$ and $\int_I \rho(s)ds = 1$. The consumer's initial endowment $\omega(> 0)$ is the total endowment (resources) of the economy. Suppose for simplicity that $n = 1$ so that $I_1 = I$ (there exists only one industry in the economy). The consumer maximizes the utility function (functional) subject to the budget constraint $\int_I p(s)x(s)ds = w\omega$. Mathematically speaking, this requires us to solve a constrained variational problem. To this end, we differentiate the constrained Lagrangian with the multiplier λ

$$\mathcal{L} = \int_I \rho(s)\log x(s)ds + \lambda\left(w\omega - \int_I p(s)x(s)ds\right)$$

in $\boldsymbol{x}(t)$ and obtain the FOC

$$\frac{d\mathcal{L}}{dx(t)} = \frac{\rho(t)}{x(t)} - \lambda p(t) = 0.$$

It follows from the FOC that $\int_I \rho(s)ds - \lambda \int_I p(s)x(s)ds = 0$; hence, $\lambda = 1/w\omega$. Therefore, the demand function for $x(t)$ is given by $x(t) = (\rho(t)\omega)(w/p(t))$. The equilibrium relative price $\hat{w}/\hat{p}(t)$ can be obtained from the firm's profit condition $\hat{p}(t)f_t(1,\omega) = \hat{w}$.

The above calculation of equilibrium for a one-consumer economy is straightforward. Our economic problem is to compute the competitive equilibrium for the multiconsumer economy. Given $\varsigma = (\varsigma_1 \dots \varsigma_n) \in [0,\omega]^n$, consider the constrained variational problem

$$P(\varsigma) : \text{Maximize} \sum_{a=1}^{m} \alpha_a u_a(x_a(t)) \text{ subject to}$$

$$\sum_{a=1}^{m} x_a(t) \leq f_t(z_t, \varsigma_i),\ t \in I_i,\ i = 1 \dots n, \text{ and } \int_I z_s ds \leq \omega,$$

where $\alpha_1 \dots \alpha_m$ are welfare weights of consumers satisfying $\alpha_a \geq 0$ and we normalize $\sum_{a=1}^{m} \alpha_a = 1$. Let $\Delta = \{\alpha = (\alpha_a) \in \mathbb{R}_+^m | \sum_{a=1}^{m} \alpha_a = 1\}$ be the unit simplex. Note that the normalization of the welfare weights is arbitrary, but each normalization determines a price normalization, as shown later. The solution of this problem is a saddle point of the constrained Lagrangian

$$\mathcal{L}_{\alpha,\varsigma}(x_a(t), z_t, p(t), w) = \sum_{a=1}^{m} \alpha_a \int_I \rho_a(s)\log x_a(s)ds$$
$$+\sum_{i=1}^{n} \int_{I_i} p(s) \left(f_s(z_s, \varsigma_i) - \sum_{a=1}^{m} x_a(s) \right) ds + w \left(\omega - \int_I z_s ds \right),$$

where $p(t)$ and w are the multipliers (*a fortiori* they will be the equilibrium prices).

The saddle point is unique given the strict concavity of the utility functional. Let the saddle point be $(\hat{x}_a(t), \hat{z}_t, \hat{p}(t), \hat{w})$ which satisfies

$$\mathcal{L}_{\alpha,\varsigma}(x_a(t), z_t, \hat{p}(t), \hat{w}) \leq \mathcal{L}_{\alpha,\varsigma}(\hat{x}_a(t), \hat{z}_t, \hat{p}(t), \hat{w})$$
$$\leq \mathcal{L}_{\alpha,\varsigma}(\hat{x}_a(t), \hat{z}_t, p(t), w)$$

for every $(x_a(t), z_t, p(t), w)$. We call these inequalities the *saddle point inequalities.* For each $\alpha \in \Delta$, define a map $\Phi : \Delta\times[0,\omega]^n \to \Delta\times[0,\omega]^n$,

$$\Phi((\alpha_a), (\varsigma_i)) = \left((\tilde{\alpha}_a), \left(\int_{I_i} \hat{z}_s ds \right) \right),$$

where $\tilde{\alpha} = (\tilde{\alpha}_a)$ is defined by

$$\tilde{\alpha}_a = \frac{\max\{0, \alpha_a + \hat{w}\omega_a - \int_I \hat{p}(s)\hat{x}_a(s)ds\}}{\sum_{a=1}^{m} \max\{0, \alpha_a + \hat{w}\omega_a - \int_I \hat{p}(s)\hat{x}_a(s)ds\}}, \quad a = 1 \ldots m.$$

Note that $\sum_{a=1}^{m} \max\{0, \alpha_a + \hat{w}\omega_a - \int_I \hat{p}(s)\hat{x}_a(s)ds\} \neq 0$, since otherwise we would have $\alpha_a + \hat{w}\omega_a - \int_I \hat{p}(s)\hat{x}_a(s)ds \leq 0$ for all $a = 1 \ldots m$. Then it follows from the first inequality that $1 = \sum_{a=1}^{m} \alpha_a \leq \int_I \hat{p}(s) \sum_{a=1}^{m} \hat{x}_a(s)ds - \hat{w}\omega_a \leq 0$, a contradiction. Hence, $\tilde{\alpha} = (\tilde{\alpha}_a) \in \Delta$ is well defined.

In the following, we compute a fixed point $(\hat{\alpha}_a, \hat{\varsigma}_i)$ of the map Φ defined by

$$\Phi((\hat{\alpha}_a), (\hat{\varsigma}_i)) = \left((\hat{\alpha}_a), \left(\int_{I_i} \hat{z}_s ds\right)\right)$$

and show that the saddle point $(\hat{x}_a(t), \hat{z}_t, \hat{p}(t), \hat{w})$ associated with the fixed point is a competitive equilibrium.

We first note that the equilibrium price vector $(\hat{p}(t), \hat{w})$ is strictly positive. Suppose $\hat{p}(t_0) = 0$ on a set of positive measure. Then defining a new allocation $y_a(t)$ by $y_a(t) = \hat{x}_a(t)$ for $t \neq t_0$, $y_a(t_0) = \hat{x}_a(t_0) + \varepsilon$ for some $\varepsilon > 0$, $\mathcal{L}_{\alpha,\varsigma}(y_a(t), z_t, \hat{p}(t), \hat{w}) > \mathcal{L}_{\alpha,\varsigma}(\hat{x}_a(t), \hat{z}_t, \hat{p}(t), \hat{w})$, contradicting the first saddle point inequality. Hence, $p(t) > 0$ a.e., and $\hat{w} > 0$ can be proved similarly.

From the saddle point inequality together with the strict positivity of prices, we conclude that the constraints of the problem $P(\varsigma)$ hold with the exact equalities,

$$\sum_{a=1}^{m} \hat{x}_a(t) = f_t\left(\hat{z}_t, \int_{I_i} \hat{z}_s ds\right) \quad \text{for } t \in I_i \text{ a.e.}, \quad \int_I \hat{z}_s ds = \omega$$

for $i = 1 \ldots n$, or the condition (DX-3) of Definition 7.7 is met.

Next, we claim $\hat{\alpha}_a > 0$ for all $a = 1 \ldots m$. If not, $\hat{\alpha}_b = 0$ for some b. It follows from the definition of $\tilde{\alpha}$ that $0 < \hat{w}\omega_a \leq \int_I \hat{p}(s)\hat{x}_a(s)ds$, hence $\hat{x}_a(t) > 0$ for all a. Given $\hat{\alpha} \in \Delta$, $\hat{\alpha}_c > 0$ for some c. Define a new allocation $(\tilde{y}_a(t))$ by $\tilde{y}_b(t) = 0$, $\tilde{y}_c(t) = \hat{x}_c(t) + \hat{x}_b(t)$ and $\tilde{y}_d(t) = \hat{x}_d(t)$ for $d \neq b, c$. Then $f_s(\hat{z}_s, \int_I \hat{z}_s ds) = \sum_{a=1}^{m} \tilde{y}_a(s)$ and

$$\begin{aligned} \mathcal{L}_{\hat{\alpha},\hat{\varsigma}}(\hat{x}_a(t), \hat{z}_t, \hat{p}(t), \hat{w}) &= \sum_{a=1}^{m} \hat{\alpha}_a u_a(\hat{x}_a(t)) \\ &< \sum_{a=1}^{m} \hat{\alpha}_a u_a(\tilde{y}_a(t)) = \mathcal{L}_{\hat{\alpha},\hat{\varsigma}}(\tilde{y}_a(t), \hat{z}_t, \hat{p}(t), \hat{w}), \end{aligned}$$

contradicting the first inequality. Therefore $0 < \hat{\alpha}_a$, hence $0 < \hat{\alpha}_a < 1$ for all a. Setting $(\alpha_a) = (\hat{\alpha}_a)$ and $(\varsigma_i) = (\int_{I_i} \hat{z}_t dt)$ in the FOC's for

$\mathcal{L}_{\alpha,\varsigma}$, we obtain

$$\frac{\partial \mathcal{L}_{\hat{\alpha},\hat{\varsigma}}}{\partial x_a(t)} = \hat{\alpha}_a \rho_a(t)/\hat{x}_a(t) - \hat{p}(t) = 0, \quad a = 1 \ldots m,$$

$$\frac{\partial \mathcal{L}_{\hat{\alpha},\hat{\varsigma}}}{\partial z_t} = \hat{p}(t) f_t\left(\hat{z}_t, \int_{I_i} \hat{z}_s ds\right) - \hat{w}\hat{z}_t = 0, \quad t \in I_i, \ i = 1 \ldots n.$$

As $\hat{\alpha}$ is a fixed point, it follows from the definition of $\tilde{\alpha}$ that

$$\hat{\alpha}_a = \frac{\hat{\alpha}_a + \left(\hat{w}\omega_a - \int_I \hat{p}(s)\hat{x}_a(s)ds\right)}{1 + \sum_{a=1}^m \left(\hat{w}\omega_a - \int_I \hat{p}(s)\hat{x}_a(s)ds\right)}, \quad a = 1 \ldots m,$$

and from this, we obtain that

$$\hat{w}\omega_a - \int_I \hat{p}(s)\hat{x}_a(s)ds$$
$$= \hat{\alpha}_a \sum_{a=1}^m \left(\hat{w}\omega_a - \int_I \hat{p}(s)\hat{x}_a(s)ds\right), \quad a = 1 \ldots m.$$

The conditions $0 < \hat{\alpha}_a < 1$ for all a imply that the budget constraint $\int_I \hat{p}(s)\hat{x}_a(s)ds = \hat{w}\omega_a$ follows.

Let $x_a(t)$ be such that $\int_I \hat{p}(s)x_a(s)ds \leq \hat{w}\omega_a$. Setting $x_b(t) = \hat{x}_b(t)$ for $b \neq a$, we obtain from the first inequality

$$\hat{\alpha}_a u_a(x_a(s)) \leq \hat{\alpha}_a u_a(\hat{x}_a(s)) + \int_I \hat{p}(s)(x_a(s) - \hat{x}_a(s))ds \leq \hat{\alpha}_a u(\hat{x}_a(s)),$$

or $u_a(x_a(s)) \leq u_a(\hat{x}_a(s))$; hence, condition (DX-1) is met. Similarly, the equilibrium condition (DX-2) follows from the saddle point inequality and $\partial_{z_t}\mathcal{L}_{\hat{\alpha},\hat{\varsigma}} = 0$.

Integrating $\partial_{x_a(t)}\mathcal{L}_{\hat{\alpha},\hat{\varsigma}} = 0$ and using the budget constraint, we have

$$\hat{\alpha}_a = \hat{w}\omega_a, \quad a = 1 \ldots m.$$

This is the celebrated *Negishi condition* [167, p. 97]; the welfare weight of a consumer is the inverse of the marginal utility of income, which is the Lagrangian multiplier of the consumer's maximization problem; see Example 7.4. As $\sum_{a=1}^{m} \hat{\alpha}_a = 1$, the equilibrium price has been normalized as follows:

$$\hat{w} = \omega^{-1}.$$

Summing $\partial_{x_a(t)} \mathcal{L}_{\hat{\alpha},\hat{\varsigma}} = 0$ over a with the help of $\hat{\alpha}_a = \hat{w}\omega_a$, $\partial_{z_t} \mathcal{L}_{\hat{\alpha},\hat{\varsigma}} = 0$ and (DX-3), we obtain

$$\hat{w} \sum_{a=1}^{m} \rho_a(t)\omega_a = \hat{p}(t) \sum_{a=1}^{m} \hat{x}_a(t) = \hat{p}(t) f_t\left(\hat{z}_t, \int_{I_i} \hat{z}_s ds\right) = \hat{w}\hat{z}_t,$$

or

$$\hat{z}_t = \sum_{a=1}^{m} \rho_a(t)\omega_a, \; t \in I.$$

For convenience, we set $\vartheta_i(\hat{z}_t) = f_t\left(\hat{z}_t, \int_{I_i} \hat{z}_s ds\right)$, $t \in I_i$. Then we have

$$\hat{p}(t) = \hat{z}_t \hat{w}/\vartheta_i(\hat{z}_t) = \hat{z}_t/\omega\vartheta_i(\hat{z}_t), \; \hat{x}_a(t) = \left(\frac{\rho_a(t)\omega_a}{\sum_{a=1}^{m} \rho_a(t)\omega_a}\right) \vartheta_i(\hat{z}_t)$$

for $t \in I_i$, $i = 1 \ldots n$, and $a = 1 \ldots m$. Our computation of the equilibrium has been completed.

Next, we consider the social optimization problem

$$P : \text{Maximize} \sum_{a=1}^{m} \alpha_a \int_I \rho_a(s) \log x_a(s) ds$$

$$\text{subject to} \sum_{a=1}^{m} x_a(t) \leq \vartheta_i(z_t), \; t \in I_i, \; i = 1 \ldots n, \text{ and}$$

$$\int_I z_s ds \leq \omega.$$

As before, $\alpha_1 \ldots \alpha_m$ are the welfare weights of consumers satisfying $\alpha_a \geq 0$ and $\sum_{a=1}^{m} \alpha_a = 1$. The solution $(\tilde{z}_t, \tilde{x}_a(t))$ of this problem is a Pareto-optimal allocation, and a saddle point of the constrained

Lagrangian

$$\mathcal{L}_\alpha(x_a(t), z_t) = \sum_{a=1}^{m} \alpha_a \int_I \rho_a(s)\log x_a(s)ds$$
$$+ \sum_{i=1}^{n} \int_{I_i} \lambda(s) \left(\vartheta_i(z_s) - \sum_{a=1}^{m} x_a(s) \right) ds + \nu \left(\omega - \int_I z_s ds \right),$$

where $\lambda(t)$ and ν are the multipliers. The FOCs for $\mathcal{L}_\alpha$ are

$$\frac{\partial \mathcal{L}_\alpha}{\partial x_a(t)} = \alpha_a \rho_a(t)/\tilde{x}_a(t) - \lambda(t) = 0, \quad a = 1 \dots m, \ t \in I,$$
$$\frac{\partial \mathcal{L}_\alpha}{\partial z_t} = \boldsymbol{\lambda}(t)\vartheta_i'(\tilde{z}_t) - \nu = 0, \quad t \in I_i, \ i = 1 \dots n,$$

where $\vartheta_i'(z_t) = f_t(1, \int_{I_i} z_s ds) + \partial_\varsigma f_t(z_t, \int_{I_i} z_s ds)$. The second term represents the externality effect of production.

The problem is this: for which value of α_a should we compare the competitive allocation $(\hat{z}_t, \hat{x}_a(t))$ and the efficient allocation $(\tilde{z}_t, \tilde{x}_a(t))$? The answer is obviously given by the Negishi condition $\hat{\alpha}_a = \hat{w}\omega_a$. We would like to emphasize that, however, from the normative point of view, this seems to be justifiable only when the initial endowments of the consumers are at least nearly identical.

Setting $\alpha_a = \hat{w}\omega$ in $\partial_{x_a(t)}\mathcal{L}_\alpha = 0$ and summing over a, we have

$$\hat{w} \sum_{a=1}^{m} \rho_a(t)\omega_a = \lambda(t) \sum_{a=1}^{m} \tilde{x}_a(t) = \lambda(t)\vartheta_i(\tilde{z}_t), \quad t \in I_i, \ i = 1 \dots n.$$

Substituting this into $\partial_{x_a(t)}\mathcal{L}_\alpha = 0$ and $\partial_{z_t}\mathcal{L}_\alpha = 0$, we obtain

$$\tilde{x}_a(t) = \left(\frac{\rho_a(t)\omega_a}{\sum_{a=1}^{m} \rho_a(t)\omega_a} \right) \vartheta_i(\tilde{z}_s),$$
$$\hat{w} \sum_{a=1}^{m} \rho_a(t)\omega_a \left(\frac{\vartheta_i'(\tilde{z}_t)}{\vartheta_i(\tilde{z}_t)} \right) = \nu, \quad t \in I_i, \ i = 1 \dots n,$$

respectively. Multiplying the last equation with $\tilde{z}_t$, integrating over I and summing over i, we obtain

$$\nu = \hat{w}\omega^{-1} \sum_{a=1}^{m} \sum_{i=1}^{n} \int_{I_i} \rho_a(s)\omega_a \left(\frac{\vartheta_i'(\tilde{z}_s)\tilde{z}_s}{\vartheta_i(\tilde{z}_s)} \right) ds.$$

We define the elasticity $\varepsilon_i(z_t)$ of firm t in industry i as

$$\varepsilon_i(z_t) = \left(\frac{\vartheta_i'(z_t)z_t}{\vartheta_i(z_t)} \right), \quad t \in I_i, \ i = 1 \ldots n.$$

Then it follows from the last three equations that

$$\tilde{z}_t = \left(\frac{\varepsilon_i(\tilde{z}_t)}{\omega^{-1} \sum_{a=1}^{m} \sum_{i=1}^{n} \int_{I_i} \rho_a(s)\omega_a \varepsilon_i(\tilde{z}_s) ds} \right)$$

$$\times \sum_{a=1}^{m} \rho_a(t)\omega_a, \quad t \in I_i, \ i = 1 \ldots n.$$

Thus, we have completed the computation of the optimal allocation $(\tilde{z}_t, \tilde{x}_a(t))$. Comparing it with the competitive allocation $(\hat{z}_t, \hat{x}_a(t))$, we can immediately deduce the following theorem which was, for economies with finitely many homogeneous commodities, first stated by Chipman [43, p. 365] for the case $m = 1$.

Theorem 7.4. *Suppose that every consumer has a Cobb–Douglas utility functional and the same amount of labor as an endowment. Then, the optimal output (with the welfare weights proportional to the endowments) of the tth product is greater than, equal to, or less than the competitive level of output, if and only if the elasticity (at the optimum)* $\varepsilon_i(\tilde{z}_t)$ *of the firm t in industry i is greater than, equal to, or less than the weighted average of the elasticities of all industries,* $\omega^{-1} \sum_{a=1}^{m} \sum_{i=1}^{n} \int_{I_i} \rho_a(s)\omega_a \varepsilon_i(\tilde{z}_s) ds$, *respectively. In particular, if all firms' elasticities are equal to the weighted average, the competitive equilibrium is Pareto-optimal.*

7.5 A Simple Model of the Difference Principle

John Rawls [202] proposed and established the difference principle as an alternative to the utilitarian principle in social justice. The *difference principle* states that

> Social and economic inequalities are to be arranged so that they are attached to the greatest benefit of the least advantaged [202, p. 83].

However, we note that Rawls did not provide any positive or convincing proof to demonstrate that the second principle was indeed implementable within any liberal or democratic society. The purpose of this section is to provide a proof that supports his conclusion.

Consider Figure 7.5, which corresponds to Figure 6 in Rawls [202, p. 76]. As far as we are aware, this is the only exposition by Rawls that elaborates upon the difference principle theoretically. Rawls explained this as follows:

> Suppose that $\mathbf{x_1}$ is the most favored representative man in the basic structure. As his expectations are increased so are the prospects of $\mathbf{x_2}$, the least advantaged man. In Figure 6 (Figure 7.5), let the curve OP represent the contribution to $\mathbf{x_2}$'s expectation made by the greater expectations of $\mathbf{x_1}$. The point O, the origin, represents the hypothetical state in which all social primary goods are distributed equally. Now the OP curve is always below the 45° line, since $\mathbf{x_1}$ is always better off. Thus, the only relevant parts of the indifference curves are

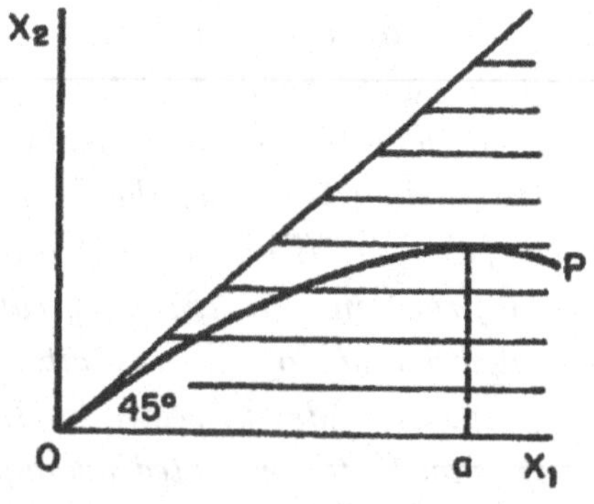

Figure 7.5. Figure 6 in [202, p. 76].

> those below this line, and for this reason the upper left-hand part of figure 6 is not drawn in. Clearly the difference principle is perfectly satisfied only when the OP curve is just tangent to the highest indifference curve it touches. In Figure 6, this is at the point a [202, p. 76].

However, the question is *where does the contribution curve OP come from?* In [202], the mechanism that generates this curve is a black box, which is precisely what we wish to explore in the remainder of this section by means of external increasing returns.

In society, two kinds of goods trade in markets. Consumption goods are consumed by all members of society with an amount denoted by x. Consumption goods determine the current level of the quality of life for citizens and lead to expectations for their future lives. The other type of good is input goods, which are used as an input (resource) to produce the consumption good and its amount denoted by z. The input good is not consumed, which means that the consumers' utilities do not depend on it. Therefore, the value of the consumption good can be identified as the utility level if we restrict ourselves within the class of monotonous and continuous utility functions. This allows us to avoid the problem of the interpersonal comparability of utilities. We can also interpret the input good as a *primary good* in Rawls's sense [202, pp. 92–93] because it is required to develop each person's life plan; i.e., advantaged citizens will use it as "money" and increase their "productivity" (see below). For simplicity, we assume that there is initially no consumption good in the society (market) and so it must be produced by the social cooperative production activity described as follows.

There are two groups of citizens (consumers). Each citizen of group 1 represents more advantaged (or talented) persons and each citizen of the group 2 represents less advantaged persons. Consumption levels of citizens 1 and 2 are denoted by x_1 and x_2, respectively, (see variables $\boldsymbol{x}_1$ and $\boldsymbol{x}_2$ in Figure 7.5). Citizens possess some amount (possibly 0) of the primary good as an initial endowment. Let $\omega > 0$ be the total amount of the primary good that initially exists in society. Suppose that there are n_1 citizens of group 1 and n_2 citizens of group 2, and assume that each member of group 1 owns the same amount of initial endowment ω_1, and similarly citizens are endowed with ω_2 in group 2; hence, we have $n_1\omega_1 + n_2\omega_2 = \omega$. Let $\theta = n_1/n_2$ be the population ratio, and we denote $\Omega_1 = \omega/n_1$ and $\Omega_2 = \omega/n_2$; then it follows that

$$\theta\Omega_1 = \theta\omega_1 + \omega_2 = \Omega_2.$$

We identify ω_1 as the index of the policy parameter through which the government arranges justice in the society. When $\omega_1 = \Omega_1$, then the advantaged person owns all, and when $\omega_1 = 0$, the least advantaged person owns all. We then search for the value of ω_1 that achieves the social state in which the difference principle is satisfied.

The reason why citizen 1 is the more "talented" or better-endowed person is that the citizen has a "production function" $f(z; \omega_1)$ that satisfies the conditions (Hom) and (Mon) in Definition 7.5, where z is the value of the input to produce the consumption good, and ω_1 is a parameter denoted ς thus far.

A natural interpretation for $f(z; \omega_1)$ is a potential ability such as natural talents. However, this production function reflects the background institutions explained in Section 1.6. Therefore, its interpretation should not be restricted to natural talents. Instead, it will include one's own environmental elements such as family and education. As to the latter, while opportunities in education are certainly open to *all* citizens in liberal societies, they must choose (and be chosen by) an appropriate school (consider schools of law, medicine, engineering, and music). The production function in this case results from a combination of one's natural talent and school education. Indeed, educational systems are a good example of institutions for information transmission or knowledge spillover of a Romer type.

When citizen 1 acts as a "producer," the citizen takes the value of ω_1 as a parameter, which means that it is given. Therefore, the citizen will subjectively produce under constant returns to scale. Here the role of "primary good" is threefold; it not only serves as an input commodity in the usual sense but also develops the ability of citizen 1 through the background institutions and enters the model as a positive external effect, and finally, it is used as "money" to purchase the consumption good. As stated, positive externalities, as well as market trades, convey the concept of reciprocity. As citizen 1 is a representative agent of the class of talented people, the share value of ω_1 included in the production function is also the value of the other members of the class represented by citizen 1. The functional form $f(z;\omega_1)$ restricts the range of the externality (mutual advantage) to the group of talented persons.

We set up a political and cooperative scheme for this society as follows. Suppose that the entire amount ω of the primary good is initially owned by citizen 2. The "government" levies a tax ω_1 on this and redistributes it to citizen 1. As there is initially no consumption good, the less endowed citizen 2 must give some amount of the primary good to the better endowed citizen 1 to help that citizen produce the consumption good. Using the remainder of the primary commodity (disposable income), citizen 2 purchases the consumption good in the market. In contrast, citizen 1 can undertake production activity only when some amount of the primary commodity $\omega_1 > 0$ is supplied by the society (government); otherwise, they can produce nothing given equation $f(z;0) = 0$. The difference principle requires ω_1 to be set, which maximizes the consumption (welfare level) of citizen 2 at the equilibrium. Obviously, this sort of "fiscal policy" is realizable in property-owning democracies and liberal socialism regimes.

For notational convenience, we set $f(1,\omega_1) = \varphi(\omega_1)$ and assume $\varphi(\omega_1)$ to be twice-continuously differentiable. We also assume that $\varphi(0) = 0$ or that no outputs can be produced without the primary good. Note that the condition (Mon) implies that $\varphi'(\omega_1) > 0$.

The market equilibrium is now determined by the following equations:

$$px_1 = p\varphi(\omega_1)z - z + \omega_1,$$

$$px_2 = \omega_2,$$

$$\theta x_1 + x_2 = \varphi(\omega_1)\Omega_2,$$

where we set the price of the primary good to be 1. By (Hom), the equilibrium price is determined as $p = 1/\varphi(\omega_1)$. The *equilibrium curve* $\Phi(x_1, x_2) = 0$ is obtained by eliminating p, ω_1, and ω_2 from the four equations which define the equilibrium. The derivatives are calculated as

$$\frac{dx_1}{d\omega_1} = \varphi'(\omega_1)\omega_1 + \varphi(\omega_1), \quad \frac{dx_2}{d\omega_1} = \varphi'(\omega_1)(\Omega_2 - \theta\omega_1) - \theta\varphi(\omega_1).$$

The fundamental properties of the equilibrium curve are stated in the following lemma.

Lemma 7.5. *The equilibrium curve* $\Phi(x_1, x_2) = 0$ *starts from the origin when* $\omega_1 = 0$, *intersects with the* 45° *line when* $\omega_1 = \omega_2$ *and ends at the* x_1*-axis,* $x_1 = \varphi(\Omega_1)\Omega_1$ *when* $\omega_1 = \Omega_1$. *Moreover, it is concave to the* x_1*-axis if*

$$\varphi''(\omega_1)\varphi(\omega_1) < 2\varphi'^2(\omega_1)$$

and has the peak at $x_1 = \varphi(\hat{\omega}_1)\hat{\omega}_1$, *where* $\hat{\omega}_1$ *satisfies*

$$\varphi'(\hat{\omega}_1)(\Omega_2 - \theta\hat{\omega}_1) - \theta\varphi(\hat{\omega}_1) = 0.$$

Proof. The first three properties are verified by just substituting 0, ω_2, and Ω_1 for ω_1 in the defining equations of the equilibrium, respectively. We notice that the intersection of the curve with the 45° line at $\omega_1 = \omega_2$ comes from the condition (Hom). Given the "profit" of citizen 1 is 0, the income needed to purchase the consumption good is solely determined by that initial endowment. Therefore, citizens 1 and 2 will purchase the same amount when $\omega_1 = \omega_2$. From $dx_1/d\omega_1$ and $dx_2/d\omega_1$, we can compute the following equations:

$$\frac{dx_2}{dx_1} = \frac{\varphi'(\omega_1)(\Omega_2 - \theta\omega_1) - \theta\varphi(\omega_1)}{\varphi'(\omega_1)\omega_1 + \varphi(\omega_1)},$$

$$\frac{d^2x_2}{dx_1^2} = \frac{(\varphi''(\omega_1)\varphi(\omega_1) - 2\varphi'^2(\omega_1))\Omega_2}{(\varphi'(\omega_1)\omega_1 + \varphi(\omega_1))^3}.$$

The value of x_1 which attains the peak of $\Phi(x_1, x_2) = 0$ and the concavity of the curve follow from those derivatives, respectively. □

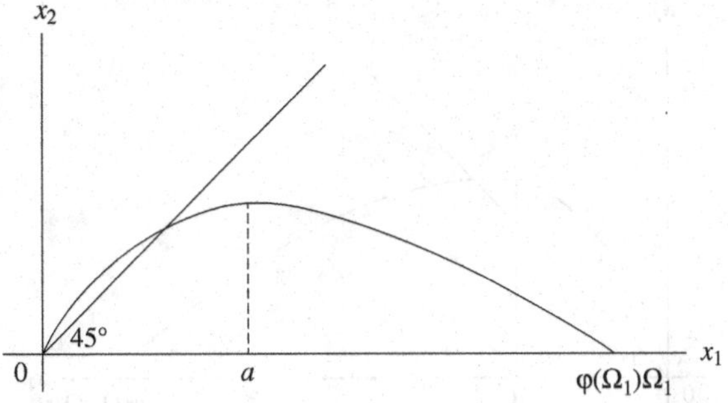

Figure 7.6. Effective reciprocity.

Note that the concavity condition $\varphi''(\omega_1)\varphi(\omega_1) < 2\varphi'^2(\omega_1)$ is mild and does not depend on θ. We define a measure of reciprocity from the derivative of the equilibrium curve on the 45° degree line.

Definition 7.8. We say that reciprocity *works effectively* or reciprocity is *effective* if, and only if, $dx_2/dx_1 > 0$ when $\omega_1 = \omega_2$.

We can readily see that reciprocity is effective if, and only if,

$$\left(\frac{\omega}{n_1+n_2}\right)\varphi'\left(\frac{\omega}{n_1+n_2}\right) > \theta\varphi\left(\frac{\omega}{n_1+n_2}\right).$$

Figure 7.6 illustrates the situation where reciprocity is effective.

Suppose for a moment that the reciprocity of the society is effective. The relation between the contribution curve OP in Rawls and the equilibrium curve is as follows. As we mentioned earlier, Rawls sets the origin as the reference point where each individual is assigned the same amount of the primary good.

> The point O, the origin, represents the hypothetical state in which all social primary goods are distributed equally ([202, p. 76]).

Thus, the point at which the equilibrium curve intersects the 45° line corresponds to the origin O in Figure 7.6, as given by Lemma 7.5, we have $\omega_1 = \omega/2$ at that point. The difference principle is applied to the equilibrium (x_1, x_2) which satisfies $\Phi(x_1, x_2) = 0$, thereby instructing the society (or the government) how to set the value of ω_1 to obtain

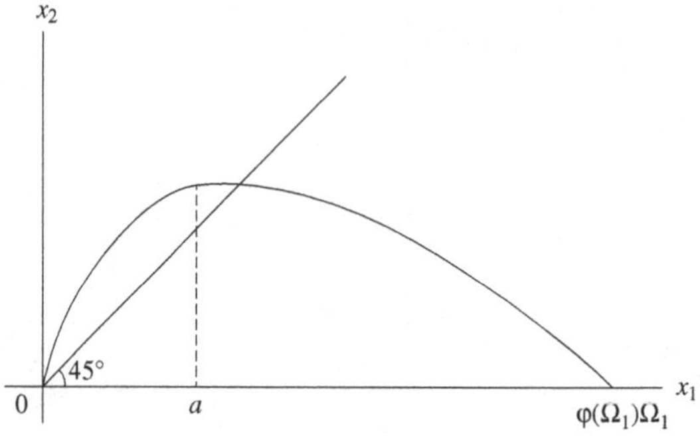

Figure 7.7. Ineffective reciprocity.

the largest value of x_2; this state can be achieved at $x_1 = \varphi(\hat{\omega}_1)\hat{\omega}_1$ of Lemma 7.5.

What about the case where reciprocity is ineffective? Figure 7.7 illustrates the case. In this case, point a in Figure 7.7 is placed to the left of the 45° line. If point a is chosen, this means that talented person 1 is "sacrificed" for the society. Obviously, this is unjust. In this case, therefore, the difference principle would order the point on the 45° line. The idea behind this observation is as follows. We always start from the reference point of $\omega_1 = \omega_2$. Any deviation from this point is admitted only when both of the more and less advantaged persons gain by this deviation, which will be possible under the condition that reciprocity is effective. In other words, the difference principle will mandate the states that are distinct from that of perfect equality only for those societies where reciprocity works sufficiently. Rawls stated

> [T]he difference principle is a strongly egalitarian condition in the sense that unless there is a distribution that makes both persons better off (limiting ourselves to the two person case for simplicity), an equal distribution is to be preferred. The [social] indifference curves [...] are actually made up of vertical and straight [horizontal] lines that intersect at right angles at the 45° line (again supposing an intertemporal and cardinal interpretations of the axes). No matter how much either person's situation is improved, there is no gain from the standpoint of the indifference principle unless the other gains also ([202, p. 76]).

We can summarize these discussions in the following proposition.

Proposition 7.4. *The difference principle mandates the allocation associated with* $\omega_1 = \hat{\omega}_1$ *on an equilibrium curve with* $x_1 > x_2$*, where* $\hat{\omega}_1$ *is a solution of the equation* $\varphi'(\hat{\omega}_1)(\Omega_2 - \theta\hat{\omega}_1) - \theta\varphi(\hat{\omega}_1) = 0$ *when reciprocity works effectively, and the allocation on the* 45° *line* ($\omega_1 = \omega/2$) *when reciprocity is ineffective.*

Note that we do not assume the concavity condition $\varphi''(\omega_1)\varphi(\omega_1) < 2\varphi'^2(\omega_1)$ in Proposition 7.4. For the following analysis, we need to distinguish two concepts of optimality. The first is the standard notion of Pareto optimality which is as follows:

Definition 7.9. A feasible allocation (x_1, x_2), which satisfies the equation $\theta x_1 + x_2 \leq \varphi(\omega_1)\Omega_2$, is said to be *globally Pareto-optimal* (or Pareto-optimal for short), if, and only if, another feasible allocation exists (y_1, y_2), such that $(x_1, x_2) < (y_1, y_2)$.

Note that globally Pareto-optimal allocations are exactly those which satisfy $\theta x_1 + x_2 = \varphi(\Omega_1)\Omega_2$. Our second concept of optimality is a local one, defined only for the equilibrium allocations:

Definition 7.10. An equilibrium allocation (x_1, x_2), which satisfies the equation $\Phi(x_1, x_2) = 0$, is said to be *locally Pareto-optimal* (or locally optimal for short), if, and only if, the derivative dx_2/dx_1 is less than or equal to 0 at (x_1, x_2).

Obviously, the allocations on the equilibrium curve over the domain $a \leq x_1 \leq \varphi(\Omega_1)\Omega_1$ are locally optimal. Hence, the allocation stipulated by the difference principle under effective reciprocity is locally optimal. Rawls was concerned with local optimality when he discussed the efficiency of allocations ([202, p. 77]). We trust that the local optimality is more appropriate than the global one for liberal societies, given the latter will generally need to compare very distant allocations. To attain an improving allocation, the society must gather all produced commodities at once and redistribute them among citizens. Although such drastic reallocations might be possible in societies with centrally planned economies, they would be unrealistic for liberal societies with market economies.

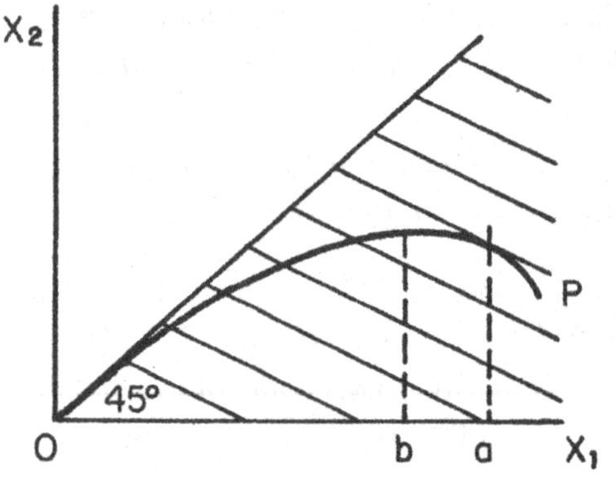

Figure 7.8. Figure 8 in [202, p. 77].

Next, we would like to identify allocations determined by the *utilitarian principle* that select the equilibrium allocations maximizing total (average) utility $\theta x_1 + x_2$.

For this, the relevant statements in Rawls are stated as follows:

> A classical utilitarian [...] is indifferent as to how a constant sum of benefits is distributed. If there are but two persons, then assuming an interpersonal cardinal interpretation of the axes, the utilitarian's indifference lines for distributions are straight lines perpendicular to the 45° line. Since, however, x_1 and x_2 are representative men, the gains to them have to be weighted by the number of persons they represent. Since presumably x_2 represents rather more persons than x_1, the indifference lines become more horizontal, as seen in Figure 8 [Figure 7.8]. The ratio [$= \theta$] of the number of advantaged to the number of disadvantaged defines the slope of these straight lines. Drawing the same contribution curve OP as before, we see that the best distribution from a utilitarian point of view is reached at the point which is beyond the point b where the OP curve reaches its maximum ([202, p. 77]).

For this case, we obtain a more precise result.

Proposition 7.5. *Suppose that the concavity condition holds. Then the average utilitarian principle orders the allocation* $(\varphi(\Omega_1)\Omega_1, 0)$ *for any* $\theta > 0$.

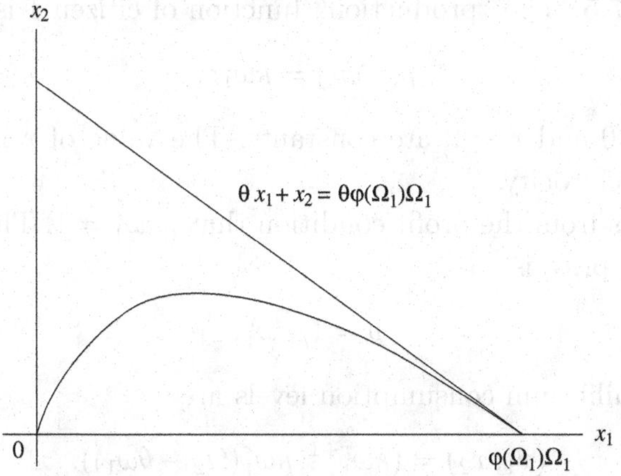

Figure 7.9. Proposition 7.5.

Proof. We first note the average utilitarian principle orders to maximize $\theta x_1 + x_2$. Set $\omega_1 = \Omega_1$ in dx_2/dx_1. Then we have $\theta\omega_1 = \Omega_2$; therefore, it follows that

$$\left.\frac{dx_2}{dx_1}\right|_{\omega_1=\Omega_1} = \frac{-\theta\varphi(\Omega_1)}{\varphi'(\Omega_1)\Omega_1 + \varphi(\Omega_1)} > -\theta.$$

Given the equilibrium curve is concave to the horizontal axis, the result follows. □

Proposition 7.5 illustrated by Figure 7.9 might look surprising, given it states that the utilitarian principle stipulates as just an allocation in which talented people monopolize all of society's wealth! It is interesting that the set of equilibrium allocations that are globally Pareto-optimal is just $(\varphi(\Omega_1)\Omega_1, 0)$ because every other point on the equilibrium curve can be moved to an optimal point on $\theta x_1 + x_2 = \theta\varphi(\Omega_1)\Omega_1$, which is located in a north-east direction. However, we cannot help but be strongly skeptical about the utilitarian principle as a principle of justice for societies irrespective of whether reciprocity is effective or not, even if they realize Pareto-optimal allocations. Indeed, Proposition 7.5 might be considered yet another reason for the justification of the difference principle over the utilitarian principle from the original discussion.

Example 7.5. The "production" function of citizen 1 is given by

$$f(z;\omega_1) = \kappa\omega_1^r z,$$

where $\kappa > 0$ and $r \geq 0$ are constants. The value of r is called the *degree* of reciprocity.

It follows from the profit condition that $p\kappa\omega_1^r = 1$. Therefore, the equilibrium price is

$$p = 1/\kappa\omega_1^r,$$

and the equilibrium consumption levels are

$$(x_1, x_2) = (\kappa\omega_1^{r+1}, \kappa\omega_1^r(\Omega_2 - \theta\omega_1)).$$

Eliminating ω_1 from this, we obtain the equilibrium curve,

$$(\theta x_1 + x_2)^{r+1} = \kappa\Omega_2^{r+1}x_1^r.$$

We then compute

$$\frac{dx_1}{d\omega_1} = \kappa(r+1)\omega_1^r, \quad \frac{dx_2}{d\omega_1} = \kappa\omega_1^{r-1}(r\Omega_2 - \theta(r+1)\omega_1).$$

From these, we obtain

$$\frac{dx_2}{dx_1} = \frac{r\Omega_2 - \theta(r+1)\omega_1}{(r+1)\omega_1},$$

and

$$\frac{d^2x_2}{dx_1^2} = \left(\frac{dx_1}{d\omega_1}\right)^{-1}\frac{d}{d\omega_1}\left(\frac{dx_2}{dx_1}\right) = \frac{-r\Omega_2}{\kappa(r+1)^2\omega_1^{r+2}} < 0.$$

When $\omega_1 = \omega_2$, $dx_2/dx_1 = (r-\theta)/(r+1)$. Hence, reciprocity is effective if, and only if, $r > \theta = n_1/n_2$. It is interesting that the smaller the proportion of the advantaged (talented) persons, the more likely the reciprocity is to be effective. When $r > \theta$, the difference principle mandates the allocation

$$(x_1, x_2) = \left(\kappa\left(\frac{r\Omega_2}{\theta(r+1)}\right)^{r+1}, \kappa(\theta/r)\left(\frac{r\Omega_2}{\theta(r+1)}\right)^{r+1}\right)$$

associated with $\omega_1 = r\Omega_2/\theta(r+1)$ on the equilibrium curve.

7.6 Notes

Section 7.1: There are so many references concerning marginal cost pricing equilibria. Among these, we cite [23, 35, 36, 45, 105]. The interested reader can consult the survey articles [34, 44] and the references therein. The idea of the proof of Theorem 7.1 is due to [36].

Section 7.2: The original paper regarding the distributive sets is, of course, Scarf [224]; see also [57]. Scarf considered a finite consumers economy, but obtained a remarkable theorem which asserts, roughly speaking, that the distributiveness of a production set is necessary for the non-emptiness of the core [224, Theorem 5, p. 426]. Oddou [183] reconsidered his results in a production economy with a measure space of consumers, but in a slightly different way from our formulation in this section which is closer to the original formulation of Scarf.

Section 7.3: We discussed the history of external increasing returns in Chapter 1. The exposition of this section entirely follows [244]. There are special technical difficulties when production includes externalities. The case of external increasing returns seems to be exceptional for escaping these problems. In the following, we present a point.

Recall the proof of Lemma 7.3 and the condition (BA-2). We could not claim that the vector $\pi \in ba$ was an equilibrium price vector, as we merely showed that the firms maximized profits at π in only one period. This is in contrast to Lemma 4.2 for exchange economies, and the reason why we could not show that

$$\pi\left(F_b\left(\boldsymbol{z}_b, \sum_{c=1}^{n} \boldsymbol{z}_c\right) - \boldsymbol{z}_b\right) \geq \pi\left(F_b\left(\boldsymbol{z}, \sum_{c=1}^{n} \boldsymbol{z}_c\right) - \boldsymbol{z}\right)$$

for every $\boldsymbol{z} \in \ell^{\infty}_{\beta+}$ is the following. Suppose on the contrary that

$$\pi\left(F_b\left(\boldsymbol{z}_b, \sum_{c=1}^{n} \boldsymbol{z}_c\right) - \boldsymbol{z}_b\right) < \pi\left(F_b\left(\boldsymbol{z}, \sum_{c=1}^{n} \boldsymbol{z}_c\right) - \boldsymbol{z}\right)$$

for some $\boldsymbol{z} \in \ell^{\infty}_{\beta+}$ and try to deduce a contradiction. Strictly speaking, we have a problem here that the price vector π belongs to ba rather than $ba_{\beta^{-1}}$, so that the profit value may not exist. We ignore this problem for the moment and assume that the value exists in order to concentrate on the problem arising from the production externalities. Then we have

$$\pi(T)\left(F_b\left(\boldsymbol{z}_b, \sum_{c=1}^{n} \boldsymbol{z}_c\right) - \boldsymbol{z}_b\right) < \pi(T)\left(F_b\left(\boldsymbol{z}, \sum_{c=1}^{n} \boldsymbol{z}_c\right) - \boldsymbol{z}\right)$$

for some T large enough.

If (the projections of) $F_b(\boldsymbol{z}_b, \sum_{c=1}^{n} \boldsymbol{z}_c)$ and $F_b(\boldsymbol{z}, \sum_{c=1}^{n} \boldsymbol{z}_c)$ are technologically feasible outputs for the finite-dimensional economy $\mathcal{E}^T_\varsigma$, we would obtain the desired contradiction. This is not so, however, and what we actually need to show is that

$$\pi(T)\left(F_b\left(\boldsymbol{z}_b(T), \sum_{c=1}^{n} \boldsymbol{z}_c(T)\right) - \boldsymbol{z}_b(T)\right)$$
$$< \pi(T)\left(F_b\left(\boldsymbol{z}, \sum_{c=1}^{n} \boldsymbol{z}_c(T)\right) - \boldsymbol{z}\right).$$

Lacking the joint continuity of the inner product $\pi \boldsymbol{z}$ in the weak* topology, this inequality does not follow.[2] Note that if the externalities do not exist in production, this problem does not occur under the constant returns to scale technology $F_b(\boldsymbol{z}_b)$. Indeed, in this case, all we have to show is that

$$0 < \pi(T)(F_b(\boldsymbol{z}) - \boldsymbol{z}) \text{ for some } T,$$

since $\pi(T)(F_b(\boldsymbol{z}_b(T)) - \boldsymbol{z}_b(T)) = 0$ for all T. This is exactly what Bewley showed in [27, Theorem 3, p. 525]. We should point out that the assumption that the technology function consists of the one-period production functions was also important. However, we can say that

[2]Hence, we do not have any problem if the externalities come into the utility functions in the condition (BA-1), if they are continuous in the weak* topology.

this assumption of the periodic structure of the production process is natural in order to study the infinite time horizon economies.

Section 7.4: The original contribution regarding the results of this section is Chipman [43]. Our formulation entirely follows [250].

In [141, p. 389], Marshall wrote

> ... By similar reasoning, it may be shown that a tax on a commodity which obeys the law of increasing return is more injurious to the consumer than if levied on one which obeys the law of constant return. For it lessens the demand and therefore the output. It thus probably increases the expenses of manufacture somewhat: sends up the price by more than the amount of the tax; and finally diminishes consumers' surplus by much more than the total payments which it brings in to the exchequer. On the other hand, a bounty on such a commodity causes so great a fall in its price to the consumer, that the consequent increase of consumers' surplus may exceed the total payments made by the State to the producers; and certainly will do so in case the law of increasing return acts at all sharply.

In light of the analysis of this section, the theoretical background of the so-called "Pigou tax" originally due to Marshall cited above seems to become clear; see also [43, Section IV]. Finally, Theorem 7.4 answers the question of "socially optimal product differentiation" raised by Lancaster [127] in the presence of the external increasing returns.

Section 7.5: For philosophical discussions on the difference principle and the theory of justice (justice as fairness), readers must consult Rawls [202, 203]. The formulation and exposition in this section entirely follow [251].

Chapter 8

Monopolistically Competitive Economies

8.1 Monopolistically Competitive Markets

A remarkable aspect of the development of markets in reality is that, on the one hand, the number of consumers has grown rapidly — which is perhaps the meaning of the "extension of the markets" — but, on the other hand, the production activities have become concentrated more and more within a small number of firms. In other words, as the markets have become large in the sense of Chapter 3, they have become more monopolistic. In 1917, Lenin wrote that

> The enormous growth of the industry and remarkably rapid process of concentration of production in ever-larger enterprises are one of the most characteristic features of capitalism. Modern censuses of production give most complete and exact data on this process.
>
> In Germany, for example, in 1882, there were only three large enterprises, i.e., those employing more than 50 workers, in every 1,000 industrial enterprises. In 1895, there were six and in 1907, there were nine. In these years, large enterprises employed 22, 30, and 37 workers, respectively, for every 100 workers employed. However, concentration of production is much more intense than the concentration of workers, as the labor in the large enterprises is much more productive than in small enterprises. This is shown by the figures available on steam engines and electric motors.

If we take what is called industry in the broad sense of the term, that is, including commerce, transport, etc., in Germany, we obtain the following picture: large-scale enterprises accounted for 30, 588 out of a total of 3, 265, 623 enterprises, that is to say, 0.9%. Yet, these large-scale enterprises employed 5,700,000 workers out of a total of 14,400,000 (39.4%) and they used 6,660,000 steam horsepower out of a total of 8,800,000 (75.3%) and 1,200,000 kilowatts of electricity out of a total of 1,500,000 (77.2%).

Thus, less than one-hundredth of the total enterprises utilized more than three-fourths of the steam and electric power! In contrast, 2,970,000 small enterprises (employing up to five workers), which represented 91% of the total enterprises, utilized only 7% of the steam and electric power. Tens of thousands of large-scale enterprises are everything; millions of small ones are nothing.

In 1907, there were 586 establishments in Germany employing 1,000 and more workers. They employed nearly one-tenth (1,380,000) of the total number of workers employed in industry and utilized *almost one-third* (32%) of the total steam and electric power employed.[1] As we shall see, money, capital, and banks make this superiority of a handful of the largest enterprises still more overwhelming, in the most literal sense of the word, as millions of small and medium firms and even some big "masters" are in fact in complete subjection to hundreds of millionaire financiers.[2]

In another advanced capitalist country, the United States of America, the growth in the concentration of production is still greater. The United States statistics focus on industry in the narrow sense of the word and group enterprises according to the value of their annual output. In 1904, large-scale enterprises with an annual output of 1,000,000 dollars and over numbered 1,900 (out of 216,180, i.e., 0.9%). They employed 1,400,000 workers (out of 5,500,000, i.e., 25.6%) and their combined annual output was valued at $5,600,000,000 (out of $14,800,000,000 i.e., 38%). Five years later, in 1909, there were 3,060 large-scale enterprises out of 268,491, i.e., 1.1%, which employed 2,000,000 workers out of 6,600,000, i.e., 30.5%, and which produced output valued at $9,000,000,000 out of $20,700,000,000, i.e., 43.8%.[3]

[1] *Annalen des Deutschen Reiches* (Annals of German Empire), 1911, Zahn, pp. 165–69.

[2] In the following chapters of his book, Lenin emphasized the concentration in the banking and financial sectors, which is out of scope in this monograph.

[3] Statistical Abstract of the United States, 1912, p. 202.

> Thus, almost half of the total production of all the enterprises in the country was conducted by a hundredth part of those enterprises! These 3,000 giant enterprises embraced 268 branches of industry. From this, it can observed that, at a certain stage of development, concentration itself leads to monopoly; the difficulty of competition and the ease with which a score or so of giant enterprises can arrive at an agreement creates a tendency towards monopoly based on the very dimensions of the enterprises. This transformation of competition into monopoly is one–if not the — most important phenomena of the modern capitalist economy, and we must deal with it in greater detail ... ([130, pp. 22–23]).

We note that Lenin's observation has been proven by the history of the twentieth century, as described in the above statements. This means that the competitive equilibrium for the production economies in Chapter 6 is seriously insufficient for the equilibrium analysis, if we expect that the equilibrium is an approximation in one sense or another of the actual state of the economy.

Therefore, our first task is to generalize the price-taking behavior to incorporate the monopolistic behavior of the firms. This has been achieved by Negishi [168]. He assumed that each firm has its *subjective (perceived or expected inverse) demand* $q^t = \gamma_b(y_b^t, p^t)$, where the commodity t is assumed to be produced by the firm b, and p^t is the current market price of the commodity t. In normal situations, the firm would expect a downward-sloping inverse demand curve that obeys the law of demand, and Negishi specified the function γ_b to be linear with respect to the output quantity y^t, $q^t = a_b(p^t)y_b^t + d_b(p^t)$, where $a_b(p^t) \leq 0$ and $d_b(p^t) \geq 0$ for all p^t, and they satisfy

$$\sum_a x_a^t = \sum_b y_b^t + \sum_a \omega_a^t \text{ implies that } q^t = p^t,$$

which means that, in equilibrium, the firm must have a correct or consistent expectation of the prevailing market price.

In particular, if the firm has a constant expectation that $q^t = p^t$, then the firm is a price taker. Hence, Negishi's formulation of the monopolistic firms contains the competitive firms as a special case. See Figures 8.1(a) and 8.1(b).

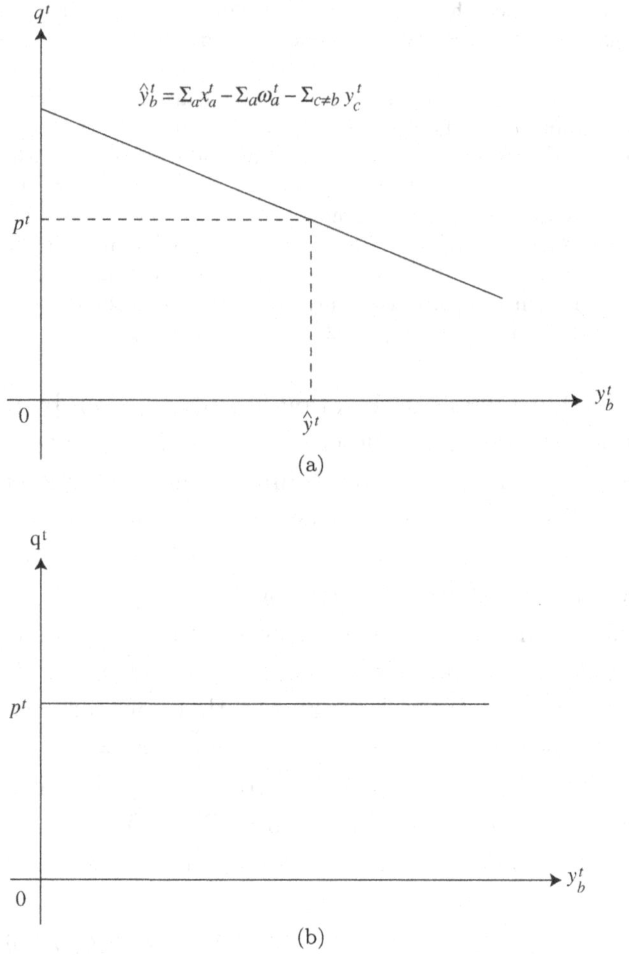

Figure 8.1. (a) Perceived demand of a monopolist. (b) Perceived demand of a competitive firm.

Generally speaking, we believe that the technologies of the monopolistic firms exhibit at least one of the following properties: (a) increasing returns to scale (Figure 6.1(c)), (b) large setup costs or fixed costs (Figure 8.2), and (c) production of differentiated commodities (see Sections 4.3 and 7.4).

The essential point is that, in cases of (a) increasing returns and (b) large setup costs, the convexity of the production sets will

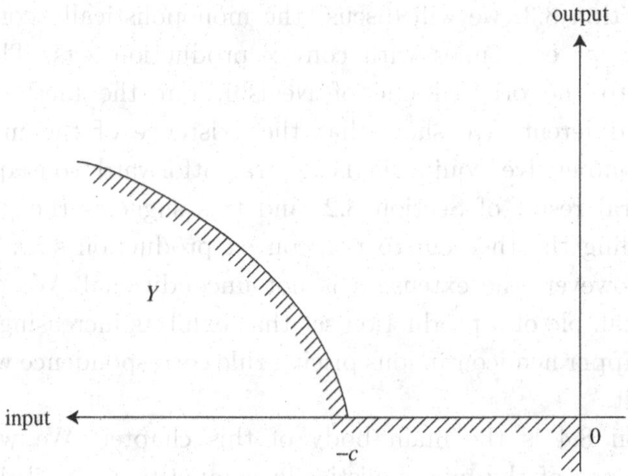

Figure 8.2. Setup cost.

be violated. Unfortunately, Negishi discussed monopolistic competition with convex production sets on a finite-dimensional commodity space. Therefore, none of the above cases are contained in his model.

In the following sections, we will discuss the Negishi-type monopolistically competitive equilibria with non-trivial setup costs, following Dehez *et al.* [58]. Section 8.2 will be devoted to developing the fundamental machinery to prove the existence of the monopolistically competitive equilibria. A basic tool is the pricing rule, which is a correspondence defined on the production set of each firm. It assigns to each (efficient) production vector a subset of the normal cone of the production set at that point. As the firms should maximize their profits at least locally even if their sets are not convex, the first-order condition at the equilibrium point is a natural condition to be stipulated. The pricing rule correspondence will also depend on the market price vector, so that Negishi's consistency condition explained above is naturally embodied in it. To apply the usual fixed point argument, the correspondence has to be non-empty, compact and convex-valued, and upper hemicontinuous. Most of our technical tasks involve guaranteeing these conditions on the pricing rule when the production sets are not convex.

In Section 8.3, we will discuss the monopolistically competitive equilibria for economies with convex production sets. The model is close to the original one of Negishi, but the method of the proof is different. We show that the existence of the monopolistically competitive equilibria is a straightforward consequence of the general result of Section 8.2, and this suggests the possibility of extending the theorem to non-convex production sets. Unfortunately, however, the extension is not unconditional. We present a simple example of a production set that exhibits increasing returns, and the upper hemicontinuous pricing rule correspondence would not exist on it.

Section 8.4 is the main body of this chapter. We will prove the existence of the monopolistically competitive equilibrium with large setup costs. The non-existence example of Section 8.3 indicates that we need an essential assumption to be imposed; for each firm, a commodity that is used as the setup cost is distinguished from the other commodities. As a consequence, the non-convex production set Y is a union of two convex sets, Y_1 and Y_2, where $Y = Y_1 \cup Y_2$. In Figure 8.3, the commodity z_1 is a variable input, z_2 is a fixed input that represents the setup cost and y is an output. To define the pricing rule correspondence on $Y = Y_1 \cup Y_2$, the most technically intricate part is the vertical face of Figure 8.3, which connects to $Y_1 \cap Y_2$. Generally speaking, the firm with a non-convex production set could not maximize the profit globally, as a supporting hyperplane would not exist. Therefore, in the equilibrium condition for such a firm, it is unavoidable that the global profit maximization is replaced by a local maximization. Moreover, if the production set is not convex, the profit can be negative even if the production set contains the origin; see Figure 7.1 in Chapter 7. Hence, the condition that the profits of the firms are non-negative should be added to the definition of the equilibrium.

To summarize: *Each firm has a downward sloping expected inverse demand. In equilibrium, each firm maximizes the expected profit locally and the profit is non-negative.*

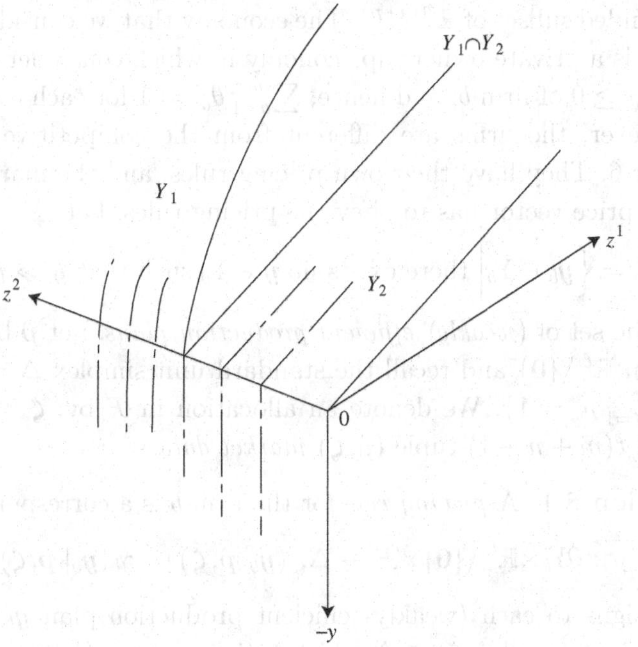

Figure 8.3. $Y = Y_1 \cup Y_2$.

8.2 Existence of Equilibria with Pricing Rules

As usual, we assume that ℓ commodities exist in the market, where $t = 1 \ldots \ell$. The consumption sector is standard enough. There exist m consumers indexed by $a = 1 \ldots m$ and the consumption set X_a of a is a closed and convex subset of $\mathbb{R}^\ell$ that is bounded from below. For every a, the preference relation $\prec_a \subset X_a \times X_a$ satisfies the basic postulates, (NTR), (CV), and (MT). The consumer a also has the initial endowment vector that satisfies (MI). There exist n firms in the economy indexed by $b = 1 \ldots n$. The production set Y_b of the firm b is a closed subset of $\mathbb{R}^\ell$ that satisfies (NP) and (FD). We also assume (BTP), which implies that the set of feasible allocations

$$\mathcal{F} = \left\{ (\boldsymbol{x}_a, \boldsymbol{y}_b) \in \prod_{a=1}^{m} X_a \times \prod_{b=1}^{n} Y_b \,\middle|\, \sum_{a=1}^{m} \boldsymbol{x}_a \leq \sum_{a=1}^{m} \omega_a + \sum_{b=1}^{n} \boldsymbol{y}_b \right\},$$

is a bounded subset of $\mathbb{R}^{\ell(m+n)}$. The economy that we consider in this chapter is a private ownership economy in which consumer a owns a share $\theta_{ab} \geq 0$ of firm b, and hence, $\sum_{a=1}^{m} \theta_{ab} = 1$ for each $b = 1 \ldots n$.

However, the firms are different from the competitive firms of Chapter 6. They have their own pricing rules, and the market equilibrium price vector has to obey the pricing rules. Let

$$\partial Y_b = \left\{ \boldsymbol{y}_b \in Y_b \middle| \text{ there exists no } \boldsymbol{y} \in Y_b \text{ such that } \boldsymbol{y} \gg \boldsymbol{y}_b \right\}.$$

∂Y_b is the set of (*weakly*) *efficient production plans*. Let $\hat{\boldsymbol{p}}$ be a price vector in $\mathbb{R}^{\ell}_{+} \backslash \{\boldsymbol{0}\}$ and recall the standard unit simplex $\Delta = \{(p^t) \in \mathbb{R}^{\ell}_{+} | \sum_{t=1}^{\ell} p^t = 1\}$. We denote an allocation in $\mathcal{F}$ by: $\hat{\boldsymbol{\zeta}} = (\hat{\boldsymbol{x}}_a, \hat{\boldsymbol{y}}_b)$ and call $\ell(m+n+1)$-tuple $(\hat{\boldsymbol{p}}, \hat{\boldsymbol{\zeta}})$ *market data*.

Definition 8.1. A *pricing rule* for the firm b is a correspondence

$$\gamma_b : \partial Y_b \times \mathbb{R}^{\ell}_{+} \backslash \{\boldsymbol{0}\} \times \mathcal{F} \to \Delta, \ (\boldsymbol{y}_b, \hat{\boldsymbol{p}}, \hat{\boldsymbol{\zeta}}) \mapsto \gamma_b(\boldsymbol{y}_b | \ \hat{\boldsymbol{p}}, \hat{\boldsymbol{\zeta}}),$$

that assigns to each (weakly) efficient production plan $\boldsymbol{y}_b \in \partial Y_b$ a set of prices $\gamma_b(\boldsymbol{y}_b | \ \hat{\boldsymbol{p}}, \hat{\boldsymbol{\zeta}}) \subset \Delta$ which satisfies, given the market data $(\hat{\boldsymbol{p}}, \hat{\boldsymbol{\zeta}})$, the following:

(PR-1): the correspondence $\gamma_b : \partial Y_b \times \mathbb{R}^{\ell}_{+} \times \mathcal{F} \to \Delta$ is non-empty, compact and convex-valued, and upper hemicontinuous,
(PR-2): $\boldsymbol{p}\boldsymbol{y}_b \geq 0$ for all $\boldsymbol{p} \in \gamma_b(\boldsymbol{y}_b | \ \hat{\boldsymbol{p}}, \hat{\boldsymbol{\zeta}})$.

The list $(X_a, \prec_a, \omega_a, \theta_{ab}, Y_b, \gamma_b)$ is called an *economy with the pricing rule* and is denoted by $\mathcal{E}_\gamma$. The fundamental equilibrium concept of this section reads as follows:

Definition 8.2. The market data $(\hat{\boldsymbol{p}}, \hat{\boldsymbol{x}}_a, \hat{\boldsymbol{y}}_b)$ are said to consist of an *equilibrium with the pricing rule* if and only if

(P-1) $\hat{\boldsymbol{p}}\hat{\boldsymbol{x}}_a \leq \hat{\boldsymbol{p}}\omega_a + \sum_{b=1}^{n} \theta_{ab}\hat{\boldsymbol{p}}\hat{\boldsymbol{y}}_b$ and
$\hat{\boldsymbol{x}}_a \succsim_a \boldsymbol{x}$ whenever $\hat{\boldsymbol{p}}\boldsymbol{x} \leq \hat{\boldsymbol{p}}\omega_a + \sum_{b=1}^{n} \theta_{ab}\hat{\boldsymbol{p}}\hat{\boldsymbol{y}}_b$, $a = 1 \ldots m$,
(P-2) $\hat{\boldsymbol{p}} \in \gamma_b(\hat{\boldsymbol{y}}_b | \ \hat{\boldsymbol{p}}, \hat{\boldsymbol{\zeta}})$, $b = 1 \ldots n$,
(P-3) $\sum_{a=1}^{m} \hat{\boldsymbol{x}}_a \leq \sum_{b=1}^{n} \hat{\boldsymbol{y}}_b + \sum_{a=1}^{m} \omega_a$.

The next result is the basis of our discussion of the monopolistically competitive equilibria.

Theorem 8.1. *Suppose that an economy* $\mathcal{E}_\gamma = (X_a, \prec_a, \omega_a, \theta_{ab}, Y_b, \gamma_b)$ *satisfies the basic postulates with (NTR), (CV), (MT) and (MI) for all* $a = 1 \ldots m$, *(NP) and (FD) for all* $b = 1 \ldots m$, *and (BTP). Then, there exists a competitive equilibrium* $(\hat{\boldsymbol{p}}, \hat{\boldsymbol{x}}_a, \hat{\boldsymbol{y}}_b)$ *for* $\mathcal{E}_\gamma$.

Proof. The idea of the proof is essentially the same as that of Theorem 7.1. As the set of feasible allocations $\mathcal{F}$ is non-empty and bounded by the (BTP) assumption, the feasible consumption and production sets, which are defined as the projections of $\mathcal{F}$ to X_a and Y_b, respectively, are bounded. Therefore, a closed cube $\hat{K}$ in $\mathbb{R}^\ell$ with the length $k > 0$ exists, centered at the origin and containing in its interior the feasible consumption and production plans, $\boldsymbol{x}_a, \boldsymbol{y}_b \subset int\hat{K}$ for $(\boldsymbol{x}_a, \boldsymbol{y}_b) \in \mathcal{F}$. Then, we define

$$\hat{X}_a = X_a \cap \hat{K},\ a = 1 \ldots m,$$

$$\hat{Y}_b = (Y_b + k\mathbf{1}) \cap \mathbb{R}^\ell_+, \text{ where } \mathbf{1} = (1 \ldots 1),\ b = 1 \ldots n.$$

Let Δ be the unit $\ell - 1$-simplex as usual and for each $\boldsymbol{r} = (r^t) \in \Delta$, we define the following:

$$\tau_b(\boldsymbol{r}) = \sup\{\tau > 0 \mid \tau\boldsymbol{r} \in \hat{Y}_b\},$$

$$h_b(\boldsymbol{r}) = \tau_b(\boldsymbol{r})\boldsymbol{r},$$

$$\partial\hat{Y}_b = \{h_b(\boldsymbol{r}) \in \hat{Y}_b \mid \boldsymbol{r} \in \Delta\},\ b = 1 \ldots n.$$

Let g denote the projection of points in $\mathbb{R}^\ell_+ \backslash \{0\}$ on Δ. Obviously, $g_b \equiv g|_{\partial\hat{Y}_b}$ and h_b are the inverses of each other, $h_b(\boldsymbol{r})$ is weakly efficient (up to $k\mathbf{1}$), or $h_b(\boldsymbol{r}) - k\mathbf{1} \in \partial Y_b$ for each $\boldsymbol{r} \in \Delta$ and we can take $k > 0$ sufficiently large that if $\boldsymbol{y}_b$ is a feasible production vector, then $\boldsymbol{y}_b + k\mathbf{1} \gg \mathbf{0}$; see Figure 7.2 of Chapter 7.

Now, we prove the following lemma.

Lemma 8.1. $g_b : \partial\hat{Y}_b \to \Delta$ *is a homeomorphism between* $\partial\hat{Y}_b$ *and* Δ *that satisfies that* $g_b(\boldsymbol{y}) \gg \mathbf{0}$ *if and only if* $\boldsymbol{y} \gg \mathbf{0}$.

Proof. The proof is the same as that of Lemma 7.1. □

Then, let us define a function η_b on Δ by $\eta_b(\boldsymbol{r}) = h_b(\boldsymbol{r}) - k\mathbf{1}$. The assumption (FD) ensures that $\eta_b(\boldsymbol{r}) \in \partial Y_b$ for all $\boldsymbol{r} \in \Delta$. Moreover, if $\eta_b(\boldsymbol{r})$ is a feasible production vector, then $h_b(\boldsymbol{r}) \gg \mathbf{0}$, and hence, $\boldsymbol{r} \gg \mathbf{0}$. Let the budget relation $\beta_a : \Delta^{n+1} \to \hat{X}_a$ and the quasi-demand relation $\hat{\phi}_a : \Delta^{n+1} \to \hat{X}_a$ be defined by

$$\beta_a(\boldsymbol{p}, \boldsymbol{r}_1 \dots \boldsymbol{r}_n) = \left\{ \boldsymbol{x} \in \hat{X}_a \,\middle|\, \boldsymbol{p}\boldsymbol{x} \leq \boldsymbol{p}\omega_a + \sum_{b=1}^{n} \theta_{ab}\boldsymbol{p}\eta_b(\boldsymbol{r}_b) \right\}$$

and $\hat{\phi}_a(\boldsymbol{p}, \boldsymbol{r}_1 \dots \boldsymbol{r}_n)$

$$= \begin{cases} \Big\{ \boldsymbol{x} \in \hat{X}_a \,|\, \boldsymbol{x} \in \beta_a(\boldsymbol{p}, \boldsymbol{r}_1 \dots \boldsymbol{r}_n) \text{ and } \boldsymbol{x} \succsim_a \boldsymbol{z} \\ \qquad \text{for all } \boldsymbol{z} \in \beta_a(\boldsymbol{p}, \boldsymbol{r}_1 \dots \boldsymbol{r}_n) \Big\} \\ \qquad \text{if } \inf \boldsymbol{p}\hat{X}_a < \boldsymbol{p}\omega_a + \sum_{b=1}^{n} \theta_{ab}\boldsymbol{p}\eta_b(\boldsymbol{r}_b), \\ \left\{ \boldsymbol{x} \in \hat{X}_a \,\middle|\, \boldsymbol{p}\boldsymbol{x} = \inf \boldsymbol{p}\hat{X}_a \right\} \text{ otherwise,} \end{cases}$$

respectively. By Proposition 2.4, the quasi-demand correspondence $\hat{\phi}_a$ is non-empty, compact and convex-valued, and upper hemicontinuous as its range $\hat{X}_a$ is compact. For each $b = 1 \dots n$, we define the correspondence $\hat{\psi}_b : \Delta^3 \to \Delta$ by

$$\hat{\psi}_b(\boldsymbol{p}, \boldsymbol{q}_b, \boldsymbol{r}_b) = \left(\frac{\max\{0, r_b^t + p^t - q_b^t\}}{\sum_{t=1}^{\ell} \max\{0, r_b^t + p^t - q_b^t\}} \right),$$

where $\boldsymbol{p} = (p^t)$, $\boldsymbol{q}_b = (q_b^t)$ and $\boldsymbol{r}_b = (r_b^t)$. Obviously, $\sum_{t=1}^{\ell} \max\{0, r_b^t + p^t - q_b^t\} \geq 1$, and hence, $\hat{\psi}_b$ is a continuous function. Here, $\boldsymbol{p}$ denotes a "market price" as opposed to $\boldsymbol{q}_b$, which denotes a "producer price."

The market price vector $\boldsymbol{p}$ is determined through the standard market correspondence $\mu : \prod_{a=1}^{m} \hat{X}_a \times \Delta^n \to \Delta$ defined by

$$\mu(\boldsymbol{x}_a, \boldsymbol{r}_b) = \left\{ \boldsymbol{p} \in \Delta \,\middle|\, \boldsymbol{r}\left(\sum_{a=1}^{m} \boldsymbol{x}_a - \sum_{b=1}^{n} \eta_b(\boldsymbol{r}_b) - \sum_{a=1}^{m} \omega_a \right) \right.$$
$$\left. \leq \boldsymbol{p}\left(\sum_{a=1}^{m} \boldsymbol{x}_a - \sum_{b=1}^{n} \eta_b(\boldsymbol{r}_b) - \sum_{a=1}^{m} \omega_a \right) \text{ for all } \boldsymbol{r} \in \Delta \right\}.$$

The continuity of η_b ensures that the correspondence μ is upper hemicontinuous with non-empty, compact, and convex values.

For each b, the producer price $\boldsymbol{q}_b$ is determined through the correspondence $\chi_b : \Delta^2 \times \prod_{a=1}^m \hat{X}_a \to \Delta$, which is defined by

$$\chi_b(\boldsymbol{r}_b, \boldsymbol{p}, \boldsymbol{x}_a) = \gamma_b(\eta_b(\boldsymbol{r}_b) |\, \boldsymbol{p}, \boldsymbol{x}_a, \eta_b(\boldsymbol{r}_b)).$$

As the correspondences γ_b are upper hemicontinuous with non-empty, compact, and convex values, and the η_b's are a continuous function, it follows from Proposition D.1 that the correspondence χ_b is upper hemicontinuous with values that are non-empty, compact, and convex.

Then, we can define a fixed point mapping that is simply the direct product of these correspondences

$$\Phi(\boldsymbol{p}, \boldsymbol{q}_b, \boldsymbol{r}_b, \boldsymbol{x}_a) = \mu(\boldsymbol{x}_a, \boldsymbol{r}_b) \times \prod_{b=1}^n \chi_b(\boldsymbol{r}_b, \boldsymbol{p}, \boldsymbol{x}_a)$$
$$\times \prod_{b=1}^n \hat{\psi}_b(\boldsymbol{p}, \boldsymbol{q}_b, \boldsymbol{r}_b) \times \prod_{a=1}^m \hat{\phi}_a(\boldsymbol{p}, \boldsymbol{r}_b).$$

Using Kakutani's fixed point theorem (Theorem D.1), the correspondence Φ has a fixed point. Let $(\hat{\boldsymbol{p}}, \hat{\boldsymbol{q}}_b, \hat{\boldsymbol{r}}_b, \hat{\boldsymbol{x}}_a)$ be the fixed point and define $\hat{\boldsymbol{y}}_b = \eta_b(\hat{\boldsymbol{r}}_b)$ and $\hat{\boldsymbol{z}} = \sum_{a=1}^m \hat{\boldsymbol{x}}_a - \sum_{a=1}^m \omega_a - \sum_{b=1}^n \hat{\boldsymbol{y}}_b$. Then, $\hat{\boldsymbol{y}}_b \in \partial Y_b$ for all b and the following conditions hold

$$\begin{aligned}
&\hat{\boldsymbol{r}}_b = \hat{\psi}_b(\hat{\boldsymbol{p}}, \hat{\boldsymbol{q}}_b, \hat{\boldsymbol{r}}_b),\ b = 1 \dots n,\\
&\hat{\boldsymbol{x}}_a \in \hat{\phi}_a(\hat{\boldsymbol{p}}, \hat{\boldsymbol{r}}_b),\ a = 1 \dots m,\\
&\boldsymbol{p}\hat{\boldsymbol{z}} \le \hat{\boldsymbol{p}}\hat{\boldsymbol{z}} \text{ for all } \boldsymbol{p} \in \Delta,\\
&\hat{\boldsymbol{q}}_b \in \chi_b(\hat{\boldsymbol{r}}_b, \hat{\boldsymbol{p}}, \hat{\boldsymbol{x}}_a),\ b = 1 \dots n.
\end{aligned}$$

Let us define $\hat{\sigma}_b = \sum_{t=1}^{\ell} \max\{0, \hat{r}_b^t + \hat{p}^t - \hat{q}_b^t\}$. Then, $\hat{\boldsymbol{r}}_b = \hat{\psi}_b(\hat{\boldsymbol{p}}, \hat{\boldsymbol{q}}_b, \hat{\boldsymbol{r}}_b)$ implies that

$$\hat{\sigma}_b \hat{r}_b^t \ge \hat{r}_b^t + \hat{p}^t - \hat{q}_b^t,\ t = 1 \dots \ell$$

for each b with the exact equality whenever $\hat{r}_b^t > 0$. Multiplying both sides of the above inequalities by $\hat{r}_b^t$ and summing over t, we have

$$(\hat{\sigma}_b - 1)\hat{\boldsymbol{r}}_b\hat{\boldsymbol{r}}_b = (\hat{\boldsymbol{p}} - \hat{\boldsymbol{q}}_b)\hat{\boldsymbol{r}}_b,$$

where $\hat{\sigma}_b \geq 1$ and $\hat{\boldsymbol{r}}_b\hat{\boldsymbol{r}}_b \geq 1/\ell$. Therefore, it follows that

$$(\hat{\boldsymbol{p}} - \hat{\boldsymbol{q}}_b)\hat{\boldsymbol{r}}_b \geq 0, \quad b = 1 \ldots n.$$

By the definition of h_b, there exists a $\hat{\tau}_b > 0$, such that

$$\hat{\tau}_b\hat{\boldsymbol{r}}_b = \hat{\boldsymbol{y}}_b + k\mathbf{1}, \quad b = 1 \ldots n.$$

Using the fact that $(\hat{\boldsymbol{p}}-\hat{\boldsymbol{q}}_b)\mathbf{1} = 0$, we obtain $\hat{\tau}_b(\hat{\boldsymbol{p}}-\hat{\boldsymbol{q}}_b)\hat{\boldsymbol{r}}_b = (\hat{\boldsymbol{p}}-\hat{\boldsymbol{q}}_b)\hat{\boldsymbol{y}}_b$, which, combined with $(\hat{\boldsymbol{p}} - \hat{\boldsymbol{q}}_b)\hat{\boldsymbol{r}}_b \geq 0$, implies that

$$\hat{\boldsymbol{p}}\hat{\boldsymbol{y}}_b \geq \hat{\boldsymbol{q}}_b\hat{\boldsymbol{y}}_b, \quad b = 1 \ldots n.$$

As $\hat{\boldsymbol{q}}_b \in \gamma_b(\hat{\boldsymbol{y}}_b |\ \hat{\boldsymbol{p}}, \hat{\boldsymbol{x}}_a, \hat{\boldsymbol{y}}_b)$, we have $\hat{\boldsymbol{q}}_b\hat{\boldsymbol{y}}_b \geq 0$. Hence, $\hat{\boldsymbol{p}}\hat{\boldsymbol{y}}_b \geq 0$ for all b and the minimum income assumption (MI) ensures that, for every a,

$$\hat{\boldsymbol{p}}\omega_a + \sum_{b=1}^{n} \theta_{ab}\hat{\boldsymbol{p}}\hat{\boldsymbol{y}}_b > \inf \boldsymbol{p}X_a.$$

From the definition of $\hat{\phi}_a$, we see that the budget inequalities

$$\hat{\boldsymbol{p}}\hat{\boldsymbol{x}}_a \leq \hat{\boldsymbol{p}}\omega_a + \sum_{b=1}^{n} \theta_{ab}\hat{\boldsymbol{p}}\hat{\boldsymbol{y}}_b$$

apply for all a. Summing over all a, we obtain $\hat{\boldsymbol{p}}\hat{\boldsymbol{z}} \leq 0$. As $\boldsymbol{p}\hat{\boldsymbol{z}} \leq \hat{\boldsymbol{p}}\hat{\boldsymbol{z}}$ for all $\boldsymbol{p} \in \Delta$, we set $\boldsymbol{p} = \boldsymbol{e}_s = (\delta_s^t)$, the sth elementary vector, $s = 1 \ldots \ell$. It follows that $\hat{\boldsymbol{z}} \leq \mathbf{0}$, and therefore, $(\hat{\boldsymbol{x}}_a, \hat{\boldsymbol{y}}_b) \in \mathcal{F}$. Consequently, we have $h_b(\hat{\boldsymbol{r}}_b) \gg \mathbf{0}$, and therefore, $\hat{\boldsymbol{r}}_b \gg \mathbf{0}$ by Lemma 8.1. Then, the inequality

$$\hat{\sigma}_b\hat{r}_b^t \geq \hat{r}_b^t + \hat{p}^t - \hat{q}_b^t, \ t = 1 \ldots \ell$$

yields $\hat{\sigma}_b = 1$ and $\hat{p}^t = \hat{q}_b^t$ for all $t = 1 \ldots \ell$ and for all $b = 1 \ldots n$. Therefore, the equilibrium condition (P-2), or $\hat{\boldsymbol{p}} \in \gamma_b(\hat{\boldsymbol{y}}_b |\ \hat{\boldsymbol{p}}, \hat{\boldsymbol{x}}_a, \hat{\boldsymbol{y}}_b)$ is established on $\hat{Y}_b$. As $\hat{\boldsymbol{p}}\omega_a + \sum_{b=1}^{n} \theta_{ab}\hat{\boldsymbol{p}}\hat{\boldsymbol{y}}_b > \inf \boldsymbol{p}X_a$ and $\hat{\boldsymbol{x}}_a \in \hat{\phi}_a(\hat{\boldsymbol{p}}, \hat{\boldsymbol{r}}_b)$ for all a, the condition (P-1) on $\hat{X}_a$ follows from the definition of $\hat{\phi}_a$. The usual limiting argument as in Theorem 2.4 or Theorem 6.2 will verify the conditions (P-1) and (P-2) on X_a and Y_b, respectively. Finally, we have already showed the condition (P-3) or $\hat{\boldsymbol{z}} \leq \mathbf{0}$. Hence, the proof of Theorem 8.1 is complete. □

8.3 Monopolistic Competition under Convex Technologies

The consumption sector is the same as that of Section 8.2. The consumer $a(= 1 \dots m)$ is characterized by the consumption set X_a, which is a convex and closed subset of $\mathbb{R}^\ell$ and it is assumed to be bounded from below. The preference relation $\prec_a \subset X_a \times X_a$ is assumed to satisfy the same conditions as in Section 8.2. The consumer a also has the initial endowment vector $\omega_a \in \mathbb{R}^\ell$. As before, we assume that for the relation between the consumption set X_a and the endowment vector ω_a, the condition (MI) holds.

For the production sector, we assume that each firm produces only one commodity, that there exist ℓ firms in the economy and that the firm t is assumed to produce the commodity $t(= 1, \dots, \ell)$. This specification is made for simplicity and is not essential; see the Notes for details. Then, the production set Y_t of the firm t is a closed and convex subset of $\mathbb{R}_- \times \dots \mathbb{R}_+ \times \dots \mathbb{R}_-$, where $\mathbb{R}_+$ is in the tth position. Below we will set up the production set Y_t which satisfies the (FD), (NP), and the (BTP) conditions.

Recall that a list $(\hat{\boldsymbol{p}}, \hat{\boldsymbol{x}}_a, \hat{\boldsymbol{y}}_t)$ of a price vector $\hat{\boldsymbol{p}}$ and a feasible allocation $\hat{\boldsymbol{\zeta}} = (\hat{\boldsymbol{x}}_a, \hat{\boldsymbol{y}}_t) \in \mathcal{F}$ is called the market data. The firm t is assumed to have a *perceived demand* function $\boldsymbol{p}_t(\boldsymbol{y}|\, \hat{\boldsymbol{p}}, \hat{\boldsymbol{\zeta}})$ defined as

$$p_t^t((y^t)|\, \hat{\boldsymbol{p}}, \hat{\boldsymbol{\zeta}}) = a_t(\hat{\boldsymbol{p}}, \hat{\boldsymbol{\zeta}}) y^t + d_t(\hat{\boldsymbol{p}}, \hat{\boldsymbol{\zeta}}),$$
$$p_t^s((y^t)|\, \hat{\boldsymbol{p}}, \hat{\boldsymbol{\zeta}}) = \hat{p}^s \text{ for } s \neq t,$$

where $a_t(\hat{\boldsymbol{p}}, \hat{\boldsymbol{\zeta}})$ is a real-valued function on $\mathbb{R}_+^\ell \times \mathcal{F}$, such that $a_t(\hat{\boldsymbol{p}}, \hat{\boldsymbol{\zeta}}) \leq 0$ for all $(\hat{\boldsymbol{p}}, \hat{\boldsymbol{\zeta}}) \in \mathbb{R}_+^\ell \times \mathcal{F}$.

The economic meaning of the perceived demand function is that the firm t expects to be able to sell the quantity y^t at the price p_t^t when the market data are $(\hat{\boldsymbol{p}}, \hat{\boldsymbol{\zeta}})$. Note that the firm t behaves competitively in the markets for the commodities other than t. In other words, it is a price taker in the markets for the input commodities such as factors of production or labor. This assumption is not necessary; see Section 8.5 for details. However, the assumption of a firm that behaves monopolistically as a seller but competitively as a buyer seems to be intuitively natural. The second term $d_t(\hat{\boldsymbol{p}}, \hat{\zeta})$ is assumed to ensure

(CN) (Consistency with observations): $p_t^t(\hat{\boldsymbol{y}}_t|\, \hat{\boldsymbol{p}}, \hat{\zeta}) \equiv \hat{p}^t$, $t = 1 \dots \ell$.

Furthermore, we require that the perceived demand function is homogeneous of degree 1 with respect to price and continuous,

(HG) (Homogeneity and continuity):

$$\begin{aligned} p_t^t(\boldsymbol{y}|\, \tau\hat{\boldsymbol{p}}, \hat{\zeta}) &= a_t(\tau\hat{\boldsymbol{p}}, \hat{\zeta})y^t + d_t(\tau\hat{\boldsymbol{p}}, \hat{\zeta}) \\ &= \tau a_t(\hat{\boldsymbol{p}}, \hat{\zeta})y^t + \tau d_t(\hat{\boldsymbol{p}}, \hat{\zeta}) = \tau p_t^t(\boldsymbol{y}|\, \hat{\boldsymbol{p}}, \hat{\zeta}) \end{aligned}$$

for all $\tau \geq 0$, and that $a_t(\hat{\boldsymbol{p}}, \hat{\zeta})$ and $d_t(\hat{\boldsymbol{p}}, \hat{\zeta})$ are continuous.

Then, the *expected profit* function of the firm t is given by

$$\pi_t(\hat{\boldsymbol{p}}, \hat{\zeta}) = a_t(\hat{\boldsymbol{p}}, \hat{\zeta})(y_t^t)^2 + d_t(\hat{\boldsymbol{p}}, \hat{\zeta})y_t^t + \sum_{s \neq t} \hat{p}^s y_t^s$$

and, from this, one obtains

$$\begin{aligned} \frac{\partial \pi_t(\hat{\boldsymbol{p}}, \hat{\zeta})}{\partial y_t^t} &= 2a_t(\hat{\boldsymbol{p}}, \hat{\zeta})y_t^t + d_t(\hat{\boldsymbol{p}}, \hat{\zeta}) \\ &= a_t(\hat{\boldsymbol{p}}, \hat{\zeta})y_t^t + p_t(\boldsymbol{y}_t|\, \hat{\boldsymbol{p}}, \hat{\zeta}) \equiv q^t. \end{aligned}$$

Let $\boldsymbol{q} = (\hat{p}^1 \dots \hat{p}^{t-1}, q^t, \hat{p}^{t+1} \dots \hat{p}^\ell)$. The first-order condition at $\boldsymbol{y}_t \in Y_t$ is that $\boldsymbol{q}\boldsymbol{y} \leq \boldsymbol{q}\boldsymbol{y}_t$ for all $\boldsymbol{y} \in Y_t$ sufficiently close to $\boldsymbol{y}_t$, or $\boldsymbol{q}$ is normal to Y_t at $\boldsymbol{y}_t$.

Let $N(\boldsymbol{y}, Y)$ be the normal cone of a set Y at $\boldsymbol{y} \in \partial Y$, which is defined by

$$N(\boldsymbol{y}, Y) = \{\boldsymbol{q} \in \mathbb{R}^\ell |\, \boldsymbol{q}\boldsymbol{z} \leq \boldsymbol{q}\boldsymbol{y} \text{ for all } \boldsymbol{z} \in Y\}.$$

If the set Y is convex, then $N(\boldsymbol{y}, Y) \neq \emptyset$ by the separation hyperplane theorem (Theorem A.1). $N(\boldsymbol{y}, Y)$ is a closed and convex cone with the vertex at the origin.

Let $\Gamma_t : \mathbb{R}_- \times \dots \mathbb{R}_+ \times \dots \mathbb{R}_- \to \mathbb{R}$ be a production function that is convex, continuously differentiable with $\Gamma_t(\mathbf{0}) = 0$ and satisfies

$D\Gamma_t(\boldsymbol{y}) = (\partial_1\Gamma_t(\boldsymbol{y})\ldots\partial_\ell\Gamma_t(\boldsymbol{y})) > \mathbf{0}$. Suppose that the tth firm's production set is defined as

$$Y_t = \{\boldsymbol{y} \in \mathbb{R}^\ell |\ \Gamma_t(\boldsymbol{y}) \leq 0\}.$$

As the function Γ_t is convex, so is Y_t as a set. We note that Y_t satisfies (NP) and (FD). The set of weakly efficient production plans is obviously $\partial Y_t = \{\boldsymbol{y} \in \mathbb{R}^\ell |\ \Gamma_t(\boldsymbol{y}) = 0\}$. As in Section 7.1, we also define the normalized gradient of $\Gamma_t(\boldsymbol{y})$ by

$$\partial\Gamma_t(\boldsymbol{y}) = \frac{D\Gamma_t(\boldsymbol{y})}{\sum_{s=1}^{\ell} \partial_s\Gamma_t(\boldsymbol{y})}.$$

Under these conditions, we obviously obtain $N(\boldsymbol{y}_t, Y_t) \cap \Delta = \{\partial\Gamma_t(\boldsymbol{y})\}$. Let $\hat{\boldsymbol{q}} = (\hat{q}^s) = \partial\Gamma_t(\boldsymbol{y})$. Then, by the homogeneity (HG) condition, the first-order condition is written as

$$p_t^t(\boldsymbol{y}|\ \tau\hat{\boldsymbol{p}}, \hat{\boldsymbol{\zeta}}) + a_t(\tau\hat{\boldsymbol{p}}, \hat{\boldsymbol{\zeta}})y_t^t = \nu_\tau \hat{q}_t,$$
$$p_t^s(\boldsymbol{y}|\ \tau\hat{\boldsymbol{p}}, \hat{\boldsymbol{\zeta}}) = \nu_\tau \hat{q}^s \ (s \neq t).$$

Summing over $s = 1\ldots\ell$, we have

$$\sum_{s=1}^{\ell} p_t^s(\boldsymbol{y}|\ \hat{\boldsymbol{p}}, \hat{\boldsymbol{\zeta}}) + a_t(\hat{\boldsymbol{p}}, \hat{\boldsymbol{\zeta}})y^t = \frac{\nu_\tau}{\tau}\sum_{s=1}^{\ell} \hat{q}_s = \frac{\nu_\tau}{\tau}.$$

To obtain $\sum_{s=1}^{\ell} p_t^s(\boldsymbol{y}|\ \hat{\boldsymbol{p}}, \hat{\boldsymbol{\zeta}}) = 1$, it is necessary that

$$\frac{\nu_\tau}{\tau} = 1 + a_t(\hat{\boldsymbol{p}}, \hat{\boldsymbol{\zeta}})y^t.$$

This yields the first-order conditions in the normalized form

$$p_t^t(\boldsymbol{y}|\ \hat{\boldsymbol{p}}, \hat{\boldsymbol{\zeta}}) = (1 + a_t(\hat{\boldsymbol{p}}, \hat{\boldsymbol{\zeta}})y_t^t)\hat{q}^t - a_t(\hat{\boldsymbol{p}}, \hat{\boldsymbol{\zeta}})y_t^t,$$
$$p_t^s(\boldsymbol{y}|\ \tau\hat{\boldsymbol{p}}, \hat{\boldsymbol{\zeta}}) = (1 + a_t(\hat{\boldsymbol{p}}, \hat{\boldsymbol{\zeta}})y_t^t)\hat{q}^s \ (s \neq t).$$

These equations can be written as

$$p_t^s(\boldsymbol{y}|\ \hat{\boldsymbol{p}}, \hat{\boldsymbol{\zeta}}) = (1 + a_t(\hat{\boldsymbol{p}}, \hat{\boldsymbol{\zeta}})y_t^t)\hat{q}^t - \delta_t^s a_t(\hat{\boldsymbol{p}}, \hat{\boldsymbol{\zeta}})y_t^t,\ s = 1\ldots\ell,$$

where $\delta_s^t = 1$ for $s = t$, and $\delta_s^t = 0$ otherwise.

For every $t(= 1 \ldots \ell)$, we define the pricing rule $\gamma_t : \partial Y_t \times \Delta \times \mathcal{F} \to \mathbb{R}^\ell_+$ by

$$\begin{aligned}\gamma_t(\boldsymbol{y}|\, \hat{\boldsymbol{p}}, \hat{\zeta}) &= (p^s)\\ &= \left(\hat{q}^s \max\{0, 1 + a_t(\hat{\boldsymbol{p}}, \hat{\zeta})y_t^t\} - \frac{\delta_t^s a_t(\hat{\boldsymbol{p}}, \hat{\zeta})y_t^t}{\max\{1, -a_t(\hat{\boldsymbol{p}}, \hat{\zeta})y_t^t\}}\right).\end{aligned}$$

Then, we can prove the following lemma.

Lemma 8.2. *Under the assumptions that $a_t(\hat{\boldsymbol{p}}, \hat{\zeta})$ is continuous and the production function Γ_t is convex with $\Gamma_t(\boldsymbol{0}) = 0$, and $D\Gamma_t(\boldsymbol{y}) > \boldsymbol{0}$ the pricing rule γ_t is a continuous function from $\partial Y_t \times \mathbb{R}^\ell_+ \times \mathcal{F}$ to Δ that satisfies the condition (PR-2). Moreover, γ_t satisfies that the first-order condition for the profit maximization or else some of the prices are non-positive (in fact zero).*

Proof. The continuity of $\gamma_t(\boldsymbol{y}|\, \hat{\boldsymbol{p}}, \hat{\zeta})$ is evident from the construction. Let $\boldsymbol{p} = (p^s) = \gamma_t(\boldsymbol{y}|\, \hat{\boldsymbol{p}}, \hat{\zeta})$. Then, by definition,

$$p^s = \hat{q}^s \max\{0, 1 + a_t(\hat{\boldsymbol{p}}, \hat{\zeta})y_t^t\} - \frac{\delta_t^s a_t(\hat{\boldsymbol{p}}, \hat{\zeta})y_t^t}{\max\{1, -a_t(\hat{\boldsymbol{p}}, \hat{\zeta})y_t^t\}}, \quad s = 1 \ldots \ell$$

for $\hat{\boldsymbol{q}} = (\hat{q}^s) = \partial\Gamma_t(\boldsymbol{y})$. If $1 + a_t(\hat{\boldsymbol{p}}, \hat{\zeta})y_t^t > 0$, then

$$p^s = (1 + a_t(\hat{\boldsymbol{p}}, \hat{\zeta})y_t^t)\hat{q}^s - \delta_t^s a_t(\hat{\boldsymbol{p}}, \hat{\zeta})y_t^t \geq 0, \; s = 1 \ldots \ell.$$

Summing over s,

$$\sum_{s=1}^{\ell} p^s = (1 + a_t(\hat{\boldsymbol{p}}, \hat{\zeta})y_t^t) \sum_{s=1}^{\ell} \hat{q}^s - a_t(\hat{\boldsymbol{p}}, \hat{\zeta})y_t^t = 1.$$

Similarly, if $1 + a_t(\hat{\boldsymbol{p}}, \hat{\zeta})y_t^t \leq 0$, then

$$p^s = \frac{-\delta_t^s a_t(\hat{\boldsymbol{p}}, \hat{\zeta})y_t^t}{-a_t(\hat{\boldsymbol{p}}, \hat{\zeta})y_t^t} = \delta_t^s \geq 0, \; s = 1 \ldots \ell,$$

therefore, $\sum_{s=1}^{\ell} p^s = 1$ or $\boldsymbol{p} \in \Delta$.

The profit function $\pi_t(\boldsymbol{y}|\, \hat{\boldsymbol{p}}, \hat{\boldsymbol{\zeta}})$ is calculated as

$$\begin{aligned}\pi_t(\boldsymbol{y}|\, \hat{\boldsymbol{p}}, \hat{\boldsymbol{\zeta}}) &= \boldsymbol{p}_t(\boldsymbol{y}|\, \hat{\boldsymbol{p}}, \hat{\boldsymbol{\zeta}})\boldsymbol{y}_t \\ &= \sum_{s=1}^{\ell}\left(\hat{q}^s y_t^s \max\{0, 1 + a_t(\hat{\boldsymbol{p}}, \hat{\boldsymbol{\zeta}})y_t^t\} - \frac{\delta_t^s a_t(\hat{\boldsymbol{p}}, \hat{\boldsymbol{\zeta}})y_t^t y_t^s}{\max\{1, -a_t(\hat{\boldsymbol{p}}, \hat{\boldsymbol{\zeta}})y_t^t\}}\right) \\ &\geq \max\{0, 1 + a_t(\hat{\boldsymbol{p}}, \hat{\boldsymbol{\zeta}})y_t^t\}\hat{\boldsymbol{q}}\hat{\boldsymbol{y}}_t \geq 0,\end{aligned}$$

as $a_t(\hat{\boldsymbol{p}}, \hat{\boldsymbol{\zeta}}) \leq 0$ and $\hat{\boldsymbol{q}}\boldsymbol{y}_t \geq 0$. This establishes $\pi_t(\boldsymbol{y}|\, \hat{\boldsymbol{p}}, \hat{\boldsymbol{\zeta}}) \geq 0$.

Finally, when $1 + a_t(\hat{\boldsymbol{p}}, \hat{\boldsymbol{\zeta}})y_t^t > 0$, then the pricing rule implements the first-order condition

$$p_t^s(\boldsymbol{y}|\, \hat{\boldsymbol{p}}, \hat{\boldsymbol{\zeta}}) = (1 + a_t(\hat{\boldsymbol{p}}, \hat{\boldsymbol{\zeta}})y_t^t)\hat{q}^t - \delta_t^s a_t(\hat{\boldsymbol{p}}, \hat{\boldsymbol{\zeta}})y_t^t \geq 0, \; s = 1 \ldots \ell.$$

When $1 + a_t(\hat{\boldsymbol{p}}, \hat{\boldsymbol{\zeta}})y_t^t \leq 0$, then

$$p_t^s(\boldsymbol{y}|\, \hat{\boldsymbol{p}}, \hat{\boldsymbol{\zeta}}) = 0 \text{ for } s \neq t.$$

This proves Lemma 8.2. □

Now, we state the definition of the monopolistically competitive equilibrium. Since we will assume that the production sets Y_t are convex for all $t = 1 \ldots \ell$, the profit maximizations are global, and it is ensured that the maximized profits are non-negative at the equilibria.

Definition 8.3. The market data $(\hat{\boldsymbol{p}}, \hat{\boldsymbol{x}}_a, \hat{\boldsymbol{y}}_t)$ are said to consist of a *monopolistically competitive equilibrium* if and only if

(MP-1) $\hat{\boldsymbol{p}}\hat{\boldsymbol{x}}_a \leq \hat{\boldsymbol{p}}\omega_a + \sum_{t=1}^{\ell} \theta_{at}\hat{\boldsymbol{p}}\hat{\boldsymbol{y}}_t$ and
$\hat{\boldsymbol{x}}_a \succsim_a \boldsymbol{x}$ whenever $\hat{\boldsymbol{p}}\boldsymbol{x} \leq \hat{\boldsymbol{p}}\omega_a + \sum_{t=1}^{\ell} \theta_{at}\hat{\boldsymbol{p}}\hat{\boldsymbol{y}}_t$, $a = 1 \ldots m$,
(MP-2) $\boldsymbol{p}_t(\boldsymbol{y}|\, \hat{\boldsymbol{p}}, \hat{\boldsymbol{\zeta}})\boldsymbol{y} \leq \hat{\boldsymbol{p}}\hat{\boldsymbol{y}}_t$ for all $\boldsymbol{y} \in Y_t$, $t = 1 \ldots \ell$,
(MP-3) $\sum_{a=1}^{m} \hat{\boldsymbol{x}}_a \leq \sum_{t=1}^{\ell} \hat{\boldsymbol{y}}_t + \sum_{a=1}^{m} \omega_a$.

The existence of the equilibrium is immediately obtained by Lemma 8.2 and the result of Section 8.2.

Theorem 8.2. *Assume all assumptions of Theorem 8.1. Suppose that the perceived demand function* $\boldsymbol{p}_t(\boldsymbol{y}|\, \hat{\boldsymbol{p}}, \hat{\boldsymbol{\zeta}})$ *and the production function* $\Gamma_t(\boldsymbol{y})$ *are described as above. Then, there exists a monopolistically competitive equilibrium.*

Proof. The proof follows from Theorem 8.1 and Lemma 8.2. □

Theorem 8.2 generalizes Theorem 6.2 in the way that it implements the monopolistic behavior of the firms. However, it is not fully satisfactory as it assumes the convexity of the production sets. As pointed out in Section 8.1, in reality, monopolistically competitive firms would undertake production activities under non-convex production technologies. However, the following example indicates that not all non-convex production sets are compatible with the pricing rule verifying (PR-1) and (PR-2) with the first-order condition as in Lemma 8.2.

Example 8.1. There are two commodities, an output y and an input z, with the prices p and r, respectively. The production set is defined by

$$Y = \{(y, z) \in \mathbb{R}_+ \times \mathbb{R}_- \mid y \leq \max\{0, -z - c\}\};$$

see Figure 8.4.

Note that $y \leq 0$ when $z \geq -c$, and $y \leq -z - c$ when $z \leq -c$. The production function $\Gamma(y, z) = y - \max\{0, -z - c\}$ is not smooth at $z = -c$ or $z = 0$, but they are irrelevant for the following analysis. The production set Y represents a technology that includes a setup cost c.

The perceived inverse demand function of the firm is defined by

$$\begin{aligned} p(y, z \mid \hat{\boldsymbol{p}}, \hat{\boldsymbol{\zeta}}) &= \hat{p} + b(y - \hat{y}),\ b < 0, \\ r(y, z \mid \hat{\boldsymbol{p}}, \hat{\boldsymbol{\zeta}}) &= \hat{r}, \end{aligned}$$

given $\hat{\boldsymbol{p}} = (\hat{p}, \hat{r})$ and $\hat{\boldsymbol{\zeta}} = (\hat{\boldsymbol{x}}_a, \hat{y}, \hat{z})$. Then, the profit function is calculated as

$$\pi(y, z \mid \hat{\boldsymbol{p}}, \hat{\boldsymbol{\zeta}}) = \begin{cases} \hat{r}z & \text{if } z \geq -c, \\ (\hat{p} + b(y - \hat{y}))y - \hat{r}(y + c) & \text{otherwise.} \end{cases}$$

Then, the first-order condition for the profit maximization, given $z \leq -c$, is given by

$$\frac{d\pi}{dy} = p(y, z \mid \hat{\boldsymbol{p}}, \hat{\boldsymbol{\zeta}}) + by - \hat{r} = 0,\ p = \hat{r} - by.$$

This yields the normalized price vector $\boldsymbol{p} = (p, r)$ given by

$$p = \frac{1 - by}{2}, \; r = \frac{1 + by}{2}.$$

The corresponding profits are

$$\pi = \left(\frac{1 - by}{2}\right) y - \left(\frac{1 + by}{2}\right)(y + c) = -by^2 - \frac{1 + by}{2} c.$$

Therefore, we see that

$$\pi \geq 0 \text{ if and only if } \hat{y} \equiv \frac{bc - \sqrt{b^2c^2 - 8bc}}{4b} \leq y \leq -\frac{1}{b}.$$

For example, if $b = -1$, then $0 \leq \hat{y} \leq 1$ for $c \leq 1$.

We wish to construct, on the boundary ∂Y of Y, a pricing rule γ, which is upper hemicontinuous (PR-1) and yields non-negative profits (PR-2) and is such that $p, r \geq 0$ or else $pr \leq 0$; see Lemma 8.2. We will show that a contradiction arises at the point $(\hat{y}, -\hat{y} - c)$, labeled a in Figure 8.4.

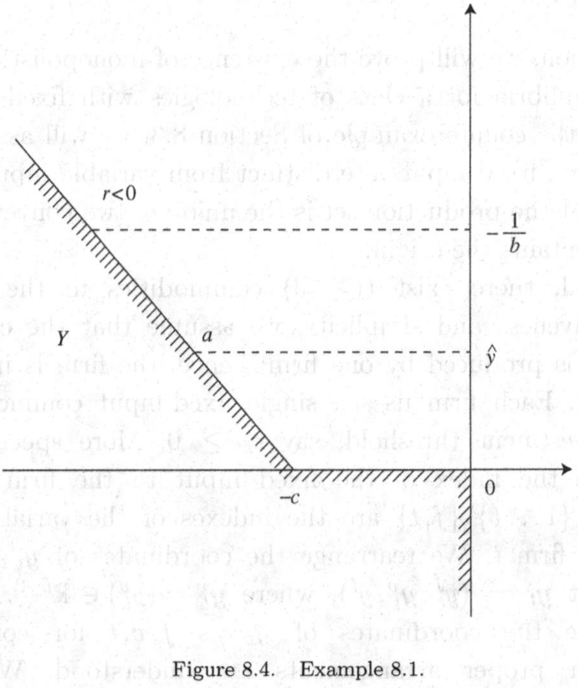

Figure 8.4. Example 8.1.

For $y \in [\hat{y}, -1/b]$, the first-order condition imposes the unique prices given by $p(\hat{y}) = (1 - b\hat{y})2^{-1}$ and $r(\hat{y}) = (1 + b\hat{y})2^{-1}$. In particular, $p(\hat{y}), r(\hat{y}) > 0$ for $\hat{y} < -1/b$. On the other hand, for $0 \leq y < \hat{y}$, the first-order condition yields negative profits, and hence, the conditions of Lemma 8.2 impose $pr \leq 0$ with $p \geq 0$ and $r \leq 0$.

Let $y_n \to \hat{y}$ as $n \to +\infty$ and $y^n < \hat{y}$ for all n. Take $(p_n, r_n) \in \gamma(y_n)$, so that $p^n \geq 0$ and $r_n \leq 0$. If $\gamma(y)$ is upper hemicontinuous, there exists a (not necessarily unique) limit $(\bar{p}, \bar{r})$, such that $\bar{p} \geq 0$ and $\bar{r} \leq 0$. As $\gamma(y)$ is upper hemicontinuous, compact, and convex-valued at $\hat{y}$, it follows that $(p_T, r_T) = \tau(p(\hat{y}), r(\hat{y})) + (1 - \tau)(\bar{p}, \bar{r}) \in \gamma(\hat{y})$ for all $\tau \in [0, 1]$. However, then $(p_T, r_T) \gg (0, 0)$ for τ close to 1. For $\tau \neq 1$, we have a strictly positive vector $(p_T, r_T) \in \gamma(\hat{y})$, which is different from $(p(\hat{y}), (\hat{y}))$. Hence, (p_T, r_T) does not obey the first-order condition. This is a contradiction and we cannot construct a pricing rule with the desired properties in this example.

8.4 Monopolistic Competition with Fixed Costs

In this section, we will prove the existence of monopolistically competitive equilibria for a class of technologies with fixed costs. On account of the counterexample of Section 8.3, we will assume that, for each firm, fixed inputs are distinct from variable inputs or outputs, so that the production set is the union of two convex sets, one of which contains the origin.

As usual, there exist $\ell(\geq 3)$ commodities in the economy. For definitiveness and simplicity, we assume that the commodity $t(= 1 \ldots \ell)$ is produced by one firm; hence, the firm is indexed by $t(= 1 \ldots \ell)$. Each firm uses a single fixed-input commodity with a fixed investment threshold, say $c_t \geq 0$. More specifically, let $f(\neq t)$ be the index of the fixed input to the firm t; hence, $v \in \boldsymbol{v} \equiv \{1 \ldots \ell\} \backslash \{f, t\}$ are the indexes of the variable inputs. Consider a firm t. We rearrange the coordinates of $\boldsymbol{y}_t \in \mathbb{R}^\ell$ and denote that $\boldsymbol{y}_t = (y_t^f, \boldsymbol{y}_t^{\boldsymbol{v}}, y_t^t)$, where $\boldsymbol{y}_t^{\boldsymbol{v}} = (y_t^v) \in \mathbb{R}^{\ell-2}$. Although we indicate the coordinates of $\boldsymbol{y}_t$ as $f, \boldsymbol{v}, t$ for convenience, below their proper arrangements are understood. We assume that the production set Y_t of the firm t is written in the following form

$$Y_t = Y_t^1 \cup Y_t^2,$$
$$Y_t^1 = \{\boldsymbol{y}_t = (y_t^f, \boldsymbol{y}_t^{\boldsymbol{v}}, y_t^t) |\ y_t^f \leq -c_t, \Gamma_t^1(\boldsymbol{y}_t^{\boldsymbol{v}}, y_t^t) \leq 0\},$$
$$Y_t^2 = \{\boldsymbol{y}_t = (y_t^f, \boldsymbol{y}_t^{\boldsymbol{v}}, y_t^t) |\ -c_t \leq y_t^f \leq 0, \Gamma_t^2(\boldsymbol{y}_t^{\boldsymbol{v}}, y_t^t) \leq 0\},$$

where $\Gamma_t^1, \Gamma_t^2 : \mathbb{R}_-^{\ell-2} \times \mathbb{R}_+ \to \mathbb{R}$ are twice differentiable, convex, and monotone functions with $\Gamma_t^i(\mathbf{0}, 0) = 0$, $i = 1, 2$. Then, the sets $\hat{Y}_t^i = \{(\boldsymbol{y}_t^{\boldsymbol{v}}, y_t^t) \in \mathbb{R}^{\ell-1} |\ \Gamma_t^i(\boldsymbol{y}_t^{\boldsymbol{v}}, y_t^t) \leq 0\}$ are closed and convex subsets of $\mathbb{R}^{\ell-1}$, and satisfy $\hat{Y}_t^i + \mathbb{R}_-^{\ell-1} \subset \hat{Y}_t^i$, $\hat{Y}_t^i \cap \mathbb{R}_+^{\ell-1} = \{\mathbf{0}\}$, $i = 1, 2$. We also assume that $\Gamma_t^1(\boldsymbol{y}^{\boldsymbol{v}}, y^t) < \Gamma_t^2(\boldsymbol{y}^{\boldsymbol{v}}, y^t)$ for all $(\boldsymbol{y}^{\boldsymbol{v}}, y^t)$. This reflects the fact that a production plan that is feasible without fixed investment remains feasible with fixed investment; see Figure 8.3.

The set of weakly efficient production plans or the boundary ∂Y_t of Y_t can be described as follows;

$$\begin{aligned}\partial Y_t = &\{\boldsymbol{y}_t = (y_t^f, \boldsymbol{y}_t^{\boldsymbol{v}}, y_t^t) |\ y_t^f < -c_t, \Gamma_t^1(\boldsymbol{y}_t^{\boldsymbol{v}}, y_t^t) = 0\}\\ &\cup\{\boldsymbol{y}_t = (y_t^f, \boldsymbol{y}_t^{\boldsymbol{v}}, y_t^t) |\ y_t^f = -c_t, \Gamma_t^1(\boldsymbol{y}_t^{\boldsymbol{v}}, y_t^t) \leq 0, \Gamma_t^2(\boldsymbol{y}_t^{\boldsymbol{v}}, y_t^t) \geq 0\}\\ &\cup\{\boldsymbol{y}_t = (y_t^f, \boldsymbol{y}_t^{\boldsymbol{v}}, y_t^t) |\ -c_t < y_t^f < 0, \Gamma_t^2(\boldsymbol{y}_t^{\boldsymbol{v}}, y_t^t) = 0\}\\ &\cup\{\boldsymbol{y}_t = (y_t^f, \boldsymbol{y}_t^{\boldsymbol{v}}, y_t^t) |\ y_t^f = 0, \Gamma_t^2(\boldsymbol{y}_t^{\boldsymbol{v}}, y_t^t) \leq 0\}.\end{aligned}$$

To define a pricing rule $\gamma_t : \partial Y_t \times \Delta \times \mathcal{F} \to \Delta$, we can rely on Lemma 8.2 for the first and the third sets of the union defining ∂Y_t. However, we have to extend that specification to cover the second and fourth sets and preserve the upper hemicontinuity at $y_t^f = -c_t$ and 0.

The definition of the *monopolistically competitive equilibrium with fixed costs* can be obtained by replacing the condition (MP-2) of Definition 8.3 with the following:

(MP-2′) there exists a neighborhood U of $\hat{\boldsymbol{y}}_t$, such that $\boldsymbol{p}_t(\boldsymbol{y}|\hat{\boldsymbol{p}}, \hat{\zeta})\boldsymbol{y} \leq \hat{\boldsymbol{p}}\hat{\boldsymbol{y}}_t$ for all $\boldsymbol{y} \in U \cap Y_t$, and $\hat{\boldsymbol{p}}\hat{\boldsymbol{y}}_t \geq 0$, $t = 1 \ldots \ell$.

As the production sets Y_t are not convex, we can expect, at best, the local profit maximization for each firm. The non-negativity of the profits is not obvious, and we add it into the definition. The fundamental existence theorem of this chapter now reads as follows:

Theorem 8.3. *Let the production set Y_t be described as above. Under the assumptions of Theorem 8.2 but the convexity of Y_t (the convexity of Γ_t), there exists a monopolistically competitive equilibrium.*

Proof. The proof of Theorem 8.3 relies on the following lemma.

Lemma 8.3. *Under the assumptions that $a_t(\hat{\boldsymbol{p}}, \hat{\boldsymbol{\zeta}})$, which was defined in Section 8.3, is continuous and that the production set Y_t is defined as above, there exists a pricing rule $\gamma_t : \partial Y_t \times \Delta \times \mathcal{F} \to \Delta$ satisfying the conditions (PR-1) and (PR-2). Moreover, γ_t satisfies the first-order condition for the profit maximization, or else some of the prices are zero (non-positive).*

Proof. The proof is constructive. A desired pricing rule is defined successively for all $\boldsymbol{y}_t = (y_t^f, \boldsymbol{y}_t^v, y_t^t) \in \partial Y_t$, such that

1. $-c_t < y_t^f < 0$,
2. $y_t^f < -c_t$,
3. $y_t^f = -c_t$,
4. $y_t^f = 0$.

The proof applies to an arbitrary firm, so we omit the subscript t (but the superscript t). Similarly, we often omit explicit reference to $(\hat{\boldsymbol{p}}, \hat{\boldsymbol{\zeta}})$. Let $\gamma^i(\boldsymbol{y}^v, y^t)$ $(i = 1, 2)$ be the pricing rules defined in Lemma 8.2 with $\hat{\boldsymbol{q}} = \partial\Gamma^i(\boldsymbol{y}^v, y^t) \in \mathbb{R}^{\ell-1}$.

1. When $-c < y^f < 0$, then $\Gamma^2(\boldsymbol{y}^v, y^t) = 0$. We set

$$\gamma((y^f, \boldsymbol{y}^v, y^t) | -c < y^f < 0, \Gamma^2(\boldsymbol{y}^v, y^t) = 0)$$
$$= \left\{ (p^f, \boldsymbol{p}^v, p^t) \in \Delta \,\middle|\, p^f = 0, (\boldsymbol{p}^v, p^t) = \gamma^2(\boldsymbol{y}^v, y^t) \right\}.$$

As Lemma 8.2 applies to $\gamma^2(\boldsymbol{y}^v, y^t)$, it also applies to $\gamma(y^f, \boldsymbol{y}^v, y^t)$.

2. When $y^f < -c$, then $\Gamma^1(\boldsymbol{y^v}, y^t) = 0$ and we define

$$\gamma((y^f, \boldsymbol{y^v}, y^t)|\, y^f < -c, \Gamma^1(\boldsymbol{y^v}, y^t) = 0)$$
$$= \left\{(p^f, \boldsymbol{p^v}, p^t) \in \Delta \,\middle|\, p^f = 0, (\boldsymbol{p^v}, p^t) = \gamma^1(\boldsymbol{y^v}, y^t)\right\}.$$

Again, Lemma 8.2 applies to $\gamma(y^f, \boldsymbol{y^v}, y^t)$, as defined above.

3. When $y^f = -c$, then $\Gamma^1(\boldsymbol{y^v}, y^t) \leq 0$ and $\boldsymbol{y} = (y^f, \boldsymbol{y^v}, y^t)$ is efficient if and only if $\Gamma^1(\boldsymbol{y^v}, y^t) = 0$. Otherwise, the first-order conditions cannot be satisfied at strictly positive price vectors, and $\boldsymbol{y}$ cannot be an equilibrium production plan when the preferences are (strictly) monotone. Then, we set $p^f = 0$ when $\Gamma^1(\boldsymbol{y^v}, y^t) < 0$ and extend the rule appropriately. To do so, it is convenient to distinguish three subcases.

3.1. Suppose that $\Gamma^1(\boldsymbol{y^v}, y^t) = 0$. Then, we set $(\boldsymbol{p^v}, p^t) = (1 - p^f)(\hat{\boldsymbol{p}}^{\boldsymbol{v}}, \hat{p}^t)$ for $(\hat{\boldsymbol{p}}^{\boldsymbol{v}}, \hat{p}^t) = \gamma^1(\boldsymbol{y^v}, y^t)$ and $p^f \in [0, (1/c)(\boldsymbol{p^v y^v} + p^t y^t)]$. This is equivalent to setting

$$p^f \in \left[0, \frac{\hat{\boldsymbol{p}}^{\boldsymbol{v}}\boldsymbol{y^v} + \hat{p}^t y^t}{c + \hat{\boldsymbol{p}}^{\boldsymbol{v}}\boldsymbol{y^v} + \hat{p}^t y^t}\right],$$

and $(\boldsymbol{p^v}, p^t) = (1 - p^f)(\hat{\boldsymbol{p}}^{\boldsymbol{v}}, \hat{p}^t)$, which defines the correspondence

$$\gamma((y^f, \boldsymbol{y^v}, y^t)|\, y^f = -c, \Gamma^1(\boldsymbol{y^v}, y^t) = 0)$$
$$= \left\{(p^f, \boldsymbol{p^v}, p^t) \in \Delta \,\middle|\, p^f \in \left[0, \frac{\hat{\boldsymbol{p}}^{\boldsymbol{v}}\boldsymbol{y^v} + \hat{p}^t y^t}{c + \hat{\boldsymbol{p}}^{\boldsymbol{v}}\boldsymbol{y^v} + \hat{p}^t y^t}\right], (\boldsymbol{p^v}, p^t) = (1 - p^f)(\hat{\boldsymbol{p}}^{\boldsymbol{v}}, \hat{p}^t)\right\},$$

which is upper hemicontinuous by Propositions D.1, D.2, and D.5, and non-empty, compact, and convex-valued. The convex values are obvious from the construction. As $p^f \leq (1/c)(\boldsymbol{p^v y^v} + p^t y^t)$, it follows that $\boldsymbol{py} \geq 0$ for each $\boldsymbol{p} = (p^f, \boldsymbol{p^v}, p^t) = \gamma(\boldsymbol{y}|\, \hat{\boldsymbol{p}}, \hat{\boldsymbol{\zeta}})$, which verifies the condition (PR-2). Hence, Lemma 8.2 applies to this case.

3.2. Suppose that $\Gamma^1(\boldsymbol{y^v}, y^t) < 0$ and $\Gamma^2(\boldsymbol{y^v}, y^t) = 0$. Then, we define

$$\begin{aligned}\gamma((y^f, \boldsymbol{y^v}, y^t)|\, y^f &= -c, \Gamma^2(\boldsymbol{y^v}, y^t) = 0)\\ &= \left\{(p^f, \boldsymbol{p^v}, p^t) \in \Delta \,\middle|\, p^f = 0, (\boldsymbol{p^v}, p^t) = \gamma^2(\boldsymbol{y^v}, y^t)\right\}.\end{aligned}$$

Again, Lemma 8.2 applies.

3.3. For the case in which $\Gamma^1(\boldsymbol{y^v}, y^t) < 0$ and $\Gamma^2(\boldsymbol{y^v}, y^t) > 0$, the construction of the pricing rule is more intricate.

Given $\boldsymbol{y} = (-c, \boldsymbol{y^v}, y^t)$, we take $d_1 > 0$ and $d_2 > 0$, such that $\Gamma^1(\boldsymbol{y^v} + d_1\mathbf{1}, y^t + d_1) = 0$ and $\Gamma^2(\boldsymbol{y^v} - d_2\mathbf{1}, y^t - d_2) = 0$, where $\mathbf{1} = (1\ldots1) \in \mathbb{R}^{\ell-2}$. Such a d_1 and d_2 exist by the continuity and monotonicity of Γ^1 and Γ^2, and obviously, they are unique. Clearly, they are continuous as functions of $(\boldsymbol{y^v}, y^t)$. Then, we define

$$\begin{aligned}\gamma((y^f, \boldsymbol{y^v}, y^t)|\, y^f &= -c,\ \Gamma^1(\boldsymbol{y^v}, y^t) < 0,\ \Gamma^2(\boldsymbol{y^v}, y^t) > 0)\\ &= \Bigg\{(p^f, \boldsymbol{p^v}, p^t) \in \Delta \,\Bigg|\, p^f = 0,\\ (\boldsymbol{p^v}, p^t) &= \frac{d_2}{d_1 + d_2}\gamma^1(\boldsymbol{y^v} + d_1\mathbf{1}, y^t + d_1)\\ &\quad + \frac{d_1}{d_1 + d_2}\gamma^2(\boldsymbol{y^v} - d_2\mathbf{1}, y^t - d_2)\Bigg\}.\end{aligned}$$

That is to say, $\gamma((y^f, \boldsymbol{y^v}, y^t)|\, y^f = -c, \Gamma^1(\boldsymbol{y^v}, y^t) < 0, \Gamma^2(\boldsymbol{y^v}, y^t) > 0)$ is a vector $(0, \boldsymbol{p^v}, p^t)$, where $(\boldsymbol{p^v}, p^t)$ is a convex combination of $\gamma^1(\boldsymbol{y^v} + d_1\mathbf{1}, y^t + d_1)$ and $\gamma^2(\boldsymbol{y^v} - d_2\mathbf{1}, y^t - d_2)$ for d_1 and d_2, as defined above. Obviously, if $(\boldsymbol{y^v_n}, y^t_n) \to (\hat{\boldsymbol{y}}^{\boldsymbol{v}}, \hat{y}^t)$ with $\Gamma^1(\hat{\boldsymbol{y}}^{\boldsymbol{v}}, \hat{y}^t) = 0$, then $\gamma(-c, \boldsymbol{y^v_n}, y^t_n) \to \gamma^1(\hat{\boldsymbol{y}}^{\boldsymbol{v}}, \hat{y}^t)$, and if $(\boldsymbol{y^v_n}, y^t_n) \to (\tilde{\boldsymbol{y}}^{\boldsymbol{v}}, \tilde{y}^t)$ with $\Gamma^2(\tilde{\boldsymbol{y}}^{\boldsymbol{v}}, \tilde{y}^t) = 0$, then $\gamma(-c, \boldsymbol{y^v_n}, y^t_n) \to \gamma^2(\tilde{\boldsymbol{y}}^{\boldsymbol{v}}, \tilde{y}^t)$. The continuity comes from that of γ^1 and γ^2. Therefore, (PR-1) is verified.

To verify (PR-2), note that it follows from Lemma 8.2 that $\hat{\boldsymbol{p}}^{\boldsymbol{v}}_i(\boldsymbol{y^v} + \sigma_i d_i\mathbf{1}) + \hat{p}^t_i(y^t + \sigma_i d_i) \geq 0$ for $(\hat{\boldsymbol{p}}^{\boldsymbol{v}}_i, \hat{p}^t_i) = \gamma^i(\boldsymbol{y^v} + \sigma_i d_i\mathbf{1}, y^t +$

$\sigma_i d_i)$, $i = 1, 2$, where $\sigma_1 = 1$ and $\sigma_2 = -1$. Hence,

$$\begin{aligned} \boldsymbol{py} = \boldsymbol{p^v y^v} + p^t y^t &= \frac{d_2}{d_1 + d_2} \\ &\quad \times \left(\hat{\boldsymbol{p}}_1^{\boldsymbol{v}}(\boldsymbol{y^v} + d_1\mathbf{1} - d_1\mathbf{1}) + \hat{p}_1^t(y^t + d_1 - d_1)\right) \\ &\quad + \frac{d_1}{d_1 + d_2}\left(\hat{\boldsymbol{p}}_2^{\boldsymbol{v}}(\boldsymbol{y^v} - d_2\mathbf{1} + d_2\mathbf{1}) + \hat{p}_2^t(y^t - d_2 + d_2)\right) \\ &\geq \frac{-d_1 d_2}{d_1 + d_2} + \frac{d_1 d_2}{d_1 + d_2} = 0. \end{aligned}$$

Therefore, (PR-2) is verified and, on account of $p^f = 0$, Lemma 8.3 holds for this case. It should also be noted that Case 3.2 is a special case of Case 3.3, which could also have been defined for $\Gamma^2(\boldsymbol{y^v}, y^t) = 0$.

4. When $y^f = 0$ and $\Gamma^2(\boldsymbol{y^v}, y^t) \leq 0$, it is convenient to distinguish two subcases again.

4.1. For $\Gamma^2(\boldsymbol{y^v}, y^t) = 0$, let

$$\begin{aligned} &\gamma((y^f, \boldsymbol{y^v}, y^t)|\, y^f = 0, \Gamma^2(\boldsymbol{y^v}, y^t) = 0) \\ &\quad = \left\{(p^f, \boldsymbol{p^v}, p^t) \in \Delta \,\middle|\, p^f \in [0,1], (\boldsymbol{p^v}, p^t) = (1 - p^f)\gamma^2(\boldsymbol{y^v}, y^t)\right\}. \end{aligned}$$

From the argument put forth out for Case 3.1, Lemma 8.2 applies to this correspondence.

4.2. For $(\boldsymbol{y^v}, y^t)$ with $\Gamma^2(\boldsymbol{y^v}, y^t) < 0$, let

$$\begin{aligned} &\gamma((y^f, \boldsymbol{y^v}, y^t)|\, y^f = 0, \Gamma^2(\boldsymbol{y^v}, y^t) < 0) \\ &\quad = \left\{\boldsymbol{p} = (p^f, \boldsymbol{p^v}, p^t) \in \Delta \,\middle|\, p^f = 1, (\boldsymbol{p^v}, p^t) = (0, 0)\right\}. \end{aligned}$$

Lemma 8.2 now applies trivially.

The correspondence $\gamma(\boldsymbol{y})$ defined on the seven regions of Figure 8.5 labeled 1, 2, 3.1, 3.2, 3.3, 4.1, and 4.2 satisfies the conditions of Lemma 8.2. Accordingly, it satisfies (PR-2) and is non-empty, compact, and convex-valued everywhere. To verify upper hemicontinuity at the common boundaries of these seven regions, we note the following:

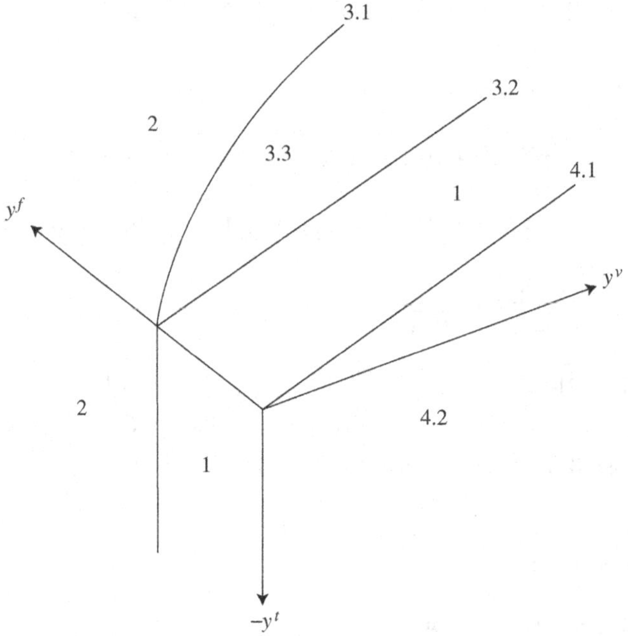

Figure 8.5. Lemma 8.3.

(1) The relevant connections (the common boundaries) concern 1 and 3.2, 1 and 4.1, 2 and 3.1, 3.1 and 3.3, 3.2 and 3.3, and 4.1 and 4.2.
(2) On 1 and 3.2, the pricing rules are identical,
(3) 1 connects to 4.1 for $p^f = 0$ in 4.1,
(4) 2 connects to 3.1 for $p^f = 0$ in 3.1,
(5) 3.1 connects to 3.3 and, hence, to 3.2 for $p^f = 0$ in 3.1 with $d_1 = 0$ in 3.3,
(6) 3.2 connects to 3.3 with $d_2 = 0$ in 3.3,
(7) 4.1 connects to 4.2 for $p^f = 1$ in 4.1.

This completes the proof of Lemma 8.3. □

The proof of Theorem 8.3 follows from Theorem 8.1 and Lemma 8.3.

8.5 Notes

Section 8.1: For a general survey of theory of monopolistic competition, the readers can consult Bénnasy [24]. See also Friedman [73].

Section 8.2: Theorem 8.1 follows Dehez and Dreze [56].

Section 8.3: As explained in the Introduction, an equilibrium theory in which firms behave monopolistically based on their perceived demand function was originally developed by Negishi [168]. Theorem 8.2 was first obtained by Negishi, but his proof did not use the pricing rule approach. As a consequence, Negishi's proof is simpler than ours, but it is difficult to see the means to extend the proof to the case of non-convex production sets.

Section 8.4: Theorem 8.3 is based on Dehez *et al.* [58]. They did not distinguish the input and output commodities and formulated the perceived demand map of the firm $j = 1 \ldots n$ as a vector-valued function

$$\boldsymbol{p}_j(\boldsymbol{y}_j | \hat{\boldsymbol{p}}, \hat{\boldsymbol{\zeta}}) = H(\hat{\boldsymbol{p}}, \hat{\boldsymbol{\zeta}})\boldsymbol{y}_j + K(\hat{\boldsymbol{p}}, \hat{\boldsymbol{\zeta}}),$$

where $H(\hat{\boldsymbol{p}}, \hat{\boldsymbol{\zeta}})$ is a negative semi-definite $\ell \times \ell$-matrix and $K(\hat{\boldsymbol{p}}, \hat{\boldsymbol{\zeta}})$ is an ℓ-vector. Our proof in the text has been significantly simplified compared with the original proof of [58] because we assume the production sets are defined by smooth production functions, and hence, every normal cone is a singleton.

Appendices

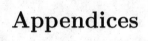

Appendix A

Convex Sets and Functions

We start with the notion of a *set* as a primitive concept and denote it as $X = \{x, y, z, \dots\}$ or $X = \{x|$ the properties that x has$\}$. "$x \in X$" means that x is an *element* of the set X, or x is contained in X. Suppose that if $x \in X$, then $x \in Y$. Then, we say that X is a *subset* of Y or the set Y includes the set X, and is denoted by $X \subset Y$. The *union* $X \cup Y$ and the *intersection* $X \cap Y$ are defined by

$$X \cup Y = \{x|\ x \in X \text{ or } x \in Y\} \text{ and } X \cap Y = \{x|\ x \in X \text{ and } x \in Y\},$$

respectively. For the family of sets $\{X_\alpha|\ \alpha \in \mathscr{A}\}$, where $\mathscr{A}$ is an arbitrary set, the union and the intersection $\cup_{\alpha\in\mathscr{A}} X_\alpha$ and $\cap_{\alpha\in\mathscr{A}} X_\alpha$ are understood in the same way. The *empty set* $\emptyset$ is the set that contains nothing, and we postulate $\emptyset \subset X$ for every set X. Let X be a subset of a set Y. The *complement* of X in Y is defined by $Y \backslash X = \{x \in Y|\ x \notin X\}$. Sometimes, we denote it by X^c when the set Y is understood. De Morgan's laws $(\cup_{\alpha\in\mathscr{A}} X_\alpha)^c = \cap_{\alpha\in\mathscr{A}} X_\alpha^c$ and $(\cap_{\alpha\in\mathscr{A}} X_\alpha)^c = \cup_{\alpha\in\mathscr{A}} X_\alpha^c$ are well known.

Let X and Y be sets. The set of ordered pairs $\{(x, y)|\ x \in X,\ y \in Y\}$ is called the *direct product* of X and Y and is denoted by $X \times Y$. The product of n sets, $X_1 \dots X_n$ is defined inductively by $\prod_{i=1}^{n} X_i = (\prod_{i=1}^{n-1} X_i) \times X_n$. The subset $\mathcal{R}$ of the product $X \times Y$ is called a *binary relation* or simply a relation of X to Y. $(x, y) \in \mathcal{R}$ is often written as $x\mathcal{R}y$. If $X = Y$, then the binary relation $\mathcal{R} \subset X \times X$ is called a *relation* on X.

A relation $\sim$ on X that satisfies

(Reflexivity) $x \sim x$ for every $x \in X$,
(Symmetricity) $x \sim y$ implies $y \sim x$ for all $x, y \in X$,
(Transitivity) $x \sim y$ and $y \sim z$ imply $x \sim z$ for all $x, y, z \in X$,

is called an *equivalence relation.* For $x \in X$, the set $[x] = \{y \in X|\ y \sim x\}$ is called an *equivalent class* represented by x. Each element belongs to a unique equivalent class, for if $x \in [y] \cap [z]$, $y \not\sim z$, then $y \sim z$ follows from $x \sim y$ and $x \sim z$ by transitivity, a contradiction. Therefore, the set X is decomposed into the set of disjoint equivalent classes denoted by $X \backslash \sim$ called the *quotient space.*

A relation on X that satisfies reflexivity and transitivity is called a *preorder relation.* The preference relation introduced in Section 2.2 is an example of a preorder relation. A preorder relation $\geq$ that satisfies

(Anti-symmetricity) $x \geq y$ and $y \geq x$ imply $x = y$

is called an *order relation.*

Let X and Y be sets and $f \subset X \times Y$ a relation of X to Y. If for every $x \in X$, there exists a unique $y \in Y$ such that $(x, y) \in f$, we call the relation f a *map(ping)* or a *function* from X to Y and denote it as $f : X \to Y$, $x \mapsto y$. In this case, the set X is called a *domain* and the set Y the *range.* We often denote the function as simply $y = f(x)$. A function $f : X \to Y$ is called *injective* or *one to one* if $x \neq y$ implies $f(x) \neq f(y)$. It is called *surjective* or *onto* if for every $y \in Y$, there exists an $x \in X$ with $y = f(x)$. When a map f is one to one and onto, it is called *bijective* and the bijective map is also called a *bijection.* The set $f(X) = \{y \in Y|\ f(x) = y \text{ for some } x \in X\}$ is called the *image* of a map f, and for a set $Z \subset Y$, the set $f^{-1}(Z) = \{x \in X|\ f(x) \in Z\}$ is called the *inverse image* of Z by f. A map f is injective if and only if $f^{-1}(\{y\})$ is a singleton (one-point set) or $\emptyset$ for every $y \in Y$, and it is surjective if and only if $f(X) = Y$. It is well known that there exists a bijection between the set of natural numbers $\mathbb{N}$ and the set of rational numbers $\mathbb{Q}$, in other words, $\mathbb{Q}$ is a *countable set.* However, there does not exist any onto map from $\mathbb{N}$ to $\mathbb{R}$, or the set $\mathbb{R}$ is *uncountable.* It is easy to verify the next proposition.

Proposition A.1. *Let X and Y be sets, and $\{X_\alpha\}$ and $\{Y_\alpha\}$ be families of subsets of X and Y, respectively. For a map $f : X \to Y$, we have*

(i) $f(\cup_\alpha X_\alpha) = \cup_\alpha f(X_\alpha)$, $f(\cap_\alpha X_\alpha) \subset \cap_\alpha f(X_\alpha)$,
(ii) $f^{-1}(\cup_\alpha X_\alpha) = \cup_\alpha f^{-1}(X_\alpha)$, $f^{-1}(\cap_\alpha X_\alpha) = \cap_\alpha f^{-1}(X_\alpha)$.

A counterexample for the equality in the second relation of (*i*) is given by $X = \{1, 2\}$, $Y = \{3, 4\}$, $X_1 = \{1\}$, and $X_2 = \{2\}$, $f(1) = f(2) = 3$. Since $X_1 \cap X_2 = \emptyset$, one has $f(X_1 \cap X_2) = \emptyset$, but $f(X_1) \cap f(X_2) = \{3\}$. Therefore, $f(X_1 \cap X_2) \neq f(X_1) \cap f(X_2)$.

For a subset X of the real line $\mathbb{R}$, $a \in \mathbb{R}$ with $a \leq x$ for all $x \in X$ is called a *lower bound* of X. The greatest lower bound is called an *infimum* and denoted $\inf X$. $a = \inf X$ is characterized as (i) $a \leq x$ for all $x \in X$, and (ii) if $c \leq x$ for all $x \in X$, then $c \leq a$. Similarly, $b \in \mathbb{R}$ with $b \geq x$ for all $x \in X$ is called an *upper bound.* The least upper bound or *supremum*, $\sup X$, is analogously defined as $\inf X$. $\sup\{x, -x\}$ is called the *absolute value* of x and denoted by $|x|$. The triangle inequality $|x + y| \leq |x| + |y|$ is well known.

The ℓ-times product of $\mathbb{R}$, $\mathbb{R} \times \cdots \times \mathbb{R}$ is denoted $\mathbb{R}^\ell$ and called the *ℓ-dimensional Euclidean space.* Hence, an element of $\mathbb{R}^\ell$ is an ℓ-tuple of real numbers $\boldsymbol{x} = (x^1 \dots x^\ell)$, $x^t \in \mathbb{R}$, $t = 1 \dots \ell$. The elements of $\mathbb{R}^\ell$ are also called *ℓ-vectors* or simply vectors. We shall sometimes denote $\mathbf{0} = (0 \dots 0)$ and $\mathbf{1} = (1 \dots 1)$. Let $\boldsymbol{e}_t$(or sometimes $\mathbf{1}_t$) $= (0 \dots 0, 1, 0 \dots 0)$ be a vector that has all 0 components but 1 at the tth coordinate. It is called the tth *elementary (or unit) vector.*

The space $\mathbb{R}^\ell$ has the order relation $\geq$ on it, which is defined by

$$(x^1 \dots x^\ell) \geq (y^1 \dots y^\ell) \text{ if and only if } x^t \geq y^t \text{ for all } t = 1 \dots \ell$$

for all $\boldsymbol{x} = (x^1 \dots x^\ell), \boldsymbol{y} = (y^1 \dots y^\ell) \in \mathbb{R}^\ell$. We will use the following related notations. $\boldsymbol{x} > \boldsymbol{y}$ if and only if $\boldsymbol{x} \geq \boldsymbol{y}$ and $\boldsymbol{x} \neq \boldsymbol{y}$. $\boldsymbol{x} \gg \boldsymbol{y}$ if and only if $x^t > y^t$ for all $t = 1 \dots \ell$. The *positive orthant* of $\mathbb{R}^\ell$ is defined by $\mathbb{R}^\ell_+ = \{\boldsymbol{x} \in \mathbb{R}^\ell | \, \boldsymbol{x} \geq \mathbf{0}\}$.

The addition and the scalar product of vectors are defined, respectively, by $(x^1 \dots x^\ell) + (y^1 \dots y^\ell) = (x^1 + y^1 \dots x^\ell + y^\ell)$ and $a(x^1 \dots x^\ell) = (ax^1 \dots ax^\ell)$ for all $\boldsymbol{x} = (x^1 \dots x^\ell), \boldsymbol{y} = (y^1 \dots y^\ell) \in \mathbb{R}^\ell$

and for all $a \in \mathbb{R}$. By these operations, the space $\mathbb{R}^\ell$ becomes a *vector space* (cf. Appendix E).

A finite set of vectors $\{\boldsymbol{x}_1 \ldots \boldsymbol{x}_n\}$ is said to be *linearly independent* if for every $a_i \in \mathbb{R}$, $i = 1 \ldots n$, $\sum_{i=1}^n a_i \boldsymbol{x}_i = \mathbf{0}$ implies $a_1 = \cdots = a_n = 0$. It can be easily seen that every $\boldsymbol{x} \in \mathbb{R}^\ell$ is written uniquely as $\boldsymbol{x} = \sum_{i=1}^\ell a_i \boldsymbol{x}_i$ by any set of ℓ linearly independent vectors $\{\boldsymbol{x}_1 \ldots \boldsymbol{x}_\ell\}$, or they consist of a *basis* of $\mathbb{R}^\ell$. Obviously, the set of elementary vectors $\{\boldsymbol{e}_1 \ldots \boldsymbol{e}_\ell\}$ is linearly independent, hence a basis of $\mathbb{R}^\ell$.

A map $f : \mathbb{R}^\ell \to \mathbb{R}^m$, $\boldsymbol{x} \mapsto \boldsymbol{y} = f(\boldsymbol{x})$ is called a *linear map* if it satisfies $f(\boldsymbol{x}+\boldsymbol{y}) = f(\boldsymbol{x})+f(\boldsymbol{y})$ for all $\boldsymbol{x}, \boldsymbol{y} \in \mathbb{R}^\ell$, and $f(a\boldsymbol{x}) = af(\boldsymbol{x})$ for all $a \in \mathbb{R}$ and $\boldsymbol{x} \in \mathbb{R}^\ell$. Let $f : \mathbb{R}^\ell \to \mathbb{R}^m$ be a linear map and write $\boldsymbol{x} = (x^s)$ and $\boldsymbol{y} = (y^t) = f(\boldsymbol{x})$. Then by the linearity,

$$(y^t) = \boldsymbol{y} = f(\boldsymbol{x}) = f\left(\sum_{s=1}^{\ell} x^s \boldsymbol{e}_s\right) = \sum_{s=1}^{\ell} x^s f(\boldsymbol{e}_s) = \left(\sum_{s=1}^{\ell} a_s^t x^s\right),$$

where $f(\boldsymbol{e}_s) = (f^t(\boldsymbol{e}_s)) = (a_s^t) \in \mathbb{R}^m$, $s = 1 \ldots \ell$. Setting $A = (a_s^t)$, the above equation can be written simply by the *matrix* form $\boldsymbol{y} = A\boldsymbol{x}$. Given an $\ell \times m$-matrix $A = (a_s^t)$, its *transpose matrix* is defined by an $m \times \ell$-matrix $A' = (a_t^s)$.

For $\boldsymbol{p} = (p^t) \in \mathbb{R}^\ell$ and $\boldsymbol{x} = (x^t) \in \mathbb{R}^\ell$, we define their *inner product* by $\boldsymbol{p}\boldsymbol{x} = \sum_{t=1}^\ell p^t x^t$. For the mathematical consistency, we should denote it as $\boldsymbol{p}'\boldsymbol{x}$, where $\boldsymbol{p}'$ is the transpose of $\boldsymbol{p}$, considering $\boldsymbol{p}$ as an $\ell \times 1$-matrix. However, we will always identify finite-dimensional vectors with their transposed counterparts for notational simplicity. We will discuss more on the linear spaces and maps in Appendices E and G.

A subset X of $\mathbb{R}^\ell$ is said to be *convex* if and only if for every $\boldsymbol{x}, \boldsymbol{y} \in X$ and for every $0 \leq \tau \leq 1$, $\tau\boldsymbol{x} + (1-\tau)\boldsymbol{y} \in X$. Clearly, if $X_1 \ldots X_m \subset X$ are convex, then $\prod_{i=1}^m X_i$, $\sum_{i=1}^m X_i$ and $\cap_{i=1}^m X_i$ are also convex. Let S be a subset of $\mathbb{R}^\ell$. The *convex hull* of the set S is defined by the intersection of all convex sets that contain S, and it is denoted by coS. It is the smallest convex subset that contains S. The following theorem (due to Minkowski) is known as the *separation hyperplane theorem.*

Theorem A.1. *Let X and Y be non-empty convex subsets of $\mathbb{R}^\ell$. If $X \cap Y = \emptyset$, then there exists a vector $\boldsymbol{p} \in \mathbb{R}^\ell$ with $\boldsymbol{p} \neq \boldsymbol{0}$ such that $\boldsymbol{px} \leq \boldsymbol{py}$ for every $\boldsymbol{x} \in X$ and every $\boldsymbol{y} \in Y$.*

A set $C \subset \mathbb{R}^\ell$ is said to be a *cone* if and only if for every $\boldsymbol{x} \in C$ and every $\tau \geq 0$, $\tau\boldsymbol{x} \in C$. For a set $X \subset \mathbb{R}^\ell$, the *polar* of X denoted by $\mathscr{P}_X$ or $\mathscr{P}(X)$ is defined by $\mathscr{P}_X = \{\boldsymbol{p} \in \mathbb{R}^\ell |\ \boldsymbol{px} \leq 0 \text{ for every } \boldsymbol{x} \in X\}$.

Proposition A.2. *Let C be a cone of $\mathbb{R}^\ell$ that is closed and convex. Then, the polar $\mathscr{P}_C$ is a closed and convex cone such that $\mathscr{P}(\mathscr{P}_C) = C$.*

A real-valued function $f : X \to \mathbb{R}$, where $X \subset \mathbb{R}^\ell$ is a convex set, is said to be *concave* if for every $\boldsymbol{x}, \boldsymbol{y} \in X$ and every $0 \leq \tau \leq 1$,

$$\tau f(\boldsymbol{x}) + (1 - \tau)f(\boldsymbol{y}) \leq f(\tau\boldsymbol{x} + (1 - \tau)\boldsymbol{y}).$$

f is called *convex* if the function $-f$ is concave. It can be easily seen that for a convex function $f : X \to \mathbb{R}$, the set $\{\boldsymbol{x} \in X |\ f(\boldsymbol{x}) \leq a\}$ is convex for every $a \in \mathbb{R}$.

Theorem A.2. *A concave function is continuous on its relative interior of the domain.*

The concepts of the *continuity* and the *relative interior* (or *relative topology*) will be explained in Appendix B. An *n-simplex* in $\mathbb{R}^\ell$ denoted s_n is the convex hull of $n + 1$ points $\{\boldsymbol{x}_0 \ldots \boldsymbol{x}_n\}$ in $\mathbb{R}^\ell$, such that $\boldsymbol{x}_1 - \boldsymbol{x}_0 \ldots \boldsymbol{x}_n - \boldsymbol{x}_0$ are linearly independent. The points $\boldsymbol{x}_i$ are called the vertices of the n-simplex s_n. If the vertices of s_n have been given a specific order, then s_n is called an *ordered simplex*. Note that linear independence does not depend upon the designation of $\boldsymbol{x}_0$ by the following proposition.

Proposition A.3. *Let $\{\boldsymbol{x}_0 \ldots \boldsymbol{x}_n\} \subset \mathbb{R}^\ell$. Then, $\boldsymbol{x}_1 - \boldsymbol{x}_0 \ldots \boldsymbol{x}_n - \boldsymbol{x}_0$ are linearly independent if and only if $a_i = b_i$ for $i = 0 \ldots n$ whenever $\sum_{i=0}^n a_i\boldsymbol{x}_i = \sum_{i=0}^n b_i\boldsymbol{x}_i$ and $\sum_{i=0}^n a_i = \sum_{i=0}^n b_i$. In this case, the points $\boldsymbol{x}_0 \ldots \boldsymbol{x}_n$ are called independent.*

Proof. Suppose that $\boldsymbol{x}_1 - \boldsymbol{x}_0, \ldots, \boldsymbol{x}_n - \boldsymbol{x}_0$ are linearly independent. If $\sum_{i=0}^n a_i\boldsymbol{x}_i = \sum_{i=0}^n b_i\boldsymbol{x}_i$ and $\sum_{i=0}^n a_i = \sum_{i=0}^n b_i$, then

$0 = \sum_{i=0}^{n}(a_i - b_i)\boldsymbol{x}_i = \sum_{i=0}^{n}(a_i - b_i)\boldsymbol{x}_i - \sum_{i=0}^{n}(a_i - b_i)\boldsymbol{x}_0 = \sum_{i=1}^{n}(a_i - b_i)(\boldsymbol{x}_i - \boldsymbol{x}_0)$. It follows from the linear independence of $\boldsymbol{x}_1-\boldsymbol{x}_0, \ldots, \boldsymbol{x}_n-\boldsymbol{x}_0$ that $a_i = b_i$ for $i = 1\ldots n$. As $\sum_{i=0}^{n}(a_i-b_i) = 0$, $a_0 = b_0$. Conversely, suppose that $a_i = b_i$ for $i = 0, \ldots, n$ whenever $\sum_{i=0}^{n} a_i\boldsymbol{x}_i = \sum_{i=0}^{n} b_i\boldsymbol{x}_i$ and $\sum_{i=0}^{n} a_i = \sum_{i=0}^{n} b_i$. If $\sum_{i=1}^{n} b_i(\boldsymbol{x}_i - \boldsymbol{x}_0) = 0$, then $\sum_{i=1}^{n} b_i\boldsymbol{x}_i = \sum_{i=1}^{n} b_i\boldsymbol{x}_0$; hence, $b_i = 0$ for $i = 1, \ldots, n$. □

By definition, the simplex s_n is a convex hull of $n+1$ points $\boldsymbol{x}_0, \ldots, \boldsymbol{x}_n$ that are independent, and Proposition A.3 ensures the *standard representation* of the n-simplex

$$s_n = \left\{ \sum_{i=0}^{n} a_i\boldsymbol{x}_i \,\middle|\, \sum_{i=0}^{n} a_i = 1,\ a_i \geq 0,\ i = 0, \ldots, n \right\}.$$

In this case, we shall often identify s_n with the set

$$\Delta_n = \left\{ (a_0 \ldots a_n) \in \mathbb{R}^{n+1} \,\middle|\, \sum_{i=0}^{n} a_i = 1,\ a_i \geq 0,\ i = 0, \ldots, n \right\},$$

which is called the *standard n-simplex*, or *unit simplex*. In this book, the unit simplex Δ_n is often denoted simply by Δ when the dimension n is understood.

Knaster, Kuratowski and Mazurkiewicz proved the following theorem, known as the *K–K–M Lemma*.

Theorem A.3. *(K–K–M) Let $A = \{a_0, a_1 \ldots a_r\}$ be a set of $r+1$ points in $\mathbb{R}^\ell$. Let $\{S_0, S_1 \ldots S_r\}$ be a set of $r+1$ closed subsets of $\mathbb{R}^\ell$, and $I = \{0, 1 \ldots r\}$. Suppose that for all $J \subset I$, the convex hull of $\{a_i\}_{i\in J}$ is contained in $\cup_{i\in J}S_i$, that is, $co\{a_i\}_{i\in J} \subset \cup_{i\in J}S_i$. Then, $\cap_{i\in I}S_i \neq \emptyset$.*

The definition of the *closed sets* will be given in Appendix B. Let $A = \{1, 2, \ldots N\}$ be a finite set and $\mathcal{A} = \{C \subset A \mid C \neq \emptyset\}$ be the set of non-empty subsets of A. We say that a family $\mathcal{B}$ of subsets of A is *balanced* if there exist non-negative weights w_C for $C \in \mathcal{B}$, such that $\sum_{C\in\mathcal{B}_a} w_C = 1$ for every $a \in A$, where $\mathcal{B}_a = \{C \in \mathcal{B} \mid a \in C\}$ is the family of sets in $\mathcal{B}$ that contain $a \in A$. For each $C \in \mathcal{B}$, we denote the *unit simplex spanned by the set C* by $\Delta^C = co\{\boldsymbol{e}_a \mid a \in C\}$, where $\boldsymbol{e}_a$

is the ath elementary vector. The generalized K–K–M lemma which is due to Shapley [231] now reads as follows:

Theorem A.4. *Let $\{F_C\}_{C \in \mathcal{A}}$ be a family of closed subsets of Δ^A, indexed by the members of $\mathcal{A}$, such that for every $D \in \mathcal{A}$, $\Delta^D \subset \cup\{F_C | \; C \subset D\}$. Then, there exists a balanced family $\mathcal{B}$ for which $\cap\{F_C | \; C \in \mathcal{B}\} \neq \emptyset$.*

We conclude this appendix by the fundamental theorem of non-linear programming.

Theorem A.5 (Kuhn–Tucker). *Let $f, g_1 \ldots g_m$ be real-valued concave functions defined on a convex set X in $\mathbb{R}^\ell$. Suppose that the Slater's condition holds, namely that*

$$\textit{there exists an } \boldsymbol{x}_0 \in X \textit{ such that } g_j(\boldsymbol{x}_0) > 0, \; j = 1 \ldots m.$$

Let $\hat{\boldsymbol{x}}$ be a point which achieves a maximum of $f(\boldsymbol{x})$ on X subject to $g_j(\boldsymbol{x}) \geq 0$, $j = 1 \ldots m$. Then, there exists a non-negative vector $\hat{\boldsymbol{\lambda}} = (\hat{\lambda}_1 \ldots \hat{\lambda}_m) \in \mathbb{R}^m_+$, such that

$$f(\boldsymbol{x}) + \hat{\boldsymbol{\lambda}} g(\boldsymbol{x}) \leq f(\hat{\boldsymbol{x}}) + \hat{\boldsymbol{\lambda}} g(\hat{\boldsymbol{x}}) \leq f(\hat{\boldsymbol{x}}) + \boldsymbol{\lambda} g(\hat{\boldsymbol{x}})$$

for every $\boldsymbol{x} \in X$ and every non-negative $\boldsymbol{\lambda} \in \mathbb{R}^m_+$.

In the above inequalities, we set $g(\boldsymbol{x}) = (g_1(\boldsymbol{x}) \ldots g_m(\boldsymbol{x}))$. In Theorem A.5, the function $\mathcal{L}(\boldsymbol{x}, \boldsymbol{\lambda}) = f(\boldsymbol{x}) + \boldsymbol{\lambda} g(\boldsymbol{x})$ is called a *Lagrangian*, the vector $\boldsymbol{\lambda}$ is the *Lagrangian multiplier*, and the point $(\hat{\boldsymbol{x}}, \hat{\boldsymbol{\lambda}}) \in X \times \mathbb{R}^m_+$ is called a *saddle point* of the Lagrangian $\mathcal{L}(\boldsymbol{x}, \boldsymbol{\lambda})$.

Appendix B

General Topology

The purpose of this appendix is to clarify the terminology and present the general topology results that are used in the text. We will not provide proofs, because they are available in standard textbooks with which students of mathematical economics would be familiar. Let X be a set. A family $\mathscr{G}$ of subsets of X is said to be *topology* on X if it satisfies

(Tp-1) $X, \emptyset \in \mathscr{G}$,
(Tp-2) if $G_1, G_2 \in \mathscr{G}$, then $G_1 \cap G_2 \in \mathscr{G}$,
(Tp-3) if $G_\alpha \in \mathscr{G}$, for all $\alpha \in \mathscr{A}$, then $\cup_{\alpha \in \mathscr{A}} G_\alpha \in \mathscr{G}$,

where $\mathscr{A}$ is an arbitrary set. In this case, we call $(X, \mathscr{G})$ a *topological space*. We often call X itself a topological space when the topology $\mathscr{G}$ is understood. Each member of $\mathscr{G}$ is called an *open set*. A subset F of X is said to be *closed* if its complement $X \backslash F$ is open. The smallest closed set containing a set $S \subset X$ is called the *closure* of S and denoted by $\bar{S}$ or clS. This means that if F is closed and $S \subset F$, then $\bar{S} \subset F$. The closure of S is characterized by the intersection of all closed sets containing S. It is clear that for every subset S of X, $S \subset \bar{S}$ and S is closed if and only if $S = \bar{S}$. A point x of the set $\bar{S}$ is called a *point of closure* of S. It is characterized such that every open set U containing x has a non-empty intersection with X, that is, if U is open and $x \in U$, then $U \cap X \neq \emptyset$.

A set $U \subset X$ containing $x \in X$ is called a *neighborhood* of x if U includes an open set to which x belongs. Hence, every open set containing x is a neighborhood of x. Let $\mathscr{U}_x$ be the set of all neighborhoods of $x \in X$. $\mathscr{U}_x$ is called the *neighborhood system* of x and satisfies

(Nh-1) if $U \in \mathscr{U}_x$, then $x \in U$,
(Nh-2) if $U, V \in \mathscr{U}_x$, then $U \cap V \in \mathscr{U}_x$,
(Nh-3) if $U \in \mathscr{U}_x$ and $U \subset V$, then $V \in \mathscr{U}_x$,
(Nh-4) if $U \in \mathscr{U}_x$, then there exists $V \in \mathscr{U}_x$, such that $V \subset U$ and $V \in \mathscr{U}_y$ for each $y \in V$.

A set $\mathscr{B}_x \subset \mathscr{U}_x$ is called a *base* or a *basis* at $x \in X$ for the neighborhood system $\mathscr{U}_x$ if for every $U \in \mathscr{U}_x$, there exists $V \in \mathscr{B}_x$ with $V \subset U$. For instance, the set of all open sets containing x is a base at x. A point x of a subset S of X is said to be an *interior point* of S if the set S includes a neighborhood of x. The set of all interior points of the set S is called the *interior* of S and denoted by $\mathring{S}$ or $intS$. It is clear that a set $G \subset X$ is open if and only if $\mathring{G} = G$. For every subset S of X, $\bar{S} \backslash \mathring{S}$ is called the *boundary* of S and denoted by ∂S or $bdryS$.

A *directed system* is a set $\mathscr{K}$ together with a relation[1] $\prec \subset \mathscr{K} \times \mathscr{K}$ denoted $(\mathscr{K}, \prec)$ that satisfies the following conditions:

(Dr-1) if $(\kappa, \mu) \in \prec$ and $(\mu, \nu) \in \prec$, then $(\kappa, \nu) \in \prec$,
(Dr-2) $(\kappa, \kappa) \in \prec$ for all $\kappa \in \mathscr{K}$,
(Dr-3) if $\kappa, \mu \in \mathscr{K}$, then there exists a $\nu \in \mathscr{K}$, such that $(\kappa, \nu) \in \prec$ and $(\mu, \nu) \in \prec$.

Note that the condition (Dr-1) is nothing but the transitivity condition and (Dr-2) the reflexivity condition. As usual, we denote $(\kappa, \mu) \in \prec$ by $\kappa \prec \mu$. For each $x \in X$, it follows from (Nh-1) and (Nh-2) that the neighborhood system $(\mathscr{U}_x, \supset)$ constitutes a direct system by the set inclusion.

[1]This relation should not be confused with the preference relation!

A *net* on a topological space X is a map of a directed system $(\mathscr{K}, \prec)$ to X. We usually denote by x_κ or $x(\kappa)$ the value of the net at $\kappa \in \mathscr{K}$ and by $\{x_\kappa\}$ the net itself. A map from $\mathbb{N}$ to X is called a *sequence*. The sequence $\{x_n\}$ is an example of the net with a directed set $(\mathbb{N}, \leq)$. A net $\{y_\sigma, \sigma \in (\mathscr{S}, \prec')\}$ is a *subnet* of a net $\{x_\kappa, \kappa \in (\mathscr{K}, \prec)\}$ if there exists a map $\phi : \mathscr{S} \to \mathscr{K}$, such that

(Sn-1) $y_\sigma = x_{\phi(\sigma)}$ for each $\sigma \in \mathscr{S}$,

(Sn-2) for each $\kappa \in \mathscr{K}$, there exists $\sigma_0 \in \mathscr{S}$, such that if $\sigma_0 \prec' \sigma$, then $\kappa \prec \phi(\sigma)$.

The subnet is often denoted as $\{x_{\kappa(\sigma)}\}$ or $\{x(\kappa_\sigma)\}$. A point $x \in X$ is said to be the *limit* of a net $\{x_\kappa\}$ or the net $\{x_\kappa\}$ converges to x if for every open neighborhood U of x, there is an $\kappa_0 \in \Lambda$ such that $x_\kappa \in U$ for all $\kappa_0 \prec \kappa$. In this case, we denote $x_\kappa \to x$ or $\lim_\kappa x_\kappa = x$. A point $x \in X$ is called a *cluster point* of the net $\{x_\kappa\}$ if for every neighborhood U of x and for every $\kappa \in \Lambda$, there exists a μ such that $\kappa \prec \mu$ and $x_\mu \in U$.

The points of closure and the cluster points are characterized by the net.

Proposition B.1. *A point $x \in X$ is a point of closure of a set S if and only if it is the limit of some net $\{x_\kappa\}$ in S. A point $x \in X$ is a cluster point of a net $\{x_\kappa\}$ if and only if some subnet of $\{x_\kappa\}$ converges to x.*

A subset D of X is said to be *dense* in X if $X \subset \bar{D}$. A topological space X is *separable* if and only if there exists a countable dense subset of X. A subset S of a topological space $(X, \mathscr{G})$ is called a *subspace* when S is endowed with the topology $\mathscr{G}_S$ whose open sets are the intersections with S of the open sets in X; $\mathscr{G}_S = \{G \cap S | \ G \in \mathscr{G}\}$. The topology $\mathscr{G}_S$ is called the *relative topology*. A set $G \in \mathscr{G}_S$ is said to be *open relative to S* or simply *open in S*.

Proposition B.2. *Every open (closed) set G in the subspace $(S, \mathscr{G}_S)$ of $(X, \mathscr{G})$ is open (closed) in X if and only if S is open (closed) in X.*

A mapping f of a topological space X to a topological space Y is said to be *continuous* if for every open set G of Y, its inverse image of f, $f^{-1}(G)$ is open in X. In terms of the net, we can state the following proposition:

Proposition B.3. *Let X and Y be topological spaces. A mapping $f : X \to Y$ is continuous if and only if for every $x \in X$ and for every net $\{x_\kappa\}$ on X converging to x, the net $\{f(x_\kappa)\}$ on Y converges to $f(x)$.*

When there exists a bijection between topological spaces X and Y which is continuous and so is its inverse, X and Y are said to be *isomorphic.* We have derived the concepts of convergence of a net, the limit, the cluster point, and so on from the field of topology. On the contrary, it is known that once the concepts of a net and its convergence are given, we can construct a unique topology such that the convergence of a net with respect to this topology coincides with the originally given convergence. In this sense, the ideas of the topology and the net are equivalent. For details, see [113, Chapter 2].

A continuous and bijective map f from a topological space $(X, \mathscr{G})$ to a topological space $(Y, \mathscr{G}')$ whose inverse f^{-1} is also continuous is called a *homeomorphism.* In this case, the set X and the set Y are said to be *homeomorphic.* Homeomorphic spaces are considered to be identical from a topological perspective.

A subset K of a topological space is said to be *compact* if for every family of open sets $\{G_\alpha\}$, $\alpha \in \mathscr{A}$, where $\mathscr{A}$ is an arbitrary set, such that $K \subset \cup_{\alpha \in \mathscr{A}} G_\alpha$, there exists a finite subfamily $\{G_{\alpha_i}\}, i = 1 \dots n$, such that $K \subset \cup_{i=1}^{n} G_{\alpha_i}$. This definition of compactness is equivalent to the definition that K is compact if and only if for every family $\{F_\alpha\}$ of closed subsets of K, such that every finite subfamily $\{F_{\alpha_i}\}$, $i = 1 \dots n$, has non-empty intersection $\cap_{i=1}^{n} F_{\alpha_i} \neq \emptyset$, its intersection is non-empty, $\cap_{\alpha \in \mathscr{A}} F_\alpha \neq \emptyset$. The compact set is also characterized by the net.

Proposition B.4. *Let X be a topological space. Then, X is compact if and only if every net in X has a cluster point.*

A topological space is called *Hausdorff space* if and only if for every $x, y \in X$ with $x \neq y$, there are disjoint open sets U and V, such that $x \in X$ and $y \in V$.

Proposition B.5. *Every closed subset of a compact space is compact. Every compact subset of a Hausdorff space is closed.*

Proposition B.6. *Let f be a continuous mapping of a compact space $(X, \mathscr{G})$ to a topological space $(Y, \mathscr{G}')$. Then, $f(X)$ is compact.*

Let X be a compact space, Y a Hausdorff space, and $f : X \to Y$ a continuous bijection. Then, every closed subset F of X is compact by Proposition B.5. Therefore, $f(F)$ is compact by Proposition B.6; hence, it is closed by Proposition B.5. This shows that $f^{-1} : Y \to X$ is also continuous. Thus, we have obtained the following proposition:

Proposition B.7. *A continuous and bijective map f of a compact space $(X, \mathscr{G})$ to a Hausdorff space $(Y, \mathscr{G}')$ is a homeomorphism.*

Let $\mathscr{A}$ be an arbitrary set, and $\{X_\alpha, \mathscr{G}_\alpha\}$ be a family of topological spaces indexed by $\alpha \in \mathscr{A}$. The set of mappings ξ from $\mathscr{A}$ to $\cup_\alpha X_\alpha$ with $\xi(\alpha) \in X_\alpha$ for all α is called the *direct product* of X_α and written $X = \prod_{\alpha \in \mathscr{A}} X_\alpha$. Consider the subsets B of X of the form $B = \prod_{\alpha \in \mathscr{A}} G_\alpha$, where $G_\alpha \in \mathscr{G}_\alpha$ and $G_\alpha = X_\alpha$ but for a finite number of values of α. Let $\mathscr{G}$ be the family of subsets of X that are unions of the sets B. Then, it is clear that $\mathscr{G}$ satisfies (Tp-1), (Tp-2), and (Tp-3) at the start of this section. The topological space $(X, \mathscr{G})$ is called the *product space* and $\mathscr{G}$ the product topology.

Theorem B.1 (Tychonoff). *The product space $X = \prod_{\alpha \in \mathscr{A}} X_\alpha$ is compact if each X_α is compact.*

A topological space X is said to be *locally compact* if every point $x \in X$ has a compact neighborhood.

Theorem B.2 (Alexandroff). *Let $(X, \mathscr{G})$ be a locally compact space. Then, there exists a compact space $(X', \mathscr{G}')$, such that X is homeomorphic to a subset of X', whose complement consists of exactly one point.*

The space X' in the above theorem is unique up to a homeomorphism. The space X' is called the *Alexandroff compactification* or the *one-point compactification* and denoted by $X' = X \cup \{\infty\}$.

An important class of topological spaces is *metric spaces*. Let X be a set. A non-negative real-valued function d on $X \times X$,

$$d : X \times X \to \mathbb{R}_+, (x, y) \mapsto d(x, y),$$

is called a *metric* on X if it satisfies the following properties for all x, y, and $z \in X$;

(Mt-1) $d(x, y) \geq 0$ and $d(x, y) = 0$ if and only if $x = y$,
(Mt-2) $d(x, y) = d(y, x)$,
(Mt-3) $d(x, y) \leq d(x, z) + d(z, y)$.

The pair consisting of the set X and a metric d on X is called a *metric space*. We often call X itself the metric space if the metric is understood. Let $x \in X$ and $\epsilon > 0$. The set $B(x, \epsilon) = \{z \in X |\ d(x, z) < \epsilon\}$ is called *a(n open) ball* or simply a ball with the center x and radius ϵ. A ball is also called a *disk*. We will sometimes denote $\bar{B}(x, \epsilon) = \{z \in X |\ d(x, z) \leq \epsilon\}$ and call it a *closed ball* or *closed disk*. The family $\mathscr{G}_d$ of subsets G of X, such that for all $x \in G$, $B(x, \epsilon) \subset G$ for some $\epsilon > 0$ satisfies (Tp-1), (Tp-2), and (Tp-3); hence, the pair $(X, \mathscr{G}_d)$ is a topological space. The topology $\mathscr{G}_d$ is called the *metric topology*. Regarding metric spaces, much of the topological discussions go smoothly in terms of sequences. For example, a sequence $\{x_n\}$ in a metric space (X, d) converges to $x \in X$ if and only if $d(x_n, x) \to 0$ as $n \to \infty$. Moreover by Proposition B.3, a mapping f of a metric space (X, d) to a metric space (Y, d') is continuous (at x) if $f(x_n) \to f(x)$ whenever $x_n \to x$, and closed sets in metric spaces are characterized by sequences.

Proposition B.8. *A subset F of a metric space (X, d) is closed if and only if for every sequence $\{x_n\}$ in F with $x_n \to x$, it follows that $x \in F$.*

Example B.1. Let $X = \mathbb{R}^\ell$ be the ℓ-dimensional Euclidean space. We can consider three metrics on $\mathbb{R}^\ell$.

$$d_1(\boldsymbol{x}, \boldsymbol{y}) = |x^1 - y^1| + \cdots + |x^\ell - y^\ell|,$$
$$d_2(\boldsymbol{x}, \boldsymbol{y}) = \sqrt{(x^1 - y^1)^2 + \cdots + (x^\ell - y^\ell)^2},$$
$$d_\infty(\boldsymbol{x}, \boldsymbol{y}) = \max\{|x^t - y^t| \mid 1 \leq t \leq \ell\}.$$

They look different from the other, but all three metrics define the same topology on $\mathbb{R}^\ell$, because we have

$$d_\infty(\boldsymbol{x}, \boldsymbol{y}) \leq d_2(\boldsymbol{x}, \boldsymbol{y}) \leq \sqrt{\ell} d_\infty(\boldsymbol{x}, \boldsymbol{y})$$

and

$$d_\infty(\boldsymbol{x}, \boldsymbol{y}) \leq d_1(\boldsymbol{x}, \boldsymbol{y}) \leq \ell d_\infty(\boldsymbol{x}, \boldsymbol{y}),$$

so that any ball with respect to d_2 is contained in a ball with respect to d_∞ and so on.

Example B.2. Consider the direct product of countably many real lines,

$$\mathbb{R} \times \mathbb{R} \times \cdots = \{(x^0, x^1 \ldots) \mid x^t \in \mathbb{R}\},$$

which is nothing but the set of all sequences of real numbers. Let $\{\boldsymbol{x}_n\} = \{(x_n^0, x_n^1 \ldots)\}$ be a sequence in the product space. Then, $\boldsymbol{x}_n \to \mathbf{0} = (0, 0 \ldots)$ in the product topology if and only if $x_n^t \to 0$ for every $t = 0, 1, \ldots$. The product topology on $\mathbb{R} \times \mathbb{R} \ldots$ is a metric topology defined by

$$d(\boldsymbol{x}, \boldsymbol{y}) = \sum_{t=0}^{+\infty} \frac{|x^t - y^t|}{2^t(|x^t - y^t| + 1)}$$

for $\boldsymbol{x} = (x^t), \boldsymbol{y} = (y^t) \in \mathbb{R} \times \mathbb{R} \times \ldots$.

As a special case of the subnets, we have a concept known as the *subsequence.* Let $\{x_n\}$ be an infinite sequence. We say that $\{x_{n_i}\}$ is a subsequence when n_i is a monotone mapping from $\mathbb{N}$ to $\mathbb{N}$, that is, the mapping $i \mapsto n_i$ which satisfies that $i < j$ implies $n_i < n_j$. Let x be a cluster point of a sequence $\{x_n\}$ in a metric space X. Then, we have a subsequence $\{x_{n_i}\}$ such that $x_{n_i} \to x$. Consider the ball $B(x, (1+i)^{-1})$, $i \in \mathbb{N}$. As x is a cluster point, there exists $x_{n_i} \in B(x, (1+i)^{-1})$ for every i. Then, $\{x_{n_i}\}$ is the desired subsequence because the radius of the balls converges to 0. The converse is easily verified; if every sequence in X has a converging subsequence, X has a cluster point. Propositions B.1 and B.4 are restated as follows:

Proposition B.9. *Let (X, d) be a metric space. Then, X is compact if and only if every sequence $\{x_n\}$ in X has a cluster point; hence, there exists a subsequence converging to a point of X.*

Let S be a subset of a metric space (X, d) and d_S be the restriction of d to S. Clearly, d_S is a metric on the set S and the topology on S derived from the metric d_S coincides with the subspace topology. Hence, we call (S, d_S) a *subspace* of (X, d).

Proposition B.10. *Every subspace of a separable metric space is separable.*

A sequence$\{x_n\}$ in a metric space (X, d) is called a *Cauchy sequence* if for every $\epsilon > 0$, there exists an $n \in \mathbb{N}$, such that $d(x_p, x_q) \leq \epsilon$ for all $p, q \geq n$. A metric space (X, d) is said to be *complete* if every Cauchy sequence converges to a point in X. It is well known that the space of real numbers $\mathbb{R}$ is complete, but the set of rational numbers $\mathbb{Q}$ is not.

Proposition B.11. *Every compact metric space is complete and separable.*

Let X be a metric space and ϵ a positive number. A finite set of points $\{x_1 \dots x_n\} \subset X$ is called an *ϵ-net* if $X \subset \cup_{i=1}^{n} B(x_i, \epsilon)$. Then, we have the following proposition:

Proposition B.12. *Let (X, d) be a compact metric space. Then, there exists an ϵ-net for every $\epsilon > 0$.*

Let $S \subset \mathbb{R}^\ell$ be a subset of the ℓ-dimensional Euclidean space $\mathbb{R}^\ell$. The set S is said to be *bounded* if there exists an $M > 0$, such that

$S \subset B(\mathbf{0}, M)$, namely that S is contained in a ball that is large enough. The following proposition is useful.

Proposition B.13. *Let $K \subset \mathbb{R}^\ell$ be a subset of the ℓ-dimensional Euclidean space $\mathbb{R}^\ell$. Then, K is compact if and only if it is closed and bounded.*

Let S, T be metric spaces and consider the real-valued function on $S \times T$,

$$f : S \times T \to \mathbb{R},\ (x, y) \mapsto f(x, y).$$

When f is continuous with respect to the product topology of $S \times T$, we sometimes say that f is *bicontinuous* or *joint continuous*, emphasizing the difference between the continuity of each variable separately. In terms of sequences, f is joint continuous if and only if $(x_n, y_n) \to (x, y)$ implies that $f(x_n, y_n) \to f(x, y)$. This is different from the statement that for each $y \in T$, $x_n \to x \in S$ implies that $f(x_n, y) \to f(x, y)$ and for each $x \in S$, $(x, y_n) \to (x, y)$ implies that $f(x, y_n) \to f(x, y)$. The joint continuity implies separate continuity for each variable, but not vice versa. The inner product on $\mathbb{R}^\ell$ is an example of a map with two variables;

$$(\boldsymbol{p}, \boldsymbol{x}) = ((p^t), (x^t)) \mapsto \boldsymbol{p}\boldsymbol{x} = \sum_{t=1}^{\ell} p^t x^t$$

and it is joint continuous. Indeed, let $(\boldsymbol{p}_n, \boldsymbol{x}_n) \to (\boldsymbol{p}, \boldsymbol{x})$. Then, we have

$$\begin{aligned} |\boldsymbol{p}_n\boldsymbol{x}_n - \boldsymbol{p}\boldsymbol{x}| &= |\boldsymbol{p}_n\boldsymbol{x}_n - \boldsymbol{p}_n\boldsymbol{x} + \boldsymbol{p}_n\boldsymbol{x} - \boldsymbol{p}\boldsymbol{x}| \\ &\leq |\boldsymbol{p}_n\boldsymbol{x}_n - \boldsymbol{p}_n\boldsymbol{x}| + |\boldsymbol{p}_n\boldsymbol{x} - \boldsymbol{p}\boldsymbol{x}| \\ &\leq \|\boldsymbol{p}_n\|\|\boldsymbol{x}_n - \boldsymbol{x}\| + \|\boldsymbol{p}_n - \boldsymbol{p}\|\|\boldsymbol{x}\| \to 0. \end{aligned}$$

In the above inequalities, we used the Cauchy–Schwartz inequality; $|\boldsymbol{x}\boldsymbol{y}| \leq \|\boldsymbol{x}\|\|\boldsymbol{y}\|$; see Appendix E. In Appendix G, we will see that the problem of joint continuity of the inner product is more subtle on infinite-dimensional spaces.

Appendix C

Some Algebraic Topology

The purpose of this appendix is to present an algebraic topological proof of Brower's fixed point theorem in a self-contained manner as much as possible. Instead of the results in this appendix, the fixed point theorem will be used in the text, and all maps between topological spaces are assumed to be continuous unless otherwise stated. The reader should note that in this appendix, G and F and so on will denote abelian groups (see below), not open or closed sets as in Appendix B.

Definition C.1. An *abelian group* $(G, +)$ is a set with the operation $+$ called *addition* which satisfies

(Ab-1) for all $x, y \in G$, $x + y = y + x$,
(Ab-2) for all $x, y, z \in G$, $(x + y) + z = x + (y + z)$,
(Ab-3) there exists an element 0, such that $x + 0 = x$,
(Ab-4) for all $x \in G$, there exists an element called the inverse $-x$ such that $x + (-x) = 0$.

Note that the element 0 in (Ab-3) is unique, for if $0'$ also satisfies (Ab-3), then $0 = 0 + 0' = 0' + 0 = 0'$. The abelian group $(G, +)$ is often simply denoted G.

A subset $H \subset G$ is called a *subgroup* of G if $x - y \in H$ for all $x, y \in H$. Let H be a subgroup of G. We can define an equivalence relation $x \sim y$ for $x, y \in G$ by $x - y \in H$. Denoting the equivalence class of x by $[x]$, the set of equivalence classes $G/H = \{[x] | \ x \in G\}$ becomes a group called the *quotient group* when $[x] + [y]$ is defined by $[x + y]$. Note that $[0] = H$ and $[-x] = -[x]$. This definition does not depend upon the particular representatives of the equivalence classes, for letting $x, u \in [x]$ and $y, v \in [y]$, $(x + y) - (u + v) = (x - u) + (y - v) \in H + H = H$, which shows that $[x + y] = [u + v]$.

Let G and H be abelian groups. A map $\psi : G \to H$ is called a *homomorphism* if $\psi(x + y) = \psi(x) + \psi(y)$ for all $x, y \in G$. It can be readily seen that for a homomorphism $\psi : G \to H$, the kernel $\psi^{-1}(0)$ is a subgroup of G, and the image $\psi(G)$ is also a subgroup of H. An injective homomorphism is called a *monomorphism* and a surjective homomorphism is an *epimorphism*. A bijective homomorphism is called an *isomorphism*. Two groups G and H are *isomorphic* if there is an isomorphism between them denoted $G \approx H$ or even $G = H$. Then, we have the following fundamental isomorphism theorem:

Theorem C.1. *Let $\psi : G \to H$ be a surjective homomorphism. Then,*

$$G/\psi^{-1}(0) \approx H.$$

The direct product of two abelian groups G and H, $\{(x, y)|\ x \in G,\ y \in H\}$ is called the *direct sum* and denoted by $G \oplus H$. The direct sum is also an abelian group when we define $(x, y) + (x', y') = (x + x', y + y')$. Let F and H be subgroups of G, and every element x of G has the *unique* representation $x = y + z$, $y \in F$ and $z \in H$. Then, it is obvious that G is isomorphic to $F \oplus H$.

Let G be an abelian group and $x \in G$. When there exists a natural number $n \in \mathbb{N}$ with $nx = 0$, we say that x is of *finite order*; otherwise, x is of *infinite order*. The order of an element x is the smallest natural number n with $nx = 0$. When there exists an element $g \in G$, such that every element $x \in G$ can be written as $x = ng$ for some $n \in \mathbb{Z}$, G is called a *cyclic group* and g is a *generator*. Obviously, if g is a generator, so is $-g$. A cyclic group may be of finite or infinite order.

Example C.1. The set of integers $\mathbb{Z}$ is obviously an abelian group of infinite order with respect to the usual addition $+$. It is easy to see that the infinite cyclic group is nothing but (strictly speaking, isomorphic to) $\mathbb{Z}$.

For $n \in \mathbb{Z}$, we define an equivalent relation $x \equiv y \pmod{n}$ on $\mathbb{Z}$ if and only if $x - y = nz$ for some $z \in \mathbb{Z}$. In this case, we say x and y are *congruent modulo n*. Let $\bar{x}$ be an equivalence class to which x belongs. The set $\mathbb{Z}_n = \{\bar{0}, \bar{1} \ldots \overline{n-1}\}$ is an abelian group of order n under the addition $\bar{x} + \bar{y} = \overline{x+y}$.

Let $n\mathbb{Z} = \{nz|\ z \in \mathbb{Z}\}$. It easy to see that $n\mathbb{Z}$ is a subgroup of $\mathbb{Z}$ and $\mathbb{Z}_n$ can be realized as a quotient group; $\mathbb{Z}_n = \mathbb{Z}/n\mathbb{Z}$.

An abelian group G is *free* if there exists a subset $B \subset G$ called a *basis*, such that every element $x \in G$ has a unique representation

$$x = \sum_{z \in B} k_z z,$$

where k_z is an integer and equal to zero for all but finitely many z in B. It is obvious that if G is a free abelian group with a basis B and H is an abelian group, then every map $\psi : B \to H$ can be uniquely extended to a homomorphism $\hat{\psi} : G \to H$ defined by $\hat{\psi}(x) = \sum_{z \in B} k_z \psi(z)$.

Let X be a topological space and Δ_n be the standard n-simplex. A *singular n-simplex* in X is a continuous function $\sigma_n : \Delta_n \to X$. The singular 0-simplices may be identified with the points in X, the singular 1-simplex with paths in X, and so on. For a singular n-simplex σ_n, we define a singular $n-1$-simplex $\partial^i \sigma_n$ in X, which is called the ith *face* of σ_n by

$$\partial^i \sigma_n(t^0 \ldots t^{n-1}) = \sigma_n(t^0 \ldots t^{i-1}, 0, t^i \ldots t^{n-1}).$$

Let X be a topological space. We define $S_n(X)$ to be the free abelian group whose basis is the set of all singular n-simplices of X. Each element σ of $S_n(X)$ is called a *singular n-chain* of X and has the form $\sigma = \sum_{\sigma_n} k_{\sigma_n} \sigma_n$, where k_{σ_n} is an integer, equal to zero for all but a finite number of σ_n. As the ith face operator ∂^i is a map from the set of singular n-simplices to the set of $n-1$-simplices, there is a unique extension to a homomorphism $\partial^i : S_n(X) \to S_{n-1}(X)$

given by

$$\partial^i\left(\sum_{\sigma_n} k_{\sigma_n}\sigma_n\right) = \sum_{\sigma_n} k_{\sigma_n}\partial^i\sigma_n.$$

Define the *boundary operator* to be the homomorphism $\partial_n : S_n(X) \to S_{n-1}(X)$ given by

$$\partial_n = \partial^0 - \partial^1 \dots (-1)^n\partial^n = \sum_{i=0}^{n}(-1)^i\partial^i.$$

The following proposition is easy but fundamental.

Proposition C.1. $\partial_{n-1} \circ \partial_n = 0$.

Proof. We indicate the position i of 0 inserted by the operator ∂^i, $\partial^i\sigma_n(t^0 \dots t^{n-1}) = \sigma_n(t^0 \dots t^{i-1}, 0^i, t^i \dots t^{n-1})$. As the boundary operator is linear, it suffices to show for simplices. Then, we have

$$\begin{aligned}
&\partial_{n-1} \circ \partial_n\sigma_n(t^0 \dots t^{n-2})\\
&= \sum_{i=0}^{n-1}(-1)^i\partial_n\sigma_n(t^0 \dots t^{i-1}, 0^i, t^i \dots t^{n-2})\\
&= \sum_{i=0}^{n-1}(-1)^i\left(\sum_{j=0}^{i}(-1)^j\sigma_n(t^0 \dots 0^j \dots 0^i \dots t^{n-2})\right.\\
&\quad\left. + \sum_{j=i+1}^{n-1}(-1)^{j+1}\sigma_n(t^0 \dots 0^i \dots 0^j \dots t^{n-2})\right)\\
&= \sum_{i>j}(-1)^{i+j}\sigma_n(t^0 \dots 0^j \dots 0^i \dots t^{n-2})\\
&\quad + \sum_{i<j}(-1)^{i+j+1}\sigma_n(t^0 \dots 0^i \dots 0^j \dots t^{n-2}) = 0.
\end{aligned}$$

□

An n-chain $\sigma \in S_n(X)$ is an *n-cycle* if $\partial\sigma = 0$. It is called an *n-boundary* if $\sigma = \partial\tau$ for some $\tau \in S_{n+1}(X)$. As $\partial_n : S_n(X) \to S_{n-1}(X)$ is a homomorphism its kernel or the set of all n-cycles is a subgroup of $S_n(X)$ denoted by $Z_n(X)$. Similarly, the image of $\partial_{n+1} : S_{n+1}(X) \to S_n(X)$ in $S_n(X)$ is the subgroup $B_n(X)$ of all n-boundaries. By Proposition C.1, $B_n(X) \subset Z_n(X)$ is a subgroup.

Definition C.2. The quotient group

$$H_n(X) = Z_n(X)/B_n(X)$$

is the nth *singular homology group* of X.

To prove Brower's fixed point theorem, we have to know the homology group of the n-sphere, $H_p(S^n)$, $p = 0, 1 \ldots$, which is simply denoted $H_*(S^n)$. The computation is not difficult, but requires some more concepts and techniques.

Let $f : X \to Y$ be a continuous function between topological spaces X and Y, and $\sigma_n : \Delta_n \to X$ a singular n-simplex in X. We can "push forward" σ_n to a singular n-simplex $f_\#(\sigma_n)$ in Y defined by $f_\#(\sigma_n) = f \circ \sigma_n$. We can extend it uniquely to a homomorphism still denoted by $f_\#$ from $S_n(X)$ to $S_n(Y)$. If $g : Y \to Z$ is also a continuous map and $\iota_X : X \to X$ is the identity map on X, then it is easy to see

$$(g\circ f)_\#(\sigma) = g_\# \circ f_\#(\sigma) \text{ and } \iota_{X\#}(\sigma) = \sigma.$$

The homomorphism $f_\# : S_n(X) \to S_n(Y)$ is a *chain map*, that is, $\partial f_\#(\sigma) = f_\#(\partial\sigma)$; in other words, the following diagram commutes

$$\begin{array}{ccc} S_n(X) & \xrightarrow{f_\#} & S_n(Y) \\ \partial\downarrow & & \downarrow\partial \\ S_{n-1}(X) & \xrightarrow{f_\#} & S_{n-1}(Y) \end{array}$$

It suffices for showing the commutativity to check for the singular simplices σ_n and the face operator ∂^i. We now have

$$f_\#\partial^i\sigma_n(t^0 \ldots t^{n-1}) = f \circ \sigma_n(t^0 \ldots 0^i \ldots t^{n-1})$$

and

$$\begin{aligned}\partial^i f_{\#}\sigma_n(t^0\dots t^{n-1}) &= f_{\#}\sigma_n(t^0\dots 0^i\dots t^{n-1})\\ &= f\circ\sigma_n(t^0\dots 0^i\dots t^{n-1}).\end{aligned}$$

The homomorphism $f_{\#}$ maps the chains to the chains and the boundaries to the boundaries; hence, it induces a homomorphism between the homology groups

$$f_* : H_*(X) \to H_*(Y).$$

Moreover, the properties of $f_{\#}$ stated above imply that

$$(g\circ f)_*(\sigma) = g_*\circ f_*(\sigma) \text{ and } \iota_{X*}(\sigma) = \sigma$$

for a continuous map $g : Y \to Z$ and the identity map $\iota_X : X \to X$. In particular, if the map $f : X \to Y$ is a homeomorphism, then there exists a continuous inverse $g : Y \to X$ with $g\circ f = \iota_X$ and $f\circ g = \iota_Y$; hence, $g_* \circ f_* = \iota_{X*}$ and $f_* \circ g_* = \iota_{Y*}$, which proves the following proposition:

Proposition C.2. *If X and Y are homeomorphic, then $H_*(X) \approx H_*(Y)$.*

Let X and Y be topological spaces and $I = [0, 1]$ the unit interval. Two maps $f_0, f_1 : X \to Y$ are said to be *homotopic* and denoted $f_0 \simeq f_1$ if there exists a continuous map

$$F : X\times I \to Y,\ F(x, 0) = f_0(x),\ F(x, 1) = f_1(x)$$

for all $x \in X$. The map F is called a *homotopy* between f_0 and f_1. It is easy to see that the homotopy relation is an equivalence relation on the set of all continuous maps from X to Y. A map $f : X \to Y$ is said to be *homotopy equivalent* if there exists a map $g : Y \to X$ with $f\circ g \simeq \iota_Y$ and $g\circ f \simeq \iota_X$. We also say that X and Y are homotopic. If X and Y are homeomorphic, then they are evidently homotopic, but not vice versa. Proposition C.2 is generalized to the following theorem:

Theorem C.2. *If X and Y are homotopic, then $H_*(X) \approx H_*(Y)$.*

For a proof, see [258, p.16] or [153, p.21]. When $A \subset X$, we can consider the identity map ι_A as the *inclusion map* $\iota_A : A \to X$ of A into X. A map $g : X \to A$ such that $g \circ \iota_A = \iota_A$ is called a *retraction* of X onto A. The retraction $g : X \to A$ is called a *deformation retraction* if $\iota_A \circ g \simeq \iota_X$. In this case, A is called a *deformation retract* of X. Obviously, the inclusion $\iota_A : A \to X$ is a homotopy equivalence when A is a deformation retract of X. Theorem C.2 implies the following corollary:

Corollary C.1. *If $A \subset X$ is a deformation retract of X, then $\iota_{A*} : H_*(A) \to H_*(X)$ is an isomorphism.*

A sequence of abelian groups and homomorphisms

$$\dots G_n \xrightarrow{f_n} G_{n-1} \xrightarrow{f_{n-1}} G_{n-2} \dots$$

is called a *chain complex* if $f_{n-1} \circ f_n = 0$. The chain complex is *exact* if $f_n(G_n) = f_{n-1}^{-1}(0)$ for all n. The exact chain complex is called an *exact sequence.* The homology group $H_*(X)$ measures the deviation of a chain complex

$$\dots S_n(X) \xrightarrow{\partial_n} S_{n-1}(X) \xrightarrow{\partial_{n-1}} S_{n-2}(X) \dots$$

from being exact, namely that, $H_n(X) = 0$ for $n > 0$ when the chain complex is exact. An exact sequence

$$0 \to F \xrightarrow{f} G \xrightarrow{g} H \to 0$$

is called *short exact.* It is easy to see that for the short exact sequence, f is injective and g is surjective.

Let X and Y be topological spaces. We can consider the direct sum of the chain complex

$$\begin{aligned} \dots S_n(X) \oplus S_n(Y) &\xrightarrow{\partial_n} S_{n-1}(X) \oplus S_{n-1}(Y) \\ &\xrightarrow{\partial_{n-1}} S_{n-2}(X) \oplus S_{n-2}(Y) \dots, \end{aligned}$$

where $\partial_n : S_n(X) \oplus S_n(Y) \to S_{n-1}(X) \oplus S_{n-1}(Y)$ is defined by $\partial_n(\sigma, \tau) \to (\partial_n \sigma, \partial_n \tau)$ for $\sigma \in S_n(X)$ and $\tau \in S_n(Y)$. This chain complex yields the homology group $H_n(S_*(X)\oplus S_*(Y))$ and it is easy to see that $H_n(S_*(X)\oplus S_*(Y)) = H_n(X)\oplus H_n(Y)$.

A continuous function

$$\pi : [0,1] \to X$$

is called a *path* between $x, y \in X$ if $\pi(0) = x$ and $\pi(1) = y$. Note that the path π is nothing but the singular 1-simplex σ_1. A topological X is decomposed into equivalence classes induced by the equivalence relation $\sim$ defined by $x \sim y$ if and only if there exists a path between x and y. Each equivalence class is called a *path component.* A topological space X is *path-wise connected* if there exists only one path component. Then, we have the following proposition:

Proposition C.3. *If $X = Y \cup Z$, where Y and Z are path components of X, then $H_*(X) = H_*(Y)\oplus H_*(Z)$.*

Proof. We denote $\sigma = \sum_{\sigma_n} k_{\sigma_n}\sigma_n \in S_n(Y)$ and $\tau = \sum_{\tau_n} k_{\tau_n}\tau_n \in S_n(Z)$. There is a natural homomorphism $\psi : S_n(Y) \oplus S_n(Z) \to S_n(X)$ given by

$$\psi\left(\sum_{\sigma_n} k_{\sigma_n}\sigma_n, \sum_{\tau_n} k_{\tau_n}\tau_n\right) = \sum_{\sigma_n} k_{\sigma_n}\sigma_n + \sum_{\tau_n} k_{\tau_n}\tau_n.$$

As the groups $S_n(Y)$ and $S_n(Z)$ are free abelian, or the representations of σ and τ are unique, ψ is injective. To observe the surjectivity, note that for any singular n-simplex $\sigma_n \in S_n(X)$, it follows that $\sigma_n \in S_n(Y)$ or $\sigma_n \in S_n(Z)$, because $\sigma_n(\Delta_n)$ is path-wise connected. Hence, we can write $\sigma = \sum_{\sigma_n} k_{\sigma_n}\sigma_n + \sum_{\tau_n} k_{\tau_n}\tau_n$ for $\sigma \in S_n(X)$; therefore, ψ is surjective. It is easy to see that ψ is a chain map between $S_*(Y)\oplus S_*(Z)$ and $S_*(X)$, so that $H_n(X)$ is isomorphic to $H_n(S_*(Y)\oplus S_*(Z)) = H_n(X)\oplus H_n(Y)$. □

Proposition C.3 can be generalized inductively to the case of finitely many path components.

We now begin to compute the homology groups for some simple spaces. Obviously, the (non-empty) simplest space is the singleton! Let us start with it and set $X = \{x\}$. Then, for each $n \geq 0$, there exists a unique singular n-simplex $\sigma_n : \Delta_n \to \{x\}$ and it is easy to verify that $\partial^i \sigma_n = \sigma_{n-1}$ ($\sigma_{-1} = 0$ is understood). Consider the chain complex

$$\ldots \to S_2(X) \to S_1(X) \to S_0(X) \to 0.$$

For every $n \geq 0$, $S_n(X)$ is an infinite cyclic group $\mathbb{Z}$ generated by σ_n. Hence, the boundary operator is calculated as

$$\partial_n \sigma_n = \sum_{i=0}^{n} (-1)^i \partial^i \sigma_n = \sum_{i=0}^{n} (-1)^i \sigma_{n-1}.$$

It follows that $\partial_{2n-1}\sigma_{2n-1} = 0$ and $\partial_{2n}\sigma_{2n} = \sigma_{2n-1}$ for $n > 0$. Applying this to the chain complex, we have $Z_n(X) = B_n(X)$ for $n > 0$. As $Z_0(X) = S_0(X) = \mathbb{Z}$ and $B_0(X) = \{0\}$, we finally obtain

$$H_n(\{x\}) = \begin{cases} \mathbb{Z} & \text{if } n = 0, \\ 0 & \text{otherwise.} \end{cases}$$

A space is *contractible* if it is homotopic to the singleton. Theorem C.2 and the above result imply the following corollary:

Corollary C.2. *If X is contractible, then*

$$H_n(X) = \begin{cases} \mathbb{Z} & \text{if } n = 0, \\ 0 & \text{otherwise.} \end{cases}$$

The ℓ-dimensional Euclidean space $\mathbb{R}^\ell$ is contractible (why?). Corollary C.2 for the case $X = \mathbb{R}^\ell$ is called the *Poincaré Lemma*.

Suppose X is path-wise connected, and consider the last three terms of the chain complex of X given by

$$S_1(X) \xrightarrow{\partial} S_0(X) \to 0.$$

Obviously, $S_0(X) = Z_0(X)$, which is the free abelian group generated by the points of X; hence, $Z_0(X) = \{z \in X |\, z = \sum_{x \in X} k_x x\}$, where k_x are integers with all, but finitely many equal to 0. Next, $S_1(X)$ is the free abelian group generated by the set of all paths (the singular

1-simplices) in X. Denoting the vertices of $\pi = \sigma_1$ by x_0 and x_1, we have

$$\partial\pi = x_1 - x_0 \in B_0(X).$$

Define a homomorphism $\xi : S_0(X) \to \mathbb{Z}$ by $\xi(\sum k_x x) = \sum k_x$. It can be readily verified that ξ is surjective whenever X is non-empty. As for any singular 1-simplex π in X, $\xi(\partial\pi) = \xi(x_1 - x_0) = 1 - 1 = 0$, we obtain that $B_0(X)$ is contained in the kernel of ξ, $B_0(X) \subset \xi^{-1}(0)$. Conversely, suppose that $\sum_{i=1}^{n} k_i x_i \in Z_0(X)$ and $\sum_{i=1}^{n} k_i = 0$. For each $x \in X$ and each $i = 1 \dots n$, we can take a singular 1-simplex $\sigma^i : [0,1] \to X$ with $\partial^0\sigma^i = x_i$ and $\partial^1\sigma^i = x$, because X is path-wise connected. For the singular 1-chain $\sum_{i=1}^{n} k_i\sigma^i$, it follows that

$$\partial\sum_{i=1}^{n} k_i\sigma^i = \sum_{i=1}^{n} k_i x_i - \left(\sum_{i=1}^{n} k_i\right)x = \sum_{i=1}^{n} k_i x_i,$$

which implies that the kernel ξ is contained in $B_0(X)$; hence, $B_0(X) = \xi^{-1}(0)$. By Theorem C.1, we have the below proposition:

Proposition C.4. *If X is path-wise connected, then $H_0(X) \approx \mathbb{Z}$.*

So far, all calculations have been derived directly from the definitions. For the remaining calculations, we require some additional tools. Let $\mathscr{U} = \{U, V\}$ be a covering of a topological space X such that $\mathring{U} \cup \mathring{V} = X$. We then have the *Mayer–Vietoris sequence*, which is the most basic tool of homology theory.

Theorem C.3. *The Mayer–Vietoris sequence*

$$\dots H_n(U \cap V) \xrightarrow{g_*} H_n(U) \oplus H_n(V) \xrightarrow{h_*} H_n(X) \xrightarrow{\delta_*} H_{n-1}(U \cap V) \dots$$

is exact, where the homomorphisms $g_*(x) = (i_*(x), -j_*(x))$ and $h_*(y,z) = k_*(y) + l_*(z)$ are defined by the inclusion maps $i : U \cap V \to U$, $j : U \cap V \to V$, $k : U \to U \cup V$, and $l : V \to U \cup V$, respectively.

The essence of the Mayer–Vietoris sequence is contained in the homomorphism δ_* called the *connecting homomorphism*; see [258, p. 22] or [153, p. 59] for details. Armed with the Mayer–Vietoris sequence, we can compute the homology group of the n-sphere S^n as follows:

First, we compute $H_1(S^1)$, the first homology group of the 1-sphere $S^1 = \{(x^0, x^1) \in \mathbb{R}^2 |\ (x^0)^2 + (x^1)^2 = 1\}$. Let $z = (0,1)$ and $z' = (0,-1)$ be the north and south poles and $x = (-1,0)$ and $y = (1,0)$ be the points on the equator. Let $U = S^1 \backslash \{z'\}$ and $V = S^1 \backslash \{z\}$. Then, the Mayer–Vietoris sequence yields an exact sequence

$$H_1(U) \oplus H_1(V) \xrightarrow{h_*} H_1(S^1) \xrightarrow{\delta_*} H_0(U \cap V) \xrightarrow{g_*} H_0(U) \oplus H_0(V).$$

As U and V are contractible, the first term is zero by Corollary C.2. Hence, δ_* is injective and $H_1(S^1)$ is isomorphic to $\delta_*(H_1(S^1)) = g_*^{-1}((0,0))$. $H_0(U \cap V)$ is isomorphic to $\mathbb{Z} \oplus \mathbb{Z}$ by Proposition C.3 and its elements can be written as $mx + ny$, where m and n are integers. According to Theorem C.3,

$$g_*(mx + ny) = (i_*(mx + ny), -j_*(mx + ny)).$$

$i_*(mx + ny) = 0$ in the homology $H_0(U)$ means that the chain $mx + ny$ is a boundary ∂z in $S_0(U)$ for some $z \in S_1(U)$. As U and V are pathwise connected, in view of the proof of Proposition C.4 or $B_0(U) = \xi^{-1}(0)$, this is the case if and only if $m + n = 0$, and similarly for j_*. Therefore, $g_*^{-1}((0,0)) = \{m(x - y) |\ m \in \mathbb{Z}\}$, or $g_*^{-1}((0,0))$ is an infinite cyclic group generated by $x - y$; hence, we conclude that

$$H_1(S^1) = \mathbb{Z}.$$

For $n > 1$, the portion of Mayer–Vietoris sequence

$$H_n(U) \oplus H_n(V) \xrightarrow{h_*} H_n(S^1) \xrightarrow{\delta_*} H_{n-1}(U \cap V)$$

has two end terms equal to zero; hence, $H_n(S^1) = 0$, which completes the computation of the homology group of S^1.

We now proceed inductively to compute the homology of

$$S^n = \left\{ (x^0 \dots x^n) \in \mathbb{R}^{n+1} \,\middle|\, \sum_{i=0}^{n} (x^i)^2 = 1 \right\}$$

for $n > 1$. We consider the inclusion $S^{n-1} = \{(x^0 \dots x^n) \in S^n |\, x^n = 0\} \subset S^n$ as the "equator" of S^n. We denote by $z = (0 \dots 0, 1)$ and $z' = (0 \dots 0, -1)$ the north and south poles of S^n. Then, by the standard stereographic projection, $S^n \backslash \{z\}$ is homeomorphic to $\mathbb{R}^n$ and similarly for $S^n \backslash \{z'\}$. Moreover, $S^n \backslash \{z, z'\}$ is homeomorphic to $\mathbb{R}^n \backslash \{0\}$ and S^{n-1} is a deformation retract of $\mathbb{R}^n \backslash \{0\}$ (why?). Now let $U = S^n \backslash \{z\}$ and $V = S^n \backslash \{z'\}$; hence, $U \cap V = S^n \backslash \{z, z'\}$. The Mayer–Vietoris sequence for this covering becomes

$$H_m(\mathbb{R}^n) \oplus H_m(\mathbb{R}^n) \xrightarrow{h_*} H_m(S^n) \xrightarrow{\delta_*} H_{m-1}(S^{n-1}) \\ \xrightarrow{g_*} H_{m-1}(\mathbb{R}^n) \oplus H_{m-1}(\mathbb{R}^n).$$

For $m > 1$, the end terms are zero so that δ_* is an isomorphism. For $m = 1$ and $n > 1$, it follows from $H_1(\mathbb{R}^n) = 0$ that δ_* is a monomorphism. Since g_* is also a monomorphism, $H_1(S^n) = 0$. This furnishes the inductive step and we conclude the following theorem:

Theorem C.4.

$$H_m(S^n) = \begin{cases} \mathbb{Z} & \text{if } m = 0, n, \\ 0 & \text{otherwise.} \end{cases}$$

Recall the n-disk (closed ball)

$$D^n = \left\{ (x^0 \dots x^{n-1}) \in \mathbb{R}^n \,\middle|\, \sum_{i=0}^{n-1} (x^i)^2 \leq 1 \right\},$$

and note that $S^{n-1} \subset D^n$ is its boundary.

Corollary C.3. *There is no retraction of D^n onto S^{n-1}.*

Proof. For $n = 1$, this is obvious, because $D^1 = [0, 1]$ is connected but $S^0 = \{0, 1\}$ is not. Suppose $n > 1$ and $g : D^n \to S^{n-1}$ is a map such that $g \circ \iota_{S^{n-1}} = \iota_{S^{n-1}}$. This implies that the following diagram is commutative.

$$\begin{array}{ccc} H_{n-1}(S^{n-1}) & \xrightarrow{\iota_*} & H_{n-1}(S_{n-1}) \\ {\scriptstyle \iota_*}\big\downarrow & & \big\uparrow{\scriptstyle g_*} \\ H_{n-1}(D^n) & = & H_{n-1}(D^n) \end{array}$$

This is impossible, because $H_{n-1}(S^{n-1}) = \mathbb{Z}$ and $H_{n-1}(D^n) = 0$. $\square$

We can finally prove Brower's fixed point theorem.

Theorem C.5. *Every continuous map $f : D^n \to D^n$ has a fixed point.*

Proof. Suppose $f : D^n \to D^n$ has no fixed points. Then, we can define a map $g : D^n \to S^{n-1}$ as follows: for $x \in D^n$, $g(x) \in S^{n-1}$ is the point at which the ray starting from $f(x)$ passing through $\{x\}$ intersects with S^{n-1}. Then, g is continuous and $g(x) = x$ for all $x \in S^{n-1}$, contradicting Corollary C.3. $\square$

Let X be a convex subset of $\mathbb{R}^\ell$ and $n+1$ be the largest number of points in X that are independent[2]. We call n the *dimension* of X and denote $n = \dim X$.

Proposition C.5. *Let X be a compact and convex subset of $\mathbb{R}^\ell$ with dimension n. Then, X is homeomorphic to D^n.*

Let $X \subset \mathbb{R}^\ell$ be a non-empty compact and convex set of Proposition C.5 and $g : X \to D^n$ be a homeomorphism. For a continuous map $f : X \to X$, the map $g \circ f \circ g^{-1} : D^n \to D^n$ has a fixed point $\hat{p} \in D^n$; $\hat{p} = g \circ f \circ g^{-1}(\hat{p})$ by Theorem C.5. As $g^{-1}(\hat{p}) = f(g^{-1}(\hat{p}))$, $\hat{x} = g^{-1}(\hat{p})$ is a fixed point of f. Then, we obtain the usual statement of the fixed point theorem.

Theorem C.6. *Let X be a non-empty compact and convex subset of $\mathbb{R}^\ell$. Then, every continuous map $f : X \to X$ has a fixed point.*

---

[2]Note that we have started from 0 in counting the number of independent points; see Appendix A.

Appendix D

Continuous Correspondences

In this appendix, we present definitions and results that are related to the concept of correspondences (see below). No proofs are provided but those concerning economics. Readers can refer Notes for their sources. Furthermore, the sets X and Y and so on in this appendix are assumed to be metric spaces unless otherwise specified.

A *correspondence* ϕ is a relation $\phi \subset X \times Y$ such that for every $x \in X$, there exists $y \in Y$ such that $(x, y) \in \phi$. For every $x \in X$, we denote the set $\{y \in Y | \, (x, y) \in \phi\}$ by $\phi(x)$ and call the *value* of ϕ at $x \in X$. It is a non-empty subset of Y. If the value $\phi(x)$ is a singleton, $\phi(x) = \{y\}$, the correspondence is nothing but a mapping. Hence, we will often denote the correspondence $\phi \subset X \times Y$ as

$$\phi : X \to Y, \; x \mapsto \phi(x).$$

Definition D.1. A correspondence $\phi : X \to Y$ is said to be *upper hemicontinuous* (u.h.c) at $x \in X$ if for every open subset V of Y that contains $\phi(x)$, there exists a neighborhood U of x such that $\phi(z) \subset V$ for every $z \in U$. The correspondence ϕ is called upper hemicontinuous if it is u.h.c. at every $x \in X$.

Let $\phi : X \to Y$ and $\psi : Y \to Z$ be correspondences. Then, we can define the composition of ϕ and ψ, $\psi \circ \phi : X \to Z$ by $\psi \circ \phi(x) = \{z \in Z | \; z \in \psi(y) \text{ for some } y \in \phi(x)\}$. We can immediately deduce the following proposition from the definition.

Proposition D.1. *Let the correspondences $\phi : X \to Y$ and $\psi : Y \to Z$ be u.h.c. Then, the composition $\psi \circ \phi : X \to Z$ is also u.h.c.*

The correspondence $\phi : X \to Y$ is said to be *closed* if its graph $\{(x, y) \in X \times Y |\, y \in \phi(x)\}$ is a closed subset of $X \times Y$.

Proposition D.2. *Let ϕ and ψ be correspondences of X to Y such that $\phi(x) \cap \psi(x) \neq \emptyset$ for all $x \in X$. Suppose that either condition (i) or (ii) holds, (i) ϕ and ψ are closed valued and u.h.c at $\bar{x} \in X$, (ii) ϕ is closed, ψ is u.h.c at $\bar{x}$ and $\psi(\bar{x})$ is compact. Then, the correspondence $x \mapsto \phi(x) \cap \psi(x)$ is u.h.c at $\bar{x}$.*

As a corollary, it follows that the closed correspondence $\phi : X \to Y$ is u.h.c if Y is compact. Moreover, in this case, ϕ is compact-valued. Let $y_n \in \phi(x)$ for all n and $y_n \to y$. As $(x, y_n) \to (x, y)$ and ϕ is closed, we have $y \in \phi(x)$. Hence, $\phi(x)$ is a closed subset of Y. As every closed subset of a compact set is compact, $\phi(x)$ is compact.

Proposition D.3. *Let the correspondence ϕ of X to Y be compact-valued and u.h.c. Then, the image $\phi(K)$ of a compact set K is compact.*

The following proposition characterizes the u.h.c correspondences by sequences.

Proposition D.4. *A compact-valued correspondence ϕ of X to Y is u.h.c at $x \in X$ if and only if for every sequence $\{x_n\}$ converging to x and every sequence $\{y_n\}$ with $y_n \in \phi(x_n)$, there exists a converging subsequence y_{n_q} with the limit y such that $y \in \phi(x)$.*

Proposition D.5. *Let $\phi_1 \ldots \phi_m$ be compact-valued and u.h.c correspondences of X to $\mathbb{R}^\ell$. Then, we have that the convex hull, $x \mapsto co\phi_i(x)$, and the sum, $x \mapsto \sum_{i=1}^m \phi_i(x)$ are compact-valued and u.h.c.*

Proposition D.6. *Let $\phi_i : X \to Y_i$ be compact-valued and u.h.c correspondences of X to Y_i, $i = 1 \ldots m$. Then, the product, $x \mapsto \prod_{i=1}^m \phi_i(x)$ is also compact-valued and u.h.c.*

The following theorem generalizes Theorem C.6 to the correspondences and certainly plays a fundamental role in mathematical economics.

Theorem D.1 (Kakutani). *Let X be a compact and convex subset of $\mathbb{R}^\ell$. If the correspondence ϕ of X to X is convex-valued and u.h.c, there exists a point $\hat{x} \in X$ such that $\hat{x} \in \phi(\hat{x})$.*

Proof. As X is compact, there exists a $1/n$-net $\{z_1 \ldots z_{k(n)}\}$ for every $n \in \mathbb{N}$ by Proposition B.12. We then define functions $\theta_i : X \to \mathbb{R}_+$ by $\theta_i(x) = \max\{0, (1/n) - d(x, z_i)\}$, $i = 1 \ldots k(n)$. As $d(x, z_i) < 1/n$ or $\theta_i(x) > 0$ for some i on X, we have $\sum_{i=1}^{k(n)} \theta_i(x) > 0$ on X. Hence, a function $g_n : X \to X$ defined by

$$g_n(x) = \sum_{i=1}^{k(n)} \frac{\theta_i(x) y_i}{\sum_{j=1}^{k(n)} \theta_j(x)} = \sum_{i=1}^{k(n)} \rho_i(x) y_i,$$

where $y_i \in \phi(z_i)$ is well defined and continuous. By Brower's fixed point theorem (Theorem C.6), we have a fixed point $x_n = g_n(x_n)$ for every n. As X is compact, we can assume without loss of generality that the sequence $\{x_n\}$ converges to a point $\hat{x} \in X$. We shall show that $\hat{x}$ is a fixed point.

Let $\epsilon > 0$ be given. As $\phi(x)$ is u.h.c., there exists a $\delta > 0$ such that $x \in B(\hat{x}, \delta)$ implies that $\phi(x) \subset B(\phi(\hat{x}), \epsilon)$, where $B(\phi(\hat{x}), \epsilon) = \phi(\hat{x}) + B(0, \epsilon)$. As $\phi(\hat{x})$ and $B(0, \epsilon)$ are convex, so is $B(\phi(\hat{x}), \epsilon)$. We can take an N so large that $(1/n) < \delta/2$ and $d(\hat{x}, x_n) < \delta/2$ for $n \geq N$. If $\rho_i(x_n) > 0$ and $n \geq N$, then $d(x_n, z_i) < \delta/2$. Hence, it follows from $d(\hat{x}, z_i) \leq d(\hat{x}, x_n) + d(x_n, z_i) = \delta$ that $z_i \in B(\hat{x}, \delta)$; thus, $y_i \in \phi(z_i) \subset B(\phi(\hat{x}), \epsilon)$.

As $x_n = g_n(x_n) = \sum_{i=1}^{k(n)} \rho_i(x_n) y_i$ and $\sum_{i=1}^{k(n)} \rho_i(x_n) = 1$, we have $x_n \in B(\phi(\hat{x}), \epsilon)$ by the convexity of $B(\phi(\hat{x}), \epsilon)$. Then, n to $+\infty$, $\hat{x} \in \overline{B(\phi(\hat{x}), \epsilon)}$, or $d(\hat{x}, B(\phi(\hat{x}), \epsilon)) = 0$. As $\epsilon > 0$ is arbitrary, it follows that $d(\hat{x}, \phi(\hat{x})) = 0$. Therefore, $\hat{x} \in \overline{\phi(\hat{x})} = \phi(\hat{x})$. This completes the proof. □

The point $\hat{x}$ such that $\hat{x} \in \phi(\hat{x})$ is called a *fixed point* of ϕ.

Definition D.2. A correspondence $\phi : X \to Y$ is said to be *lower hemicontinuous* (l.h.c) at $x \in X$ if for every open set G of Y such that $\phi(x) \cap G \neq \emptyset$, there exists a neighborhood U of x such that $\phi(z) \cap G \neq \emptyset$ for every $z \in U$. The correspondence ϕ is called lower hemicontinuous if it is l.h.c. at every $x \in X$.

When a correspondence $\phi : X \to Y$ is upper hemicontinuous (at $x \in X$) and lower hemicontinuous (at $x \in X$), then it is called *continuous* (at $x \in X$). The following theorem is used to guarantee the upper hemicontinuity of the individual demand correspondences or the supply correspondences of competitive firms.

Theorem D.2 (Berge). *Let β be a nonempty, compact-valued and continuous correspondence from X to Y, and let $f : X \times Y \to \mathbb{R}$ be a continuous function. Then, we have*

(a) *the function $m : X \to \mathbb{R}$, $x \mapsto \max\{f(x,y) | \, y \in \beta(x)\}$ is continuous,*
(b) *the correspondence $x \mapsto \{y \in \beta(x) | \, f(x,y) = m(x)\}$ is nonempty, compact-valued and upper hemicontinuous.*

For a metric space X, let $\{S_n\}$ be a sequence of subsets of X.

Definition D.3. A *topological limes inferior* $L_i(S_n)$ is a subset of X such that $x \in L_i(S_n)$ if and only if for every neighborhood U of x, there is an integer N such that $U \cap S_n \neq \emptyset$ for all $n \geq N$. A *topological limes superior* $L_s(S_n)$ is a subset of X such that $x \in L_s(S_n)$ if and only if for every neighborhood U of x, there are infinitely many n such that $U \cap S_n \neq \emptyset$.

The following proposition is immediate from the definitions.

Proposition D.7. *For every sequence $(F_n)_{n \in \mathbb{N}}$ of subsets of X, the following properties hold.*

(a) *$L_i(F_n)$ and $L_s(F_n)$ are closed (possibly empty) and $L_i(S_n) \subset L_s(S_n)$,*
(b) *$x \in L_i(F_n)$ if and only if there exists an integer $N \in \mathbb{N}$ and a sequence $(x_n)_{n \in \mathbb{N}}$ with $x_n \in F_n$ for all $n \geq N$ and $x_n \to x$,*

(c) $x \in L_s(F_n)$ *if and only if there exists a subsequence* (F_{n_q}) *from which one can choose an element* $x_{n_q} \in F_{n_q}$ *for every* n_q *such that* $x_{n_q} \to x$.

The following proposition is often useful for examining whether a correspondence is lower hemicontinuous.

Proposition D.8. *If a correspondence* $\phi : X \to Y$ *is l.h.c at* x, *then it follows that* $\phi(x) \subset L_i(\phi(x))$ *for every sequence* $\{x_n\}$ *converging to* x. *Conversely, if* $\emptyset \neq \phi(x) \subset L_s(\phi(x))$ *for every sequence* $\{x_n\}$ *converging to* x, *then the correspondence* ϕ *is l.h.c at* x.

From Proposition D.7, we can easily obtain the below proposition:

Proposition D.9. *Let* $\phi_1 \ldots \phi_m$ *be l.h.c. correspondences of* X *to* $\mathbb{R}^\ell$. *Then, we have that the convex hull,* $x \mapsto co\phi_i(x)$, *and the sum,* $x \mapsto \sum_{i=1}^m \phi_i(x)$ *are also l.h.c.*

Proposition D.10. *Let* $\phi_i : X \to Y_i$ *be l.h.c. correspondences of* X *to* Y_i, $i = 1 \ldots m$. *Then, the product,* $x \mapsto \prod_{i=1}^m \phi_i(x)$ *is l.h.c.*

As usual, let $\Delta = \{\boldsymbol{p} \in \mathbb{R}^\ell |\ p^t \geq 0,\ \sum_{t=1}^\ell p^t = 1\}$ be the unit $\ell - 1$ simplex. The following theorem which is known as the *Gale–Nikaido–Debreu lemma* or simply *Gale–Nikaido lemma* is fundamental to proving the existence of equilibria.

Theorem D.3 (Gale–Nikaido). *Let* X *be a compact and convex subset of* $\mathbb{R}^\ell$, *and* P *a closed and convex subset of* Δ. *Let* $Z : P \to X$ *be an upper hemicontinuous correspondence for every* $\boldsymbol{p} \in P$ *and* $Z(\boldsymbol{p})$ *be a non-empty convex set that satisfies* $\boldsymbol{p}Z(\boldsymbol{p}) \leq 0$ *for every* $\boldsymbol{p} \in P$. *Then, there exist* $\boldsymbol{p}^* \in P$ *with* $Z(\boldsymbol{p}^*) \cap \mathscr{P}_P \neq \emptyset$, *where* $\mathscr{P}_P$ *is the polar of* P, *or there exists a vector* $\boldsymbol{z}^* \in Z(\boldsymbol{p}^*)$ *such that* $\boldsymbol{q}\boldsymbol{z}^* \leq 0$ *for every* $\boldsymbol{q} \in P$. *If* $P = \Delta$, *then* $\boldsymbol{z}^* \leq \boldsymbol{0}$.

Proof. Define a map $M : X \to P$ by

$$M(\boldsymbol{z}) = \{\boldsymbol{p} \in P |\ \boldsymbol{p}\boldsymbol{z} \geq \boldsymbol{q}\boldsymbol{z} \text{ for every } \boldsymbol{q} \in P\},$$

and then we can construct the correspondence

$$M \times Z : P \times X \to P \times X,\ (\boldsymbol{p}, \boldsymbol{z}) \mapsto M(\boldsymbol{z}) \times Z(\boldsymbol{p}).$$

First, we shall show that the correspondence M is upper hemicontinuous. As a closed subset of a compact set Δ, P is compact. Hence, it is enough to show that M is closed by Proposition D.2. Take a sequence $\{(\boldsymbol{p}_n, \boldsymbol{z}_n)\}$ with $\boldsymbol{p}_n \in M(\boldsymbol{z}_n)$ for all n and $(\boldsymbol{p}_n, \boldsymbol{z}_n) \to (\boldsymbol{p}, \boldsymbol{z})$. We want to show that $\boldsymbol{p} \in M(\boldsymbol{z})$. Suppose not. Then, $\boldsymbol{pz} < \boldsymbol{qz}$ for some $\boldsymbol{q} \in P$. As $(\boldsymbol{p}_n, \boldsymbol{z}_n) \to (\boldsymbol{p}, \boldsymbol{z})$, we have $\boldsymbol{p}_n\boldsymbol{z}_n \to \boldsymbol{pz}$; hence, $\boldsymbol{p}_n\boldsymbol{z}_n < \boldsymbol{q}\boldsymbol{z}_n$ for n large enough. This contradicts $\boldsymbol{p}_n \in M(\boldsymbol{z}_n)$. This proves that M is upper hemicontinuous. Hence, so is the product map $M \times Z$ by Proposition D.6. Therefore, $M \times Z$ is an upper hemicontinuous correspondence from a compact and convex set $X \times P$ to itself. It is easy to check that $M \times Z$ is convex-valued. By the Kakutani fixed point theorem (Theorem D.1), there exists a fixed point $(\boldsymbol{p}^*, \boldsymbol{z}^*) \in M(\boldsymbol{z}^*) \times Z(\boldsymbol{p}^*)$, or $\boldsymbol{p}^* \in M(\boldsymbol{z}^*)$ and $\boldsymbol{z}^* \in Z(\boldsymbol{p}^*)$. We have $\boldsymbol{p}^*\boldsymbol{z}^* \leq 0$, because $\boldsymbol{z}^* \in Z(\boldsymbol{p}^*)$. By the definition of the correspondence M, we have

$$\boldsymbol{q}\boldsymbol{z}^* \leq \boldsymbol{p}^*\boldsymbol{z}^* \leq 0 \text{ for every } \boldsymbol{q} \in P.$$

This proves the first part of the theorem. Now let $P = \Delta$. Taking $\boldsymbol{q} = \boldsymbol{e}_t (= (0 \ldots 1 \ldots 0))$ in the above inequality, one obtains $\boldsymbol{z}^t \leq 0$ for $t = 1 \ldots \ell$. This proves the latter part of the theorem. □

Theorem D.3 was strengthened by Hildenbrand [94] such that the exact equality between the supply and demand holds when Walras' law is met by the exact equality and the boundary condition (see (iii) below) is additionally imposed. Let $\mathring{\Delta} = \{\boldsymbol{p} \in \mathbb{R}^\ell |\ \sum_{t=1}^{\ell} p^t = 1,\ p^t > 0,\ t = 1 \ldots \ell\}$ be the interior of the simplex.

Theorem D.4 (Hildenbrand). *Let Z be a correspondence of $\mathring{\Delta}$ into $\mathbb{R}^\ell$ that satisfies the following properties:*

(i) *for every strictly positive price vector $\mathbf{p} \gg \mathbf{0}$, $\mathbf{p}Z(\mathbf{p}) = 0$,*
(ii) *the correspondence Z is compact and convex-valued, bounded from below and u.h.c.,*
(iii) *if a sequence $\{\mathbf{p}_n\}$ in $\mathring{\Delta}$ converges to $\mathbf{p} \in \partial\Delta = \{\mathbf{p} = (p^t) \in \Delta |\ p^t = 0$ for some $t\}$, then $\inf\{\sum_{t=1}^{\ell} z^t |\ \boldsymbol{z} = (z^t) \in Z(\mathbf{p})\} > 0$ for n large enough.*

Then, there exists a vector $\mathbf{p}^* \gg \mathbf{0}$ such that $\mathbf{0} \in Z(\mathbf{p}^*)$.

Proof. For $n = 1, 2 \dots$, define

$$\Delta_n = \left\{ \boldsymbol{p} = (p^t) \in \mathbb{R}^\ell \,\middle|\, \sum_{t=1}^{\ell} p^t = 1,\ p^t \geq 1/n,\ t = 1 \dots \ell \right\}.$$

By Proposition D.3, the set $Z(\Delta_n)$ is compact. Hence, we can apply Theorem D.3 and there exist vectors $\boldsymbol{p}_n \in \Delta_n$ and $\boldsymbol{z}_n \in \mathbb{R}^\ell$ such that (1) $\boldsymbol{z}_n \in Z(\boldsymbol{p}_n)$ and (2) $\boldsymbol{q}\boldsymbol{z}_n \leq 0$ for every $\boldsymbol{q} \in \Delta_n$. It suffices to show that $\boldsymbol{z}_n = \boldsymbol{0}$ for some n. As Δ is compact, we can assume that $\boldsymbol{p}_n \to \boldsymbol{p} \in \Delta$. We now claim that $\boldsymbol{p} \gg \boldsymbol{0}$. Otherwise, it would follow from assumption (iii) that $\sum_{t=1}^{\ell} z_n^t > 0$ for n large enough, because $\boldsymbol{z}_n \in Z(\boldsymbol{p}_n)$. Setting $\boldsymbol{q} = (1/\ell \dots 1/\ell)$ in property (2), one obtains $\sum_{t=1}^{\ell} z_n^t \leq 0$, a contradiction. Finally, $\boldsymbol{p} \gg \boldsymbol{0}$ implies that $\boldsymbol{z}_n = \boldsymbol{0}$ for n large enough. Indeed, let N be such that Δ_N contains $\boldsymbol{p}$ in its interior. By assumption (i), we have $\boldsymbol{p}_n\boldsymbol{z}_n = 0$. As $\boldsymbol{p}_n \in \mathring{\Delta}_N$ for n large enough, it follows from the property (2) that $\boldsymbol{z}_n = \boldsymbol{0}$. □

Note that in Theorem D.4, we restrict the domain of the correspondence Z to the interior of the price simplex, and the range of Z is not restricted to a compact set.

For a metric space (X, d), let $\mathcal{K}(X)$ be the set of all non-empty compact subsets of (X, d). For every E and $F \in \mathcal{K}(X)$, we define the *Hausdorff distance* δ by

$$\delta(E, F) = \inf\{\epsilon \in [0, \infty) |\ E \subset \bar{B}(F, \epsilon) \text{ and } F \subset \bar{B}(E, \epsilon)\},$$

where $\bar{B}(E, \epsilon)$ is the (closed) ϵ-neighborhood of E,

$$\bar{B}(E, \epsilon) = \{x \in X |\ \inf_{z \in E} d(x, z) \leq \epsilon\}.$$

Then, one can show that δ satisfies the condition (Mt-1), (Mt-2), and (Mt-3) of Appendix B; hence, the space $(\mathcal{K}(X), \delta)$ is a metric space.

Proposition D.11. *The metric space* $(\mathcal{K}(X), \delta)$ *has the following properties:*

(a) *the Hausdorff distance topology on* $\mathcal{K}(X)$ *depends only on the topology of* X *and not on the particular metric on* X,

(b) *if* X *is separable, then so is* $(\mathcal{K}(X), \delta)$,

(c) *if X is separable and locally compact, then $(\mathcal{K}(X), \delta)$ is complete,*
(d) *if X is compact, then so is $(\mathcal{K}(X), \delta)$.*

A subset F of X is called the *closed limit* of a sequence $(F_n)_{n \in \mathbb{N}}$ if $L_i(F_n) = F = L_s(F_n)$. Let (X, d) be a compact metric space and let $(\mathcal{F}_0, \delta)$ be the set of all non-empty closed subsets of X with the topology of the Hausdorff distance. Then, the metric space $(\mathcal{F}_0, \delta)$ is compact by Proposition D.11. Moreover, we have the following proposition:

Proposition D.12. *A sequence (F_n) converges to F in $(\mathcal{F}_0, \delta)$ if and only if $L_i(F_n) = F = L_s(F_n)$. Every open set of $(\mathcal{F}_0, \delta)$ can be written as a union of the sets of the form*

$$B(G; G_1 \ldots G_k) = \{F \in \mathcal{F}_0 |\ F \subset G \text{ and } F \cap G_i \neq \emptyset,\ i = 1 \ldots k\},$$

where $G, G_1 \ldots G_k$ are open sets of X.

Let $\mathcal{K}_c$ be the family of all compact and convex subsets of $\mathbb{R}^\ell$. Then, by Proposition D.11, $\mathcal{K}_c$ is metrizable by the Hausdorff distance δ. Furthermore, Rådström proved the following theorem:

Theorem D.5. *The space $(\mathcal{K}_c, \delta)$ can be embedded as a convex cone in a real normed space L in such a way that*

(a) *the embedding is isometric,*
(b) *the addition in L induces an addition in $\mathcal{K}_c$,*
(c) *the multiplication by non-negative scalars in L induces the same operation in $\mathcal{K}_c$.*

For a metric space (X, d), let $\mathcal{F}(X)$ be the set of all closed subsets of X. For a compact subset K of X and a finite family $G_1 \ldots G_k$ of open subsets of X, consider subsets of $\mathcal{F}(X)$ that are of the form

$$[K; G_1 \ldots G_k] \equiv \{F \in \mathcal{F}(X) |\ F \cap K = \emptyset \text{ and } F \cap G_i \neq \emptyset,\ i = 1 \ldots k\}.$$

As a finite intersection of sets of this form is again of this form, the family of arbitrary unions of these sets satisfies the conditions (Tp-1), (Tp-2), and (Tp-3) of Appendix B; hence, the family is a topology on the set $\mathcal{F}(X)$. This topology is called the *topology of closed convergence* and denoted by τ_c. Then, one can prove the following theorem:

Theorem D.6. *Let (X, d) be a locally compact and separable metric space. Then, the set $\mathcal{F}(X)$ of all closed subsets of X endowed with the topology of closed convergence, $(\mathcal{F}(X), \tau_c)$ is a compact metrizable space. A sequence $(F_n)_{n \in \mathbb{N}}$ converges to F if and only if $L_i(F_n) = F = L_s(F_n)$.*

For a metric space (X, d), we can endow the set $\mathcal{K}(X)$ with both the Hausdorff distance topology and the topology of closed convergence. If X is not compact, then the topological space $(\mathcal{K}(X), \delta)$ is distinct from the topological space $(\mathcal{K}(X), \tau_c)$. The topology induced by the Hausdorff distance δ is finer than the topology of closed convergence τ_c.

Appendix E

Differential Calculus and Manifolds

Definition E.1. An abelian group (see Definition C.1) $(L, +)$ is called a *vector space* or *linear space* over $\mathbb{R}$ if we have a function $\cdot : \mathbb{R} \times L \to L$ that satisfies the following conditions:

(Vc-1) $a \cdot (\boldsymbol{x} + \boldsymbol{y}) = a \cdot \boldsymbol{x} + a \cdot \boldsymbol{y}$ for all $a \in \mathbb{R}$ and all $\boldsymbol{x}, \boldsymbol{y} \in L$,
(Vc-2) $(a + b) \cdot \boldsymbol{x} = a \cdot \boldsymbol{x} + b \cdot \boldsymbol{x}$ for all $a, b \in \mathbb{R}$ and all $\boldsymbol{x} \in L$,
(Vc-3) $a(b \cdot \boldsymbol{x}) = (ab) \cdot \boldsymbol{x}$ for all $a, b \in \mathbb{R}$ and all $\boldsymbol{x} \in L$,
(Vc-4) $0 \cdot \boldsymbol{x} = \boldsymbol{0}$, and $1 \cdot \boldsymbol{x} = \boldsymbol{x}$ for all $\boldsymbol{x} \in L$.

The function $\cdot$ the multiplication by scalars. In the following, we will often omit the dot for the scalar multiplication. We have $\boldsymbol{x} + (-1)\boldsymbol{x} = 1 \cdot \boldsymbol{x} + (-1)\boldsymbol{x} = (1-1)\boldsymbol{x} = 0\boldsymbol{x} = \boldsymbol{0}$; hence, the element $(-1)\boldsymbol{x}$ is the inverse of $\boldsymbol{x}$ and written as $-\boldsymbol{x}$. When the space has a basis $\{\boldsymbol{b}_1, \ldots, \boldsymbol{b}_n\}$, then the *dimension* of the space is n. For n-dimensional spaces, obviously there do not exist more than n linearly independent vectors. If there exist n linearly independent vectors for every $n \in \mathbb{N}$, the space is called *infinite dimensional.*

Definition E.2. A non-negative real-valued function $\|\cdot\|$ defined on a vector space L is called a *norm* if it satisfies

(Nr-1) $\|\boldsymbol{x}\| = \boldsymbol{0}$ if and only if $\boldsymbol{x} = \boldsymbol{0}$,
(Nr-2) $\|\boldsymbol{x} + \boldsymbol{y}\| \leq \|\boldsymbol{x}\| + \|\boldsymbol{y}\|$ for all $\boldsymbol{x}, \boldsymbol{y} \in L$,
(Nr-3) $\|a\boldsymbol{x}\| = |a|\|\boldsymbol{x}\|$ for all $a \in \mathbb{R}$ and for all $\boldsymbol{x}, \boldsymbol{y} \in L$.

A normed vector space becomes a metric space, and hence, a topological space with the norm topology if we define a metric d by

$d(\boldsymbol{x}, \boldsymbol{y}) = \|\boldsymbol{x} - \boldsymbol{y}\|$. When a normed vector space is complete in this metric, it is called a *Banach space.*

Example E.1. Let $L = \mathbb{R}^\ell$, the ℓ-dimensional Euclidean space. We can consider three norms on $\mathbb{R}^\ell$.

$$\|\boldsymbol{x}\|_1 = |x^1| + \cdots + |x^\ell|,$$

$$\|\boldsymbol{x}\|_2 = \sqrt{(x^1)^2 + \cdots + (x^\ell)^2},$$

$$\|\boldsymbol{x}\|_\infty = \max\{|x^t| \mid 1 \leq t \leq \ell\}.$$

We observed in Appendix B that these norms are equivalent in the sense that they induce the same topology on $\mathbb{R}^\ell$. The finite-dimensional spaces are extended to the infinite-dimensional spaces.

Example E.2. The set of sequences $\boldsymbol{x} = (x^t)$ satisfying $\sum_{t=0}^\infty |x^t|^p < +\infty$ for $1 \leq p < +\infty$ is a Banach space with the norm

$$\|x\|_p = \left(\sum_{t=0}^\infty |x^t|^p\right)^{1/p}.$$

The conditions (Nr-1) and (Nr-2) can be easily verified. The condition (n-3) follows from the Minkowski's inequality

$$\left(\sum_{t=0}^\infty |x^t + y^t|^p\right)^{1/p} \leq \left(\sum_{t=0}^\infty |x^t|^p\right)^{1/p} + \left(\sum_{t=0}^\infty |y^t|^p\right)^{1/p},$$

which can be proved for every integer p with $1 \leq p < +\infty$. Moreover, we can show that the space ℓ^p is complete and separable with respect to this norm topology; hence, it is a Banach space. In particular, for $p = 1, 2$, the norms of the ℓ^1 and ℓ^2 spaces are clearly generalizations of the first and second norms of Example E.1 to infinite-dimensional spaces of sequences.

The third norm of Example E.1 is extended to the norm $\|\boldsymbol{x}\|_\infty = \sup_{t\geq 0} |x^t|$ for a sequence $\boldsymbol{x} = (x^t)$. The space with this norm is called the ℓ^∞ space and defined by

$$\ell^\infty = \left\{ \boldsymbol{x} = (x^t) \,\middle|\, \sup_{t\geq 0} |x^t| < +\infty \right\}.$$

The conditions (Nr-1), (Nr-2), and (Nr-3) can be easily verified and one can prove that the space ℓ^∞ is complete with respect to this norm. Hence, the space ℓ^∞ is also a Banach space. Unfortunately, it is not separable. We will discuss these spaces further in Appendix G.

An important class of the Banach spaces is the Hilbert spaces that possess the inner product between any two vectors of the space.

Definition E.3. A *pre-Hilbert space* is a vector space L together with an *inner product* $(\boldsymbol{x}, \boldsymbol{y}) \mapsto \langle \boldsymbol{x}, \boldsymbol{y} \rangle$ that satisfies

(Ip-1) $\langle \boldsymbol{x}, \boldsymbol{y} \rangle = \langle \boldsymbol{y}, \boldsymbol{x} \rangle$ for all $\boldsymbol{x}, \boldsymbol{y} \in L$,
(Ip-2) $\langle \boldsymbol{x} + \boldsymbol{y}, \boldsymbol{z} \rangle = \langle \boldsymbol{x}, \boldsymbol{z} \rangle + \langle \boldsymbol{y}, \boldsymbol{z} \rangle$ for all $\boldsymbol{x}, \boldsymbol{y}, \boldsymbol{z} \in L$,
(Ip-3) $\langle a\boldsymbol{x}, \boldsymbol{y} \rangle = a\langle \boldsymbol{x}, \boldsymbol{y} \rangle$ for all $\boldsymbol{x}, \boldsymbol{y} \in L$ and for all $a \in \mathbb{R}$,
(Ip-4) $\langle \boldsymbol{x}, \boldsymbol{x} \rangle \geq 0$ and $\langle \boldsymbol{x}, \boldsymbol{x} \rangle = 0$ if and only if $\boldsymbol{x} = \boldsymbol{0}$.

For a pre-Hilbert space L, we can define a norm $\|\boldsymbol{x}\|$ by $\|\boldsymbol{x}\| = (\langle \boldsymbol{x}, \boldsymbol{x} \rangle)^{1/2}$ for $\boldsymbol{x} \in L$, as shall be shown below. We then have the following proposition:

Proposition E.1 (The Cauchy–Schwarz inequality). *For all $\boldsymbol{x}$, $\boldsymbol{y}$ in a pre-Hilbert space L, it follows that $|\langle \boldsymbol{x}, \boldsymbol{y} \rangle| \leq \|\boldsymbol{x}\|\|\boldsymbol{y}\|$. The equality holds if and only if $\boldsymbol{x} = a\boldsymbol{y}$ for some $a \in \mathbb{R}$, or $\boldsymbol{x} = \boldsymbol{0}$.*

Let L be a pre-Hilbert space. For any $\boldsymbol{x}, \boldsymbol{y} \in L$, we have by (Ip-1), (Ip-2), and Proposition E.1,

$$\begin{aligned}
\|\boldsymbol{x} + \boldsymbol{y}\|^2 &= \langle \boldsymbol{x} + \boldsymbol{y}, \boldsymbol{x} + \boldsymbol{y} \rangle \\
&= \langle \boldsymbol{x}, \boldsymbol{x} \rangle + \langle \boldsymbol{x}, \boldsymbol{y} \rangle + \langle \boldsymbol{y}, \boldsymbol{x} \rangle + \langle \boldsymbol{y}, \boldsymbol{y} \rangle \\
&\leq \|\boldsymbol{x}\|^2 + 2|\langle \boldsymbol{x}, \boldsymbol{y} \rangle| + \|\boldsymbol{y}\|^2 \\
&\leq \|\boldsymbol{x}\|^2 + 2\|\boldsymbol{x}\|\|\boldsymbol{y}\| + \|\boldsymbol{y}\|^2 = (\|\boldsymbol{x}\| + \|\boldsymbol{y}\|)^2,
\end{aligned}$$

hence, $\|\boldsymbol{x} + \boldsymbol{y}\| \leq \|\boldsymbol{x}\| + \|\boldsymbol{y}\|$. Therefore, if L is a pre-Hilbert space, then the map $\|\cdot\| : L \to \mathbb{R}_+$ defined by $\|\boldsymbol{x}\| = \langle \boldsymbol{x}, \boldsymbol{x} \rangle^{1/2}$ is indeed a norm on L. A pre-Hilbert space is called a *Hilbert space* if it is complete in this norm. A finite-dimensional Euclidean space $\mathbb{R}^\ell$ is of course an example of the Hilbert spaces, but the inner product of $\boldsymbol{x}$ and $\boldsymbol{y}$ of $\mathbb{R}^\ell$ was denoted simply as $\boldsymbol{xy}$ rather than $\langle \boldsymbol{x}, \boldsymbol{y} \rangle$. In Example E.2, the space ℓ^2 is also a Hilbert space that has an inner product defined by $\langle \boldsymbol{x}, \boldsymbol{y} \rangle = \sum_{t=0}^{\infty} x^t y^t$ for $\boldsymbol{x} = (x^t)$ and $\boldsymbol{y} = (y^t)$ in ℓ^2.

Let L and M be Banach spaces. A map $f : L \to M$ is said to be *linear* if

$$f(a\boldsymbol{x} + b\boldsymbol{y}) = af(\boldsymbol{x}) + bf(\boldsymbol{y})$$

for every $\boldsymbol{x}, \boldsymbol{y} \in L$ and every $a, b \in \mathbb{R}$. The space L is called the *domain* of f. When $M = \mathbb{R}$, the linear map $f : L \to \mathbb{R}$ is called a *linear functional.* The idea of linearity is generalized to that of a *multilinear map.* Let $L_1, \ldots, L_n, M$ be Banach spaces. A map $g : L_1 \times \ldots \times L_n \to M$ is said to be n *multilinear* if $g(\boldsymbol{x}_1 \ldots \boldsymbol{x}_n)$ is linear in each variable separately. For instance, the linearity in the first variable means that

$$g(a\boldsymbol{x}_1 + b\boldsymbol{y}_1 \ldots \boldsymbol{x}_n) = ag(\boldsymbol{x}_1 \ldots \boldsymbol{x}_n) + bg(\boldsymbol{y}_1 \ldots \boldsymbol{x}_n).$$

The n multilinear map g is continuous if and only if there exists $C > 0$ such that

$$\|g(\boldsymbol{x}_1 \ldots \boldsymbol{x}_n)\| \le C\|\boldsymbol{x}_1\| \ldots \|\boldsymbol{x}_n\| \text{ for all } \boldsymbol{x}_k \in L_k, k = 1 \ldots n.$$

The space of continuous n multilinear maps of $L_1 \times \cdots \times L_n$ to M is denoted by $\mathscr{L}(L_1 \ldots L_n, M)$. When $L_1 = \cdots = L_n = L$, it is denoted by $\mathscr{L}^n(L, M)$. When $n = 1$, we usually write $\mathscr{L}^1(L, M) = \mathscr{L}(L, M)$. This is nothing but the space of continuous linear maps of L to M. In particular, if $M = \mathbb{R}$, then $\mathscr{L}(L, \mathbb{R})$ is often denoted as L^* and called the *dual space* of L.

The space $\mathscr{L}^n(L, M)$ is obviously a linear space and we can endow a norm on it that is called the *operator norm* defined by

$$\|g\| = \sup\left\{ \frac{\|g(\boldsymbol{x}_1 \ldots \boldsymbol{x}_n)\|}{\|\boldsymbol{x}_1\| \ldots \|\boldsymbol{x}_n\|} \,\middle|\, \boldsymbol{x}_1, \ldots, \boldsymbol{x}_n \ne \boldsymbol{0} \right\}.$$

We can prove easily that the space $\mathscr{L}^n(L, M)$ is complete if M is. Moreover, we can show that $\mathscr{L}(L, \mathscr{L}^{n-1}(L, M)) = \mathscr{L}^n(L, M)$. Indeed, let $f \in \mathscr{L}(L, \mathscr{L}^{n-1}(L, M))$. As $f(\boldsymbol{x}_1) \in \mathscr{L}^{n-1}(L, M)$, we can define a map $g : L^n \to M$ by $g(\boldsymbol{x}_1 \ldots \boldsymbol{x}_n) = f(\boldsymbol{x}_1)(\boldsymbol{x}_2 \ldots \boldsymbol{x}_n)$. Obviously, the map $f \mapsto g$ is bijective and linear, and it is easily verified that $\|f\| = \|g\|$; hence, $\mathscr{L}(L, \mathscr{L}^{n-1}(L, M))$ and $\mathscr{L}^n(L, M)$ are *isomorphic*:

Definition E.4. Banach spaces L and M are said to be *isomorphic* if there exists a bijective (namely one to one and onto) map $f \in \mathscr{L}(L, M)$ with $f^{-1} \in \mathscr{L}(M, L)$.

Two isomorphic Banach spaces are considered to be the same space. Hence, if L and M are isomorphic, we often write $L \approx M$ or even $L = M$.

Let L and M be Banach spaces, and $f : L \to M$ is a map from L to M. We say that the map f is open if and only if $f(U)$ is open in M whenever U is open in L.

Theorem E.1 (Open mapping theorem). *Let L and M be Banach spaces and suppose $f \in \mathscr{L}(L, M)$ is onto. Then, f is an open mapping.*

Therefore, if a linear map $f \in \mathscr{L}(L, M)$ is one to one and onto, it is an isomorphism (Banach isomorphism theorem). The convex subsets of a normed space L and the concavity of the real-valued functions defined on a convex subset of L are defined in the same way as the case of $\mathbb{R}^\ell$ (see Appendix A). The separation hyperplane theorem (Theorem A.1) is extended to Banach spaces.

Theorem E.2 (Hahn–Banach). *Let X and Y be non-empty convex subsets of a Banach space L, and $\mathring{X} \neq \emptyset$ or $\mathring{Y} \neq \emptyset$. If $X \cap Y = \emptyset$, then there exists a vector $\boldsymbol{p} \in L^*$ with $\boldsymbol{p} \neq \boldsymbol{0}$ such that $\boldsymbol{px} \leq \boldsymbol{py}$ for every $\boldsymbol{x} \in X$ and every $\boldsymbol{y} \in Y$.*

Theorem A.5 is generalized to the case of Banach spaces.

Theorem E.3 (Kuhn–Tucker). *Let L be a Banach space and let f be a real-valued concave function defined on a convex set X of L,* *and $g_1 \dots g_m$ be real-valued concave functions defined on X. Suppose that Slater's condition holds, namely that there exists an $\boldsymbol{x}_0 \in X$ such that $g_j(\boldsymbol{x}_0) > 0$, $j = 1 \dots m$. Then, a point $\hat{\boldsymbol{x}}$ achieves a maximum of $f(\boldsymbol{x})$ on X subject to $g_j(\boldsymbol{x}) \geq 0$, $j = 1 \dots m$ if and only if there exists a non-negative vector $\hat{\boldsymbol{\lambda}} = (\hat{\lambda}_1 \dots \hat{\lambda}_m) \geq \mathbf{0}$ such that*

$$\mathcal{L}(\boldsymbol{x}, \hat{\boldsymbol{\lambda}}) \leq \mathcal{L}(\hat{\boldsymbol{x}}, \hat{\boldsymbol{\lambda}}) \leq \mathcal{L}(\hat{\boldsymbol{x}}, \boldsymbol{\lambda})$$

for every $\boldsymbol{x} \in X$ and every $\boldsymbol{\lambda} \geq \mathbf{0}$, where $\mathcal{L}(\boldsymbol{x}, \boldsymbol{\lambda}) = f(\boldsymbol{x}) + \boldsymbol{\lambda} g(\boldsymbol{x}) = f(\boldsymbol{x}) + \sum_{j=1}^{m} \lambda_j g_j(\boldsymbol{x})$ is the Lagrangian.

When $L = \mathbb{R}^\ell$ and $M = \mathbb{R}^m$, the linear map $f : L \to M$ is of course represented by the matrix $A = (a_s^t)$, $a_s^t = f^t(\boldsymbol{e}_s)$, $s = 1 \dots \ell$, $t = 1 \dots m$. The transpose matrix $A' = (b_s^t)$, $b_s^t = a_t^s$ is extended to the *adjoint map*.

Proposition E.2. *Let L and M be Banach spaces. For each $f \in \mathscr{L}(L, M)$, there exists a unique $f^* \in \mathscr{L}(M^*, L^*)$ called the adjoint of f satisfying $\mathbf{p}(f(\boldsymbol{x})) = f^*(\mathbf{p})(\boldsymbol{x})$ for all $\boldsymbol{x} \in L$ and all $\mathbf{p} \in M^*$, and $\|f\| = \|f^*\|$.*

Let H be a Hilbert space with the inner product $\langle \boldsymbol{x}, \boldsymbol{y} \rangle$ for $\boldsymbol{x}, \boldsymbol{y} \in H$. Then, we have the Riesz' lemma.

Theorem E.4 (Riesz' Lemma). *For each $\mathbf{p} \in H^*$, there exists $\boldsymbol{y} \in H$ such that $\|\mathbf{p}\| = \|\boldsymbol{y}\|$ and $\mathbf{p}(\boldsymbol{x}) = \langle \boldsymbol{y}, \boldsymbol{x} \rangle$ for all $\boldsymbol{x} \in H$.*

As a corollary, it follows that $H = H^*$, and the condition of the adjoint map of Proposition E.2 is written as $\langle \boldsymbol{y}, f(\boldsymbol{x}) \rangle = \langle f^*(\boldsymbol{y}), \boldsymbol{x} \rangle$, which might look more transparent.

For $f \in \mathscr{L}(L, M)$, the *kernel* (or the *null space*) and the *range* of f are denoted by $\mathscr{N}(f) = \{\boldsymbol{x} \in L \mid f(\boldsymbol{x}) = \mathbf{0}\} (= f^{-1}(\mathbf{0}))$ and $\mathscr{R}(f) = \{\boldsymbol{y} \in M \mid f(\boldsymbol{x}) = \boldsymbol{y} \text{ for some } \boldsymbol{x} \in L\}$, respectively. Let M be a subspace of a Banach space L. The *annihilator* $M^\perp$ of M is defined by

$$M^\perp = \{\mathbf{p} \in L^* \mid \mathbf{p}(\boldsymbol{x}) = 0 \text{ for all } \boldsymbol{x} \in M\}.$$

Similarly, for a subspace N of L^*, the annihilator $N^\perp$ of N is defined by

$$N^\perp = \{\boldsymbol{x} \in L|\ \boldsymbol{p}(\boldsymbol{x}) = 0 \text{ for all } \boldsymbol{p} \in N\}.$$

Proposition E.3. *Let L and M be Banach spaces and $f \in \mathscr{L}(L, M)$. Then, $\mathscr{N}(f) = \mathscr{R}(f^*)^\perp$ and $\mathscr{N}(f^*) = \mathscr{R}(f)^\perp$.*

Let L be a Banach space and M a linear subspace of L. We can define an equivalence relation on L by $\boldsymbol{x} \equiv \boldsymbol{y}$ if and only if $\boldsymbol{x} - \boldsymbol{y} \in M$. Let $[\boldsymbol{x}]$ be the equivalence class containing $\boldsymbol{x} \in L$. The set of all equivalent classes are called the *quotient vector space* and denoted by L/M. We can define the sum and the scalar multiplication on L/M by $[\boldsymbol{x}] + [\boldsymbol{y}] = [\boldsymbol{x} + \boldsymbol{y}]$ and $a[\boldsymbol{x}] = [a\boldsymbol{x}]$ for $a \in \mathbb{R}$, and it can be easily shown that L/M is a vector space. When *dimension* $L/M = n$, we say that M is a subspace of L with the *codimension* n. The null space of $\boldsymbol{p} \in L^*$, $\mathscr{N}(\boldsymbol{p}) = \{\boldsymbol{x} \in L|\ \boldsymbol{p}(\boldsymbol{x}) = 0\}$ is an example of a subspace of L with the codimension 1, which is called a *hyperplane* perpendicular to $\boldsymbol{p}$.

As in the finite-dimensional spaces, we can approximate a general map between Banach spaces by a linear map (derivative). If this procedure is possible, the approximated map is called *smooth* and one can get a lot of information from the approximating linear map, which is generally simpler (because it is linear!) than the original map.

Definition E.5. Let U be an open subset of a Banach space L. A map f from U to a Banach Space M is called (*Frechet*) *differentiable* at $x \in U$ if there is a continuous linear map $Df \in \mathscr{L}(L, M)$ such that

$$\lim_{\boldsymbol{h} \to \boldsymbol{0}} \frac{\|f(\boldsymbol{x} + \boldsymbol{h}) - f(\boldsymbol{x}) - Df(\boldsymbol{x})\|}{\|\boldsymbol{h}\|} = 0.$$

The linear map $Df(\boldsymbol{x}) \in \mathscr{L}(L, M)$ is called the derivative of f at $\boldsymbol{x}$.

An advantage of this definition of the derivative is that it is "coordinate free"; hence, it can be applied to the case of general Banach spaces. When we use the coordinates of $L = \mathbb{R}^\ell$, $\boldsymbol{x} = (x^1 \dots x^\ell)$ and

$M = \mathbb{R}^m$, $f(\boldsymbol{x}) = (f^1(\boldsymbol{x}) \dots f^m(\boldsymbol{x}))$, the derivative can be written in standard matrix form as follows:

$$Df(\boldsymbol{x}) = \begin{pmatrix} \partial_1 f^1(\boldsymbol{x}) \dots \partial_\ell f^1(\boldsymbol{x}) \\ \dots\dots\dots\dots\dots\dots \\ \partial_1 f^m(\boldsymbol{x}) \dots \partial_\ell f^m(\boldsymbol{x}) \end{pmatrix},$$

where $\partial_s f^t(\boldsymbol{x}) = \partial f^t(\boldsymbol{x})/\partial x^s$. This $\ell \times m$ matrix is often called the *Jacobian matrix* of f.

For every integer $r \geq 0$, the rth derivative $D^r f(\boldsymbol{x})$ of f at $x \in U$ is defined inductively

$$D^r f(x) \equiv D(D^{r-1} f)(x) : U \to \mathscr{L}(L, \mathscr{L}^{r-1}(L, M)) \approx \mathscr{L}^r(L, M),$$

which maps $\boldsymbol{x}$ at U to an r-multilinear map of L to M. A map f is said to be of *class* C^r at $\boldsymbol{x} \in U$ if this map is continuous. When $L = \mathbb{R}^\ell$ and $M = \mathbb{R}^m$, it is equivalent for every partial derivative (in the usual sense)

$$\frac{\partial^n f^t}{\partial x^{s_1} \dots \partial x^{s_n}}(\boldsymbol{x}), \quad 1 \leq s_1 \dots s_n \leq \ell, \quad 1 \leq t \leq m, \quad 1 \leq n \leq r$$

to exist and be continuous.

The fundamental property of the derivative is that it is linear.

Proposition E.4. *Let L, M be Banach spaces, and $U \subset L$ an open set. Let a be a real number. If $f, g : U \to M$ are of class C^r, then af and $f + g$ are also of class C^r and*

$$D^r(f(\boldsymbol{x}) + g(\boldsymbol{x})) = D^r f(\boldsymbol{x}) + D^r g(\boldsymbol{x}) \text{ and } D^r(af)(\boldsymbol{x}) = aD^r f(\boldsymbol{x}).$$

Proposition E.5 (Chain rule). *Let L, M, N be Banach spaces, $U \subset L$ and $V \subset M$ are open, and map $f : U \to V$ and $g : V \to N$ which are of Class C^1. Then, we have*

$$D(g \circ f)(\boldsymbol{x}) = Dg(f(\boldsymbol{x})) \circ Df(\boldsymbol{x}).$$

Let σ be a permutation of $\{1 \dots r\}$, or one to one and onto a map of $\{1 \dots r\}$ to itself. For $g \in \mathscr{L}^r(L, M)$, we define $\sigma g(\boldsymbol{x}_1 \dots \boldsymbol{x}_r) = g(\sigma(\boldsymbol{x}_1) \dots \sigma(\boldsymbol{x}_r))$. When $\sigma g = g$ for all σ, we say that the r multilinear map g is *symmetric*. Then, we have the following proposition.

Proposition E.6 (Euler). *Let L, M be Banach spaces, and $U \subset L$ be an open set. If $f : U \to M$ is of class C^r, then $D^r f \in \mathscr{L}^r(L, M)$ is symmetric.*

Proposition E.7 (Leibniz rule). *Let L, M_1, M_2, and N be Banach spaces and $U \subset L$ an open set. Suppose $f : U \to M_1$ and $g : U \to M_2$ are of class C^r, and $\psi \in \mathscr{L}(M_1 \times M_2, N)$. Let $f \times g : U \to M_1 \times M_2$, $f \times g(\boldsymbol{x}) = (f(\boldsymbol{x}), g(\boldsymbol{x}))$ and $\psi(f, g) = \psi \circ (f \times g)$. Then, $\psi(f, g)$ is of class C^r by the Leibniz rule*

$$D\psi(f, g)(\boldsymbol{x})\boldsymbol{h} = \psi(Df(\boldsymbol{x})\boldsymbol{h}, g(\boldsymbol{x})) + \psi(f(\boldsymbol{x}), Dg(\boldsymbol{x})\boldsymbol{h}).$$

Proposition E.8 (Taylor's formula). *Let L, M be Banach spaces, and $U \subset L$ be an open set. If $f : U \to M$ is of class C^r, then we have*

$$f(\boldsymbol{x} + \boldsymbol{h}) = \sum_{k=0}^{r} \frac{D^k f(\boldsymbol{x})\boldsymbol{h}^k}{k!} + R(\boldsymbol{x}, \boldsymbol{h})\boldsymbol{h}^r,$$

where $\boldsymbol{h}^k = (\boldsymbol{h} \dots \boldsymbol{h}) \in L^k$ and $R(\boldsymbol{x}, \boldsymbol{h})$ is given by

$$R(\boldsymbol{x}, \boldsymbol{h}) = \int_0^1 \frac{(1-t)^{r-1}}{(r-1)!}(D^r f(\boldsymbol{x} + t\boldsymbol{h}) - D^r f(\boldsymbol{x}))dt,$$

and $\boldsymbol{h}$ is assumed to be so small that $\{\boldsymbol{x} + t\boldsymbol{h} |\, 0 \le t \le 1\} \subset U$.

Note that in the above formula, $R(\boldsymbol{x}, \boldsymbol{h})$ is continuous and $R(\boldsymbol{x}, \boldsymbol{0}) = 0$. A related concept of the derivative is explained in the following definition:

Definition E.6. Let U be an open subset of L and let M be a Banach space. We say that a map $f : U \to M$ has a derivative in the direction $\boldsymbol{h} \in L$ at $\boldsymbol{x}$ if

$$\frac{d}{dt} f(\boldsymbol{x} + t\boldsymbol{h})_{t=0}$$

exists. We call this element of M the *Gateaux derivative* at $x \in U$, and if it exists everywhere in U, we say that the map f is *Gateaux differentiable*.

The Fréchet differentiability is stronger than the Gateaux differentiability according to the following proposition:

Proposition E.9. *If f is Fréchet differentiable at $\boldsymbol{x}$, then the Gateaux derivatives of f at $\boldsymbol{x}$ exist, and they are given by*

$$\frac{d}{dt} f(\boldsymbol{x} + t\boldsymbol{h})_{t=0} = Df(\boldsymbol{x})\boldsymbol{h}.$$

We can also give a "coordinate-free" definition of the partial derivative. Let $L = L_1 \times L_2$, where L_1 and L_2 are Banach spaces, and $U \subset L$ is an open subset of L. For a map of class C^r $f : U \to M$, the *partial derivative* with respect to L_1 is defined by

$$\partial_1 f(\boldsymbol{x}) : L_1 \to M, \ \boldsymbol{v} \mapsto Df(\boldsymbol{x})(\boldsymbol{v}, \boldsymbol{0}), \ \boldsymbol{v} \in L_1.$$

The partial derivative with respect to L_2 is defined similarly. The following theorem is of central importance in the differential calculus on manifolds; see [1, p. 121] for details.

Theorem E.5 (Implicit function theorem). *Let $U \subset L$, $V \subset M$ be open and $f : U \times V \to M$ be of class C^r $(r \geq 1)$. For some $\boldsymbol{x}_0 \in U$ and $\boldsymbol{y}_0 \in V$, assume that $\partial_2 f(\boldsymbol{x}_0, \boldsymbol{y}_0) : M \to M$ is an isomorphism. Then, there are neighborhoods $U_0 \subset U$ of $\boldsymbol{x}_0$ and $W \subset M$ of $f(\boldsymbol{x}_0, \boldsymbol{y}_0)$, and a unique C^r map $g : U_0 \times W \to V$ such that for all $(\boldsymbol{x}, \boldsymbol{z}) \in U_0 \times W$,*

$$f(\boldsymbol{x}, g(\boldsymbol{x}, \boldsymbol{z})) = \boldsymbol{z}.$$

We now give the definition of the (finite-dimensional) differentiable manifolds.

Definition E.7. A Hausdorff topological space $\mathcal{M}$ is an m-dimensional *manifold* if there exists an open cover $\{U_\alpha\}$ of $\mathcal{M}$ and local isomorphisms ϕ_α on U_α to $\mathbb{R}^m$ such that $\phi_\beta \circ \phi_\alpha^{-1} : \phi_\alpha(U_\alpha \cap U_\beta) \to \phi_\beta(U_\alpha \cap U_\beta)$ is bijective and of class C^r for each α and β; see the following diagram.

$$\begin{array}{ccc} U_\alpha & \longrightarrow & U_\beta \\ {\scriptstyle \phi_\alpha}\downarrow & & \downarrow{\scriptstyle \phi_\beta} \\ \mathbb{R}^m & \longrightarrow & \mathbb{R}^m \end{array}$$

As $(\phi_\beta \circ \phi_\alpha^{-1})^{-1} = \phi_\alpha \circ \phi_\beta^{-1} : \phi_\beta(U_\alpha \cap U_\beta) \to \phi_\alpha(U_\alpha \cap U_\beta)$, the inverse of $\phi_\beta \circ \phi_\alpha^{-1}$ is also of class C^r, so that it is C^r-diffeomorphism. (U_α, ϕ_α) is called the *chart* and the family of all charts is called an *atlas*.

Example E.3. The m-dimensional sphere

$$S^m = \left\{ \boldsymbol{x} = (x^0 \ldots x^m) \in \mathbb{R}^{m+1} \,\middle|\, \sqrt{(x^0)^2 + \cdots + (x^m)^2} = 1 \right\}$$

is a manifold. Indeed, define $2(m+1)$ open sets $U_\pm^t$, $t = 0, 1 \ldots m$ on S^m by

$$U_+^t = \{(x^0 \ldots x^m) \in S^m |\, x^t > 0\} \quad \text{and}$$
$$U_-^t = \{(x^0 \ldots x^m) \in S^m |\, x^t < 0\},$$

and $2(m+1)$ maps $\phi_\pm^t$ on $U_\pm^t$ to the open disk

$$B(0,1) = \left\{ \boldsymbol{y} = (y^1 \ldots y^m) \in \mathbb{R}^m \,\middle|\, \|\boldsymbol{y}\| = \sqrt{(y^1)^2 + \cdots + (y^m)^2} < 1 \right\}$$

by

$$\phi_\pm^t(x^0 \ldots x^m) = (x^0 \ldots x^{t-1}, x^{t+1}, \ldots x^m).$$

Then, it is easy to see that

$$\begin{aligned}
&\phi_\pm^s \circ (\phi_\pm^t)^{-1}(y^1 \ldots y^m) \\
&\quad = \left(y^1 \ldots y^{s-1}, y^{s+1} \ldots y^{t-1}, \pm\sqrt{1 - \|\boldsymbol{y}\|^2}, y^t \ldots y^m \right),
\end{aligned}$$

which is obviously smooth.

Let $f : \mathcal{M} \to \mathcal{N}$ be a map from a manifold $\mathcal{M}$ to a manifold $\mathcal{N}$. The map f is said to be of *class* C^r, if the map $\psi_\beta \circ f \circ \phi_\alpha^{-1}$ from $\phi_\alpha(U_\alpha)$ to $\psi_\beta(V_\beta)$ is of class C^r, where (U_α, ϕ_α) and (V_β, ψ_β) are charts of $\mathcal{M}$ and $\mathcal{N}$, respectively. This definition does not depend on the choice of the coordinate charts. Indeed, let (U_γ, ϕ_γ) and (V_δ, ψ_δ) be another chart of $\mathcal{M}$ and $\mathcal{N}$, respectively. As $\psi_\delta \circ f \circ \phi_\gamma^{-1} = (\psi_\delta \circ \psi_\beta^{-1}) \circ (\psi_\beta \circ f \circ \phi_\alpha^{-1}) \circ (\phi_\alpha \circ \phi_\gamma^{-1})$ and $\psi_\delta \circ \psi_\beta^{-1}$ and $\phi_\alpha \circ \phi_\gamma^{-1}$ are of class C^r by definition, $\psi_\delta \circ f \circ \phi_\gamma^{-1}$ is of class C^r.

As a special case, a C^r *curve* through $p \in \mathcal{M}$ is a C^r map from $(-\epsilon, \epsilon)$ to $\mathcal{M}$ such that $c(0) = p$. Two C^r curves c and d are called equivalent if and only if

$$\dot{c}(0) \left(\equiv \frac{d(\phi_\alpha \circ c)}{dt}(0) \right) = \dot{d}(0),$$

where (U_α, ϕ_α) is a chart of $\mathcal{M}$ such that $p \in U_\alpha$. This definition of the equivalence relation is also independent of the choice of the coordinate chart. Let (U_β, ϕ_β) be another chart with $p \in U_\beta$. From $\phi_\beta \circ c = (\phi_\beta \circ \phi_\alpha^{-1}) \circ (\phi_\alpha \circ c)$ and $\phi_\beta \circ d = (\phi_\beta \circ \phi_\alpha^{-1}) \circ (\phi_\alpha \circ d)$, it follows that

$$\frac{d(\phi_\beta \circ c)}{dt} = D(\phi_\beta \circ \phi_\alpha^{-1}) \frac{d(\phi_\alpha \circ c)}{dt} \quad \text{and}$$
$$\frac{d(\phi_\beta \circ d)}{dt} = D(\phi_\beta \circ \phi_\alpha^{-1}) \frac{d(\phi_\alpha \circ d)}{dt}.$$

As $D(\phi_\beta \circ \phi_\alpha^{-1})$ is a linear isomorphism, $d(\phi_\beta \circ c)/dt = d(\phi_\beta \circ d)/dt$ if and only if $d(\phi_\alpha \circ c)/dt = d(\phi_\alpha \circ d)/dt$. The equivalence class is denoted as $[c]_p$ and called a *tangent vector* at p. Let $T_p\mathcal{M}$ be the set of all tangent vectors at p and call it the *tangent space* at p. It is easy to see that $T_p\mathcal{M}$ is an m-dimensional vector space. Let (U_α, ϕ_α) be a coordinate chart with $p \in U_\alpha$. Without loss of generality, we may assume that $\phi_\alpha(p) = \mathbf{0} \in \mathbb{R}^m$. Define the smooth curves $c_i(t) = \phi_\alpha^{-1}(0 \ldots 0, t, 0 \ldots 0)$, where t is at the ith coordinate. Then, the curves $[c_i]_p$, $i = 1 \ldots m$, constitute a basis of $T_p\mathcal{M}$.

Let f be a C^r map from an m-dimensional manifold $\mathcal{M}$ to an n-dimensional manifold $\mathcal{N}$. The *tangent map* (derivative) at $p \in \mathcal{M}$ of a C^r map f is a linear map $Df(p) : T_p\mathcal{M} \to T_{f(p)}\mathcal{N}$ defined by

$$Df(p)([c]_p) = [f \circ c]_{f(p)}.$$

We have to check that this definition is independent of the choice of curves representing the equivalence class. Let c_1 and c_2 be two curves such that $[c_1]_p = [c_2]_p$. This means that $d(\phi_\alpha \circ c_1)/dt = d(\phi_\alpha \circ c_2)/dt$, where (U_α, ϕ_α) is a chart on $\mathcal{M}$ with $p \in U_\alpha$. We want to show that $[f \circ c_1]_{f(p)} = [f \circ c_2]_{f(p)}$. As $\psi_\gamma \circ f \circ c_i = (\psi_\gamma \circ f \circ \phi_\alpha^{-1}) \circ (\phi_\alpha \circ c_i)$, $i = 1, 2$, where (V_γ, ψ_γ) is a chart on $\mathcal{N}$ with $f(p) \in V_\gamma$, we have

$d(\psi_\gamma \circ f \circ c_i)/dt = D(\psi_\gamma \circ f \circ \phi_\alpha^{-1}) d(\phi_\alpha \circ c_i)/dt$, $i = 1, 2$. As $d(\phi_\alpha \circ c_1)/dt = d(\phi_\alpha \circ c_2)/dt$, we get

$$\frac{d(\psi_\gamma \circ f \circ c_1)}{dt}(0) = \frac{d(\psi_\gamma \circ f \circ c_2)}{dt}(0).$$

Let $f : \mathcal{M} \to \mathcal{N}$ be a smooth (C^r) map between an m-dimensional manifold $\mathcal{M}$ and an n-dimensional manifold $\mathcal{N}$. A point $q \in \mathcal{N}$ is a *regular value* of f (or f is *transversal to* $\{q\}$) if for every $p \in f^{-1}(q)$, $Df(p) : T_p\mathcal{M} \to T_{f(p)}\mathcal{N}$ is surjective (onto). Note that when $m < n$, a point q is a regular value only if $q \notin f(\mathcal{M})$. When dimension $m = n$, $q \in \mathcal{N}$ is a regular value if and only if $Df(p)$ is an isomorphism between $T_p\mathcal{M}$ and $T_q\mathcal{N}$ at every $p \in f^{-1}(q)$. A point of $\mathcal{N}$ that is not a regular value is called a *critical value*.

Theorem E.6 (Regular value theorem). *Let $f : \mathcal{M} \to \mathcal{N}$ be a smooth (C^r) map between an m-dimensional manifold $\mathcal{M}$ and an n-dimensional manifold $\mathcal{N}$. Suppose that $m \geq n$ and $q \in \mathcal{N}$ is a regular value of f. Then, $f^{-1}(q)$ is a submanifold of $\mathcal{M}$ with dimension $f^{-1}(q) = m - n$.*

Let $H^m = \{(x^1 \dots x^m) \in \mathbb{R}^m |\ x^m \geq 0\}$ be the *m-dimensional half space.* The boundary ∂H^m is defined as $\partial H^m = \{(x^1 \dots x^m) \in H^m |\ x^m = 0\}$. A map $\phi : H^m \to \mathbb{R}^m$ is said to be of class C^r if for every $\boldsymbol{x} \in H^m$, there exists an open neighborhood $U \subset \mathbb{R}^m$ of $\boldsymbol{x}$ and C^r map $\Phi : U \to \mathbb{R}^m$ such that $\Phi = \phi$ on $U \cap H^m$. The concept of manifolds can be extended to those with a boundary.

Definition E.8. A Hausdorff topological space $\mathcal{M}$ is an m-dimensional *manifold with boundary* if there exist an open cover $\{U_\alpha\}$ of $\mathcal{M}$ and local isomorphisms ϕ_α on U_α to H^m such that $\phi_\beta \circ \phi_\alpha^{-1} : \phi_\alpha(U_\alpha \cap U_\beta) \to \phi_\beta(U_\alpha \cap U_\beta)$ is bijective and of class C^r for each α and β.

The *boundary* $\partial\mathcal{M}$ is the set of points $p \in \mathcal{M}$ with $\phi_\alpha(p) \in \partial H^m$ when $p \in U_\alpha$. The symbol $\partial\mathcal{M}$ should not be confused with the boundary of a topological space (Appendix B). Usually, they are not identical; the boundary of a manifold is always a part of the manifold, but the boundary of a topological space may not be. When we discuss a differentiable manifold $\mathcal{M}$, we always denote by $bdry\mathcal{M}$ the set of boundary points of $\mathcal{M}$ as a topological space. It is not hard to show for $p \in \partial\mathcal{M}$, $\phi_\beta(p) \in \partial H^m$ whenever $p \in U_\beta$ and that $\partial\mathcal{M}$ is an $m-1$-dimensional manifold (without boundary), $\mathcal{M}\backslash\partial\mathcal{M}$ is an m-dimensional manifold, and the tangent space $T_p\mathcal{M}$ is a well-defined m-dimensional vector space even for $p \in \partial\mathcal{M}$.

The regular-value theorem is extended to the manifolds with boundary.

Theorem E.7. *Let $f : \mathcal{M} \to \mathcal{N}$ be a smooth (C^r) map, where $\mathcal{M}$ is an m-dimensional manifold with a boundary and $\mathcal{N}$ is an n-dimensional manifold. Suppose that $m > n$ and $q \in \mathcal{N}$ is a regular value of f and its restriction $f|_{\partial\mathcal{M}}$. Then, $f^{-1}(q)$ is a manifold of $\mathcal{M}$ with boundary of dimension $m-n$ and $\partial f^{-1}(q) = \partial\mathcal{M} \cap f^{-1}(q)$.*

A subset R of $\mathbb{R}^\ell$ is called a *rectangular solid* if it is of the form $R = \{(x^1 \dots x^\ell) |\ a^t \le x^t \le b^t,\ t = 1 \dots \ell\}$ for vectors $\boldsymbol{a} = (a^1 \dots a^\ell)$ and $\boldsymbol{b} = (b^1 \dots b^\ell)$ with $a^t \le b^t$ for all $t = 1 \dots \ell$. The *volume* of the rectangular solid R is defined by

$$\mathrm{vol}R = \prod_{t=1}^{\ell}(b^t - a^t).$$

A subset $A \subset \mathbb{R}^\ell$ is said to have *measure zero* if for every $\epsilon > 0$, there exist countably many rectangular solids $R_1, R_2 \dots$ such that $A \subset \cup_{j=1}^{\infty} R_j$ and $\sum_{j=1}^{\infty} \mathrm{vol}R_j < \epsilon$ (see Appendix F for more of measure theory).

Theorem E.8 (Sard). *Let $f : \mathcal{M} \to V$ be a C^r map where $\mathcal{M}$ is an m-dimensional manifold and V is an open subset of $\mathbb{R}^\ell$ (hence, a manifold of dimension ℓ). If $r > \max\{0, m-\ell\}$, then the set of regular values of f has measure zero.*

We can provide an alternative proof of Brower's fixed point theorem C.5. First, we show there is no *smooth* retraction from D^n onto S^{n-1} (cf. Corollary C.3). Suppose there was a retraction ρ of D^n onto S^{n-1}. By Sard's theorem E.8, we can take a regular value $q \in S^{n-1}$

of ρ, and as q is also a regular value of the identity map $\rho|_{S^{n-1}}$, it follows from Theorem E.7 that $\rho^{-1}(q)$ is a one-dimensional manifold with boundary $\rho^{-1}(q) \cap \partial\mathcal{M} = \{q\}$. As $\rho^{-1}(q)$ is a closed subset of a compact set D^n, it is compact. It is known that the only compact one-dimensional manifolds are finite disjoint unions of circles and segments (e.g., [164, pp. 55–57]), so that $\partial\rho^{-1}(q)$ must consist of an even number of points, a contradiction. Then, in exactly the same manner as the proof of Theorem C.5, we can show that any *smooth* map of D^n to itself has a fixed point.

We now prove Theorem C.5. Suppose there exists a *continuous* map $f : D^n \to D^n$ with $f(\boldsymbol{x}) \neq \boldsymbol{x}$ for all $\boldsymbol{x} \in D^n$. Then, the continuous function $\|f(\boldsymbol{x}) - \boldsymbol{x}\|$ must take a minimum value $\mu > 0$ on D^n. By the Weierstrass approximation theorem (e.g., [63, p. 133]), there is a polynomial function $g : \mathbb{R}^n \to \mathbb{R}^n$ with $\|f(\boldsymbol{x}) - g(\boldsymbol{x})\| < \mu/2 \equiv \mu'$ for all $\boldsymbol{x} \in D^n$. Let $h : \mathbb{R}^n \to \mathbb{R}^n$ be defined by $h(\boldsymbol{x}) = g(\boldsymbol{x})/(1 + \mu')$. Then, $h(D^n) \subset D^n$ and $\|f(\boldsymbol{x}) - h(\boldsymbol{x})\| < \mu$ for all $\boldsymbol{x} \in D^n$; hence, $h(\boldsymbol{x}) \neq \boldsymbol{x}$ for all $\boldsymbol{x} \in D^n$. This is a contradiction, because polynomial functions are smooth.

Appendix F

Measure and Integration

Let A be a set and $\mathcal{A}$ a collection of subsets of A. Then, $\mathcal{A}$ is called a *σ-algebra* if it satisfies

(Sm-1) $A \in \mathcal{A}$,
(Sm-2) if $B \in \mathcal{A}$, then $B \backslash A \in \mathcal{A}$,
(Sm-3) if $B_1, B_2 \ldots \in \mathcal{A}$, then $\cup_{n=1}^{\infty} B_n \in \mathcal{A}$.

By (Sm-1) and (Sm-2), it is clear that $\emptyset \in \mathcal{A}$, and if $B_1, B_2 \ldots \in \mathcal{A}$, then $\cap_{n=1}^{\infty} B_n \in \mathcal{A}$. The pair $(A, \mathcal{A})$ is called a *measurable space*, and each element A of $\mathcal{A}$ is called a *measurable set*. If $\mathcal{C}$ is a collection of subsets of A, the smallest σ-algebra containing $\mathcal{C}$ is called the *σ-algebra generated by* $\mathcal{C}$ and denoted by $\sigma(\mathcal{C})$.

Let $\lambda : \mathcal{A} \to \mathbb{R} \cup \{+\infty\} \cup \{-\infty\}$ be a set function on a σ-algebra $\mathcal{A}$. We say that λ is *finitely additive* if and only if

$$\lambda\left(\cup_{n=1}^{N} B_n\right) = \sum_{n=1}^{N} \lambda(B_n)$$

for all pairwise disjoint measurable sets $B_1 \ldots B_N \in \mathcal{A}$. The set function λ is called *countably additive* if and only if

$$\lambda\left(\cup_{n=1}^{\infty} B_n\right) = \sum_{n=1}^{\infty} \lambda(B_n)$$

for all collections of countably many pairwise disjoint measurable sets $B_1, B_2 \ldots \in \mathcal{A}$.

A countably additive set function on a measurable space $(A, \mathcal{A})$ is called a *signed measure*. It is called a *measure* if it is non-negative or $\lambda(A) \geq 0$ for every $A \in \mathcal{A}$. If in addition it satisfies $\lambda(A) = 1$, then we say that λ is a *probability measure*. The triple $(A, \mathcal{A}, \lambda)$ consisting of a set A, an σ-algebra $\mathcal{A}$, and a measure λ on $(A, \mathcal{A})$ is called a *measure space*.

Example F.1. Let $A = \mathbb{N} = \{0, 1, \ldots\}$ be the set of non-negative integers. Let $\mathcal{A}$ be the set of all subsets of $\mathbb{N}$ and define the set function λ on $\mathcal{A}$ by $\lambda(B) = \sharp B =$ the number of elements in B. Then, it is easy to see that $\mathcal{A}$ is an σ-algebra and λ is a measure. The measure λ is called the *counting measure*.

We say that a condition C is said to hold *almost everywhere*, and write it as C a.e., if there exists a measurable set of λ measure 0 such that the condition C holds outside of the set. A subset of a set of measure 0 is often called the *null* set. A set $B \subset A$ such that $A \backslash B$ is null is called a set of *full measure* in A.

Let $(A, \mathcal{A}, \lambda)$ be a measure space. If the σ-algebra $\mathcal{A}$ contains all λ-null sets, we say that the measure space $(A, \mathcal{A}, \lambda)$ is *complete*. Let $\mathcal{N}_0$ be the set of all null sets of the measure space $(A, \mathcal{A}, \lambda)$, and let $\mathcal{A}_\lambda$ be the σ-algebra generated by the family $\mathcal{A} \cup \mathcal{N}_0$. It can be shown that $\mathcal{A}_\lambda = \{B \cup N | B \in \mathcal{A},\ N \in \mathcal{N}_0\}$. We can define the measure $\bar{\lambda}$ on the measurable space $(A, \mathcal{A}_\lambda)$ by $\bar{\lambda}(A \cup N) = \lambda(A)$. Then, the measure space $(A, \mathcal{A}_\lambda, \bar{\lambda})$ is called the *completion* of the measure space $(A, \mathcal{A}, \lambda)$.

Let $A = X$ be a topological space. The *Borel σ-algebra* on X, denoted by $\mathcal{B}(X)$, is the σ-algebra generated by the open sets of X. Each element B of $\mathcal{B}(X)$ is called a *Borel set*. The set of all signed measures on $(X, \mathcal{B}(X))$ is denoted by $ca(X)$ and the set of all measures on $(X, \mathcal{B}(X))$ by $\mathscr{M}(X)$. $\mu \in ca(K)$ is called a (signed) *Borel measure*.

Proposition F.1. *For every subspace Y of a metric space X, $\mathcal{B}(Y) = \{B \cap Y | \ B \in \mathcal{B}(X)\}$. In particular, if $Y \in \mathcal{B}(X)$, then $\mathcal{B}(Y) = \{B \subset Y | \ B \in \mathcal{B}(X)\}$.*

Proposition F.2. *Let X be a separable metric space. Then for every measurable set B of measure space $(X, \mathcal{B}(X), \lambda)$, there exists a closed set $C_B \subset B$ satisfying $\lambda(C_B) = \lambda(B)$ and for any closed set $C \subset B$ with $\lambda(C) = \lambda(B)$, $C_B \subset C$.*

The closed set C_B in Proposition F.2 is called the *support* of B and denoted by *support*(B).

Example F.2. Let $A = \mathbb{R}$ and $\mathcal{A}$ be the σ-algebra generated by the family of all intervals of the form $(a, b]$, $a \leq b$. It can be easily shown that $\mathcal{A} = \mathcal{B}(\mathbb{R})$. For an interval $(a, b] \in \mathcal{A}$, set $\lambda_0((a, b]) = b - a$. Then, λ_0 can be extended as a unique measure on $\mathcal{A}$. It is called the *Borel–Lebesgue measure* on $\mathbb{R}$.

Example F.3. The measure λ_0 defined in Example 2 is not complete. The completion $\bar{\lambda}$ of λ_0 is called the *Lebesgue measure* on $A = \mathbb{R}$.

Let $(A_1, \mathcal{A}_1)$ and $(A_2, \mathcal{A}_2)$ be two measurable sets. The *product σ-algebra* $\mathcal{A}_1 \times \mathcal{A}_2$ is the σ-algebra on the set $A_1 \times A_2$ which is generated by all sets of the form $A_1 \times A_2$, $A_1 \in \mathcal{A}_1$ and $A_2 \in \mathcal{A}_2$. The measurable space $(A_1 \times A_2, \mathcal{A}_1 \times \mathcal{A}_2)$ is called the *product measurable space.* The product measurable space $(\prod_{i=1}^{m} A_i, \prod_{i=1}^{m} \mathcal{A}_i)$ for the measurable spaces $(A_1, \mathcal{A}_1) \ldots (A_m, \mathcal{A}_m)$ is also inductively defined.

Proposition F.3. *Let A_1 and A_2 be separable metric spaces. Then, we have $\mathcal{B}(A_1 \times A_2) = \mathcal{B}(A_1) \times \mathcal{B}_2(A_2)$.*

For a measure space $(A_1 \times A_2, \mathcal{B}(A_1 \times A_2), \lambda)$, the *marginal* λ_{A_1} of λ to A_1 is defined to be a measure on A_1 given by $\lambda_{A_1}(B) = \lambda(B \times A_2)$ for every $B \in \mathcal{B}(A_1)$. The marginal λ_{A_2} is similarly defined.

Let $(A, \mathcal{A}, \lambda)$ be a measure space. The set $A \in \mathcal{A}$ with $\lambda(A) > 0$ is called an *atom* if for all $B \subset A$, it follows that $\lambda(B) = \lambda(A)$ or $\lambda(B) = 0$. The measure space $(A, \mathcal{A}, \lambda)$ (or sometimes the measure λ itself) is said to be *atomless* if it has no atoms.

Theorem F.1 (Lyapunov). *Let $\lambda_1 \ldots \lambda_m$ be atomless measures on a measurable space $(A, \mathcal{A})$. Then, the set $\{(\lambda_1(A) \ldots \lambda_m(A)) \in \mathbb{R}^m \mid A \in \mathcal{A}\}$ is a closed and convex subset in $\mathbb{R}^m$.*

The concept of an atomless measure space is strengthened to that of a *saturated* measure space.

Definition F.1. Let X and Y be complete and separable metric spaces. A measure space $(A, \mathcal{A}, \lambda)$ is said to satisfy the *saturation property* for a measure $\mu \in \mathscr{M}(X \times Y)$ if for every measurable map $f : A \to X$ with $\lambda \circ f^{-1} = \mu_X$, there exists a measurable map $g : A \to Y$ such that $\lambda \circ (f, g)^{-1} = \mu$.

A probability space $(A, \mathcal{A}, \lambda)$ is called *saturated* or *super-atomless* if $(A, \mathcal{A}, \lambda)$ is atomless and for every pair of complete and separable metric spaces (X, Y), it satisfies the saturation property for every $\mu \in \mathscr{M}(X \times Y)$.

Let $(A_1, \mathcal{A}_1)$ and $(A_2, \mathcal{A}_2)$ be measurable spaces. A mapping $f : A_1 \to A_2$ is said to be *measurable* if $f^{-1}(B) \in \mathcal{A}_1$ for every $B \in \mathcal{A}_2$. Then, it is sufficient for the mapping f to be measurable that $f^{-1}(B) \in \mathcal{A}_1$ for every $B \in \mathcal{C}$, where $\mathcal{C}$ is a family of subsets of A_2 that generates $\mathcal{A}_2$, that is to say, $\mathcal{A}_2 = \sigma(\mathcal{C})$. Indeed, because the set $\{B \in \mathcal{A}_2 | f^{-1}(B) \in \mathcal{A}_1\}$ is an σ-algebra containing $\mathcal{C}$, it coincides with $\mathcal{A}_2$. In particular, if A_1 and A_2 are metric spaces, every continuous function is measurable with respect to $(A_1, \mathcal{B}(A_1))$ and $(A_2, \mathcal{B}(A_2))$. When $A_2 = X$ is a metric space, a measurable function $f : (A_1, \mathcal{A}_1) \to (X, \mathcal{B}(X))$ is called *Borel measurable.* The following proposition is clear from the definition.

Proposition F.4. *If two mappings* $f : (A_1, \mathcal{A}_1) \to (A_2, \mathcal{A}_2)$ *and* $g : (A_2, \mathcal{A}_2) \to (A_3, \mathcal{A}_3)$ *are both measurable, then the composition* $g \circ f : (A_1, \mathcal{A}_1) \to (A_3, \mathcal{A}_3)$ *is also measurable.*

Proposition F.5. *Let* f *and* g *be Borel measurable functions of* $(A, \mathcal{A})$ *to* $(\mathbb{R}, \mathcal{B}(\mathbb{R}))$ *and* $(f_n)_{n \in \mathbb{N}}$ *a sequence of Borel measurable mappings of* $(A, \mathcal{A})$ *to* $(\mathbb{R}, \mathcal{B}(\mathbb{R}))$*. Then, we have*

(a) *the functions defined by* $a \mapsto f(a) \cdot g(a)$ *and* $a \mapsto sup\{f(a), g(a)\}$ *are Borel measurable,*
(b) *if* $\lim_{n \to \infty} f_n(a)$ *exists for every* $a \in A$*, then the function defined by* $a \mapsto lim_{n \to \infty} f_n(a)$ *is Borel measurable.*

Proposition F.6. *Let $f_1 \dots f_m$ be measurable mappings of a measurable space $(A, \mathcal{A})$ to measurable spaces $(A_1, \mathcal{A}_1) \dots (A_m, \mathcal{A}_m)$, respectively. Then, the mapping defined by $a \mapsto (f_1(a) \dots f_m(a))$ of $(A, \mathcal{A})$ to $(\prod_{i=1}^m A_i, \prod_{i=1}^m \mathcal{A}_i)$ is measurable.*

A function $f : (A, \mathcal{A}) \to (\mathbb{R}, \mathcal{B}(\mathbb{R}))$ is called *simple* if it can be written in the form $f(a) = \sum_{i=1}^m x_i \mathbf{1}_{A_i}(a)$, where $x_i \in \mathbb{R}$, $A_i \in \mathcal{A}$, $i = 1 \dots m$, and $\mathbf{1}_A$ is the indicator function of the set A, which is defined by

$$\mathbf{1}_A(a) = \begin{cases} 1 & \text{for } a \in A, \\ 0 & \text{otherwise.} \end{cases}$$

It is clear that every simple function is measurable.

Proposition F.7. *Every measurable function f is the point-wise limit of a sequence of simple functions, namely that there exists a sequence $(f_n)_{n \in \mathbb{N}}$ of measurable functions such that $f_n(a) \to f(a)$ for every $a \in A$.*

Let $(A, \mathcal{A}, \lambda)$ be a measure space. We now define the *integral* of a Borel measurable function $f : A \to \mathbb{R}$.

Step 1: When the function f is simple and written as $f(a) = \sum_{i=1}^m x_i \mathbf{1}_{A_i}(a)$, we define

$$\int_A f(a) d\lambda = \sum_{i=1}^m x_i \lambda(A_i),$$

as long as both $+\infty$ and $-\infty$ together do not appear at once in the sum of the right-hand side. When they do, we say that the integral does not exist.

Step 2: When the function f is non-negative and Borel measurable, we define

$$\int_A f(a) d\lambda = \sup \left\{ \int_A s(a) d\lambda \,\middle|\, s(\cdot) \text{ is simple and } 0 \le s(\cdot) \le f(\cdot) \right\}.$$

Note that this definition agrees with the definition of Step 1 when the function is simple.

Step 3: When f is an arbitrary measurable function, let $f^+(a) = \sup\{f(a), 0\}$ and $f^-(a) = \sup\{-f(a), 0\}$. Then, we have $f(a) = f^+(a) - f^-(a)$, $|f(a)| = f^+(a) + f^-(a)$. By Proposition F.5, the functions $f^+(\cdot)$ and $f^-(\cdot)$ are both measurable. We define

$$\int_A f(a)d\lambda = \int_A f^+(a)d\lambda - \int_A f^-(a)d\lambda,$$

if at least one of the two terms on the right-hand side is not ∞. Otherwise, we say that the integral does not exist. If both the terms are finite, the function f is said to be *integrable* (or λ*-integrable*).

For a measurable set $B \in \mathcal{A}$, we define the integral of f over B as

$$\int_B f(a)d\lambda = \int_A f(a)\mathbf{1}_B(a)d\lambda.$$

As the integrand $f \cdot \mathbf{1}_B$ is measurable by Proposition F.5, this definition is legitimate. The following proposition follows easily from the definition of the integration.

Proposition F.8. *Let f and g be integrable functions of a measure space $(A, \mathcal{A}, \lambda)$ to $\mathbb{R}$.*

(a) *If $f(a) \le g(a)$ a.e., then we have $\int_A f(a)d\lambda \le \int_A g(a)d\lambda$,*
(b) *$\int_A cf(a)d\lambda = c\int_A g(a)d\lambda$ for every $c \in \mathbb{R}$, and*
(c) *$\int_A (f(a) + g(a))d\lambda = \int_A f(a)d\lambda + \int_A g(a)d\lambda$.*

and we can prove that

Theorem F.2 (Fatou's lemma). *Let f and $f_1, f_2 \ldots$ be Borel measurable functions of a measure space $(A, \mathcal{A}, \lambda)$ to $\mathbb{R}$. Then, we have*

(a) *if $f(a) \le f_n(a)$ a.e., for all n, where $-\infty < \int_A f(a)d\lambda$, then $\int_A \liminf_{n\to\infty} f_n(a)d\lambda \le \liminf_{n\to\infty} \int_A f_n(a)d\lambda$,*

(b) *if* $f_n(a) \leq f(a)$ *a.e., for all* n*, where* $\int_A f(a)d\lambda < +\infty$*, then* $\limsup_{n\to\infty} \int_A f_n(a)d\lambda \leq \int_A \limsup_{n\to\infty} f_n(a)d\lambda$.

Let $(f_n)_{n\in\mathbb{N}}$ be a sequence of Borel measurable functions of a measure space $(A, \mathcal{A}, \lambda)$ to $\mathbb{R}$. The sequence (f_n) is said to *converge almost everywhere* to a Borel measurable function f (written $f_n \to f$ a.e.) if $\lambda(\{a \in A | \lim_{n\to\infty} f_n(a) \neq f(a)\}) = 0$. From Theorem F.2, we can deduce the monotone convergence theorem as follows:

Theorem F.3 (Monotone convergence theorem). *Let* f*,* g *and* $f_1, f_2 \ldots$ *be Borel measurable functions of a measure space* $(A, \mathcal{A}, \lambda)$ *to* $\mathbb{R}$*. Then, we have*

(a) *if* $g(a) \leq f_n(a)$ *a.e., for all* n*, where* $-\infty < \int_A g(a)d\lambda$*, and* $f_n \uparrow f$ *a.e., then* $\int_A f_n(a)d\lambda \uparrow \int_A f(a)d\lambda$*,*
(b) *if* $f_n(a) \leq g(a)$ *for all* n*, where* $\int_A g(a)d\lambda < +\infty$*, and* $f_n \downarrow f$ *a.e., then* $\int_A f_n(a)d\lambda \downarrow \int_A f(a)d\lambda$*.*

Theorem F.4 (Dominated convergence theorem). *Let* f*,* g *and* $f_1, f_2 \ldots$ *be Borel measurable functions of a measure space* $(A, \mathcal{A}, \lambda)$ *to* $\mathbb{R}$*. If* $|f_n(a)| \leq g(a)$ *a.e., for all* n*, where* $\int_A g(a)d\lambda < +\infty$ *and* $f_n \to f$ *a.e., then* f *is integrable and* $\int_A f_n(a)d\lambda \to \int_A f(a)d\lambda$*.*

A sequence of Borel measurable functions $(f_n)_{n\in\mathbb{N}}$ of a measure space $(A, \mathcal{A}, \lambda)$ to $\mathbb{R}$ is said to *converge in measure* to a measurable function f if and only if for every $\epsilon > 0$, $\lambda(\{a \in A| \, |f_n(a) - f(a)| \geq \epsilon\}) \to 0$ as $n \to \infty$.

Theorem F.5. *Let the measure* λ *be finite, or* $\lambda(A) < +\infty$*. Then, a sequence of measurable functions on* $(A, \mathcal{A}, \lambda)$ *that converges almost everywhere to* f *converges in measure to* f*.*

Let f be a map from a measure space $(A, \mathcal{A}, \lambda)$ to a metric space X and assume that f is Borel measurable. In terms of probability theory, f is called a *random element.* When $M = \mathbb{R}^\ell$, it is called a *random vector* and when $M = \mathbb{R}$, a *random variable.* The *distribution* of f on the measurable space $(X, \mathcal{B}(X))$ is defined by $\nu = \lambda \circ f^{-1}$, or $\nu(B) = \lambda(\{a \in A| \, f(a) \in B\})$ for every $B \in \mathcal{B}(X)$.

A sequence of measures $(\mu_n) \subset ca(X)$ on a metric space X *converges weakly* to $\mu \in ca(X)$ if and only if $\int_X g(x)d\mu_n \to \int_X g(x)d\mu$ for every bounded and continuous function $g : X \to \mathbb{R}$.

A sequence $(f_n)_{n\in\mathbb{N}}$ of Borel measurable functions of a measure space $(A, \mathcal{A}, \lambda)$ to a metric space X is said to *converge in distribution* to a measurable function f if and only if the sequence of distributions $\nu_n = \lambda \circ f_n^{-1}$ converges weakly to the distribution $\nu = \lambda \circ f^{-1}$.

Theorem F.6. *Let $(A, \mathcal{A}, \lambda)$ be a complete measure space, X and Y complete and separable metric spaces, (f_n) and (g_n) sequences of measurable maps of A to X and Y, respectively, such that $\lambda \circ f_n^{-1}$ and $\lambda \circ g_n^{-1}$ converge weakly to $\mu \in \mathscr{M}(X)$ and $\nu \in \mathscr{M}(Y)$, respectively. Then there exists a subsequence of $\lambda \circ (f_n, g_n)^{-1}$ converging weakly to a measure $\rho \in \mathscr{M}(X \times Y)$ such that $\rho_X = \mu$ and $\rho_Y = \nu$.*

Theorem F.7 (Skorokhod). *Let X be a separable metric space and (μ_n) a sequence of measures on X converging weakly to a measure $\mu \in \mathscr{M}(X)$. Then, there exists a measure space $(A, \mathcal{A}, \lambda)$ and measurable mappings f and f_n, $n = 1, 2, \ldots$ of A to X such that $\mu = \lambda \circ f^{-1}$, $\mu_n = \lambda \circ f_n^{-1}$, and $f_n(a) \to f(a)$ a.e. Furthermore, if X is a complete and separable metric space, then the measure space $(A, \mathcal{A}, \lambda)$ can be chosen for the unit interval with the Lebesgue measure.*

Theorem F.8. *Let f and $f_1, f_2 \ldots$ be Borel measurable mappings of a measure space $(A, \mathcal{A}, \lambda)$ to a separable metric space (X, d). Then, the mapping $a \mapsto d(f_n(a), f(a))$ of $(A, \mathcal{A})$ to $(\mathbb{R}, \mathcal{B}(\mathbb{R}))$ is Borel measurable.*

Furthermore, if the sequence $\{d(f_n(\cdot), f(\cdot))\}_{n\in\mathbb{N}}$ converges in measure to 0, then the sequence $(f_n)_{n\in\mathbb{N}}$ converges in distribution to f.

A measure λ on a measurable space $(A, \mathcal{A})$ is said to be *σ-finite* if $A = \cup_{n=1}^{\infty} A_n$, where $A_n \in \mathcal{A}$ for all n and $\lambda(A_n) < \infty$. A countably additive set function μ on a measurable space $(A, \mathcal{A})$ is said to be *absolutely continuous* with respect to a measure λ if for every $B \in \mathcal{A}$, $\lambda(B) = 0$ implies that $\mu(B) = 0$.

Theorem F.9 (Radon–Nikodym). *Let λ be a σ-finite measure and μ a countably additive set function on a measure space $(A, \mathcal{A}, \lambda)$. Assume that μ is absolutely continuous with respect to the measure λ. Then, there exists a Borel measurable function $g : A \to \mathbb{R}$ such that*

$$\mu(B) = \int_B g(a)d\lambda$$

for every $B \in \mathcal{A}$.

Let $(A_1, \mathcal{A}_1, \lambda_1)$ and $(A_2, \mathcal{A}_2, \lambda_2)$ be σ-finite measure spaces. Consider the product measurable space $(A, \mathcal{A}) = (A_1 \times A_2, \mathcal{A}_1 \times \mathcal{A}_2)$. For each measurable set $B \in \mathcal{A} = \mathcal{A}_1 \times \mathcal{A}_2$, the *section* $B(a_1)$ of the set B at $a_1 \in B_1$ is defined by $B(a_1) = \{a_2 \in B_2 | \ (a_1, a_2) \in B\}$. Define the set function λ on the product measurable space $(A, \mathcal{A})$ by

$$\lambda(B) = \int_{A_1} \lambda_2(B(a_1))d\lambda_1.$$

Then, one can show that $\int_{A_1} \lambda_2(B(a_1))d\lambda_1 = \int_{A_2} \lambda_1(B(a_2))d\lambda_2$ and λ is a unique measure on $(A, \mathcal{A})$ such that $\lambda(B_1 \times B_2) = \lambda_1(B_1)\lambda_2(B_2)$ for all $B_1 \in \mathcal{A}_1$ and $B_2 \in \mathcal{A}_2$. The measure λ is called the *product measure* of λ_1 and λ_2 and written as $\lambda = \lambda_1 \times \lambda_2$. Then, we have the next theorem.

Theorem F.10 (Fubini). *Let f be a nonnegative integrable function defined on the product measure space $(A, \mathcal{A}, \lambda) = (A_1 \times A_2, \mathcal{A}_1 \times \mathcal{A}_2, \lambda_1 \times \lambda_2)$. Then, the following equality holds.*

$$\begin{aligned}\int_A f(a_1, a_2)d\lambda &= \int_{A_1} \left(\int_{A_2} f(a_1, a_2)d\lambda_2 \right) d\lambda_1 \\ &= \int_{A_2} \left(\int_{A_1} f(a_1, a_2)d\lambda_1 \right) d\lambda_2.\end{aligned}$$

From now on, we shall consider the integration theory for the correspondences. In so doing, we start to consider the mapping f of $(A, \mathcal{A}, \lambda)$ to the ℓ-dimensional Euclidean space $\mathbb{R}^\ell$. Let $(A, \mathcal{A}, \lambda)$ be a measure space and $f : (A, \mathcal{A}, \lambda) \to (\mathbb{R}^\ell, \mathcal{B}(\mathbb{R}^\ell))$ be a Borel measurable function. The function $a \mapsto (f^1(a) \ldots f^\ell(a))$ is said to be integrable if each coordinate function $f^t : A \to \mathbb{R}$, $t = 1 \ldots \ell$ is integrable. The integral $\int_A f(a) d\lambda$ is then defined by

$$\int_A f(a) d\lambda = \left(\int_A f^1(a) d\lambda \ldots \int_A f^\ell(a) d\lambda \right).$$

The set of integrable mappings of $(A, \mathcal{A}, \lambda)$ to $(\mathbb{R}^\ell, \mathcal{B}(\mathbb{R}^\ell))$ is denoted by $\mathcal{L}(A, \mathcal{A}, \lambda)$ or $\mathcal{L}(A, \mathcal{A}, \lambda;\ \mathbb{R}^\ell)$ when the range $\mathbb{R}^\ell$ should be denoted. We now define the integration of a correspondence $\phi : A \to \mathbb{R}^\ell$ as follows:

Definition F.2. The set

$$\left\{ \int_A f(a) d\lambda \in \mathbb{R}^\ell \,\middle|\, f \in \mathcal{L}(A, \mathcal{A}, \lambda),\ f(a) \in \phi(a) \text{ a.e.} \right\}$$

is called the *integral* of the correspondence ϕ and denoted by $\int_A \phi(a) d\lambda$.

The following theorem ensures that this definition makes sense for a large class of correspondences, namely that $\int_A \phi(a) d\lambda \neq \emptyset$ for sufficiently many ϕs.

Theorem F.11 (Measurable selection theorem). *Let ϕ be a correspondence of a measure space $(A, \mathcal{A}, \lambda)$ to a complete and separable metric space X with a measurable graph, that is to say,*

$$\text{Graph}(\phi) = \{(a, \xi) \in A \times X |\ \xi \in \phi(a)\} \in \mathcal{A} \times \mathcal{B}(X).$$

Then, there exists a Borel measurable function f of A to X such that $f(a) \in \phi(a)$ a.e.

A correspondence ϕ of $(A, \mathcal{A}, \lambda)$ to $(\mathbb{R}^\ell, \mathcal{B}(\mathbb{R}^\ell))$ is said to be *integrably bounded* if there exists an integrable function g of $(A, \mathcal{A}, \lambda)$ to $\mathbb{R}_+$ such that $\|\phi(a)\| \leq g(a)$ a.e. Then, by Theorem F.11 and Proposition F.8, it is immediate that $\int_A \phi(a) d\lambda \neq \emptyset$ if the correspondence ϕ is integrably bounded and has a measurable graph. The following related propositions will also be used in the text.

Proposition F.9. *Let ϕ be a closed valued correspondence of a measurable space $(A, \mathcal{A})$ to a complete and separable metric space X that* *is measurable in the sense that*

$$\{a \in A|\ \phi(a) \cap F \neq \emptyset\} \in \mathcal{A}$$

for every closed subset F of X. Then, there exists a countable family of measurable mappings $\{f_n\}_{n \in \mathbb{N}}$ of $(A, \mathcal{A})$ to X such that $\phi(a) = \overline{\{f_n(a)|n \in \mathbb{N}\}}$ for every $a \in A$.

Proposition F.10. *Let ϕ be a correspondence of a measurable space $(A, \mathcal{A})$ to $\mathbb{R}^\ell$ with a measurable graph. Then, we have*

(a) *if a function $h : (A, \mathcal{A}) \to (\mathbb{R}^\ell, \mathcal{B}(\mathbb{R}^\ell))$ is Borel measurable, then the correspondence $a \mapsto \phi(a) + h(a)$ has a measurable graph,*
(b) *if g is a measurable mapping of a measurable space $(S, \mathcal{S})$ to $(A, \mathcal{A})$, then the correspondence $\phi \circ g : S \to \mathbb{R}^\ell$, $s \mapsto \phi(g(s))$ has a measurable graph.*

Proposition F.11. *Let ϕ be a correspondence of a measure space $(A, \mathcal{A}, \lambda)$ to $\mathbb{R}^\ell$ with a measurable graph and $\int_A \phi(a)d\lambda \neq \emptyset$. Then, for every vector $\boldsymbol{p} \in \mathbb{R}^\ell$, it follows that*

$$\sup\left\{\boldsymbol{pz} \in \mathbb{R}\,\middle|\, \boldsymbol{z} \in \int_A \phi(a)d\lambda\right\} = \int_A \sup\{\boldsymbol{px} \in \mathbb{R}|\ \boldsymbol{x} \in \phi(a)\}d\lambda.$$

The following theorem is deduced from Lyapunov's theorem (Theorem F.1).

Theorem F.12. *Let ϕ be a correspondence of an atomless measure space $(A, \mathcal{A}, \lambda)$ to $\mathbb{R}^\ell$. Then, the integral $\int_A \phi(a)d\lambda$ is a convex subset of $\mathbb{R}^\ell$.*

Let X be a metric space and consider a sequence of maps $f_n : A \to X$. For each $a \in A$, if $L_s(f_n(a)) \neq \emptyset$, we can define a map $L_s(f_n(\cdot)) : A \to X, a \mapsto L_s(f_n(a))$ (for the definition of L_s, see Appendix D). The sequence of measurable functions (f_n) on A to $\mathbb{R}^\ell$ is said to be *uniformly integrable* if there exists an integrable function $g : A \to \mathbb{R}_+$ such that $\|f_n(a)\| \leq g(a)$ a.e, for all n. The following theorem is known as the ℓ-dimensional version of Fatou's lemma (Theorem F.2).

Theorem F.13 (Fatou's lemma in ℓ dimensions). *Let $(f_n(a))$ be a sequence of integrable functions of a measure space $(A, \mathcal{A}, \lambda)$ to $\mathbb{R}^\ell_+$ such that $\lim \int_A f_n(a)d\lambda$ exists. Then, there exists an integrable function $f : (A, \mathcal{A}, \lambda) \to \mathbb{R}^\ell_+$ such that*

(a) $f(a) \in L_s(f_n(a))$ *a.e. in* A,
(b) $\int_A f(a)d\lambda \leq \lim_{n\to\infty} \int_A f_n(a)d\lambda$.

Moreover, if the sequence $(f_n(a))$ is uniformly integrable and if the set $\{f_n(a) |\ n \in \mathbb{N}\}$ is bounded a.e., then there exists a measurable selection f of $L_s(f_n)$ such that

$$\int_A f(a)d\lambda = \lim_{n\to\infty} \int_A f_n(a)d\lambda.$$

Conversely, if for every function f of A to $\mathbb{R}^\ell$ with properties (a) and (b) it follows that $\int_A f(a)d\lambda = \lim_{n\to\infty} \int_A f_n(a)d\lambda$, then the sequence (f_n) is uniformly integrable.

From Theorem F.13, we can deduce the following Theorem.

Theorem F.14. *Let $(\phi)_{n\in\mathbb{N}}$ be a sequence of correspondences of a measure space $(A, \mathcal{A}, \lambda)$ to $\mathbb{R}^\ell_+$ such that there exists a sequence of functions $g_n : (A, \mathcal{A}, \lambda) \to \mathbb{R}^\ell_+$ such that (i) $\phi_n(a) \leq g_n(a)$ a.e. (ii) the sequence (g_n) is uniformly integrable and the set $\{g_n(a) |\ n \in \mathbb{N}\}$ is bounded a.e. Then, it follows that*

$$L_s\left(\int_A \phi_n(a)d\lambda\right) \subset \int_A L_s(\phi_n(a))d\lambda.$$

From Theorem F.14, we can easily obtain the following corollary:

Corollary F.1. *If the correspondence ϕ of a measure space $(A, \mathcal{A}, \lambda)$ to $\mathbb{R}^\ell$ such that $b \leq \phi(a)$ a.e. for some $b \in \mathbb{R}^\ell$ is closed valued and integrably bounded, then the integral $\int_A \phi(a)d\lambda$ is compact.*

Proof. As ϕ is bounded from below, we can assume without loss of generality that $\phi(a) \subset \mathbb{R}^\ell_+$ a.e. Define a sequence of measurable correspondences $(\phi_n)_{n\in\mathbb{N}}$ by $\phi_n(a) = \phi(a)$, $n = 1, 2, \ldots$ Then, $L_s(\phi_n(a)) = \phi(a)$ for all n, because $\phi(a)$ is a closed set. Hence, by

Theorem F.14, it follows that $L_s\left(\int_A \phi(a)d\lambda\right) \subset \int_A \phi(a)d\lambda$, in other words, every limit point of $\int_A \phi(a)d\lambda$ belongs to $\int_A \phi(a)d\lambda$. □

Furthermore, we also have the following corollary:

Corollary F.2. *Let (X,d) be a metric space. Suppose that a correspondence $\phi : A \times X \to \mathbb{R}^\ell_+$ satisfies the following conditions:*

(i) *there exists an integrable function $g : (A, \mathcal{A}, \lambda) \to \mathbb{R}_+$ with $\|\phi(a, \boldsymbol{z})\| \leq g(a)$ a.e. for every $\boldsymbol{z} \in X$,*
(ii) *the correspondence $\phi(a, \cdot) : X \to \mathbb{R}^\ell$ is closed at $\boldsymbol{z} \in X$ a.e.*

Then, the relation $\boldsymbol{z} \mapsto \int_A \phi(\cdot, \boldsymbol{z})d\lambda$ is closed at $\boldsymbol{z}$.

Proof. We need to show that for every sequence $(\boldsymbol{z}_n)_{n\in\mathbb{N}}$ converging to $\boldsymbol{z}$, we have

$$L_s\left(\int_A \phi(\cdot, \boldsymbol{z}_n)d\lambda\right) \subset \int_A \phi(\cdot, \boldsymbol{z})d\lambda.$$

As $\phi(a, \cdot)$ is closed at $\boldsymbol{z} \in X$, $L_s(\phi(\cdot, \boldsymbol{z}_n)) \subset \phi(\cdot, \boldsymbol{z})$. Therefore, by Theorem F.14, $L_s\left(\int_A \phi(\cdot, \boldsymbol{z}_n)d\lambda\right) \subset \int_A L_s(\phi(\cdot, \boldsymbol{z}_n))d\lambda \subset \int_A \phi(\cdot, \boldsymbol{z})d\lambda$. □

Radon–Nikodym's theorem (Theorem F.9) is extended to correspondences as follows:

Theorem F.15 (Radon–Nikodym for correspondences). *Let $(A, \mathcal{A}, \lambda)$ be a measure space and Φ a correspondence of $\mathcal{A}$ to $\mathbb{R}^\ell$ with the properties*

(i) *$\Phi\left(\cup_{n=1}^\infty B_n\right) = \sum_{n=1}^\infty \Phi(B_n)$ for every pairwise disjoint sequence $(B_n)_{n=1}^\infty$ of $\mathcal{A}$,*
(ii) *$\Phi(B)$ is a convex set for every $B \in \mathcal{A}$,*
(iii) *$\Phi(B) = \{\mathbf{0}\}$ if $\lambda(B) = 0$.*

Then, there exists a convex-valued correspondence of ϕ of $(A, \mathcal{A}, \lambda)$ to $\mathbb{R}^\ell$ such that

(a) *$\int_B \phi(a)d\lambda \subset \Phi(B)$ and $\overline{\int_B \phi(a)d\lambda} = \overline{\Phi(B)}$ for every $B \in \mathcal{A}$,*
(b) *the correspondence ϕ is measurable in the sense that*

$$\{a \in A | \phi(a) \cap F \neq \emptyset\} \in \mathcal{A} \text{ for every closed subset } F \text{ of } \mathbb{R}^\ell.$$

Appendix G

Topological Vector Spaces

Let L be a (real) vector space. Recall that a norm on L is a function denoted by $\|\cdot\|$ of L to $\mathbb{R}_+$ satisfying (Nr-1), (Nr-2), and (Nr-3) of Appendix E. The pair $(L, \|\cdot\|)$ is called a *normed space*. As usual, we often call L a normed space if the norm is understood. Recall that a normed space is a metric space defined by $d(\boldsymbol{x}, \boldsymbol{y}) = \|\boldsymbol{x} - \boldsymbol{y}\|$ and it is called a *Banach space* if it is complete (that is, every Cauchy sequence in L is convergent to a point of L) with respect to this metric induced from the norm.

Theorem G.1 (Theorem of completion). *Let L be a normed space that is not complete. Then, L is isometrically isomorphic to a dense linear subspace of a Banach space $\tilde{L}$, that is, there exists a one to one and onto mapping ι of L to a dense linear subspace of $\tilde{L}$ such that*

$$\iota(a\boldsymbol{x} + b\boldsymbol{y}) = a\iota(\boldsymbol{x}) + b\iota(\boldsymbol{y}), \ \|\iota(\boldsymbol{x})\|_{\tilde{L}} = \|\boldsymbol{x}\|_L$$

for all $\boldsymbol{x}$, $\boldsymbol{y} \in L$ and all a, $b \in \mathbb{R}$.

A *topological vector space* is a vector space L with a Hausdorff topology such that the addition $+ : L \times L \to L$ and the scalar multiplication $\cdot : \mathbb{R} \times L \to L$ are continuous. Then, the normed vector space is a topological vector space with the metric topology induced by the norm. For if $\boldsymbol{x}_n \to \boldsymbol{x} \in L$, $\boldsymbol{y}_n \to \boldsymbol{y} \in L$ and $a_n \to a \in \mathbb{R}$, then $\|(\boldsymbol{x}_n + \boldsymbol{y}_n) - (\boldsymbol{x} + \boldsymbol{y})\| \le \|\boldsymbol{x}_n - \boldsymbol{x}\| + \|\boldsymbol{y}_n - \boldsymbol{y}\| \to 0$, and $\|a_n\boldsymbol{x}_n - a\boldsymbol{x}\| \le \|a_n\boldsymbol{x}_n - a_n\boldsymbol{x}\| + \|a_n\boldsymbol{x} - a\boldsymbol{x}\| \le |a_n|\|\boldsymbol{x}_n - \boldsymbol{x}\| + |a_n - a|\|\boldsymbol{x}\| \to 0$.

A topological vector space is *locally convex* if every neighborhood of the origin contains a convex neighborhood of the origin. A subset V of a topological vector space is *symmetric* if $V = -V$.

The concept of linear functionals on Banach spaces is generalized to topological vector spaces. A *linear functional* on a topological vector space is a real-valued linear function, and the set of continuous linear functionals on L is called the *dual space* of L and denoted by L^*. The dual space L^* naturally has the vector space structure; if L is a normed space, then L^* is also a normed space by the operator norm defined by $\|\boldsymbol{p}\| = \sup\{|\boldsymbol{px}| \,|\, \|\boldsymbol{x}\| \leq 1\}$ for $\boldsymbol{p} \in L^*$. Obviously, this definition is equivalent to $\|\boldsymbol{p}\| = \sup\left\{|\boldsymbol{p}(\boldsymbol{x}/\|\boldsymbol{x}\|)| \,\middle|\, \boldsymbol{x} \neq \boldsymbol{0}\right\}$.

Proposition G.1. *For a normed space L, a linear functional $\boldsymbol{p} : L \to \mathbb{R}$ is continuous if and only if it is bounded, or $\|\boldsymbol{p}\| < +\infty$.*

Let L and M be topological vector spaces. A *dual pairing* is an ordered pair of vector spaces (L, M) and a real-valued bilinear function on $L \times M$ (see Appendix E for the definition of the multilinear map). For $\boldsymbol{x} \in L$ and $\boldsymbol{p} \in M$, the value of the bilinear functional is denoted by $\boldsymbol{px}$, or $\boldsymbol{p}(\boldsymbol{x})$.[3] Given a dual pairing (L, M), there exists a weakest topology on L such that $L^* = M$, which is denoted by $\sigma(L, M)$. The $\sigma(L, M)$-topology is characterized by the net as follows:

Proposition G.2. *A net $(\boldsymbol{x}_\kappa)$ in L converges to $\boldsymbol{x} \in L$ with respect to $\sigma(L, M)$-topology if and only if for every $\boldsymbol{p} \in M$, $\boldsymbol{px}_\kappa \to \boldsymbol{px}$.*

If L is already a topological vector space, then there exists naturally a dual pairing (L, L^*). As by definition the $\sigma(L, L^*)$-topology is as weak or weaker than the original topology on L, the $\sigma(L, L^*)$-topology is often called the *weak topology*. The $\sigma(L^*, L)$-topology on L^* is called the *weak* topology*.

However, given a dual pairing (L, M), there is a strongest topology on L such that $L^* = M$. It is called the *Mackey topology* and

[3] In mathematical literatures, it is usually denoted $\langle \boldsymbol{x}, \boldsymbol{p} \rangle$ or $\boldsymbol{xp}$. Our notation of the reverse order is more convenient for economic applications.

denoted by $\tau(L, M)$. The following theorem characterizes the Mackey topology by nets.

Proposition G.3. *A net $(\boldsymbol{x}_\kappa)$ in L converges to $\boldsymbol{x} \in L$ with respect to $\tau(L, M)$-topology if and only if for every $\sigma(M, L)$-compact, convex and circled subset C of M, $\sup\{|\boldsymbol{p}\boldsymbol{x}_\kappa - \boldsymbol{p}\boldsymbol{x}| | \boldsymbol{p} \in C\} \to 0$, where a set C is circled if and only if $c \in C$ implies that $rc \in C$ for every $-1 \leq r \leq 1$.*

A fundamental result on the weak and the Mackey topology is as follows:

Theorem G.2 (Mackey). *If (L, M) is a dual pairing, every convex subset of L is closed in the $\sigma(L, M)$-topology if and only if it is closed in the $\tau(L, M)$-topology.*

The next result, which is known as Alaoglu's theorem, is also a fundamental theorem with respect to the weak* topology.

Theorem G.3 (Alaoglu). *If L is a locally convex topological vector space and V is an open symmetric neighborhood of $\mathbf{0}$, then the subset of L^*, $B = \{\boldsymbol{p} \in L^* | \, |\boldsymbol{p}\boldsymbol{x}| \leq 1$ for every $\boldsymbol{x} \in V\}$ is compact in the $\sigma(L^*, L)$-topology.*

In Appendix E, we defined for every integer p with $1 \leq p < +\infty$, the spaces ℓ^p and the space ℓ^∞.

Proposition G.4. *For an integer with $1 \leq p < +\infty$, the dual space ℓ^{p*} of ℓ^p is isomorphic to the space ℓ^q, where q is an integer such that $\frac{1}{p} + \frac{1}{q} = 1$ when $p > 1$, and we set $q = +\infty$ when $p = 1$.*

Then, we can consider the $\sigma(\ell^\infty, \ell^1)$ and $\tau(\ell^\infty, \ell^1)$ topologies on the space ℓ^∞, and the $\sigma(\ell^1, \ell^\infty)$ and $\tau(\ell^1, \ell^\infty)$ topologies on the space ℓ^1, respectively. The following propositions are particular to the case of ℓ^∞ (and ℓ^1), but useful.

Proposition G.5. *Let B be a (norm) bounded subset of ℓ^∞. Then, on the set B, the Mackey topology $\tau(\ell^\infty, \ell^1)$ coincides with the product topology, or the topology of the coordinate-wise convergence.*

As the product topology is weaker than the $\sigma(\ell^\infty, \ell^1)$-topology, the product topology and the $\sigma(\ell^\infty, \ell^1)$ and $\tau(\ell^\infty, \ell^1)$ topologies are all equal on bounded subsets of ℓ^∞. Moreover, we have the following proposition:

Proposition G.6. *Let Π be a $\sigma(\ell^1, \ell^\infty)$ compact subset of ℓ^1. Then, the paring map from $\ell^\infty \times \Pi$ to $\mathbb{R}$, $(\boldsymbol{x}, \boldsymbol{p}) \mapsto \boldsymbol{p}\boldsymbol{x}$ is jointly continuous with respect to $\tau(\ell^\infty, \ell^1) \times \sigma(\ell^1, \ell^\infty)$ topology on $\ell^\infty \times \Pi$.*

Consider the measurable space $(\mathrm{N}, 2^{\mathrm{N}})$, or the set of non-negative integers in which every subset is a measurable set. Let ba be the set of finitely additive set functions with bounded variation, namely that

$$ba = \left\{ \pi : 2^{\mathrm{N}} \to \mathbb{R} \,\middle|\, \sup_{E \subset \mathrm{N}} |\pi(E)| < +\infty, \right.$$
$$\left. \pi(E \cup F) = \pi(E) + \pi(F) \text{ whenever } E \cap F = \emptyset \right\}.$$

Then, we can show that the space ba is a Banach space with the norm

$$\|\pi\| = \sup \left\{ \sum_{i=1}^{n} |\pi(E_i)| \,\middle|\, E_i \cap E_j = \emptyset \text{ for } i \neq j,\ E_i \subset \mathrm{N},\ n \in \mathrm{N} \right\}.$$

Proposition G.7. *The dual space ℓ^∞ is isomorphic to the space ba, or $\ell^{\infty *} = ba$.*

We can generalize these spaces as follows. Let $\boldsymbol{\beta} = (\beta^0, \beta^1, \ldots)$ be a sequence of strictly positive real numbers; $\beta^t > 0$ for all $t \in \mathrm{N}$. For an integer p with $1 \leq p < +\infty$, we define the "weighted" ℓ^p space by

$$\ell^p_{\boldsymbol{\beta}} = \left\{ \boldsymbol{x} = (x^t) \,\middle|\, \sum_{t=0}^{\infty} |\beta^t x^t|^p < +\infty \right\}$$

and for $p = \infty$, the "weighted"-ℓ^∞ space $\ell^\infty_{\boldsymbol{\beta}}$ is given analogously by

$$\ell^\infty_{\boldsymbol{\beta}} = \left\{ \boldsymbol{x} = (x^t) \,\middle|\, \sup_{t \geq 0} |\beta^t x^t| < +\infty \right\}$$

with the norms on these spaces obviously defined.

Finally, we define the "weighted" *ba* space by

$$ba_{\beta} = \left\{ \pi : 2^{\mathbb{N}} \to \mathbb{R} \,\middle|\, \sup_{E \subset \mathbb{N}} \int_E \beta d|\pi| < +\infty, \right.$$
$$\left. \pi(E \cup F) = \pi(E) + \pi(F) \text{ whenever } E \cap F = \emptyset \right\},$$

where the integral $\int_E \beta d|\pi|$ is defined in a similar way as the integral of the countably additive measure. The norm on the space ba_{β} is also obviously endowed. These weighted spaces are Banach spaces. Indeed, it can be easily verified that the space ℓ^p_{β} is isometrically isomorphic to the space ℓ^p for $1 \leq p \leq +\infty$, and the space ba_{β} is also isometrically isomorphic to the space *ba*.

Proposition G.8. *For an integer with $1 \leq p < +\infty$, the dual space ℓ^{p*}_{β} of ℓ^p_{β} is isomorphic to the space $\ell^q_{\beta^{-1}}$, where $\beta^{-1} = (1/\beta^0, 1/\beta^1 \ldots)$ and q is an integer such that $\frac{1}{p} + \frac{1}{q} = 1$ when $p > 1$, and we set $q = +\infty$ when $p = 1$.*

Proposition G.9. *The dual space of ℓ^{∞}_{β} is isomorphic to the space $ba_{\beta^{-1}}$, or $\ell^{\infty *}_{\beta} = ba_{\beta^{-1}}$.*

The spaces ℓ^p and *ba* can be extended to the spaces of (measurable) functions on a measure space $(A, \mathcal{A}, \lambda)$. The space $L^p(A, \mathcal{A}, \lambda)$ for $1 \leq p < +\infty$ is defined as the set of all Borel measurable functions f such that $\int_A |f(a)|^p d\lambda < +\infty$.

For a Borel measurable mapping $g : A \to \mathbb{R}$, we define the essential supremum of $|g|$ as

$$ess\sup\, |g| = \inf\{c \geq 0 |\ \lambda(\{a \in A |\ |g(a)| > c\}) = 0\}.$$

The space $L^{\infty}(A, \mathcal{A}, \lambda)$ is the set of all Borel measurable functions with finite essential supremum.

Unfortunately, the "norm" of f, $\|f\| = \left(\int_A |f(a)|^p d\lambda\right)^{1/p}$ for $f \in L^p(A, \mathcal{A}, \lambda)$, and $\|f\| = ess\sup|f|$ for $f \in L^{\infty}(A, \mathcal{A}, \lambda)$ are not exactly norms but *semi-norms*, because even if $\|f - g\| = 0$, it may be the case that $f(a) \neq g(a)$ on a set of measures 0. We can get the "true" norm by considering $f \in L^p(A, \mathcal{A}, \lambda)$ for $1 \leq p \leq +\infty$ as an equivalence class determined by the equivalence relation $f \sim g$ if and only if $f = g$ a.e. In this case, we denote the set of the equivalence classes by $\mathcal{L}^p(A, \mathcal{A}, \lambda)$ for $1 \leq p \leq +\infty$.

The space $ba(A, \mathcal{A}, \lambda)$ is defined as the collection of finitely additive set functions that are absolutely continuous with respect to λ and have a bounded (finite) total variation norm. Here, the total variation norm of $\pi : \mathcal{A} \to \mathbb{R}$ is given by

$$\|\pi\| = \sup \left\{ \sum_{i=1}^{n} |\pi(E_i)| \,\middle|\, E_i \cap E_j = \emptyset \text{ for } i \neq j, E_i \in \mathcal{A}, \right.$$
$$\left. i = 1 \ldots n, n \in \mathbb{N} \right\}.$$

The Radon–Nikodym theorem (Theorem F.9) states that the natural embedding from $L_1(A, \mathcal{A}, \lambda)$ to $ba(A, \mathcal{A}, \lambda)$ identifies $L_1(A, \mathcal{A}, \lambda)$ with the set of countably additive set functions in $ba(A, \mathcal{A}, \lambda)$. The set of all finitely additive set functions on $(A, \mathcal{A})$ that is denoted by $ba(A, \mathcal{A})$ endowed with the total variation norm given above is defined similarly.

The set function $\pi \in ba(A, \mathcal{A})$ is said to be *purely finitely additive* if $\rho = 0$ whenever $\rho \in ba(A, \mathcal{A})$ is countably additive and $0 \leq \rho \leq \pi$.

Theorem G.4 (Yosida–Hewitt). *If $\pi \in ba(A, \mathcal{A})$ and $\pi \geq 0$, then there exist set functions $\pi_c \geq 0$ and $\pi_p \geq 0$ in $ba(A, \mathcal{A})$ such that π_c is countably additive and π_p is purely finitely additive and satisfy $\pi = \pi_c + \pi_p$. This decomposition is unique.*

Let (K, d) be a compact metric space. Recall that $\mathcal{B}(K)$ is the set of all Borel measurable subsets of K (Appendix F), and the set of all bounded signed measures on $(K, \mathcal{B}(K))$ is denoted by $ca(K)$.

Proposition G.10. *Let $C(K)$ be the set of all continuous functions on K. Then, the dual space of $C(K)$ is $ca(K)$, or $C^*(K) = ca(K)$ by the pairing defined by*

$$\boldsymbol{px} = \int_K p(t) d\boldsymbol{x} \text{ for } \boldsymbol{p} = p(t) \in C(K) \text{ and } \boldsymbol{x} \in ca(K).$$

Therefore, by definition, the weak* topology on $ca(K)$ is the topology of point-wise convergence on $C(K)$, or a net $(\boldsymbol{x}_\kappa)$ on $ca(K)$ converges to $\boldsymbol{x} \in ca(K)$ if and only if $\boldsymbol{p}\boldsymbol{x}_\kappa \equiv \int_K p(t)d\boldsymbol{x}_\kappa \to \int_K p(t)d\boldsymbol{x}$ for every $\boldsymbol{p} = p(t) \in C(K)$. In Appendix F, weak* convergent nets (sequences) on a compact set were called weakly convergent nets (sequences), and Theorem G.3 is restated as

Proposition G.11. *The norm-bounded subsets of $ca(K)$ are compact and metrizable in the weak* topology.*

By Proposition B.11, the compact metric space K is separable. Hence, there exists a countable dense subset of K. For $t \in K$, the *Dirac measure* δ_t is defined by

$$\delta_t(B) = \begin{cases} 1 & \text{if } t \in B, \\ 0 & \text{otherwise.} \end{cases}$$

The following proposition is a basic tool for approximating an infinite-dimensional economy by a sequence of finite-dimensional subeconomies.

Proposition G.12. *Let $\{t_n\}$ be a countable dense subset of K. Then, $ca(K)$ is the closure of the set of finite linear combinations of the Dirac measures δ_{t_n}.*

Let (K, d) be a compact metric space and consider a sequence of pairs of closed subsets K_n of K and continuous and non-negative functions $\boldsymbol{p}_n = p_n(t) \in C_+(K_n)$ on K_n, $(K_n, \boldsymbol{p}_n)$. For $\boldsymbol{p} = p(t) \in C_+(K)$, we write $(K_n, \boldsymbol{p}_n) \to (K, \boldsymbol{p})$ if and only if $K_n \to K$ in the topology of closed convergence (Appendix D) and for all subsequences n_k, t_{n_k} with $t_{n_k} \in K_{n_k}$ and $t_{n_k} \to t$, we have $p_{n_k}(t_{n_k}) \to p(t)$. Then, we have the following proposition:

Proposition G.13. *Suppose $(K_n, \boldsymbol{p}_n) \to (K, \boldsymbol{p})$ and for a bounded sequence $\boldsymbol{x}_n$ with $support(\boldsymbol{x}_n) \subset K_n$, $\boldsymbol{x}_n \to \boldsymbol{x}$ in the weak* topology. Then, $\boldsymbol{p}_n\boldsymbol{x}_n \to \boldsymbol{p}\boldsymbol{x}$.*

The sequence $(K_n, \boldsymbol{p}_n)$ of closed subsets K_n of K and $\boldsymbol{p}_n = p_n(t) \in C(K_n)$ is said to be *equicontinuous* if and only if for all $\epsilon > 0$, there exists a $\delta > 0$ such that for all $n \in \mathbb{N}$ and all $t, s \in K$ with $d(t, s) < \delta$, it follows that $|p_n(t) - p_n(s)| < \epsilon$.

Proposition G.14. *Let K_n, $\boldsymbol{p}_n$ be as above with $K_n \subset K_{n+1}$ and $K_n \to K$ in the topology of closed convergence. If $(K_n, \boldsymbol{p}_n)$ is equicontinuous and the $\boldsymbol{p}_n$ are uniformly bounded, there exists a subsequence n_k and a continuous function $\boldsymbol{p} \in C(K)$ such that $(K_{n_k}, \boldsymbol{p}_{n_k}) \to (K, \boldsymbol{p})$.*

The examples of Banach spaces that have been given so far are equipped with the *order relation* which is compatible with the vector space structure. In general, an *ordered vector space* is a vector space L endowed with a reflexive, transitive, and antisymmetric relation $\leq$ that satisfies the following conditions:

(Or-1) if $\boldsymbol{x} \leq \boldsymbol{y}$ and $a \in \mathbb{R}_+$, then $a\boldsymbol{x} \leq a\boldsymbol{y}$,
(Or-2) if $\boldsymbol{x} \leq \boldsymbol{y}$, then $\boldsymbol{x} + \boldsymbol{z} \leq \boldsymbol{y} + \boldsymbol{z}$ for each $\boldsymbol{z} \in L$.

Let L be an ordered vector space. We define the *positive cone* L_+ by $L_+ = \{\boldsymbol{x} \in L |\ \boldsymbol{0} \leq \boldsymbol{x}\}$. For instance, the positive cone $ca_+(K)$ is nothing but $\mathscr{M}(K)$, the set of all Borel measures on K. Obviously, the positive cone L_+ is convex and a *proper cone*, or if $\boldsymbol{x} \in L_+ \cap -L_+$, then $\boldsymbol{x} = \boldsymbol{0}$. If a vector space L has a proper convex cone C, then we can make L an ordered vector space by giving the order structure defined by $\boldsymbol{x} \leq \boldsymbol{y}$ if and only if $\boldsymbol{y} - \boldsymbol{x} \in C$. An ordered vector space L is called an *ordered topological vector space* if it is a topological vector space and the positive cone L_+ is closed. Note that if L is an ordered topological vector space, then the dual space L^* is also ordered, with positive cone $L^*_+ = \{\boldsymbol{p} \in L^* |\ 0 \leq \boldsymbol{p}\boldsymbol{x}$ for every $\boldsymbol{x} \in L_+\}$. A linear functional $\boldsymbol{p} \in L^*_+$ is called *positive*. Moreover, the positive cone L^*_+ is evidently $\sigma(L^*, L)$-closed. For $\boldsymbol{x}, \boldsymbol{y} \in L$, we can define the *order interval* by $[\boldsymbol{x}, \boldsymbol{y}] = \{\boldsymbol{z} \in L |\ \boldsymbol{x} \leq \boldsymbol{z} \leq \boldsymbol{y}\}$.

Let $S \subset L$. An element $\boldsymbol{y} \in L$ is an *upper bound* of S if and only if $\boldsymbol{x} \leq \boldsymbol{y}$ for every $\boldsymbol{x} \in S$. An element $\sup S$ is called the *supremum* (or a *least upper bound*) if it is an upper bound of S and $\sup S \leq \boldsymbol{y}$ for every upper bound $\boldsymbol{y}$ of S. Similarly, $\boldsymbol{z}$ is a *lower bound* of S if $\boldsymbol{z} \leq \boldsymbol{x}$ for every $\boldsymbol{x} \in S$. $\inf S$ is the *infimum* (*greatest lower bound*) if it is a lower bound of S and $\boldsymbol{z} \leq \inf S$ for every lower bound $\boldsymbol{z}$ of S. We usually write $\boldsymbol{x} \vee \boldsymbol{y}$ rather than $\sup\{\boldsymbol{x}, \boldsymbol{y}\}$ and $\boldsymbol{x} \wedge \boldsymbol{y}$ rather than $\inf\{\boldsymbol{x}, \boldsymbol{y}\}$. If every pair $\boldsymbol{x}, \boldsymbol{y}$ of an ordered vector space L has the supremum $\boldsymbol{x} \vee \boldsymbol{y}$ and the infimum $\boldsymbol{x} \wedge \boldsymbol{y}$, then we call L a *vector lattice* or *Riesz space*. An ordered topological vector space is a *topological vector lattice* if it is a vector lattice and the lattice operations are (uniformly) continuous. We write $\boldsymbol{x}_+ = \boldsymbol{x} \vee \boldsymbol{0}$ and $\boldsymbol{x}_- = (-\boldsymbol{x}) \vee \boldsymbol{0}$ and call these the *positive part* and the *negative part* of $\boldsymbol{x}$, respectively. Then, $\boldsymbol{x} = \boldsymbol{x}_+ - \boldsymbol{x}_-$, and we write $|\boldsymbol{x}| = \boldsymbol{x}_+ + \boldsymbol{x}_-$ and call it the *absolute value* of $\boldsymbol{x}$. We say that a subset $M \subset L$ is solid if $|\boldsymbol{x}| \leq |\boldsymbol{y}|$ and $\boldsymbol{y} \in M$ imply that $\boldsymbol{x} \in M$. A fundamental property of the vector lattices is the *Riesz decomposition property.*

Theorem G.5 (Riesz decomposition property). *Let L be a vector lattice and let $\boldsymbol{x}_1 \dots \boldsymbol{x}_n$, $\boldsymbol{z}$ be positive elements of L such that $\boldsymbol{z} \leq \sum_{i=1}^n \boldsymbol{x}_i$. Then, there exist positive elements $\boldsymbol{z}_1 \dots \boldsymbol{z}_n$ of L such that $\boldsymbol{z} = \sum_{i=1}^n \boldsymbol{z}_i$ and $\boldsymbol{z}_i \leq \boldsymbol{x}_i$ for each i.*

Hahn-Banach theorem E.2 is generalized to topological vector spaces.

Theorem G.6. *Let X and Y be nonempty convex subsets of a locally convex topological vector space L, and $\mathring{X} \neq \emptyset$ or $\mathring{Y} \neq \emptyset$. If $X \cap Y = \emptyset$, then there exists a nonzero vector $\boldsymbol{p} \in L^*$ such that $\boldsymbol{px} \leq \boldsymbol{py}$ for every $\boldsymbol{x} \in X$ and every $\boldsymbol{y} \in Y$.*

A topological vector lattice L is called a *Banach lattice* if it is a Banach space and satisfies $\|\boldsymbol{x}\| \leq \|\boldsymbol{y}\|$ whenever $|\boldsymbol{x}| \leq |\boldsymbol{y}|$. The previous spaces given above are all examples of Banach lattices.

Let $(A, \mathcal{A}, \lambda)$ be a finite-measure space, L a Banach space, and L^* its norm dual space under the duality $\langle \boldsymbol{p}, \boldsymbol{z} \rangle$ or $\boldsymbol{pz}$ for short, where $\boldsymbol{z} \in L$ and $\boldsymbol{p} \in L^*$. A map $f : A \to L$ is said to be *strongly measurable* if there exists a sequence $\{f_n\}$ of simple functions with $\|f_n(a) - f(a)\| \to 0$ a.e. A strongly measurable function f is said to be *Bochner integrable* if $\int_A \|f_n(a) - f(a)\| d\lambda \to 0$. In this case, we denote $\int_A f(a) d\lambda = \lim_{n\to\infty} \int_A f_n(a) d\lambda$ and call $\int_A f(a) d\lambda$ the *Bochner integral.* It is well known ([62, Theorem 2, p. 45]) that a strongly measurable map f is Bochner integrable if and only if $\int_A \|f(a)\| d\lambda < \infty$.

A map $f : A \to L$ is said to be *weakly measurable* if for each $\boldsymbol{p} \in L^*$, $\boldsymbol{p}f(a)$ is measurable. A weakly measurable map $f(a)$ is said to be *Pettis integrable* if there exists an element $\xi_f \in L$ such that for each $\boldsymbol{p} \in L^*$, $\boldsymbol{p}\xi_f = \int_A \boldsymbol{p}f(a)d\lambda$. The vector ξ_f is denoted by $\int_A f(a)d\lambda$ and called a *Pettis integral* of f.

If f is strongly measurable, it is weakly measurable. To see this, take $\boldsymbol{p} \in L^*$. As f is strongly measurable, there exists a sequence $\{f_n\}$ of simple functions converging to f in the norm topology for almost all $a \in A$. Hence, $\boldsymbol{p}f_n(a) \to \boldsymbol{p}f(a)$ a.e. As all f_n are simple, they are Borel measurable for all n. Therefore, by Proposition F.5, $\boldsymbol{p}f(a)$ is Borel measurable.

We say that $f : A \to L$ is *λ-almost separably valued* if there exists a set N of λ-measure 0 such that $\{f(a) \in L |\ a \in A \backslash N\}$ is separable. The following theorem is considered to be a generalization of Proposition F.7 to infinite-dimensional spaces.

Theorem G.7 (Pettis). *Suppose $f : A \to L$ is weakly measurable and λ-almost separably valued. Then, f is strongly measurable. Furthermore, the approximating simple functions can be taken as*

$$\{f_n(a) |\ a \in A, n \in \mathbb{N}\} \subset \{f(a) |\ a \in A\} \cup \{\mathbf{0}\}.$$

A map $f : A \to L^*$ is said to be *weak* measurable* if for each $\boldsymbol{z} \in L$, $\boldsymbol{z}f(a)$ is measurable. A weak* measurable map $f(a)$ is said to be *Gelfand integrable* if there exists an element $\pi_f \in L^*$ such that for each $\boldsymbol{z} \in L$, $\boldsymbol{z}\pi_f = \int_A \boldsymbol{z}f(a)\boldsymbol{z}d$. The vector π_f is denoted by $\int_A f(a)d\lambda$ and called a *Gelfand integral* of f.

Theorem G.8. *Let L be a Banach space and L^* its norm dual space. If $f : A \to L^*$ is weak* measurable and $\boldsymbol{z}f(a)$ is an integrable function for all $\boldsymbol{z} \in L$, then f is Gelfand integrable.*

Theorem G.9. *Let $\{f_n\}$ be a sequence of Gelfand integrable functions from A to L^* that converges a.e., to f in the weak* topology. Then, it follows that $\int_A f_n(a)d\lambda \to \int_A f(a)d\lambda$ in the weak* topology.*

For instance, let $L = \ell^1$ and $M = L^* = \ell^\infty$. A map $f : A \to \ell^\infty$ is weak* measurable if for each $\boldsymbol{p} \in \ell^1$, $\boldsymbol{p}f(a)$ is measurable. The

Gelfand integrable of f is an element $\boldsymbol{x}_f \in \ell^\infty$ such that for each $\boldsymbol{p} \in \ell^1$, $\boldsymbol{p}\boldsymbol{x}_f = \int_A \boldsymbol{p} f(a) d\lambda$.

In general, let L be a locally convex topological vector space and L^* its dual space. A map $f : A \to L$ is said to be *weakly measurable* if for each $\boldsymbol{p} \in L^*$, $\boldsymbol{p} f(a)$ is measurable. A weakly measurable map $f(a)$ is said to be *Pettis integrable* if there exists an element $\int_A f(a) d\lambda \in L$ such that for each $\boldsymbol{p} \in L^*$, $\boldsymbol{p} \int_A f(a) d\lambda = \int_A \boldsymbol{p} f(a) d\lambda$.

The following theorem was proved by Khan *et al.* [118], which is an infinite dimensional version of Fatou's lemma; see also [83].

Theorem G.10. *Let $(A, \mathcal{A}, \lambda)$ be a complete and finite measure space that is saturated, and L^* be the dual space of a separable Banach space L. Let $f_n : A \to L^*$ be a sequence of Gelfand integrable mappings from A to L^* such that there exists an integrable function $g(a)$ with $\sup_n \|f_n(a)\| \le g(a)$ a.e. Then, there exists a Gelfand integrable map $f : A \to L^*$ with*

$$\int_A f(a) d\lambda \in L_s\left(\int_A f_n(a) d\lambda\right) \text{ and } f(a) \in L_s(f_n(a)) \text{ a.e.}$$

Let $\phi : A \to L$ be a correspondence. The integral of the correspondence ϕ is defined similarly as in the finite-dimensional case:

$$\int_A \phi(a) d\lambda = \left\{ \int_A f(a) d\lambda \in L \,\middle|\, f \in \mathcal{L}(A, \mathcal{A}, \lambda),\ f(a) \in \phi(a) \text{ a.e.} \right\}.$$

Podczeck *et al.* [193, 242] proved the following theorem:

Theorem G.11. *Let $(A, \mathcal{A}, \lambda)$ be a finite-measure space, and L^* be the dual space of a separable Banach space L. Then, the following are equivalent.*

(i) *the Gelfand integral $\int_A \phi(a) d\lambda$ is convex for every correspondence $\phi : A \to L^*$,*
(ii) *the measure space $(A, \mathcal{A}, \lambda)$ is saturated.*

Remark G.1. Recall that a topological space is a *Suslin space* if and only if it is the image of a continuous map from a complete and separable metric space (Polish space). Hence all Suslin spaces are separable. A separable Banach space L is a Suslin space, and it is known that its dual space L^* endowed with the weak* topology is also a Suslin space. It follows from $\ell^\infty = (\ell^1)^*$ that ℓ^∞ with the weak* topology $\sigma(\ell^\infty, \ell^1)$ is a Suslin space.

A map from a measure space $(A, \mathcal{A}, \lambda)$ to a locally convex Suslin space is Borel measurable if and only if it is weakly measurable [254, Theorem 1]. As ℓ^∞ is locally convex in the $\sigma(\ell^\infty, \ell^1)$-topology, the Borel measurability and the weak* measurability of a map $f : (A, \mathcal{A}, \lambda) \to (\ell^\infty, \mathcal{B}(\ell^\infty))$ are equivalent.

Appendix H

Notes

Appendix A: Most of the materials in this section can be found in standard textbooks on point set topology and linear algebra, for example, see Nikaido [174, Chapter 1] or Takayama [253, Chapter 0]. For the separation hyper-plane theorem (Theorem A.1) and K–K–M lemmas (Theorems A.3 and A.4), we recommend Ichiishi [99]. He also presented a clear and systematic account for Brower's and Kakutani's fixed point theorems. For Theorem A.2, see Rockafellar [212, p. 84]. For the Kuhn–Tucker theorem (Theorem A.5), see [253, Chapter 1].

Appendix B: All of the materials in Appendix B are standard. The basic references are Kelly [113] and Royden [216].

Appendix C: Readable accounts for the singular homology theory can be found in Massey [153] or Vick [258]. A rigorous and comprehensive treatment of algebraic topology can be found in Spanier [238].

Appendix D: A classical textbook for correspondences is Berge [25]. Much of the exposition in Appendix D follows Hildenbrand [94]. The proof of Kakutani's fixed point Theorem D.1 follows [172]. For applications of Kakutani's theorem and related propositions, see [99, Chapter 3]. Theorem D.4 follows [94]. Theorem D.5 follows Rådström [196].

Appendix E: For the coordinate-free definition of the differentiable manifolds including infinite-dimensional ones, see Abraham *et al.* [1] or Lang [128]. For Banach space theory including Hilbert spaces and linear mappings on them, see Rudin [217] or Yosida [273]. For the infinite-dimensional version of the Kuhn–Tucker theorem, see Luenberger [135].

Clear and readable textbooks of manifold theory, including in particular the regular value theorem and Sard's theorem, are Milnor [164] and Guilemmin and Polack [85]. The proof of Brower's fixed point theorem based on Sard's theorem is due to [164].

Appendix F: Basic references are Halmos [86] and Royden [216]. Ash [15] is also recommended. For integration theory for correspondences, see [94]. Theorem F.6 is due to Keisler and Sun [112]. The definition F.1 that defines the saturated measure space in terms of the saturation property is from [112]. In the following, we provide a more intrinsic definition in terms of the Maharam type. For references on saturated measure spaces, see Notes of Chapter 5 in this book.

A *measure algebra* is a pair $(\mathcal{A}, \lambda)$, where $\mathcal{A}$ is a *Boolean σ-algebra* with binary operations $\wedge$ and $\vee$, a unary operation ^{c} and λ is a real-valued function satisfying the following conditions: (i) $\lambda(B) = 0$ if and only if $B = \emptyset$, where $\emptyset = A^c$ and $A = \emptyset^c$ are the smallest and the largest elements in $\mathcal{A}$, respectively; (ii) $\lambda(\vee_{n=1}^{\infty} E_n) = \sum_{n=1}^{\infty} \lambda(E_n)$ for every sequence $\{E_n\}$ in $\mathcal{A}$ with $E_n \cap E_m = \emptyset$ whenever $m \neq n$. A map $\Phi : \mathcal{A} \to \mathcal{B}$ between measure algebras $(\mathcal{A}, \lambda)$ and $(\mathcal{B}, \mu)$ is called *homomorphism* if it is one to one, $\Phi(A^c) = \Phi(A)^c$, $\Phi(A \vee B) = \Phi(A) \vee \Phi(B)$ and $\lambda(A) = \mu(\Phi(A))$. Measure algebras $(\mathcal{A}, \lambda)$ and $(\mathcal{B}, \mu)$ are *isomorphic* if there exists a homomorphism which is onto.

A subalgebra of $\mathcal{A}$ is a subset of $\mathcal{A}$ that contains A and is closed under the Boolean operation $\wedge$, $\vee$ and ^{c}. The order $\leq$ on $\mathcal{A}$ is given by $B \leq C$ if and only if $B = B \wedge C$. A subalgebra $\mathcal{U}$ of $\mathcal{A}$ is *order-closed* with respect to $\leq$ if any non-empty upward directed subsets of $\mathcal{U}$ with their supremum in $\mathcal{A}$ have their supremum in $\mathcal{U}$. A subset $\mathcal{U} \subset \mathcal{A}$ *completely generates* $\mathcal{A}$ if the smallest order closed subalgebra in $\mathcal{A}$ containing $\mathcal{U}$ is $\mathcal{A}$ itself. The *Maharam type* of $(\mathcal{A}, \lambda)$ is the smallest cardinal of any subset $\mathcal{U}$ that completely generates $\mathcal{A}$.

Let $(A, \mathcal{A}, \lambda)$ be a finite-measure space. We define an equivalence relation on $\mathcal{A}$ by $E \sim F$ if and only if $\lambda(E \Delta F) = 0$, where

$E \Delta F = (E \wedge F^c) \vee (E^c \wedge F)$. The quotient space is denoted by $\hat{A} = A/\sim$. The equivalence class represented by $E \in \mathcal{A}$ is denoted $\hat{E}$. Then, the lattice operation and the unary operation ^{c} is defined naturally on $\hat{A}$, $\hat{E} \vee \hat{F} = \widehat{E \cup F}$ (union), $\hat{E} \wedge \hat{F} = \widehat{E \cap F}$ (intersection), and $\hat{E}^c = \widehat{(E^c)}$ (complement). The pair $(\hat{A}, \hat{\lambda})$ is a measure algebra associated with $(A, \mathcal{A}, \lambda)$, where $\hat{\lambda}(\hat{E}) = \lambda(E)$. Moreover, $(\hat{A}, \hat{\lambda})$ becomes a complete metric space by the metric $\rho(E, F) = \lambda(E \Delta F)$ (see [2, Lemma 13.13]). The measure algebra $(\hat{A}, \hat{\lambda})$ is separable if it is a separable metric space. The *Maharam type* of $(A, \mathcal{A}, \lambda)$ is defined to be that of $(\hat{A}, \hat{\lambda})$.

Let $\mathcal{A}_E = \{A \cap E | \ A \in \mathcal{A}\}$ be the sub-σ algebra of $\mathcal{A}$ restricted to $E \in \mathcal{A}$. We denote the restriction of λ to $\mathcal{A}_E$ by λ_E, or $\lambda_E(B) = \lambda(B)$ for every $B \in \mathcal{A}_E$. A finite-measure space $(A, \mathcal{A}, \lambda)$ is *Maharam-type homogeneous* if for every $E \in \mathcal{A}$ with $\lambda(E) > 0$, the Maharam type of $(E, \mathcal{A}_E, \lambda_E)$ is equal to $(A, \mathcal{A}, \lambda)$. The following fact is well known (e.g., [193, p. 838])

Fact H.1. *A finite-measure space $(A, \mathcal{A}, \lambda)$ is atomless if and only if for every $E \in \mathcal{A}$ with $\lambda(E) > 0$, the Maharam type of $(E, \mathcal{A}_E, \lambda_E)$ is infinite.*

This fact motivates the following definition.

Definition H.1. A finite-measure space $(A, \mathcal{A}, \lambda)$ is *saturated* (or *super-atomless*) if for every $E \in \mathcal{A}$ with $\lambda(E) > 0$, the Maharam type of $(E, \mathcal{A}_E, \lambda_E)$ is uncountable. A measure λ is saturated if $(A, \mathcal{A}, \lambda)$ is saturated.

Appendix G: Basic reference is Dunford and Schwartz [65]. See also [216, 217, 273]. Propositions G.5 and G.6 can be found in Bewley [30]. Propositions G.13 and G.14 are from Mas-Colell [145].

For the theory of Banach lattices including Theorem G.5, see Shaefer [227]. For Pettis' theorem (Theorem G.7), see for example, [273, p. 131] and [62]. The latter part on the range of functions is not explicitly stated in [273], but a careful examination of that proof will validate it. For the integration theory of vector-valued maps or correspondences in general, we refer [62, 269].

Bibliography

[1] Abraham, R., Marsden, J.E. and Ratiu, T. (1988). *Manifolds, Tensor Analysis, and Applications*, Springer, Berlin and New York (2nd edn., 1991).

[2] Aliprantis, C.D. and Border, K.C. (2006). *Infinite Dimensional Analysis: A Hitchhiker's Guide*, 3rd edn., Springer, Berlin.

[3] Aliprantis, C.D., Brown, D. and Burkinshaw, O. (1987). "Edgeworth equilibria", *Econometrica* 55, 1109–1137.

[4] Anderson, R. (1978). "An elementary core equivalence theorem", *Econometrica* 46, 1483–1487.

[5] Araujo, A. (1987). "The non-existence of smooth demand in general Banach spaces", *Journal of Mathematical Economics* 17, 1–11.

[6] Araujo, A., Martins-da-Rocha, V.F. and Monteiro, P.K. (2004). "Equilibria in reflexive Banach lattices with a continuum of agents", *Economic Theory* 24, 469–492.

[7] Armstrong, T.E. (1984). "The core-Walras equivalence", *Journal of Economic Theory* 33, 116–151.

[8] Armstrong, T.E. and Richter, M.K. (1986). "Existence of nonatomic core-Walras allocations", *Journal of Economic Theory* 38, 137–159.

[9] Arrow, K.J. (1973). "Some ordinalist–utilitarian notes on Rawls's Theory of Justice", *The Journal of Philosophy* 70, 245–263.

[10] Arrow, K.J. and Debreu, G. (1954). "Existence of an equilibrium for a competitive economy", *Econometrica* 22, 265–290.

[11] Arrow, K.J. and Hahn, F. (1971). *General Equilibrium Analysis*, North Holland, Amsterdam.

[12] Arrow, K.J., Block, H.D. and Hurwicz, L. (1959). "On the stability of the competitive equilibrium II", *Econometrica* 27, 82–109.

[13] Arrow, K.J. and Hurwicz, L. (1958). "On the stability of the competitive equilibrium", *Econometrica* 26, 522–552.

[14] Ash, R. (1972). *Real Analysis and Probability*, Academic Press, New York, San Fransisco and London.

[15] Aumann, R.J. (1964). "Markets with a continuum of traders", *Econometrica* 32, 39–50.

[16] Aumann, R.J. (1965). "Integrals of set-valued functions", *Journal of Mathematical Analysis and Applications* 12, 1–12.

[17] Aumann, R.J. (1966). "Existence of competitive equilibria in markets with a continuum of traders", *Econometrica* 34, 1–17.

[18] Balasko, Y. (1997a). "Pareto optima, welfare weights, and smooth equilibrium analysis", *Journal of Economic Dynamics and Control* 21, 473–503.

[19] Balasko, Y. (1997b). "Equilibrium analysis of the infinite horizon model with smooth discounted utility functions", *Journal of Economic Dynamics and Control* 21, 783–829.

[20] Balasko, Y. (1997c). "The natural projection approach to the infinite horizon model", *Journal of Mathematical Economics* 27, 251–265.

[21] Balasko, Y. (2016). *Foundations of the Theory of General Equilibrium*, World Scientific, New Jersey and Singapore.

[22] Balder, E. and Sambucini, A.L. (2005). "Fatou's lemma for multifunctions with unbounded values in a dual space", *Journal of Convex Analysis* 12, 383–395.

[23] Beato, P. (1982). "The existence of equilibria of marginal cost pricing equilibria with increasing returns", *Quarterly Journal of Economics* 389, 669–688.

[24] Bennasy, J.-P. (1991). "Monopolistic competition", in *Handbook of Mathematical Economics IV*, Hildenbrand, H. and Sonnenschein, H. (eds.), North Holland, Amsterdam and New York.

[25] Berge, C. (1963). *Topological Spaces*, Oliver and Boyde, Edinburgh. Reprinted (1997), Dover, New York.

[26] Bergstrom, T.C. (1976). "How to discard 'free disposability' at no cost", *Journal of Mathematical Economics* 3, 131–134.

[27] Bewley, T.F. (1972). "Existence of equilibria with infinitely many commodities", *Journal of Economic Theory* 4, 514–540.

[28] Bewley, T.F. (1973). "The equality of the core and the set of equilibria in economies with infinitely many commodities and continuum of traders", *International Economic Review* 14, 383–393.

[29] Bewley, T.F. (1982). "An Integration of equilibrium theory and turnpike theory", *Journal of Mathematical Economics* 10, 233–267.

[30] Bewley, T.F. (1991). "A very weak theorem on the existence of equilibria in atomless economies with infinitely many commodities", in

Equilibrium Theory in Infinite Dimensional Spaces Khan, M.A. and Yannelis, N. (eds.), Springer-Verlag, Berlin and New York.

[31] Bewley, T.F. (2007). *General Equilibrium, Overlapping Generations Models, and Optimal Growth Theory*, Harvard UP, Cambridge, Massachusetts.

[32] Boem, V.H. (1973). "Firms and market equilibria in a private ownership economy", *Zeitschrift für Nationalöconomie* 33, 87–102.

[33] Boem, V.H. and Lèvine, P. (1979). "Temporary equilibria with quantity rationing", *Review of Economic Studies* 46, 361–377.

[34] Brown, D.J. (1991). "Equilibrium analysis with non-convex technologies", in *Handbook of Mathematical Economics IV*, Hildenbrand, H. and Sonnenschein, H. (eds.), North Holland, Amsterdam and New York.

[35] Brown, D.J. and Heal, G.M. (1979). "Equity, efficiency and increasing returns", *Review of Economic Studies* 46, 571–585.

[36] Brown, D.J., Heal, G.M., Khan, M.A. and Vohra, R. (1986). "On a general existence theorem for marginal cost pricing equilibria", *Journal of Economic Theory* 38, 371–379.

[37] Caratheodory, C. (1939). "Die Homomorphieen von Somen und die Multiplication von Inhaltsfunionen", *Ann. Scuola Norm. Sup. Pisa Cl. Sci.* 8(2), 8–130.

[38] Carmona, G. and Podczeck, K. (2009). "On the existence of pure-strategy equilibria in large games", *Journal of Economic Theory* 144, 1300–1319.

[39] Castaneda, M.A. and Marton, J. (2008). "A model of commodity differentiation with indivisibilities and production", *Economic Theory* 34, 85–106.

[40] Chamberlin, E.H. (1933). *The Theory of Monopolistic Competition*, Harvard UP, Cambridge, Massachusetts.

[41] Chichilnisky, G. and Zhou, Y. (1998). "Smooth infinite economies", *Journal of Mathematical Economics* 29, 27–42.

[42] Chipman, J.S. (1965). "A survey of the theory of international trade: Part II, the Neoclassical Theory", *Econometrica* 33, 685–760.

[43] Chipman, J.S. (1970). "External economies of scale and competitive equilibrium", *Quarterly Journal of Economics* 84, 347–385.

[44] Cornet, B. (1988). "General equilibrium theory and increasing returns: Presentation", *Journal of Mathematical Economics* 17, 103–118.

[45] Cornet, B. (1990). "Existence of equilibrium in economies with increasing returns", in *Contributions to Economics and Operations Research*, Cornet, B. and Tulkens, H. (eds.), The XXth Anniversary of CORE, MIT Press, Cambridge, Massachusetts.

[46] Cornet, B. and Medecin, J.P. (2002). "Fatou's lemma for Gelfand integrable mappings", *Positivity* 6, 297–315.

[47] Cournot, A. (1838). *Recherches sur les Principes Mathématiques de la Théorie des Richesses*, in 1929 translated as *Researches into the Mathematical Principles of the Theory of Wealth*, Macmillan, New York.

[48] Dana, R.-A. (1995). "An extension of Milleron, Mitjushin and Polterovich's results", *Journal of Mathematical Economics* 24, 259–269.

[49] Debreu, G. (1952). "A social equilibrium existence theorem", *Proceedings of the National Academy of Sciences* 38, 886–893.

[50] Debreu, G. (1954). "Valuation equilibrium and Pareto optimal", *Proceedings of the National Academy of Sciences* 40, 588–592.

[51] Debreu, G. (1959). *Theory of Value*, Wiley, New York.

[52] Debreu, G. (1967). "Preference functions on measure spaces of economic agents", *Econometrica* 35, 111–122.

[53] Debreu, G. (1970). "Economies with a finite set of equilibria", *Econometrica* 38, 387–392.

[54] Debreu, G. (1974). "Excess demand functions", *Journal of Mathematical Economics* 1, 15–21.

[55] Debreu, G. and Scarf, H. (1963). "A limit theorem on the core of an economy", *International Economic Review* 4, 235–246.

[56] Dehez, P. and Dreze, J.H. (1988a). "Competitive equilibria with quantity taking producers and increasing returns to scales", *Journal of Mathematical Economics* 17, 209–230.

[57] Dehez, P. and Dreze, J.H. (1988b). "Distributive production sets and equilibria with increasing returns", *Journal of Mathematical Economics* 17, 231–248

[58] Dehez, P., Dreze, J.H. and Suzuki, T. (2003). "Imperfect competition à la Negishi, also with fixed costs", *Journal of Mathematical Economics* 39, 219–237.

[59] Dellacherie, C. and Mayer, P.A. (1978). *Probabilities and Potential*, North Holland, Amsterdam.

[60] Dierker, E. (1972). "Two remarks on the number of equilibria of an economy", *Econometrica* 50, 867–881.

[61] Dierker, E. (1974). *Topological Methods in Walrasian Economics*, Lecture Notes in Economics Mathematical Systems 92, Springer, Berlin.

[62] Diestel, J. and Uhl, J.J. (1977). *Vector Measures*, Mathematical Surveys and Monographs 15, American Mathematical Society, Province, Rhode Island.

[63] Dieudonné, J. (1960). *Foundations of Modern Analysis*, Academic Press, New York.

[64] Dirac, P.A.M. (1958). *The Principles of Quantum Mechanics*, 4th edn., Oxford University Press, London.

[65] Dunford, N. and Schwartz, J. (1988). *Linear Operators, Part I: General Theory* (Wiley Classics Library), Wiley, New York.

[66] Edgeworth, F.Y. (1881). *Mathematical Psychics: An Essay on the Application of Mathematics to the Moral Sciences*, Routledge and Kegan Paul, London, updated version (2019). Wentworth Press, New York.

[67] Edgeworth, F.Y. (1905). "Review of *A Geometrical Political Economy* by Henry Cunynghame", *Economic Journal* 15, 62–71.

[68] Edgeworth, F.Y. (1925). *Papers Relating to Political Economies, I–III*, Macmillan and Co Ltd, London, 2nd printing Burt Franklin, New York.

[69] Fajardo, S. and Keisler, H.J. (2002). *Model Theory of Stochastic Processes*, A K Peters, Ltd., Natick.

[70] Florenzano, M.J. (1983). "On the existence of equilibria in economies with an infinite dimensional commodity space", *Journal of Mathematical Economics* 12, 207–219.

[71] Florenzano, M.J. (2003). *General Equilibrium Analysis: Existence and Optimality*, Springer, Berlin and New York.

[72] Fremlin, W. (2002). *Measure Theory*, volume 3, Parts I and II: Measure Algebra, 2nd edn, Torres Fremlin, Colchester.

[73] Friedman, J. (1982). "Oligopoly theory", in *Handbook of Mathematical Economics II*,' Arrow, K. and Intrigator, M. (eds.), North Holland, Amsterdam and New York.

[74] Gale, D. and Nikaido, H. (1965). "The Jacobian matrix and global univalence of mappings", *Mathematische Annalen* 159, 81–93.

[75] Gale, D. and Mas-Colell, A. (1975). "An equilibrium existence theorem for a general equilibrium model without ordered preferences", *Journal of Mathematical Economics* 2, 9–16.

[76] Gale, D. and Mas-Colell, A. (1979). "Corrections to "An equilibrium existence theorem for a general equilibrium model without ordered preferences" ", *Journal of Mathematical Economics* 6, 297–8.

[77] Gabszewicz, J.J. and Vial, J.P. (1972). "Oligopoly à la Cournot in a general equilibrium analysis", *Journal of Economic Theory* 4, 381–400.

[78] Graham, F.D. (1923). "Some aspects of production further considered", *Quarterly Journal of Economics* 37, 199–227.

[79] Graham, F.D. (1925). "Some fallacies in the interpretation of social cost: A reply", *Quarterly Journal of Economics* 39, 324–330.

[80] Granmondt J.M. (1977). "Temporary general equilibrium theory", *Econometrica* 45, 535–572.

[81] Granmondt J.M. and Hildenbrand, W. (1976). "Stochastic process of temporary equilibria", *Journal of Mathematical Economics* 1, 247–277.

[82] Greenberg, J., Shitovitz, B. and Vial, J.P. (1979). "Existence of equilibria in atomless production economies with price dependent preferences", *Journal of Mathematical Economics* 6, 31–41.

[83] Greinecker, M. and Podczeck, K. (2017). "An exact Fatou's lemma for Gelfand integrals by means of Young measure theory", *Journal of Convex Analysis* 24, 621–64.

[84] Greinecker, M. and Podczeck, K. (2017). "Core equivalence with differentiated commodities", *Journal of Mathematical Economics* 73, 54–67.

[85] Guillemin, V. and Pollack, A. (1974). *Differential Topology*, Prentice Hall, Englewood Cliffs, New Jersey.

[86] Halmos, P. (1974). *Measure Theory*, Reprint, Springer-Verlag, Berlin and New York.

[87] Halmos, P.R. and Neumann, J.v. (1942). "Operator methods in classical mechanics II", *Annals of Mathematics* 43, 332–350.

[88] Hara, C. (2005). "Existence of equilibria in economies with bads", *Econometrica* 73, 647–658.

[89] Harrod, R.H. (1967). "Increasing returns", in *Monopolistic Competition Theory*: Studies in Impact; Essays in Honor of Edward H. Chamberlin, Robert E. Kuenne (ed.), John Wiley and Sons, Inc, New York.

[90] Hart, O.D. (1979). "Monopolistic competition in a large economy with differentiated commodities," *Review of Economic Studies* 45, 1–30.

[91] Hart, S., Hildenbrand, W. and Kohlberg, E. (1974). "On equilibrium allocations as distributions on the commodity space", *Journal of Mathematical Economics* 1, 159–166.

[92] Hart, S. and Kohlberg, E. (1974). "On equally distributed correspondences," *Journal of Mathematical Economics* 1, 167–174.

[93] Hicks, J.R. (1939). *Value and Capital*, Oxford UP, Oxford.

[94] Hildenbrand, W. (1974). *Core and Equilibria of a Large Economy*, Princeton UP, Princeton, New Jersey.

[95] Hildenbrand, W. (1983). "On the law of demand", *Econometrica* 51, 997–1020.

[96] Hildenbrand, W. and Kirman, A. (1986). *Introduction to Equilibrium Analysis*, North Holland, Amsterdam and New York.

[97] Hoover, D. and Keisler, H.J. (1984). "Adapted probability distributions", *Transactions of American Mathematical Society* 286, 159–201.

[98] Hotelling, H. (1929). "Stability in competition", *Economic Journal* 39, 41–57.

[99] Ichiishi, T. (1983). *Game Theory for Economic Analysis*, Academic Press, London and New York.

[100] Ichiishi, T. and Quinzi, M. (1983). "Decentralization for the core of a production economy with increasing returns", *International Economic Review* 24, 397–412.

[101] Jevons, W.S. (1871). *The Theory of Political Economy*, Macmillan, London and New York.

[102] Jones, L. (1983). "Existence of equilibria with infinitely many consumers and infinitely many commodities", *Journal of Mathematical Economics* 12, 119–138.

[103] Jones, L. (1984). "A competitive model of commodity differentiation", *Econometrica* 52, 507–530.

[104] Kajii, A. (1988). "Note on equilibria without ordered preferences in topological vector lattices", *Economics Letters* 27, 1–4.

[105] Kamiya, K. (1988). "Existence and uniqueness of equilibria with increasing returns", *Journal of Mathematical Economics* 17, 149–178.

[106] Kehoe, T.J. (1980). "An index theorem for general equilibrium models with production", *Econometrica* 48, 1211–1232.

[107] Kehoe, T.J. (1982). "Regular production economies", *Journal of Mathematical Economics* 10, 147–176.

[108] Kehoe, T.J. (1983). "Regularity and index theory for economies with smooth production technologies", *Econometrica* 51, 895–917.

[109] Kehoe, T.J. and Levine, D.K. (1985). "Comparative statistics and perfect foresight in infinite horizon economies", *Econometrica* 53, 433–453.

[110] Kehoe, T.J., Levine, D.K., Mas-Colell, A. and Zame, W.R. (1989). "Determinacy of equilibrium in large-square economies", *Journal of Mathematical Economics* 52, 231–263.

[111] Keisler, H.J. and Sun, Y.N. (2002). "Loeb Measures and Borel Algebras", in *Reuniting the Antipodes — Constructive and Nonstandard Views of the Continuum.* Berger, U., Osswald, H. and Schuster, P. (eds.), Proc. of Symp. in San Servolo/Venice, Italy, Kluwer, pp. 111–118.

[112] Keisler, H.J. and Sun, Y.N. (2009). "Why saturated probability spaces are necessary", *Advances in Mathematics* 221, 1584–1607.

[113] Kelly, J.L. (1975). *General Topology*, Springer-Verlag, Berlin and New York.

[114] Khan, M.A., Rath, K.P. and Sun, Y.N. (1997). "On the existence of pure strategy equilibria in games with a continuum of players", *Journal of Economic Theory* 76, 13–46.

[115] Khan, M.A. and Sagara, N. (2013). "Maharam-types and Lyapunov's theorem for vector measures on Banach spaces", *Illinois Journal of Mathematics* 57, 145–169.

[116] Khan, M.A. and Sagara, N. (2014). "Weak sequential convergence in $L^1(\mu, X)$ and an exact version of Fatou's lemma", *Journal of Mathematical Analysis and Applications* 412, 554–563.

[117] Khan, M.A. and Sagara, N. (2015). "Maharam-types and Lyapunov's theorem for vector measures on locally convex spaces with control measures", *Journal of Convex Analysis* 22, 647–672.

[118] Khan, M.A., Sagara, N. and Suzuki, T. (2016). "An exact Fatou's lemma for Gelfand integrals: a characterization of the Fatou property", *Positivity* 20, 343–354.

[119] Khan, M.A. and Sagara, N. (2017). "Fatou's Lemma, Galerkin Approximations and the Existence of Walrasian Equilibria in Infinite Dimensions", *Pure and Applied Functional Analysis* 2, 317–355.

[120] Khan, M.A. and Suzuki, T. (2016). "On differentiated and indivisible commodities: an expository re-framing of Mas-Colell's 1975 model", *Advances in Mathematical Economics* 20, 103–128.

[121] Khan, M.A. and Sun, Y.N. (1999). "Non-cooperative games on hyperfinite Loeb spaces", *Journal of Mathematical Economics* 31, 455–492.

[122] Khan, M.A. and Yamazaki, A. (1981). "On the cores of economies with indivisible commodities and a continuum of traders", *Journal of Economic Theory* 24, 218–225.

[123] Khan, M.A. and Yannelis, N.C. (1991). "Equilibria in markets with a continuum of agents and commodities", in *Equilibrium Theory in Infinite Dimensional Spaces*, Khan, M.A. and Yannelis, N.C. (eds.), Springer-Verlag, Berlin and New York.

[124] Knight, F. (1924). "Some fallacies in the interpretation of social cost", *Quarterly Journal of Economics* 38, 582–606.

[125] Knight, F. (1925). "On decreasing cost and comparative cost: A rejoinder", *Quarterly Journal of Economics* 39, 331–333.

[126] Koopmans, T.C. (1957). *Three Essays on the Sate of Economic Science*, McGraw-Hill, New York.

[127] Lancaster, K. (1971). *Consumer Demand: A New Approach*, Columbia UP, New York.

[128] Lang, S. (1972). *Differential Manifolds*, Addison-Wesley, Reading, Massachusetts.

[129] Lee, S. (2013). "Competitive equilibrium with an atomless measure space of agents and infinite dimensional commodity space without convex and complete preferences", *Hitotsubashi Journal of Economics* 54, 221–230.

[130] Lenin, V.I. (1917). *Imperialism, the Highest Stage of Capitalism*, English translation (1947), Foreign Language Publishing House, Moscow.

[131] Lipsey, R.G. (1960). "The theory of customs unions: a general survey", *Economic Journal* 70, 496–513 (reprinted in A.E.A., *Readings in International Economics*, Homewood, Ill: Richard D.Irwin, Inc., 1968).

[132] Loeb, P.A. (1975). "Conversion from nonstandard to standard measure spaces and applications in probability theory", *Transactions of American Mathematical Society* 211, 113–122.

[133] Loeb, P.A. and Sun, Y.N. (2007). "A general Fatou lemma", *Advances in Mathematics* 213, 741–762.

[134] Loeb, P.A. and Sun, Y.N. (2009). "Purification and saturation", *Proceedings of American Mathematical Society* 137, 2719–2724.

[135] Luenberger, D.G. (1969). *Optimization by Vector Space Method*, Wiley, New York.

[136] Magill, M. (1981). "An equilibrium existence theorem", *Journal of Mathematical Analysis and Applications*, 84, 162–169.

[137] Maharam, D. (1942). "On homogeneous measure algebras", *Proceedings of the National Academy of Sciences* 28, 108–111.

[138] Maharam, D. (1950). "Decompositions of measure algebras and spaces", *Transactions of American Mathematical Society* 69, 142–160.

[139] Mantel, R. (1974). "On the characterization of excess demand", *Journal of Mathematical Economics* 7, 348–53.

[140] Mantel, R. (1979). "Equilibrio con rendimiento crecientes a escala," *Anales de la Asociation Argentine de Economia Politica* 1, 271–283.

[141] Marshall, A. (1890). *Principles of Economics*, 8th edn. (1920), Macmillan, London and New York.

[142] Martin-da-Rocha, V.F. (2003). "Equilibria in large economies with a separable Banach commodity space and non-ordered preferences", *Journal of Mathematical Economics* 39, 863–889.

[143] Martin-da-Rocha, V.F. (2004). "Equilibrium in large economies with differentiated commodities", *Economic Theory* 23, 529–552.

[144] Mas-Colell, A. (1974). "An equilibrium existence theorem without complete or transitive preferences", *Journal of Mathematical Economics* 1, 237–246.

[145] Mas-Colell, A. (1975). "A model of equilibrium with differentiated commodities", *Journal of Mathematical Economics* 2, 263–296.

[146] Mas-Colell, A. (1977). "Indivisible commodities and general equilibrium theory", *Journal of Economic Theory* 16, 443–456.

[147] Mas-Colell, A. (1983). "Walrasian equilibria as limits of noncooperative equilibria, part I: Mixed strategies", *Journal of Economic Theory* 30, 153–170.

[148] Mas-Colell, A. (1984). "On a theorem of Schmeidler", *Journal of Mathematical Economics* 13, 201–206.

[149] Mas-Colell, A. (1985). *The Theory of General Economic Equilibrium: A Differentiable Approach*, Cambridge UP, Cambridge, England.

[150] Mas-Colell, A. (1986). "The price equilibrium existence problem in topological vector lattices", *Econometrica* 54, 1039–1054.

[151] Mas-Colell, A., Winston, M.P. and Green, J. (1995). *Micro Economic Theory*, Oxford UP, New York and Oxford.

[152] Mas-Colell, A. and Zame, W.R. (1991). "Equilibrium theory in infinite dimensional spaces," in *Handbook of Mathematical Economics IV*, Hildenbrand, H. and Sonnenschein, H. (eds.), North Holland, Amsterdam and New York.

[153] Massey, W. (1980). *Singular Homology Theory*, Springer, Berlin and New York.

[154] McKenzie, L. (1954). "On equilibrium in Graham's model of world trade and other competitive systems", *Econometrica* 22, 147–161.

[155] McKenzie, L. (1955). "Competitive equilibrium with dependent consumer preferences", *Second Symposium on Linear Programming*, Washington: National Bureau of Standards and Department of Air Force, 277–294.

[156] McKenzie, L. (1959). "On the existence of general equilibrium for a competitive market", *Econometrica* 27, 54–71.

[157] McKenzie, L. (1956–7). "Demand theory without utility index", *Review of Economic Studies* 24, 185–189.

[158] McKenzie, L. (1960). "Stability of equilibrium and the value of positive excess demand", *Econometrica* 28, 606–617.

[159] McKenzie, L. (1981). "The classical theorem of competitive equilibrium", *Econometrica* 49, 819–841.

[160] McKenzie, L. (2002). *Classical General Equilibrium Theory*, MIT Press, Cambridge, Massachusetts.

[161] Mead, J.E. (1952). *A Geometry of International Trade*, George Allen and Unwin Ltd, London.

[162] Menger, C. (1871). *Grundsätze der Volkswirtschaftslehre*, Wilhelm Braumüller, Wien.

[163] Mertens, J.-F. (1991). "An equivalence theorem for the core of an economy with commodity space L_∞-$\tau(L_\infty, L_1)$", in *Equilibrium Theory in Infinite Dimensional Spaces* Khan, M.A. and Yannelis, N.C. (eds.), Springer-Verlag, Berlin and New York.

[164] Milnor, J.W. (1965). *Topology from the Differentiable Viewpoint*, Virginia UP, Charlottesville.

[165] Moore, J.C. (1975). "The existence of "compensated equilibrium" and the structure of the Pareto efficiency frontier", *International Economic Review* 16, 267–300.

[166] Nash, J. (1950). "Equilibrium points in n-person games", *Proceedings of the National Academy of Sciences* 36, 48–49.

[167] Negishi, T. (1960). "Welfare economics and existence of an equilibrium for a competitive economy", *Metroeconomica* 12, 92–97.

[168] Negishi, T. (1961). "Monopolistic competition and general equilibrium", *Review of Economic Studies* 28, 196–201.

[169] Neumann, J.v. (1937). "*Über ein Ökonomisches Gleichungssystem und eine Verallgemeinerung des Browerschen Fixpunktsatzes*", *Ergebnisse eines Mathematischen Kolloquiums* 8, 73–83.

[170] Neumann, J.v. (1945). "A model of general economic equilibrium", *Review of Economic Studies* 13, 1–9 (English translation of Neumann, 1937).

[171] Newman, P. (1960). "The erosion of Marshall's theory of value", *Quarterly Journal of Economics*, 74, 587–600.

[172] Nikaido, H. (1954). "*Zusatz und Berichtigung für mine Mitteilung "Zum Beweis der Verallgeminerung des Fixpunktsatzes*"", *Kodai Mathematical Seminar Reports* 6(1).

[173] Nikaido, H. (1956). "On the classical multilateral exchange problem", *Metroeconomica* 8, 135–145.

[174] Nikaido, H. (1968). *Convex Structures and Economic Analysis*, Academic Press, New York.

[175] Nishimura, K. (1978). "A further remark on the number of equilibria of an economy", *International Economic Review* 19, 679–685.

[176] Nishino, H. (1971). "On the occurrence and existence of competitive equilibria", *Keio Economic Studies* 8, 33–67.

[177] Noguchi, M. (1997). "Economies with a continuum of consumers, a continuum of suppliers, and an infinite dimensional commodity space", *Journal of Mathematical Economics* 27, 1–21.

[178] Noguchi, M. (1997). "Economies with a continuum of agents with the commodity-price pairing (ℓ^∞, ℓ^1)", *Journal of Mathematical Economics* 8, 265–287.

[179] Noguchi, M. (2009). "Existence of Nash equilibria in large games", *Journal of Mathematical Economics* 45, 168–184.

[180] Noguchi, M. and Zame, W. (2006). "Competitive Markets with Externalities", *Theoretical Economics* **1**, 143–166.

[181] Novshek, W. (1980). "Cournot equilibrium with free entry", *Review of Economic Studies* 47, 473–486.

[182] Novshek, W. and Sonnenschein, H. (1983). "Walrasian equilibria as limits of noncooperative equilibria, part II: pure Strategies", *Journal of Economic Theory* 30, 171–187.

[183] Oddou, C. (1976). "Theorems d'existence et d'equivalence pour des economies avec production", *Econometrica* 44, 265–282.

[184] Ostroy, J.M. (1984). "On the existence of Walrasian equilibrium in large-square economies ", *Journal of Mathematical Economics*, 13, 143–164.

[185] Ostroy, J.M. and Zame, W. (1994). "Nonatomic economies and the boundaries of perfect competition", *Econometrica* 62, 593–633.

[186] Pareto, V. (1909). *Manuel d'Economie Politique*, Giard, Paris.

[187] Parthasarathy, K.P. (1967). *Probability Measures on Metric Spaces*, Academic Press, New York and London.

[188] Peleg, B. and Yaari, M. (1969). "Markets with countably many commodities", *International Economic Review* 11, 369–370.

[189] Pigou, A.C. (1920). *The Economics of Welfare*, Macmillan and Co. Ltd, London (2nd edn., 1924; 3rd edn., 1929; 4th edn., 1932).

[190] Podczeck, K. (1997). "Markets with infinitely many commodities and a continuum of agents with non-convex preferences", *Economic Theory* 9, 385–426.

[191] Podczeck, K. (1998). "Quasi-equilibrium and equilibrium in a large production economy with differentiated commodities", in *Functional Analysis and Economic Theory*, Abramovich, Y., Yannelis, N.C. and Avgerinos, E. (eds.), Berlin, Heidelberg & New York, Springer.

[192] Podczeck, K. (2005). "On core-Walras equivalence in Banach lattices", *Journal of Mathematical Economics* 41, 764–792.

[193] Podczeck, K. (2008). "On the convexity and compactness of the integral of a Banach space valued correspondence", *Journal of Mathematical Economics* 44, 836–852.

[194] Podczeck, K. (2009). "On purification of measure valued maps", *Economic Theory* 38, 399–418.

[195] Quinzi, M. (1982). "An existence theorem for the core of a productive economy with increasing returns", *Journal of Economic Theory* 28, 32–50.

[196] Rådström, H. (1952). "An embedding theorem for spaces of convex sets", *Proceedings of American Mathematical Society* 3, 165–169.

[197] Rajan, A.V. (1997). "Generic properties of the core and equilibria of pure exchange economies", *Journal of Mathematical Economics* 27, 471–486.

[198] Ramsey, F.P. (1928). "A mathematical theory of savings", *Economic Journal* 38, 543–559.

[199] Rashid, S. (1983). "Equilibrium points of nonatomic games: Asymptotic results", *Economic Letters* 12, 7–10.

[200] Rath, K.P. (1992). "A direct proof of the existence of pure strategy equilibria in games with a continuum of players", *Economic Theory* 2, 427–433.

[201] Rath, K.P., Sun, Y.N. and Yamashige, S. (1995). "The non-existence of symmetric equilibria in anonymous games with compact action spaces", *Journal of Mathematical Economics* 24, 231–346.

[202] Rawls, J. (1971). *A Theory of Justice*, Harvard UP, Cambridge, Massachusetts.

[203] Rawls, J. (2001). *Justice as Fairness: A Restatement*, Harvard UP, Cambridge, Massachusetts.

[204] Richard, S.F. (1989). "A new approach to production equilibria in vector lattices", *Journal of Mathematical Economics* 18, 41–56.

[205] Richter, H. (1963). "Verallgemeinerung eines in der Statistik Benö"tigten Statze der Massthorie", *Mathematische Annalen* 150, 85–90.

[206] Roberts, J. and Sonnenschein, H. (1977). "On the foundations of the theory of monopolistic competition", *Econometrica* 45, 101–113.

[207] Roberts, K. (1980). "The limit points of monopolistic competition", *Journal of Economic Theory* 22, 256–278.

[208] Robertson, D.H. (1924a). "Those empty boxes", *Quarterly Journal of Economics*, 34, 16–30.

[209] Robertson, D.H. (1930). "The trees of the forest", *Economic Journal*, 40, 80–89.

[210] Robertson, D.H. (1957). *Lectures on Economic Principles* Staples Press, London.

[211] Robinson, J. (1933). *Economics of Imperfect Competition*, Macmillan, London.

[212] Rockafellar, R.T. (1970). *Convex Analysis*, Princeton UP, Princeton, New Jersey.

[213] Rogers, C.A. *et al.* (eds.) (1980). *Analytic Sets*, Academic Press, London and New York.

[214] Romer, P. (1986). "Increasing returns and economic growth", *Journal of Political Economy* 94, 1002–1037.

[215] Rosen, S. (1974). "Hedonic prices and implicit markets, product differentiation in pure competition", *Journal of Political Economy* 82, 34–55.

[216] Royden, H.L. (1988). *Real Analysis*, Macmillan, New York.

[217] Rudin, W. (1991). *Functional Analysis*, McGraw-Hill, New York.

[218] Rustichini, A. and Yannelis, N.C. (1991). "What is perfect competition?" in *Equilibrium Theory in Infinite Dimensional Spaces*, Khan, M.A. and Yannelis, N.C. (eds.), Springer-Verlag, Berlin and New York.

[219] Rustichini, A. and Yannelis, N.C. (1993). "Commodity pair desirability and the core equivalence theorem", in *General Equilibrium, Growth and Trade II*, Becker, R., Boldrin, M., Jones, R. and Thomson, W. (eds.), Academic Press, San Diego, pp. 150–167.

[220] Samuelson, P.A. (1947). *Foundations of Economic Analysis*, Harvard UP, Cambridge.

[221] Samuelson, P.A. (1948). "International trade and the equalization of factor prices", *Economic Journal* 58, 163–184.

[222] Samuelson, P.A. (1953–4). "Prices of factors and goods in general equilibrium", *Review of Economic Studies* 21, 1–20.

[223] Scarf, H. (1967). "The core of an n-person game", *Econometrica* 35, 50–69.

[224] Scarf, H. (1986). "Notes on the core of a productive economy", in *Contributions to Mathematical Economics: In Honor of Gerald Debreu* Hildenbrand, W. and Mas-Colell, A. (eds.), North Holland, Amsterdam and New York.

[225] Schmeidler, D. (1969). "Competitive equilibria in markets with a continuum of traders and incomplete preferences", *Econometrica* 37, 578–585.

[226] Schmeidler, D. (1973). "Equilibrium points of nonatomic games", *Journal of Statistical Physics* 17, 295–300.

[227] Shaefer, H.H. (1974). *Banach Lattices and Positive Operators*, Springer-Verlag, Berlin and New York.

[228] Shafer, W. and Sonnenschein, H. (1975). "Equilibrium in abstract economies without ordered preferences", *Journal of Mathematical Economics* 2, 345–348.

[229] Shafer, W. and Sonnenschein, H. (1976). "Equilibrium with externalities, commodity taxation and lump sum transfers", *International Economic Review* 17, 601–611.

[230] Shannon, C. and Zame, W.R. (2002). "Quadratic concavity and determinacy of equilibrium", *Econometrica* 70, 631–662.

[231] Shapley, L. (1973). "On balanced games without side payments", in *Mathematical Programming*, Hu. T.C. and Robinson, S.M. (eds.), Academic Press, New York.

[232] Shitovitz, B. (1973). "Oligopoly in markets with a continuum of traders", *Econometrica* 41, 467–501.

[233] Smith, A. (1776/1976). *The Wealth of Nations* (reprint) Cannan, E. (ed.), University of Chicago Press, Chicago.

[234] Sonderman, D. (1974). "Economies of scale and equilibria in coalition production economies", *Journal of Economic Theory* 8, 259–291.

[235] Sonnenschein, H. (1971). "Demand theory without transitive preference with applications to the theory of competitive equilibrium", In *Preference, Utility and Demand*, Chipman, J., Hurwicz, L., Richter, M. and Sonnenschein, H. (eds.), Harcourt, Brace, Jovanovich, New York.

[236] Sonnenschein, H. (1972). "Market excess demand functions", *Econometrica* 40, 549–563.

[237] Sonnenschein, H. (1973). "Do Walras identity and continuity characterizes the class of community excess demand functions?", *Journal of Economic Theory* 6, 345–354.

[238] Spanier, E.H. (1966). *Algebraic Topology*, Springer-Verlag, Berlin and New York.

[239] Sraffa, P. (1926). "The laws of return under competitive conditions", *Economic Journal* 36, 535–550.

[240] Starr, R.M. (2011). *General Equilibrium Theory*, Cambridge UP, Cambridge, England.

[241] Sun, Y.N. (1997). "Integration of correspondences on Loeb spaces", *Transactions of American Mathematical Society* 349, 129–153.

[242] Sun, Y.N. and Yannelis, N.C. (2008). "Saturation and the integration of Banach valued correspondences", *Journal of Mathematical Economics* 44, 861–865.

[243] Suzuki, T. (1995). "Nonconvex production economies", *Journal of Economic Theory* 66, 158–177.

[244] Suzuki, T. (1996). "Intertemporal general equilibrium model with external increasing returns", *Journal of Economic Theory* 69, 117–133.

[245] Suzuki, T. (2000). "Monopolistically competitive equilibria with differentiated commodities", *Economic Theory* 16, 259–275.

[246] Suzuki, T. (2013a). "Core and competitive equilibria for a coalitional exchange economy with infinite time horizon", *Journal of Mathematical Economics* 49, 234–244.

[247] Suzuki, T. (2013b). "Competitive equilibria of a large exchange economy on the commodity space ℓ^∞", *Advances in Mathematical Economics* 17, 1–19.

[248] Suzuki, T. (2013c). "An elementary proof of an infinite dimensional Fatou's lemma with an application to market equilibrium analysis", *Journal of Pure and Applied Mathematics* 10, 159–182.

[249] Suzuki, T. (2016). "A coalitional production economy with infinitely many indivisible commodities", *Economic Theory Bulletin* 4(1), 35–52.

[250] Suzuki, T. (2017). "Welfare analysis of a market model with external increasing returns and differentiated commodities", *Theoretical Economics Letters* 7, 63–78.

[251] Suzuki, T. (2018). "A simple model of the difference principle", *Theoretical Economics Letters* 8, 1869–1888.

[252] Suzuki, T. (2020). "On large individualized and distributionalized exchange economies with infinitely many commodities", forthcoming in *Pure and Applied Functional Analysis.*

[253] Takayama, A. (1986). *Mathematical Economics*, Cambridge UP, Cambridge, England.

[254] Thomas, G.E.F. (1975). "Integration of functions with values in locally convex Suslin spaces", *Transactions of American Mathematical Society* 212, 61–81.

[255] Tourky, R and Yannelis, N.C. (2001). "Markets with more agents than commodities: Aumann's "hidden" assumption", *Journal of Economic Theory* 101, 189–221.

[256] Toussaint, T. (1985). "On the existence of equilibria in economies with infinitely many commodities", *Journal of Economic Theory*, 13, 98–115.

[257] Uzawa, H. (1962). "Competitive equilibrium and fixed point theorems II: Walras existence theorem and Brower's fixed point theorem", *Economic Studies Quarterly* 13, 59–62.

[258] Vick, J. (1991). *Homology Theory*, 2nd edn., Springer, Berlin and New York.

[259] Vind, K. (1964). "Edgeworth-allocations in an exchange economy with many traders", *International Economic Review* 5, 165–177.

[260] Wald, A. (1933–4). "*Über die Eindeutige Positive Löbarkeit der Neuen Produktions–Gleichungen*", *Ergebnisse eines Mathematischen Kolloquiums* 6, 12–20.

[261] Wald, A. (1934–5). "Über die Produktionsgleichungen der Ökonomische Wertlehre", *Ergebnisse eines Mathematischen Kolloquiums* 7, 1–6.

[262] Wald, A. (1936). "Über einige Gleichungssysteme der Mathematschen Ökonomie", *Zeitschrift für Nationalökonomie* 7, 637–670.

[263] Wald, A. (1951). "On some systems of equations of mathematical economics", *Econometrica* 19, 368–403 (English translation of Wald (1936)).

[264] Walras, L. (1874, 1877). *Elements d'Economie Politique Pure*, Lausanne: L Corbaz. In 1954 translated by W. Jaffé as *Elements of Pure Economics*, Homewood, IL: Richard D. Irwin.

[265] Yamazaki, A. (1978). "An equilibrium existence theorem without convexity assumptions", *Econometrica* 46, 541–555.

[266] Yamazaki, A. (1981). "Diversified consumption characteristics and conditionally dispersed endowment distribution: regularizing effect and existence of equilibria", *Econometrica* 49, 639–654.

[267] Yamazaki, A. (1986). *Foundations of Mathematical Economics* (in Japanese), Sobun-sha, Tokyo.

[268] Yannelis, N.C. (1985). "On a market equilibrium theorem with an infinite number of commodities", *Journal of Mathematical Analysis and Applications* 108, 595–599.

[269] Yannelis, N.C. (1991). "Integration of Banach-valued correspondences", in *Equilibrium Theory in Infinite Dimensional Spaces*, Khan, M.A. and Yannelis, N.C. (eds.), Springer-Verlag, Berlin and New York.

[270] Yannelis, N.C. and Probhakar, N.D. (1983). "Existence of maximal elements and equilibria in linear topological spaces", *Journal of Mathematical Economics* 12, 233–245.

[271] Yannelis, N.C. and Zame, W.R. (1986). "Equilibria in Banach lattices without ordered preferences", *Journal of Mathematical Economics* 15, 75–110.

[272] Yano, M. (1984). "The turnpike of dynamic equilibrium paths and its insensitivity to initial conditions", *Journal of Mathematical Economics* 13, 235–254.

[273] Yosida, K. (1968). *Functional Analysis*, Springer-Verlag, Berlin and New York (6th edn., 1980).

[274] Yosida, K. and Hewitt, E. (1952). "Finitely additive measures", *Transactions of American Mathematical Society* 72, 46–66.

[275] Young, A. (1913). "Pigou's wealth and welfare", *Quarterly Journal of Economics* 27, 672–686.

[276] Young, A. (1928). "Increasing returns and economic progress", *Economic Journal* 38, 527–542.

[277] Zame, W.R. (1987). "Competitive equilibria in production economies with an infinite dimensional commodity space", *Econometrica* 55, 1075–1108.

Index

D

Printed in the United States
by Baker & Taylor Publisher Services

Printed in the United States
by Baker & Taylor Publisher Services